AF884887

ANSWERS TO THE BIGGEST QUESTIONS EVER ASKED

An amazing exploration of many scientific discoveries that have changed our view of many of our most cherished beliefs

Gregg D Thompson

ISBN: 978-0-6454669-8-0

Copyright © Dreamtech Designs & Productions Pty Ltd & Gregg Thompson 2020. All rights are reserved. The moral right of Gregg Thompson to be identified as the author of this work has been asserted.

Answers to the BIGGEST QUESTIONS Ever Asked contains content material which is owned by or licensed to and are the copyright of Dreamtech Designs & Productions Pty Ltd.

The content material includes, but is not limited to, text, diagrams, design, layout, look, appearance, logos, trademarks and graphics. Without limiting the rights under copyright laws, you are not permitted to reproduce the content material, documents or information in the Answers to the BIGGEST QUESTIONS Ever Asked for any purpose or for use by any third party without the express written consent of Answers to the BIGGEST QUESTIONS Ever Asked. In particular you are not permitted to reproduce, republish, upload, download, transmit or store electronically or otherwise or distribute any part of Answers to the BIGGEST QUESTIONS Ever Asked, whether for commercial exploitation or otherwise, other than as specifically licensed by us upon purchase of the Answers to the BIGGEST QUESTIONS Ever Asked. Purchase of the Answers to the BIGGEST QUESTIONS Ever Asked is for use by the purchaser only.

Limited consent is hereby granted to enable content from the Answers to the BIGGEST QUESTIONS Ever Asked to be used for quotation of brief passages in reviews. We reserve the right to revoke any consent as we see fit. Please contact us to request any further consents.

We expressly reserve the right to take action against you if you breach any of these terms. You may be liable to criminal prosecution and civil claims for damages.

Although the author and publisher have taken all reasonable care in preparing this book, no warranty is given about the accuracy or completeness of its content, nor for any content linked or referred to in it, nor for any postings, message boards, Internet links, online commentary or the like. No responsibility is accepted and all liability arising from its use is disclaimed to the maximum extent permitted by law.

The author and the publisher have, to the best of their knowledge, refrained from using any copyrighted material unlawfully or without consent and have properly attributed such rights where appropriate. If you find any unindicated material protected by copyright, the copyright could not be determined. In the case of any unintentional copyright violation the author will remedy that by removing the material or properly attributing it. Kindly contact us on: info@ theultimateguidetostargazing.com

First published by Dreamtech Designs & Productions Pty Ltd 2021

DEDICATION

I dedicate this book to my wife Deborah for her loyal support and encouragement that she gave me over the many years it took to do the research, find the imagery, and then do the many rewrites. I also dedicate this book to my supportive son Tane for helping to publish and market this work.

ACKNOWLEDGMENTS

I give special thanks to the following people who offered valuable suggestions and comments when reading through the manuscript or parts thereof. They were my wife Deborah Kelly, my son Tane Thompson, Nick Parfitt, John O'Sullivan, Greg Bock, James Bryan, Olivia King, Robert Stanford, Joe Gjerek, Dan McDonald.

I wish to thank Joe and Rebecca Gjerek, and Nicole Brooke who worked with me on producing the book cover art and a number of diagrams.

I must thank my high school physics teacher Mr Thompson (no relation) and my chemistry teacher Mr McAlpine for instilling in me at an early age the value of scientific methodology, how to make accurate observations, and how to conduct useful experiments. Importantly, they enriched my thrill of discovery. I apologize to them for missing morning tea breaks due to them discussing my questions about science with me. My tertiary education was not in science so I spent a lot of time reading numerous papers about breakthroughs at the leading edge of scientific enquiry. I also thank the scientists and journal reporters who advanced my understanding of the nature of so many things.

I am indebted to the following people with whom I have had numerous informative discussions. These include professional and advanced amateur astronomers, scientists in biology, genetics, IT, and nanotechnology. I have also had very interesting conversations with psychologists, neurologists, engineers, and architects as well as experts in law and commerce, the entertainment industry, and the arts. These people opened my eyes to possibilities I was not aware of. They led me to researching and testing a very broad range of exciting subjects and challenging concepts.

I must also thank the many documentary makers and the authors of numerous science books that greatly expanded my horizons. They helped me satisfy my lust for knowledge about the big picture of things.

I must give thanks to the people at NASA and the ESA for the truly incredible work they have done to keep on making the most amazing discoveries. The extraordinarily beautiful images they have captured of the universe with Hubble Space Telescope and their many space probes, has been way beyond expectations, and truly out of this world!

ABOUT THE AUTHOR

It would be difficult to imagine anyone more interested in all fields of science than Gregg Thompson. His interest in science commenced in his early childhood before he knew what science was about. In his teens, he became intrigued with astronomy, space science, engineering, physics, chemistry, geology, meteorology, and biology.

In his 20s his interests expanded into photography, the arts, designing and building special effects entertainment and edu-tainment attractions. He authored many articles on astronomical subjects and made many television appearances about special current phenomena.

In 1989, Cambridge University Press UK published his work entitled ***'The Supernova Search Charts and Handbook'***, which he co-authored with another amateur astronomer, James Bryan Jnr at the University of Texas. This book took 12 years to develop accurate visual maps of star fields around 300 of the brightest galaxies. This allowed amateur astronomers to make visual discoveries of supernovas that randomly occur in galaxies *as they were occurring.* This was critical to enable professional astronomers to determine the characteristics of supernovas' elements and their light curves. Early detection by amateurs enabled professional astronomers to determine the size and age of the universe. For the work and innovation Gregg put into this project, Gregg was awarded the Amateur Achievement Award from the Astronomical Society of the Pacific in the US - the world's largest body of professional and advanced amateur astronomers. He shared the award with his colleague Rev Robert Evans who discovered many supernovas visually with the help of Gregg's charts and a larger photo album he produced of another thousand more distant galaxies.

In 1992, the editor of Australia's largest publishing house, Weldon Publishing attended one of Gregg's themed stargazing shows in which he used special-effects, projection systems and large telescopes. Being impressed with Gregg's knowledge of the night sky, his astronomical facts, and the way in which he conveyed his enthusiasm for astronomy, the editor asked him to write a popular book on observational astronomy. It was titled '***The Australian Guide to Stargazing***.' It was published in 1993 and sold in New Zealand and South Africa. It received much marketing from print and radio media interest. It became a top-seller. With new additions and reprints, it sold consistently for 25 years.

In 2012, Gregg decided to write what was going to be his final book "**The Ultimate Guide to Stargazing**.' It would encapsulate **all** the truly incredible discoveries that have been made in astronomy and its associated sciences. He planned this to take two years, but there was so much to cover that it took until 2020 to finish it! It was broken into four Volumes. Vol 1 encompasses for amateur astronomers every aspect of observing the universe with the naked eye, binoculars, and telescopes. It also included astrophotography and how to build the best types of observatories. Vol 2 covers observing and understanding our solar systems' workings and the extraordinary discoveries made on numerous worlds. Vol 3 comprises observing and appreciating the structures and evolution of amazing star clusters, all types of nebulas, and galaxies in deep space. Vol 4 explores the most stunning discoveries ever made about how the universe evolved. Gregg's lust for photography made him go to extraordinary lengths to include the most outstanding images taken by the world's best astrophotographers, space telescopes, space probes, and the largest ground-based observatories. He prepared excellent diagrams that make complex concepts easy to understand. And he included many interesting personal anecdotes and unanswered questions related to recent discoveries. Readers ask how a single author could have written so deeply about so many subjects. To cover all this was a mammoth task. This was exacerbated by him also starting to write another book in 2016 about the **Answers to the Biggest Questions Ever Asked,** some of which were prompted by his guide to stargazing research.

In his 30s and 40s, he started to explore exciting new discoveries in anthropology, archeology, and history. He said he loved learning so much that he begrudged having to spend precious time eating and sleeping so frequently! But he thoroughly enjoyed time with his family and his many friends. During the decades, his interests expanded into movie-making, sound production, graphics, marketing, business management, and project management. He invented many forms of high-ambiance special effects for theme parks, edutainment attractions, nightclubs, relaxation centers, restaurants, and unique signage. He used these talents in this area to create his INFINITY experience (www.infinitygc.com.au) and SPACE WALKER edu-tainment attraction.

Throughout his life, Gregg was fascinated with the theories of politics, economics, psychology, and the paranormal. He investigated many UFO sightings and assumed paranormal events. Having considerable artistic talent, in 1985 he designed and sold large numbers of his Halley's Comet chart internationally and also a planisphere to find it at night. As a businessman, he became an accomplished entrepreneur. He served as the managing director of several companies that he founded.

In his 50s, he was very interested in the tremendous implications of quantum physics, chaos theory, and probability theory. The more he learned, the more he realized how everything he was interested in seemed to be connected at various levels. Of all the sciences, astronomy excited him the most because it encompassed the Big Picture of everything. The two most important sciences that control every basic aspect of our universe are physics and chemistry, for which Gregg had a natural aptitude.

Over the last couple of decades, his relentless research into the evolution of life made it very likely that conditions should occur on many other worlds across the universe for intelligent life to evolve. He realized that most intelligences would have evolved well before our planet even formed, so they would be much more advanced than we are. He conceived fascinating outcomes from this that he has included in this book.

He also became aware of the truly extraordinary exponential explosion of human knowledge that had occurred over the last century since we started to develop high technology. He did in-depth research as to where this would take humanity. He could see the enormous implications of the high-speed evolution of computer intelligence, robotics, and other many other technologies. Gregg could see that this would very quickly change our way of life, and the speed of human evolution, and for that matter, our entire solar system.

This shocked him because he could see that this would quickly lead to an extremely exciting upside but also a very dangerous downside. Many discoveries overturned our simplistic scientific beliefs that we assumed were rock solid. The implications of this were immense.

Gregg's desire to share what he was discovered with others was the driver for him writing this book. He knew that very few science and general authors had his broad interests, and therefore this should make his book should close to being unique. The decades of research he has done would save readers the time it takes to find and consume all the information included here. After that, one has to vet it for accuracy. It would also save them not having to bring all these subjects together to see the Big Picture of nature throughout the universe at all levels that he had discovered. Maybe, most readers would be as enthusiastic about learning about these subjects as he is, and this would excite them to do further research themselves.

In his quest to discover at least some of the Big Picture of Everything before his years ran out, he also wanted to write this book for others because the answers he found not only surprised him, but many shocked him, while others excited him. He knew the book would do the same for many intelligent readers who were prepared to question their current beliefs when quality information demanded they do so. He is well aware that many of the book's answers will have other major discoveries that impact them shortly. This will change our outlook on them.

The universe has an unfathomable depth that we are always diving deeper into. With future technology's ability to make us far more intelligent and have unlimited memory capacity due to being connected to the internet, Gregg says *"We cannot imagine what the extraordinary things we will know in the very near future, or what truly incredible experiences we will have."*

Gregg hopes that he has provided a smorgasbord of thought-provoking concepts across an extensive range of subjects that will excite many readers to want to know more. He also hopes that his readers will enjoy learning about answers he has provided to some of the biggest questions we have ever asked.

CONTENTS

OUR AMAZING UNIVERSE

Chapter 3

Chapter 4

Chapter 5

Chapter 10

SPACE TRAVEL

Chapter 11

Chapter 14

TECHNOLOGICAL EVOLUTION

Chapter 15

INTELLIGENCE AND CONSCIOUSNESS

MINDBENDING QUESTIONS

WHAT IS REAL?

FINAL COMMENTS

HOW TO USE THIS BOOK

The Table of Contents allows you to jump from one chapter or section to another.

You can **Bookmark** parts of the book, add **Notes**, **Underline**, or **Highlight** sections of the text.

You can use the '**Search**' function to find certain words or phrases.

Many images are large enough to zoom into them to reveal much more detail than is possible with a traditional book. Those with a magnifier icon can be zoomed into the most. To see them at their very best, read the book on a large monitor or iPad. To view two-page spread images, select 2 page spread option in 'View'.

There are **impressive 3D images**, which are identified by this red/blue-green glasses icon. They can be viewed by using inexpensive red/blue-green anaglyph glasses.

Navigate pages and chapters by using 'Page Thumbnails' on the left of your screen.

The '**Read Aloud**' function is available on some ebook Apps and devices. They can convert the text into **different languages**.

WHAT YOU SHOULD KNOW ABOUT THIS BOOK

It requires **no specialist knowledge of science or mathematics.**

Names, technical terms, and important words are shown in bold so they can be located quickly when scanning the text.

The text in many sections have **summaries of a section shown in blue.**

The author's **anecdotes are shown in navy blue.**

Challenging questions for readers to contemplate are shown in **violet**.

Throughout the text, are, **information 'boxes'.** These have a blue background. They provide additional interesting information that is indirectly related to the topic of discussion.

Most quantitative information is presented as rounded off numbers because many measures and quantities are estimates. Rounding off numbers makes them easier to remember.

Measurements are given in Metric units with Imperial units following in brackets. Kilometers are abbreviated to **km**, meters to **m**, millimeters to **mm**, miles to **mi**. Feet are denoted by **'**, and inches by **"**. Kilograms are abbreviated to **kg**, and pounds to **lb**. Temperature is abbreviated to **C** for Centigrade, **F** for Fahrenheit, and **K** for Kelvin.

To make the text flow better and to make it easier to read, it is not interspersed with **references and footnotes**. In the **Appendix 1, there are sources that supports the conclusions in each chapter. These include recommended reading, websites, and documentaries.** There are two other informative **appendices**.

Chapter 1

WHY THE QUESTIONS IN THIS BOOK ARE SO IMPORTANT

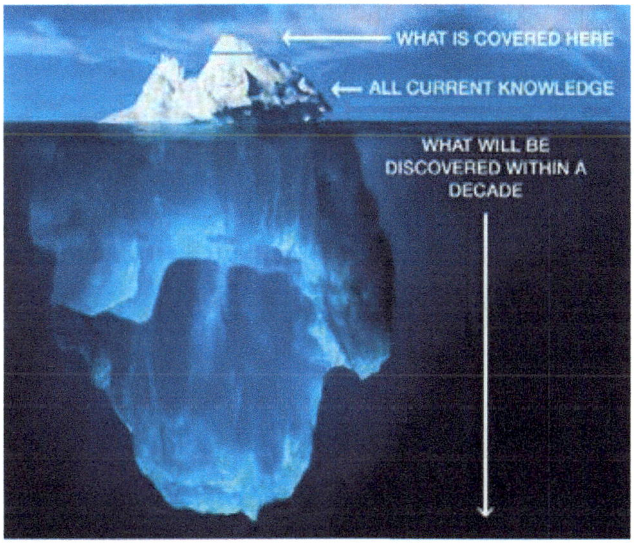

All that **we have learned over the last million years can be likened to a snowdrift at the very tip of an enormous iceberg where 92% of its mass lies below the surface. The submerged region symbolizes discoveries and concepts that we do not suspect and which are far beyond our present intelligence. For every subject we have learned about, an enormous, possibly near infinite body of knowledge lies hidden below what we think we know. We are only just starting to become aware of this.**

(Text that is in ***blue*** *is a summary of a part of each section. Sentences, names, or words that are important are in* ***bold****. The author's anecdotes are shown in* ***navy blue****. Questions that science needs to answer are in violet.)*

The aim of this book is to *blow your mind* by sharing with you the truly astonishing knowledge that we humans have learned over the last century. Much of what we are now discovering goes against what we had naively assumed to be unquestionable facts. Our past assumptions were simplistic and naïve due to us having a severe lack of knowledge compared to that available today. These stunning revelations challenge many of the beliefs we have been taught. Many will give you a '*Wow!*' reaction or a "*Surely not*" response. The answers to these really big questions have only recently been uncovered by scientists working at the leading edge of scientific inquiry. In earlier decades, we would have thought these discoveries to be impossible and rejected them. When we delve into these subjects, it becomes apparent that **there is a reason for everything, and that everything in the universe is interconnected in some way or other**. **What is this telling us?**

If you love learning what science has discovered as I do, you may know something about the most recent amazing discoveries that I have included in this book, but for most readers, most of them will be new because these discoveries are spread across many fields.

If you want to know about truly incredible discoveries made in astronomy, then read my ebooks in the four volumes of **'*The Ultimate Guide to Stargazing*'**. It overflows with stunning, high-resolution images of numerous astronomical bodies and events.

Recent discoveries are very exciting because they give us a much deeper and detailed appreciation of all that nature involves, from the smallest particles to the universe itself. These discoveries are being made increasingly faster, so for many readers, having this summary of them in one book is like finding a mine overflowing with exotic jewels.

Learning about these discoveries has been a wonderful, life-long journey for me. Hopefully, you will find them as exciting as I have.

Now, **fasten your seatbelt! This read is going to feel like an intellectual roller-coaster ride with unexpected rises and falls with twists and turns. The concepts will be very exciting for readers who do not mind going outside their comfort zone. If you are a person who likes to contemplate leading-edge scientific concepts, then you will enjoy the exhilarating opportunity to stretch your mind to its elastic limit. However, new knowledge that challenges one's beliefs can feel quite threatening, as I found it to be - at least initially. So try to have an open mind because most of what we now take for granted, most people in earlier decades would not have believed.**

Never in the history of humanity has there been a time to consider such intellectually stimulating concepts. Take your time to read them *slowly* enough to take in their importance and to think through their implications. Give yourself some weeks or months for it to sink in. Make time to do your own research and re-read the sections that interest you the most. Upon re-reading them, you will find that you have forgotten much of what you read initially, but after a second or third read, you will find these discoveries and concepts to be much more thought-provoking, easier to comprehend, and easier to remember.

Because there is so much to unveil in this book, I have restricted each subject to just the most salient points. They cover just the very tips of each iceberg of knowledge. Use the Internet and Appendix 1 to dive deeper into each subject of interest.

A LIFETIME OF RESEARCH AND DISCUSSIONS

Through my interests in all the physical sciences, the social sciences, and the arts, I've enjoyed a lifetime of discussions about numerous subjects with many scientists, science writers, science fiction authors, psychologists, researchers, philosophers, knowledgeable business people, historians, engineers, economists, and those in government.

I was born with an insatiable appetite for discovering how things work and why, even as a young boy. The more I learned, the more I wanted to know. Some of my friends call me a "*polymath relic from the 18th C's Age of the Enlightenment*". The Enlightenment was a time when most scientists were polymaths because they were interested in most sciences, although they specialized in one. Being a polymath and realizing that everything is connected to everything else in one way or another is why I have covered such a broad range of subjects.

The Internet gives us the ability to research across numerous fields, but because this takes a great deal of time, few people have the opportunity to do this. Also, we are swamped with information, much of which is false, so it also takes considerable time to extract what is real. To have decades of research condensed into this book should be a big advantage to most readers.

This book covers the really big questions that help us gain a 'big picture' understanding of our place in the universe.

The chapters on the technological evolution of humans are integral to us evaluating how other intelligences might evolve when they pass through our present level of evolution, presuming that they exist.

For decades, many people have long been keen to know if other solar systems exist. We now know that other planets exist around nearly all stars. People want to know what other planets might be like and what are the chances of there being other worlds like Earth. As an astronomer, **the question I am most frequently asked is, "*Is there life out there? And if so, what are the chances of there being other intelligent beings?*" There is always strong interest in whether UFOs could be spacecraft piloted by intelligent beings from other star systems.** People also want to know if we might detect radio communications from intelligent aliens. They ask if it might be possible to travel faster than the speed of light. They are also keen to know if we will build fleets of Star Trek-like interstellar spaceships that have warp-drives and teleporters so future humans can explore the galaxy. And I am often asked, "When will the first manned mission go to Mars?" With technology advancing so quickly, people are curious about what the future holds. Many of the answers to all these questions are covered, but most are not what we would have expected in earlier times.

Scientific discoveries are now coming so fast in every field that it is impossible to keep up with them! Discoveries in one field of science frequently have flow-on effects that impact other sciences and future technological development. Major discoveries can cause a paradigm shift in our thinking, as you will see. Many **questions that were impossible to answer just decades ago are now being answered due to science amassing an enormous amount of new knowledge ever more quickly.** New discoveries often raise questions that we have not previously thought to ask.

Supercomputers are rapidly becoming increasingly more intelligent, allowing them to quickly analyze very complex issues. Their ability to calculate is far beyond the capabilities of even teams of the smartest humans. Supercomputing is allowing us to move forward ever faster in almost every field. We are about to give birth to the first mega-intelligences that will be enormously more advanced than us - and today's supercomputers!

In order to write the summaries herein on each subject, I have done a great deal of research over many decades to seek out evidence, both for and against each subject. The answers for the majority of subjects are based on convincing evidence, but for some, the science is still in its formative stage, so present conclusions could change in the future.

UNCERTAINTY AND SPECULATION

In the vein of the late great Carl Sagan in his very popular TV series COSMOS, I will speculate from time to time about where new discoveries and inventions may be taking us and pose questions about what implications they might have.

Because most of the subjects are at the leading edge of our awareness, they provide fertile ground for speculation. Whilst much speculation turns out to be wrong or only partially right, it is useful for pushing the boundaries of science ahead. Speculation drives new explanations and theories, which can then be refined and tested to see if they are real. To progress, we must dare to fail. **Be aware that, at the leading edge of science, our conclusions are often our best *guesses* derived from what knowledge is currently available.** A conjecture can appear to fit the limited information we have available at the time, but as we amass more knowledge, a more accurate picture emerges, and this can often be quite different from our early assumptions.

So often, science finds that with each new discovery, **there is another layer of complexity underlying the previous layer,** and this raises a whole set of new questions. As an example, when astronauts went to the Moon, scientists had ten major questions about the Moon that they hoped they could find answers for. Most of those questions were answered, but there were so many unexpected discoveries that they ended up with over a hundred new questions in the years ahead! In science, the more we learn, the more we want to know.

All knowledge has associated uncertainties and biases, but too often, we forget this. The greater the uncertainty, the more speculative a hypothesis becomes. And correspondingly, the smaller the uncertainty, the more likely the hypothesis will be proven to be true (within defined limits of certainty). **As we learn more, we gain new insights into a bigger picture of nature. We often find that an existing theory becomes a subset of a larger one that is more encompassing, more complex, and which explains the observed data more comprehensively.** As an example, **Newton's Theory of Gravity** works very well until it has to deal with large gravitational fields, tight timeframes, and high velocities. In these cases, we must use **Einstein's theories of Relativity** that handle extreme conditions. We may soon find that the Theory of Relativity will become a subset of a new theory that is even more all-encompassing.

Science takes us into new frontiers where its pioneers go deep into the unknown, often unprepared for the unexpected things they discover. They do not always have the right tools to delve deeper into their newly-discovered mysteries; hence, engineers have to develop new technologies to help them test their theories to find out which one best explains what they have discovered. Few people realize that **without our ingenious engineers, we would not have gotten very far. They have done far more for humanity than any other class of people.**

When dealing with discoveries at the leading edge of science, there can be contradictory conclusions drawn by different researchers. Different interpretations of the data can be used by scientists to draw speculative conclusions that turn out to be wrong. This can be due to a lack of sufficient, good-quality data. Flawed conclusions can occur due to poor measurements, unrecognized instrumentation malfunctions, and a researcher's bias. Some scientists take a small morsel of questionable 'evidence' and portray it as a feast fit for a king in order to gain short-lived notoriety, monetary gain, or to push their political or idealistic agenda. To be scientific, scientists should always remain skeptical until there is very substantial supporting evidence for a theory. Postulations must be tested by rigorous and credible peer review. Today, reputations are sometimes built on headline-catching news releases and 'celebrity science' instead of good quality evidence. Reputations based on this are often destroyed when the truth emerges. Unfortunately nowadays, scientific conclusions in too many sciences are being bought off for political reasons and monetary gain by major corporations.

When there are contradictory conclusions, this means that one or all conclusions will turn out to be wrong. It is always prudent to keep an open mind until a sizable body of quality data is available to determine what is most likely true. Remember that, **in science, there are very few things that are 100% correct. There is always room for fine-tuning or a new theory** that explains the evidence better.

As we keep learning more, things become clearer, more exciting, and much more complex. What we are now learning about reality is often stranger than science fiction. Major new, unexpected discoveries can give us shivers up our spine, especially when the evidence goes against much of what we had assumed to be true.

THE GROWTH OF KNOWLEDGE AND TECHNOLOGY

We should remember that **most of the major astronomical discoveries that we accept today as fact were not thought possible even just a few decades ago.** As you read the chapters in this book, you may become skeptical about how much future technology will change our lives and our beliefs in the coming decades. But think about how much we have achieved over just this last century and how much this has changed what we believed back then. Most of us forget about this. Contemplate this: **at the present exponential growth of information and the advancement of technology, within just 40 years our knowledge is projected to increase by a staggering *one trillion times!! That's one thousand, billion times!*** This is beyond human comprehension. No one can imagine how much a 100 times growth rate will have on humanity and our planet, let alone a trillion. It will transform our lives beyond anything we can imagine, and it will affect our entire solar system.

Consider the following to see how rapidly we advanced in the 20th C. The 21st C will be exponentially faster!!

For nearly everyone throughout all of human history, the world has been much the same for a hundred generations due to us having had a fairly slow progression from the Stone Age up to the 17th century, admittedly marked by some significant peaks and troughs. But in *just the 20th C*, those of us in the developed world went from a low-tech civilization to a high-tech one! **It is astonishing that the vast majority of all our recent technological advancements have occurred in just one human lifespan! Stop and think about how truly amazing that is.** Incredibly, we went from horse and cart overland transportation at the beginning of the 20th C to having a jet-set lifestyle and the internet. We are now entering a space-faring civilization. *That is remarkable!* So try not to reject what the future holds because it will change dramatically faster.

In only 50 years, we went from wood fires for home stoves and heating, to coal power stations to make **electricity** for heating and light. After using candles for lighting for thousands of years, we suddenly had electric lighting systems, hot water on tap, and air-conditioned environments! In the latter half of the 20th C, many cities became powered by **atomic energy**.

In the early 1900s, medicine was primitive: there were no pain-relief drugs and nothing in the way of anesthetics or antibiotics. But throughout the 20th C, medical science developed these things, and we now have **keyhole surgery, robotic neurosurgery, and bionics**! In just a century, we dramatically improved our standard of living, our health and increased our average lifespan by a couple of decades! Today, we have sci-fi-like intelligent **computer**

technology, genetic engineering, and the beginnings of **nanotechnology**. These technologies will deliver outcomes beyond our wildest dreams - and maybe our worst nightmares in just the next couple of decades.

When electronics were invented, they allowed the unimaginable to evolve. For communication at the beginning of the 20th C, we went from sending hand-written letters by post in horse-driven coaches to instant Morse code telegraph! At that time, it was unthinkable to imagine that a satellite would be invented that could beam information everywhere across the planet instantly, and it could also determine our location precisely on the Earth's surface using GPS. Our satellites observe the weather over the whole Earth, and on the Sun as well in incredible detail. No one 100 years ago could have imagined that!

By the beginning of the 21st C, intelligent **smartphones** were invented. **In the palm of our hands, we now have a very powerful computer connected to another amazing invention, the Internet.** These technologies permit us to access enormous amounts of information as well as radio, television, music, entertainment, GPS, and all sorts of Apps. This can be done almost anywhere on the globe! No sci-fi writers ever imagined this. And on top of that, our smartphones are combined with an amazing camera, which allows us to take and share still images and video instantly! Today, events anywhere in the world are constantly televised into our homes, offices, and onto our smartphones and laptops wherever we are, often as they occur! No one in the early 20th C would have believed that these things could be developed, let alone that quickly, so don't disbelieve what our future holds in future chapters, even though it will be far more extreme than what our past has been.

The ENIAC computer was the first 'fast' computer of the mid-20th C. Incredibly, our small smartphones are far more powerful computers than this first major computer, which filled a large room. For this computer to compete with a single smartphone would require an area larger than the USA to be covered in ENIAC computers! They would require an enormous amount of energy, yet a small battery powers our smartphones! What will tomorrow bring?

No one could have envisaged that an innovation by one man could have produced a technology that was so far-reaching in its impact as the **World Wide Web** has been.

If all this could happen in just a few decades, it's difficult to predict where future technology will take us in just the next 20 years, let alone in the next 20 years. It's impossible to know what we will learn or what we will believe in the future. Despite this, many of the chapters ahead will reveal many incredible things that will occur surprisingly fast.

Who could have imagined in the early 20th C that **mass production** would drive down the cost of goods and food dramatically while at the same time, improving the quality? The ability to distribute goods and food around the whole world quickly before it goes off is beyond anything we could have imagined a century ago. Our ability to keep food fresh much longer through the invention of refrigeration and airtight packaging has been a revolution that we could not have imagined. Generally speaking, for millennia, most ordinary people had a scarcity of food, and they had very few goods, yet by the middle of the 20th C, the average person in most western countries (and now even in most third world countries) has a relative abundance of food! This is the first time in humankind's history that there has been a sustained abundance for most people.

A majority of people who have been lucky enough to have lived since the 1950s in the developed world have experienced lives of **unprecedented security, wealth, convenience, health, information, entertainment, travel, and happiness. Even most people in the lower classes have a life that surpasses what the richest kings have had since the beginning of civilization! That is an enormous achievement – all achieved through the development of science and technology. The near future will deliver far more!**

Because the growth in technology has been incre mental in the past, we have been able to adjust to it, but this will not be the case in the near future. In the coming 20 years, it's safe to assume that we will invent far more things than all that has been invented in the entire history of mankind!

The speed at which new inventions and discoveries are coming along will change our lives increasingly more quickly to a point where it will become impossible for anyone to keep up with them. We will need intelligent robots or computers to work most things for us. If we could be transported to the latter half of the 21st C, we would find a world that is unrecognizable. It would be the same as placing a caveman into a modern city. He would be completely overwhelmed by everything – huge buildings, bridges, helicopters, commercial airliners, subways, ships, cars, elevators, neon and video screen advertising, computers, robots, electricity, music, glass, video screens, the noise, millions of strangely dressed people, strange animals and so on. He would find all these things impossible to understand. He would not be able to understand how cars move along roads, especially as he would not know about the invention of the wheel! (If you look at a car going along a road, it is not obvious that the wheels are rotating: the car appears to be magically sliding along.)

Most of the time, we think little of our technological advancements because we live mostly in the 'now'.

If we could go back to the beginning of the 20th C and tell people back then what they would achieve in just 50 years, let alone 100 years, we would be ridiculed, ignored, or put in a home for the insane! It would be the same today if someone from 20 years in the future were to tell us what our future holds: most people would not believe them, despite what they would be telling us would be true. Keep this in mind when you read the chapters of this book about future technological change.

I have recounted these examples of our advances over the last century to help you realize that we will advance much more than we have in the past century in just the next two decades! This will become obvious when we discuss the evolution of technology, our own evolution, and the evolution of mega-intelligent computers.

Rapid change will occur for all technologically-advancing intelligences throughout the universe. This must surely affect the evolution of the universe itself.

THE ADVANCEMENT OF ASTRONOMY OVER THE LAST 100 YEARS

Consider the astonishing advancement that has occurred in just one science - astronomy. At the turn of the 20th C, our knowledge of astronomy was extremely limited. We had virtually no idea what the Moon was like. No one back then could imagine that men would walk and drive over the Moon in the 70s or that we would send robotic space probes to numerous worlds in the solar system and discover things that were considered to be impossible. It is hard to believe that at the outset of the 20th C, almost nothing was known about the formation and evolution of stars and planets. We had no awareness back then that black holes and gravity lenses existed or that there were other galaxies! And nor did we have any idea about how enormous the universe was. In those days, the small portion of the Milky Way that we could observe was thought to be the entire universe! Quantum physics and Relativity were not even conceived, so their mind-bending implications were not suspected. In the coming years, we will have super-high technology telescopes that will advance our knowledge of the universe at a rapidly accelerating rate. Such telescopes were well beyond the imagination of astronomers just 50 years ago.

HUMANITY'S EXPANDING KNOWLEDGE

Humanity's growth in knowledge could be likened to putting together an enormous 4-dimensional jigsaw puzzle that has length, breadth, depth, and the dimension of time. But, unlike real-life jigsaw puzzles that have only two dimensions, length and breadth, this imaginary puzzle has no convenient, box-top picture to use as a guide for us to assemble parts of the jigsaw of what we have learned. Science has taught us to look for pieces of nature's jigsaw that fit together. In doing so, it has allowed us to form numerous, scattered, small pictures of all the information that exists in our universe. The more we learn about each subject that we know about, the more the picture of those subjects expands until some of the pieces join up with other parts to give us a broader view of particular aspects of nature. We are often like children trying to force some jigsaw pieces into places where they do not belong. Eventually, we find the right pieces that do fit. This allows us to make other discoveries to understand how the universe works and to see ever more beauty and wonder in all that nature has created.

Our knowledge is not just expanding in two dimensions, as is the case with a two-dimensional jigsaw puzzle laid out on a flat table: it is growing in three dimensions i.e. in length and breadth as well as upwards and downwards. And because we are piecing together more and more about the past back to the beginning of time and projecting into the future as well, our knowledge jigsaw is also growing outward into the 4th dimension of time as well. This makes our 4D knowledge jigsaw analogy immeasurably more interesting than a 2D one made of timber or cardboard pieces.

We now realize that every aspect of nature is far more complex than we initially assumed it to be. Therefore, we are frequently revising our theories to make them more accurately represent what we observe in nature. We know that uncertainty is with us every step of the way, and the more we learn, the more we realize how much more there is to learn. This is an exciting aspect of science.

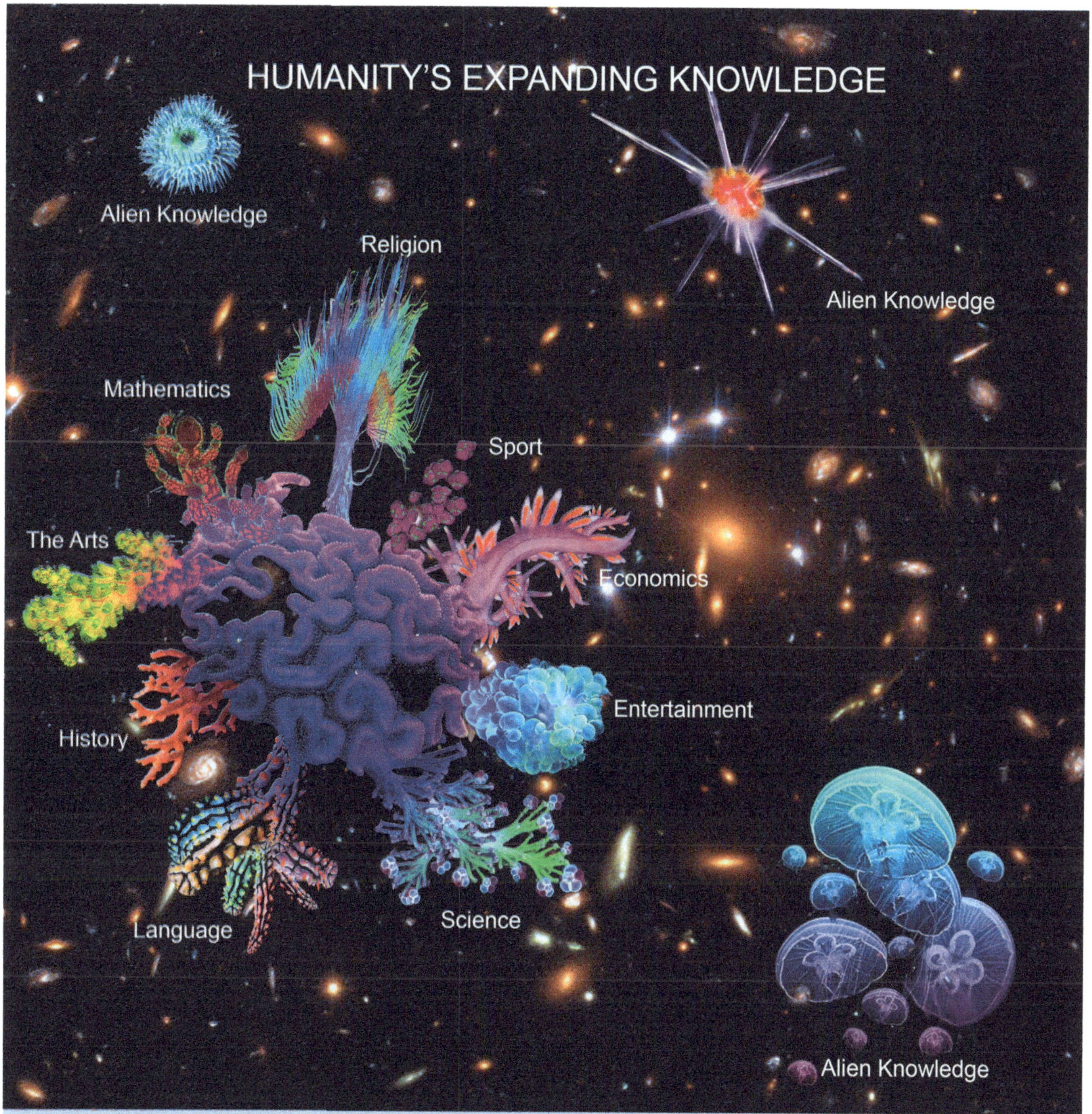

This illustration attempts to visualize what humanity's expanding knowledge might look like if we could see it from the perspective of it expanding into the universe. Areas of darkness are holes in our knowledge. Major regions of knowledge growth are shown as strange forms growing outward from our basic survival and reproduction knowledge, which is shown in purple in the middle. This is ancient knowledge handed down through our genes from the evolution of millions of life forms that have come before us. Surrounding this are strange 'tentacles' depicting different forms of knowledge. There are many more that are behind the central form. In the background, there are other knowledge explosions of alien intelligences that are also expanding into the cosmos. Their forms are very different to ours because they know things that we don't, and they do not know things that we do. Some will have orderly forms, while others will have strange structures. **What form do you think our future expanding knowledge explosion will take?** Credit: Gregg Thompson, Rebecca Gjerek

Scientific discoveries are throwing open endless doorways into a universe-sized maze full of new worlds and new levels of awareness.

If we could view our growing 'knowledge explosion' from outside ourselves from, say, a fifth dimension, we might be able to see it rapidly expanding into the universe. It might look something like a weirdly-shaped, giant amoeba with many different tentacles growing ever faster. The visualization of this in this info box shows a branch of knowledge as a distinctive tentacle. The edges of these tentacles represent the leading edge of our current knowledge in each branch. New discoveries appear like strange buds that are starting to bloom. Where our knowledge is incorrect, these buds are disfigured, and where we have no knowledge, there are dark voids. **If we could see our knowledge explosion from higher dimensions, would it be lopsided or riddled with holes?** We have no way of knowing how different our knowledge explosion would be to those of other intelligences that have different senses, different environments in which they evolved, different technologies, different minds, and possibly different forms of mathematics and science. **Could other intelligences also have different forms of logic that make them think very differently to us?** If they do, we would not be able to understand them.

In time, the knowledge explosions of many intelligences would grow large enough that they would extend to others and some would merge while others would reject them. Mergers would give them an explosive burst of growth to create new forms of knowledge. Most intelligences will become extinct as has 99% of all life forms on Earth. In many cases, the knowledge of now extinct intelligences might have been passed on to whatever they evolved into. Our supercomputers will retain all our knowledge and pass it onto their mega-intelligent descendants.

REVOLUTIONS IN SCIENCE

Throughout history, we humans have continued to gain new insights into nature through discoveries that have changed our beliefs about the universe, and recently about ourselves.

When good, new ideas emerge and gain supporting evidence, they typically spread. Once they attain mass acceptance, there is a paradigm shift into a new orthodoxy. This has occurred many times over the last few centuries. For example, in 1543, **Nicolaus Copernicus' Theory of Planetary Motions** explained how the planets revolved around the Sun. This seemed insane when it was first proposed because almost everyone thought that it was obvious that the Earth was stationary and flat. How could something as massive as the Earth move around the Sun's small disc in the sky? But as the evidence mounted for this 'crazy' idea, it eventually became accepted as being true! Similarly, no one knew why things fell to Earth, or why the planets and comets moved around the solar system as they did – until the 1680s when **Isaac Newton** brilliantly explained this with his **Theory of Gravity**.

When **Charles Darwin** came up with his incredibly insightful **Theory of Evolution** in the 1850s, all the confusing data about life forms and fossils that had been amassed with the infant sciences of paleontology and geology suddenly made sense. Darwin's theory explained fossils, how life evolved, and why there was such a diversity of life on Earth.

Albert Einstein's genius allowed him to develop the most far-reaching concept of all - his **Theory of General Relativity**. In the 1900s, he showed us that space and time were not fixed: they were, in essence, 'elastic'! Under infinite gravity, and at the speed of light, both space and time contract to zero! This illogical idea was proven to be true and eventually widely accepted. At the same time, **Max Planck** brilliantly conceived the foundations of his **Theory of Quantum Mechanics**, which explained the seemingly bizarre workings of the subatomic universe.

In 1915, the visionary geophysicist **Alfred Wegener** put forward his **Theory of Continental Drift,** which proposed that the world's massive, solid continents were constantly moving very slowly across Earth's surface. They were pulling the mid-ocean seabed apart, and in other regions, making continents crash into one another to form great mountain ranges and volcanoes. This finally explained why fossilized coral reefs were found on the tops of mountains. This made **Leonardo di Vinci** very curious. Like the previous examples, this idea also seemed impossible, but once the evidence grew ever larger, it proved Wegener's 'crazy' idea to be true, so it was eventually accepted as fact.

These few examples of uncovering the laws of nature caused major paradigm changes in our beliefs. They seemed ridiculous when they were first proposed, but they have now become well-established cornerstones of science that are understood even by school children. There are numerous other examples of **such ground-breaking discoveries, and each year there will be increasingly more, so keep this in mind when you read about the latest discoveries and concepts in this book that may seem ridiculous because they challenge our present-day beliefs. Science is littered with 'Eureka!' Discoveries like these. They will grow increasingly more numerous in the near future.**

Given how much newly discovered knowledge this book contains, you might find it beneficial to read it more slowly than you usually would in order that the concepts and their implications have time to sink in. Our primitive brains typically need time to compute the implications of big, new ideas.

As history has shown us, comprehending the frontiers of knowledge can often take decades. In the distant past, it often took centuries and even millennia before new theories could be proven and accepted. Ground-breaking discoveries and new theories require a slow, torturous route to prove them to be true before they are accepted. However, this process is happening much faster nowadays with the invention of supercomputers and new, leading-edge technologies.

In 1906, Albert Einstein said, "*If scientists can't explain their concept simply, then they don't understand it well enough.*"

"The most beautiful thing we can experience is the mysterious. It is the source of all true art in science. He to whom this emotion is a stranger and who cannot pause to wonder the mysterious in awe is as good as dead. His eyes are closed. We are in the position of a little child entering a huge library filled with books in many languages. The child knows someone must have written them, but the child does not know what they say, as he does not yet understand the languages in which they are written. The child suspects a mysterious arrangement in the order of the books but does not know what it is, but he is eager to learn."

- Albert Einstein

(If you understand this analogy, you will understand why Einstein was a genius. You will experience this throughout this book.)

UNDERSTANDING EXPONENTIAL GROWTH

The term ***'exponential growth'*** is used in several sections. Those who are not familiar with this term, it means that a growth rate is not linear and constant but rather it grows ever larger in each consecutive interval of time, becoming extraordinarily great very quickly. Exponential growth is a runaway increase that occurs with things like cell growth and the populations of locusts, flies, and rabbits - when the conditions are suitable. It also occurs in finance with compound interest and when an atomic bomb begins to explode.

A good demonstration of how fast exponential growth can occur is to place one grain of wheat on the first square of a chessboard and thereafter double the number of grains on each consecutive square. That is, place one grain on the first square, two on the second, then four on the third, eight on the fourth, sixteen on the fifth, and thirty-two on the sixth, and so on. This is exponential growth. For a challenge, calculate how many grains there will be in the 64th square.

We humans can relate to **linear growth** such as 1, 2, 3, 4, 5 etc, or 2, 4, 6, 8, 10, 12, 14, 16, 18, 20 etc. because this type of growth is what we are used to. Seldom do we experience exponential growth. When we do, it is usually limited after a period of time by a condition that stops its growth, so most the time we don't see it become extreme and cover the planet. With an explosion of particular life forms such as flies or locusts, their exponential growth is typically stopped when they run out of food, or its breeding cycle stops due to changes in the climate. Because of this, there can never be an indefinite exponential growth of any life form on Earth. If there was, it would very quickly cover the entire planet.

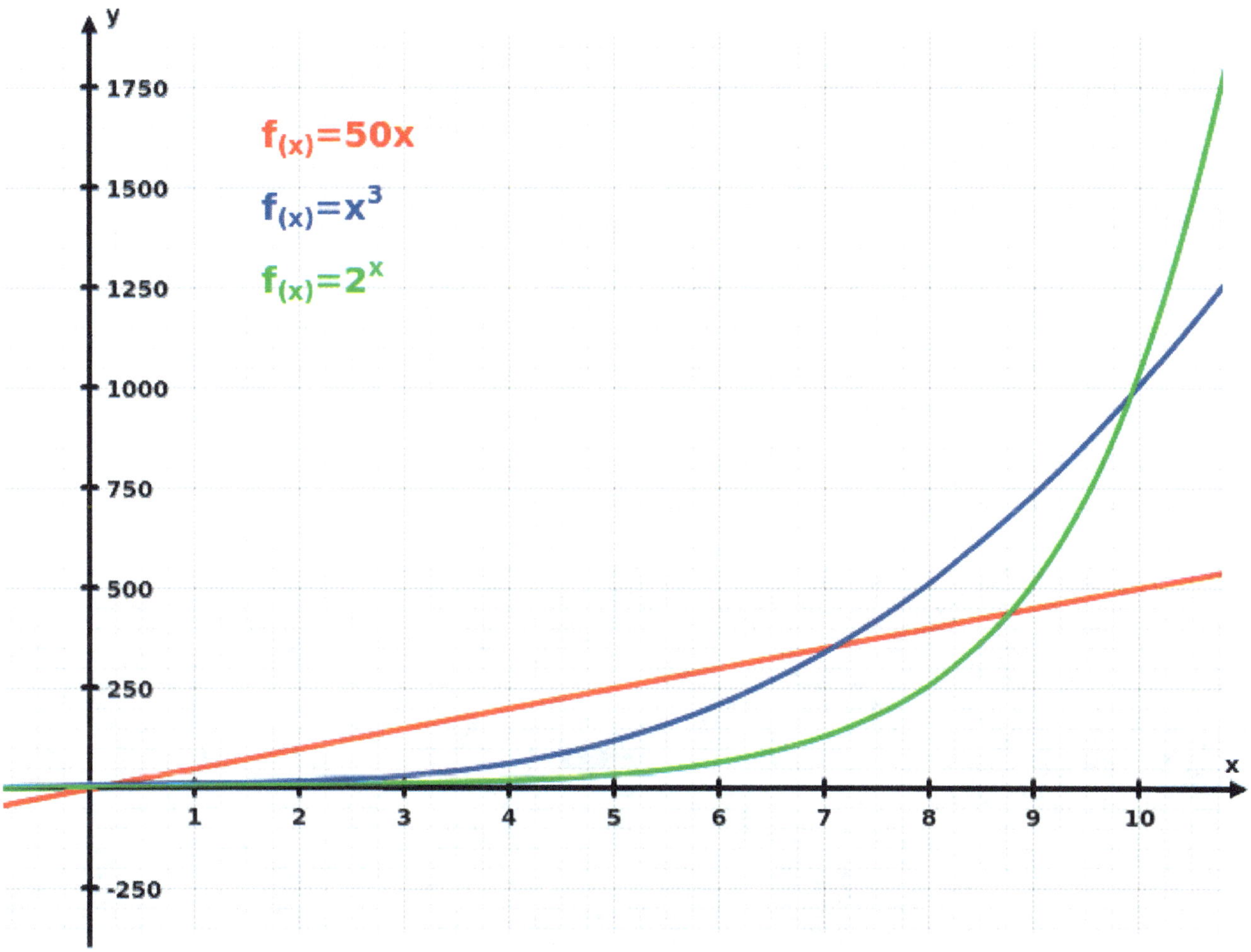

Red = Linear growth Blue = Cubic growth Green = Exponential Growth

On the graph above, the exponential curve is the green one. It starts out rising imperceptibly slowly, appearing nearly flat up to near 3 on the horizontal axis. It then starts to increase slightly towards 5, after which it grows ever faster. When it approaches 12, it becomes ever closer to vertical! Few humans are accustomed to such a fast rate of change. We are typically only familiar with linear change, which is shown in red.

Exponential growth can be a doubling in growth such as 2, 4, 8, 16, 32, 64, 128, 256, 512, 1024 etc. It grows exceptionally fast once it gets going due to it doubling at each interval. Some growths can be faster than doubling by growing at say 3 or 4.5 times or more.

(By the way, the answer to how many grains of rice are on the 64th square of a chessboard with exponential growth is 18 quintillion, 446 quadrillion, 744 trillion, 73 billion, 709 million, 551 thousand, 615!!! Expressed as an exponent, it is to the power of 263. The rate of growth becomes so large halfway through that the first square of the second half of the chessboard contains twice as many grains than the entire first half! The total amount of wheat grains on the 64th square would be 1,654 times Earth's total global wheat production!)

When we talk about the exponential growth of technology, it is very hard for most people to appreciate just how rapidly it is growing in such short timeframes. Our unfamiliarity with this rate of growth makes most people think it can't happen, but disbelief does not mean it will not occur. Computers will quickly become exponentially more intelligent and capable. Later in this chapter, you will learn where this will lead.

MEGA-INTELLIGENCE

A decade ago I coined the term 'mega-intelligence' to refer to a computer-based intelligence that is far superior to all of humanity and which evolves at an exponential rate. Such an intelligence is expected to evolve by around 2030 or before. (See Chapter 17) It will be unimaginably more intelligent and faster at computing than our present Artificial Intelligent (AI) computers.

DISCUSSIONS ABOUT THE BIG QUESTIONS

In the following chapters, we will look at answers to many of the BIG questions that are challenging humanity today. These thought-provoking subjects make for excellent conversations when camping out and relaxing under the stars. Innovative interior environments are also good for stimulating deep discussions, as you will see in the following box section.

While sitting around a log fire under the stars or looking at a view from a mountain chatting about the big concepts in this book can be an exciting and memorable experience.

INNOVATIVE ENVIRONMENTS FOR CREATIVE THINKING

The subjects in the following chapters provide excellent fuel for some deep and far-reaching discussions about many aspects of the universe, and the future, especially if you have friends who are curious, interested in science, and who like to think about concepts that are 'outside the box.' New ideas often emerge from such deliberations and this frequently generates further exciting ideas that encourage further discussion. For such discussions, it really helps to have them in comfortable, stimulating environments that create the right mood to enhance creative thinking. Such environments can be peaceful places in nature, or they can be innovative interiors like the ones I created to enhance one's train of thought.

When I was in my early 20s, I designed and built a large, three-level home on a hilltop for my family. Being a futurist and loving innovative architectural and interior design, between the 1970s and the 1980s, I created a number of special, high-ambiance environments in our home and gardens. We found that they created excellent recluses for relaxation and interesting conversations with friends, especially when we held 'deep thought' nights that went into the wee hours of the morning. On these occasions, I would invite up to 20 well-educated friends who were a mix of Ph.D. students, scientists, professors, business people, entrepreneurs, astronomers, physicists, chemists, geologists, biologists, medical researchers, architects, engineers, and others who were interested in the diverse subjects that we would cover. We all had a love of delving into a myriad of topics, such as some of those that are covered in this book. In view of this, some of my friends thought that most readers would be quite interested in the environments I built into my home because they were so different to anything that anyone had ever experienced, so they encouraged me to describe them as they considered them to be well ahead of their time back then, and even today. You will need a good imagination to visualize them!

One cherished environment was the **Conversation 'Worm Lounge' room**. Entry to this space was through a circular 'hole' in one wall. The room's subdued lighting utilized two spots; one projected a diffuse spot of magenta light onto one white wall and the other a rich deep blue onto an adjoining wall. The other two walls were floor to ceiling glass that provided a view across the city in one direction and to mountain ranges all around in the adjoining wall. The floor was covered in a very deep, plush ultramarine blue shag carpet, and the ceiling was the same color. For seating, guests reclined into very comfortable, spongy, giant intertwining 'worm' lounges, the main one being 500 mm in diameter and the other 400mm. Their upholstery was in different widths of blues, magenta, lilac, and purple. As you reclined into the larger one, it molded to your body form. A smaller worm curved around them to act as a headrest and a footrest. Guests could sit in these free-flowing 'worms' at different heights and in a variety of poses. The lounges wound around the perimeter of the room, except where the entry was and to a deck overlooking the view. The worms lounges made this room's atmosphere very relaxing, plush, cozy, and futuristic. Internally-lit, translucent, circular coffee tables glowed softly in the room's subtle ambient light. On one coffee table sat a mirrored **infinity cube** art piece that I created. It displayed what looked like an infinite number of colored stars orbiting inside it to infinity! While the interior appeared to go to infinity, the outside was clearly finite in size! The other table had some fascinating mind puzzles. In one corner, there was a large **'kinetic sculpture'** that was a floor-to-ceiling glass multi-bubble chamber that I designed and built. It released illuminated bubbles of all sizes from its base. They rose in slow motion through a series of vertical chambers that were invisible. These chambers contained transparent fluids, each with different viscosities to slow the rising of the bubbles. The densest chamber contained the largest bubbles so they would rise as slowly as the other chambers. Light reflected off the domes of the bubbles, making them look like molten, chrome jellyfish. All the bubbles would rise slowly, giving the room a tranquil atmosphere. The bubbles were internally lit from below and above with different colors. This kinetic art piece was mesmerizing to watch.

The house also had another highly atmospheric lounging environment that was affectionately called **The Music Room** or **Tomo's Holodeck** (as per Star Trek). It was built in what was essentially a concrete bunker on the lower level of the house. This room was earth-integrated into the side of the hilltop. The earth-integration design provided extraordinarily good acoustics and excellent soundproofing. Having two walls that were earth-integrated kept the room cool in summer and warm in winter. It featured a very special, fluid-filled flexible, and slightly stretchy Butynol membrane bladder that created the room's **'floatation' flooring system.** This acted like a giant, super-strong waterbed. It allowed the floor to undulate like a wave, and act as a very comfortable, lie-around lounge on which we would recline against large, soft, spherical 'planet-like' cushions of varying sizes. A soft tubular headrest encircled the perimeter of the room for those who wanted to lie around the outside. The floor covering was a deep, white, fur carpet in which artificial 'glow worms' or 'stars' could be made to glow softly amongst the tufts of fur when the room was in total darkness. These stars could also be made to glow on the walls and across the white parachute ceiling giving the room a magical effect. In this ambiance, the walls seemed to disappear, making the room appear dimensionless. It was as if we were in a large glow-worm cave or out in space amongst the stars. We were no longer in an earthly place, so it was excellent for contemplation. It was a stunning effect for listening to music like Pink Floyd, space music, or ultra-groovy albums.

Large, round, white balloons filled with just enough helium to give the ballons zero buoyancy so they floated through the room's 'sky' to great effect. When images of pop stars or dreamlike atmospheric motifs were projected onto them, the images appeared 3D as they drifted slowly overhead giving one goosebumps. Clouds of purple mist would occasionally cover the waveform floor to make us feel like we were floating on surreal clouds. When the room was in darkness mode, 'alien death rays' (lasers) beamed from rotating crystals. They slowly scanned the clouds for humans 'hiding' in the mist. It was as if we were in some futuristic movie. At other times, glow-in-the-dark space-art consisting of stars, nebulas, and galaxies fluoresced on the walls in low light, making it appear as if we were on a magic carpet ride through the Milky Way. This was a good mode in which to have deep discussions with atmospheric music. The room was designed to change to any color or brightness, so it could create many different moods. Hidden behind the white parachute ceiling and elsewhere was a diversity of special effects lighting that could change the ambiance of the room from a tranquil environment to a raging exciting nightclub. When all other lighting was off, strobe lighting made the movement of people dancing look like a surreal stop-frame movie. With no competing lighting as is the case in nightclubs, the effect was 100%, making it extraordinary. It was so powerful it was hard to believe it was real.

A computer-controlled audio system permitted an amazing variety of sound fields recorded from around the world. There were ones suited to all activities and types of music. For instance, auditorily, we could be on an open plain, or in a huge monastery or, in a cave, in a top disco, or an open-air arena. The sophisticated AV system could be made to appear as a distant city at night with a skyline of skyscrapers. A mirrored wall behind the console made the room feel extra-large. When required, the opposite wall would become a full wall movie theatre screen where we could project natural environment scenes, documentaries, movies, or music videos. It was also used for projecting images that supported the subjects we were discussing. **Patrick Moore**, the British astronomer who presented the longest-running show on stargazing in the world, said that he thought this room was like something out of a science fiction movie that had become real.

I developed this room as a prototype for very modern, up-market homes where the owners wanted the ultimate in-home entertainment. These rooms were also designed for high-end international hotels where the ultimate in private entertainment suites were desired for their richest guests. These rooms were called '**Cocooners**' because people felt like they could cocoon themselves away from reality in these dreamlike environments. These suites had surround projection capability to create the impression of being immersed in a variety of landscapes with accompanying sound fields. These ranged from moonrise over a Pacific island beach at night, to a waterfall in a rainforest, sunset over beautiful mountain farmland, an alien planet landscape, a starry night in a desert, or a volcano erupting at twilight etc. The suites also incorporated an internally-illuminated, solid dance floor at one end with traditional upmarket lounges and a bar. This mini disco area had computer-controlled special effects lighting that sensed the mood of the music being played to provide the perfect lighting ambiance. These suites incorporated a hi-tech special effects spa, and shower with special lighting. The floatation floor doubled as a very large bed with sheeting that automatically pulled out when required. There was a second more traditional modern bedroom with dreamy lighting.

Our home also incorporated a **'Kinetic Lightscape' Relaxation Lounging** area, which, by day, doubled as a playroom for my children. At night, it featured a long, translucent wall that extended from the floor to the ceiling. Inside it, soft-edged, colorful images drifted around randomly before they eventually zoomed away into the distance. Many people felt that the Lightscape was as mesmerizing as watching a giant aquarium. Relaxing music emanated from the wall. The wall's lighting and sound effects could interact with people as they approached it, or when they touched it. When psychologists tested subjects in this room to determine its relaxation effect, they found that the results showed that it was much more effective for relaxing people than any other known method of relaxation such as meditation, relaxation music, yoga, or natural settings.

Although the next environment I will describe was not a lounge, I mention it here because guests were especially thrilled when visiting my '**Infinity Loo.**' It was a place for contemplation! Once inside this small room, it appeared to go to infinity in every direction! Its mirrored walls, floor, and ceiling were embedded with tiny colored 'star' lights that gave the impression that one was sitting in deep space when the main white light was turned off. A transparent sign on the back of the door ironically and comically said, *'You have entered the 5th dimension! Small, isn't it?'* The brightness of the stars could be dimmed to a level where the walls disappeared, so you felt like you were in space. For fun, a hidden microphone converted different sound frequencies into different colors of starlight that varied in brightness with the intensity of the sound. Very high sound frequencies produced blue stars while mid-high frequencies produced green ones. Mid to low frequencies produced yellow and deep base sounds made the stars glow red. You might be able to imagine the colors that were produced by people using the loo. Girls made it shimmer mostly blue and boys green. Some visitors worked out that if they sang, this made their 'universe' go wild with changing colors. When using the loo under white light, it took me six months to not feel as if I was in the main street of a city with innumerable clones of myself looking at me from every possible direction! The Infinity Loo became so popular at parties that I had to install a fake video camera in the corner of the ceiling. A sign told users that if they stayed in there for more than 3 minutes, the video camera would come on and broadcast its vision to those in the courtyard. To ensure this was believed, three minutes after the door was closed, the fake camera's red light would flash, and it would move slowly back and forth. This ensured a speedy turnover of this very popular amenity. Hysterically, when using the loo, some girls would hold hands through the door because they were told that once the door was closed that it was difficult to find one's way back to reality! This evolved into the unique very successful INFINITY experiential attraction (See www.infinitygc.com.au.)

For coffee breaks at night, we enjoyed retreating to our subtly lit and affectionately named **'the Glow Worm Rainforest Courtyard'.** Fiber optic lighting ran through the trees and into small 'caves' that existed between the moss and lichen-covered large boulders. These acted as a retainer wall in one area. Other large rocks were scattered throughout the area under the trees to act as seats and to add atmosphere.

The courtyard was paved in one-hundred-year-old colorful bricks, handmade by convicts. They were

Left: Looking sideways in the Infinity Loo with coloured lights changing to different sound frequencies.
Right: Looking down under white light.

all different, like individual mini artforms. They meandered between boulders and rainforest gardens. A canopy of large shade trees spread over outdoor dining tables, lounge seating, and the tropical undergrowth. At night, candles and flares added a lot of atmosphere. All the bedrooms and living rooms on this level opened out onto this very pleasant, relaxing courtyard that attracted many visits from birds and the natural fauna that lived in the surrounding bushland. We had long conversations over supper breaks at midnight while sitting on **log seating around the fire pit's log fire** while roasting marshmallows and sausages. This overlooked a natural-looking 'pond' that was actually a deep, dark blue **swimming pool**. Large boulders overhung parts of its edges, and some were in the water separating the wading pool from the main pool. For some coffee breaks, we would retreat to my '**Rooftop Lounge**' on the third level, which was on a large deck outside my observatory. Here we could relax while chatting under the stars and looking out over the city to the north and the Scenic Rim of mountains to the south. It was a great place to talk while watching electrical storms brewing that produced brilliant displays of lightning.

Many international, interstate, and local astronomers, as well as friends and business people who visited my home and observatory, enjoyed their experiences in these environments. Visitors often made complimentary comments to the effect that they *"felt like they had passed through a porthole to glimpse the future!"* In the decades ahead, such environments will become very common. With advances in modern technology, they should be far superior to what I was able to create back then. Now that technology is rapidly decreasing the cost of many things, such environments will be inexpensive.

Due to the highly unusual, low-lighting effects used in my home, it was impossible in those pre-digital-camera days to take surround photographs or video using film to capture the diverse ambiances of these softly lit environments. The sound fields I used created powerful atmospheres that supported the innovative physical features and the high ambiance lighting effects. The sound fields set the scene, but they were not able to be captured in photographs. One's ability to visualize these environments in one's mind, as I hope you are doing, is probably the best way to imagine what they were like.

In case you were wondering, my home had a more traditional but still modern family room near the kitchen and dining room. The bedrooms were similarly modern but comparatively traditional with views on one side over the distant landscape while the other side opened out into the rainforest courtyard. In a wildflower patch in a forested area, my children had an adventure playground with a ropes course through the tall trees. There was a large sandpit and a cubby house amongst the large boulders. They enjoyed an exceptionally long 18 m slide down a hillside and a swing with very long cables that allowed one to swing very high. Many of our adult friends enjoyed the slide and the swing as much as the children did! There was a grassy oval on which to play games, fly gliders, and kick a ball around.

SIMPLE YET CHALLENGING PUZZLES

The following puzzles are indicative of how we scientists try to explain certain natural phenomena in unrealistic, complex ways when nature often solves problems in simple ways.

❶

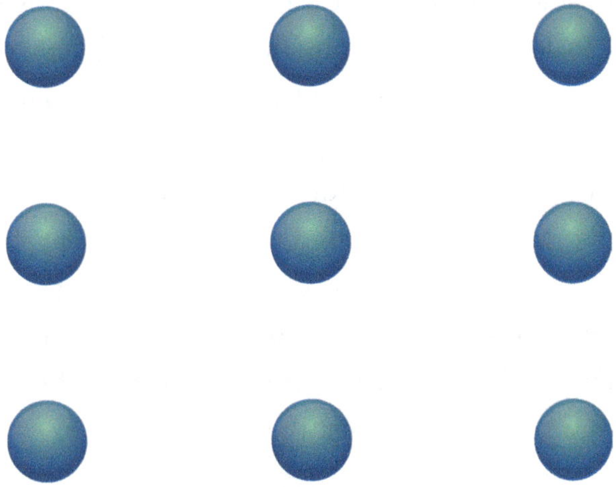

Using a pencil, draw 4 straight lines that connect all 9 dots without lifting your pencil.

> ***Clue to 1*** *: The solution involves thinking laterally so that you do not limit yourself unnecessarily.*

❷

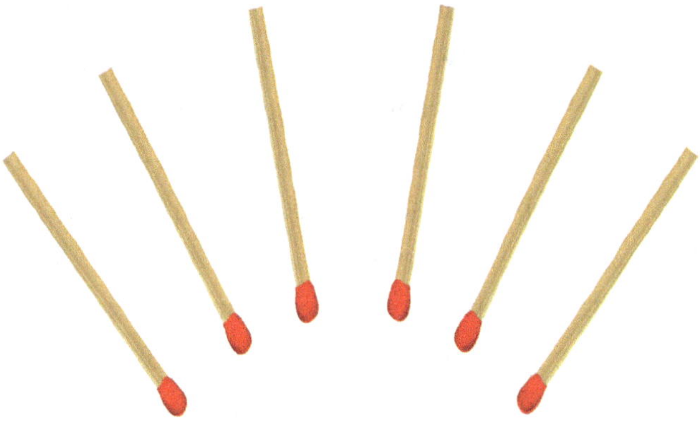

Arrange 6 matchsticks to make 4 equilateral triangles.

> ***Clue to 2:*** *Again, do not restrict yourself in dimensions unnecessarily.*

❸

A EF HIKLMN T VWX

BCD G J OPQRSU

Where does the 'Z' go, and WHY? You must say 'why' otherwise, you could be just guessing.

> *Clue: This is another one of those problems that we typically look too hard for a complicated solution when it's not needed. Young children are often able to work this it out more quickly than intelligent adults because adults have too many complex options to consider.*

AN IMAGINATIVE ANALOGY TO MODERN-DAY EXPLORERS IN SCIENCE

Try to imagine the following concept because it has no doubt been real for many intrepid explorers across the ages.

I once wondered how the following experience would feel to an Australian Aborigine whose tribe lived in Central Australia thousands of years ago, where the environment was a mostly flat, dry, partial desert. One day, a deep-thinking, young, curious individual decided that he wanted to see what was beyond the distant horizon he could see from a mountain. He decided to go on a very long 'walkabout' journey on foot to find out. But as far as he walked, day after day, month after month, he could never get to the end of his never-ending horizon: it just continued to be far away. He thought, surely it must come to an end one day if he goes far enough. He walked south for a long time – so far that the southern stars had moved noticeably higher in the sky. One day, to his amazement, he finally reached a great cliff that was the end of his world! This shook him because he realized that he had come to the edge of his universe and a new, strange universe lay before him. He had reached the Great Australian Bight.

He wanted to go back home to tell his tribe that he discovered that the world had an end! He wondered if he was the only one that the Dreamtime spirit gods had ever shown this to. He wondered what the strange, flat, deep blue world was that lay beyond the end of his world?

At the edge of his world, the land fell away abruptly. Far below was a strange, deep blue void, the likes of which he had never seen before. It was another type of space that had its own horizon that was very far away! This was an amazing thought to him. It had a strange form of matter in it that seemed to undulate. He had no idea what it was. At its edge, there were rises in it that turned into large white frothy forms as they rolled towards the edge of this universe and then disappeared. They made a loud noise as they curled over and crashed. He wandered for kilometers along the edge of his earth, trying to think about what this universe could be. Finally, he found a collapsed part of the cliff that allowed him to make his way down to study this blue phenomenon.

As he got closer to it, it looked like water, but he had never seen such a vast expanse of it before. He walked across the soft white sandy material at its edge. He decided to touch these huge waves that were far larger than any he had ever seen before in shallow lakes. They made a thunderous noise which frightened him. He played with the frothy white material. As he was thirsty, he took a mouthful of the blue liquid and immediately spat it out. What sort of water was this, he wondered? How could there be so much of it? And how could it go on forever? How deep was it? What is in it? Why was it there? He wondered if it had an end to it, but how would he ever know?

His great discovery had answered his question as to what lay beyond the horizon, but it had raised many more questions! His great discovery was totally beyond his comprehension. He had no means by which to interpret its significance.

After a long journey, our intrepid explorer discovered the edge of the world at the Great Australian Bight.

Does all this sound familiar? We are often in this position when we make ground-breaking discoveries that make us question what we think we know about the big picture of nature and the universe. Does our universe go on forever, or does it change beyond the universe we can see?

Going back to our story, try to visualize the following scenario. Imagine someone from today came down from the sky in an old Bell helicopter with a glass bubble cabin, and he decided to land near our hero. Our intrepid explorer would think the pilot was a Dreamtime god from the sky. The pilot beckoned him to come and sit with him in his glass bubble. Our aboriginal explorer was overwhelmed that this spirit god had come to see him. He wondered whether he would praise him or scold him for finding the secret edge of the world.

He wanted to enter the strange and frighteningly loud flying bird, but he was terrified. If he got into the helicopter, would he ever be brought back? But with encouragement from the pilot to enter the helicopter his extreme curiosity overcame his fear. He was taken up into the 3rd dimension! This was beyond what he or any of his tribe could ever imagine. He was absolutely mind-blown by this exhilarating experience. He could see forever across both his land universe and the new ocean universe. Had he become a Dreamtime spirit that was able to look down on his land and see what seemed like all of it. He was one with the clouds! It shocked him that he was traveling at speeds that were incomprehensible to him. How was this possible, he wondered?

How could he ever share his incredible, magical experience of discovery and flying with a Dreamtime spirit with anyone? His tribe would think he had gone insane from eating too many hallucinogenic cacti. He knew his tribe would shun him for telling lies, yet he wanted so much to share what he had learned with his family and fellow tribe members. This is similar to today for some of those at the leading edge of science who make astounding discoveries that seem impossible to others who do not have their vision.

Little did he know that in the future, when he finally returned home, he would marry a lovely girl who would believe him, and their children would also believe their father's story. When his children were 9 and 11, he decided to take his family on a long walkabout to see the edge of the world. After months of traveling, they saw that there really was an end to their world, and they could see that what he had told them was the truth. They decided to live there because there was plentiful seafood, and many lizards, birds, and kangaroos to eat. They loved swimming in the protected beaches by the sea and exploring the caves, many of which were filled with clear, deep freshwater. Along the bottom of the cliffs, there were many caves for them to live in. He and his family were blessed by living in what they considered to be heaven. In time, they discovered that there were other small Aboriginal tribes further along the coast that also lived by the sea, so they had friends.

Our hero never suspected that his land in Central Australia would have an end. In this new land by the sea, there were no mountains or rivers and very few trees. He discovered that many kinds of small and large fish, turtles, seals, and huge whales lived in the ocean universe. Little did he know that if he had our technology that was far more advanced than his spear, he would be able to go into the water-universe and discover that it is filled with an incredible diversity of strange animals, canyons, volcanoes, and huge plains. And at its bottom, whole ecosystems of strange life forms thrived in total darkness and under extreme pressure.

Many aspects of this story parallel those that have happened to many of our leading scientists and explorers over many millennia. They had realistic visions of some aspects of nature that others could not envisage. This story is a good analogy for what we are starting to discover of the big picture of the universe. Imagine if we could ever find a way to see past the horizon of our observable universe? If so, **how different might it be, and what might we discover that is beyond our imagination.**

This analogy is closely related to how, even today, we too are confronted with conceptual challenges that are beyond our knowledge limit, our intellectual grasp, and our technology. We experience this when we delve into the inexplicable quantum universe and when we peer into the depths of intergalactic space to discover that the universe has an unexpected filamentary structure. We have now reached the boundary of time by observing the edge of our observable universe, and we cannot fathom where the matter that created our universe came from. We are completely bewildered by what type of space exists inside a black hole. It's unfathomable to us how our universe could have so many finely tuned parameters to allow it to exist. Now that we are delving into the workings of our brain and into the extremely complex chemical structures of molecular biology, we are confronted with ever more mysteries at this new boundary to our knowledge. All these ambiguities are beyond our present ability to solve them, but they will be solved in time.

In the very near future, we are about to discover incredible new concepts that will take us far beyond what we presently think reality is. This book will allow you to glimpse some of these concepts. **How will people handle these new concepts about the nature or the many things that go against our present beliefs?**

ANSWERS TO SIMPLE YET CHALLENGING PUZZLES

Answer to Puzzle 1:

Think outside the square, go beyond the dots.

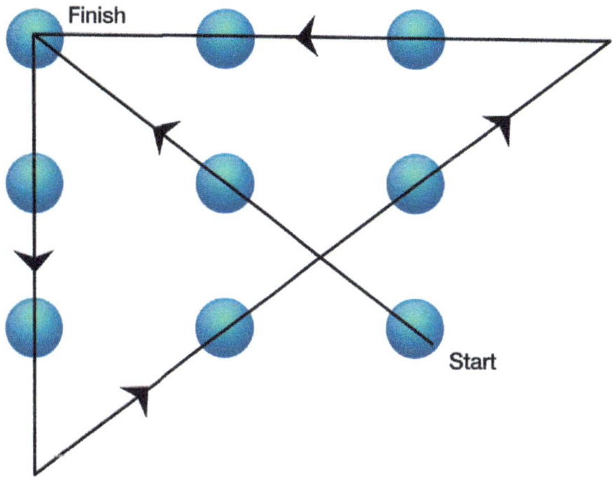

Answer to Puzzle 2:

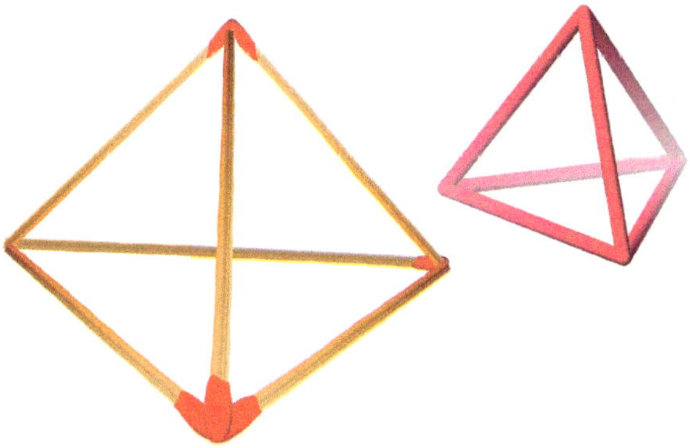

Go into the 3rd Dimension to form a three sided pyramid known as a tetrahedron.

Answer to Puzzle 3:

The 'Z' goes at the end of the top line because all the letters above the line are all made of straight lines, while those below have curves.

WARNING! CHALLENGING CONCEPTS FOLLOW

A number of the following chapters may be confronting for readers who have strong beliefs or desires that they do not want to be challenged. If you are such a person, **you are advised *not* to read the chapters listed below**, because the information in them could be upsetting to you. The information in these parts is based on:

1. **scientific discoveries** that I, and many scientists, initially found to be unexpected and alarming because they went against what we have been taught and believed to be true,
2. **logical deductions based on evidence and fact**, and
3. **the impact that the rapid advancement of technology will have on humanity's future.**

The chapters that could be upsetting for those with fixed beliefs or wishes are:

- **Chapter 6** – about gods,
- **Chapter 10** – about aliens and UFOs,
- **Chapter 11** – the plausibility of humans exploring Mars,
- **Chapter 12** – the practicality of space travel across our galaxy,
- **Chapter 13** – how realistic space warps and teleporters are,
- **Chapter 16** – how quickly computers might take control,
- **Chapter 20** – what the impact of robots will be,
- **Chapter 23** – whether the future is predetermined,
- **Chapter 24** – whether we have free will,
- **Chapter 25** – whether we consciously make our own decisions,
- **Chapter 26** – to what extent do our genes control who we are,
- **Chapter 30** – whether everything in the universe is programmed,
- **Chapter 31** – are paranormal occurrences real.

When confronted with concepts that go against our ingrained beliefs, we typically reject them out-of-hand. Even though I am naturally open-minded, I reacted this way when I was first confronted with many of these discoveries (and others not included here). As I matured, I learned to broaden my beliefs about what could be possible because, over time, evidence built up to the point that forced me to research subjects deeply. To my astonishment, I found that there was often considerable, and in some cases, overwhelming evidence supported very challenging new theories, extraordinary natural events, and discoveries that had seemed illogical or implausible.

Before considering the probability of something being true, it's important to seek out a broad range of views and interpretations from *reliable* sources that have little reason to be biased. Unfortunately, we all have genetically programmed biases built into our thinking that we cannot easily undo. (See Chapter 27.)

There are few absolutes, so we must consider new information from the perspective of how *probable* it may be. I require strong evidence before I am prepared to consider that something has a high probability of being right, or partially right. Very few things have a probability of being 100%

true because the more we learn, the more we realize that we do not have the whole picture. **There is always a deeper layer of knowledge underlying what we know.** As well, presumptions are often made on insufficient data, and this causes us to jump to erroneous conclusions.

I have had decades to absorb new surprising discoveries across many fields of science. This has given me time to come to grips with confronting evidence. But for readers that are confronted with startling new concepts and evidence for the first time, accepting this can be very difficult. It's important to keep an open mind and to take one's time to think through such things. Over time, we soften our objections and become more prepared to consider broader possibilities. Exercising one's mind by thinking deeply can be as hard to do, just as it is to do hard physical exercise. **Remember that most of what we know today was considered impossible just a century ago.** And also remember that we currently know very little compared to what we will know in the decades to come. The more we learn, the more we realize that our beliefs have been very simplistic and often completely wrong, so we are forced to broaden our perspective on many issues.

When people are confronted with information that goes against the current paradigm of the day, they often want to shoot the messenger. Whilst I sometimes express my own conclusions, I am mostly presenting research from credible sources. It is up to readers to make up their own minds after they have done further research.

It seems that the more we learn, the stranger and more complex the universe becomes. This is immensely exciting. It challenges us to think beyond our encrusted old beliefs, which are mostly based on simplistic thinking and a severe lack of knowledge.

We are now able to learn so much about numerous aspects of our world, the universe, and our selves that we cannot help but wonder if there will ever be an end to the seemingly endless discoveries that can be made. New insights will turn many of our current beliefs upside down. This is why science is so enthralling to inquisitive minds.

"Learn from yesterday, live for today,
and hope for tomorrow.
The most important thing is to not stop questioning."

– Albert Einstein

For readers who are prepared to keep an open mind to challenging new possibilities, read on.

OUR AMAZING UNIVERSE

Chapter 2

HOW DID THE UNIVERSE EVOLVE and WHAT IS IT MADE OF?

The following is a short summary of how matter evolved into increasingly more complex forms to create life and, ultimately, intelligence.

To understand how the universe works, it is necessary to realize that **gravity is the basic organizing force in the cosmos**. It is crucial in creating the structure of everything in the universe. It is an incredibly feeble force on small scales compared to the other three forces (electromagnetism and the strong and weak nuclear forces) because it is 10 billion, billion, billion, billion times weaker. Gravity is a force that acts over large scales on everything in the universe, whereas the other forces only act over comparatively short distances. Gravity is what causes mass to collect together to form stars, planets, galaxies, and huge superclusters of galaxies. But **gravity is not actually a force! It occurs as a result of mass (matter) curving space and slowing time,** just as Einstein predicted. There are many proofs for this.

WHAT OCCURRED AFTER THE BIG BANG?

At the birth of the universe, it was so hot that it is thought that only energy existed. It is postulated that almost immediately after the universe came into existence, it must have experienced an incredibly fast inflation to spread out its force fields exceptionally evenly. **Inflation** is the only way physicists can *presently* explain how the universe did not become lumpy. This extremely rapid expansion of space caused the embryonic universe to expand at an astonishing 100 trillion, trillion, trillion times(!!!) - that's 10^{26} times, or 10 with 26 zeros after it, in a minute fraction of a second! It went from being as small as a proton in an atom to the size of a grain of sand, or maybe a grapefruit in an instant. (There is currently no cause postulated for inflation.) Afterward, it expanded at a much slower rate that was nevertheless much faster than the speed of light! (Space can expand at any speed, but matter cannot exceed the speed of light.)

The expansion caused the universe to cool from an infinite, hot temperature. This allowed a multitude of subatomic particles to form. It is thought that there was no matter until three of the energy fields 'froze out' into extremely tiny particles known as **quarks**. The expansion continued very quickly, continuing to cool the universe, thereby allowing six types of quarks to combine in many different arrangements to form around a thousand different, small, **sub-atomic particles.** The main ones being protons, neutrons, electrons, and photons. As the temperature continued to drop, these particles combined to form the first **atoms** – mostly hydrogen gas with a little helium. Where the most gas existed, after about half a billion years gravity drew these elements together to form **stars** and their **planets.** There had to be the equivalent of hills and valleys in the gravity field for matter to collect in certain places more densely. But no one has explained what could have caused that.

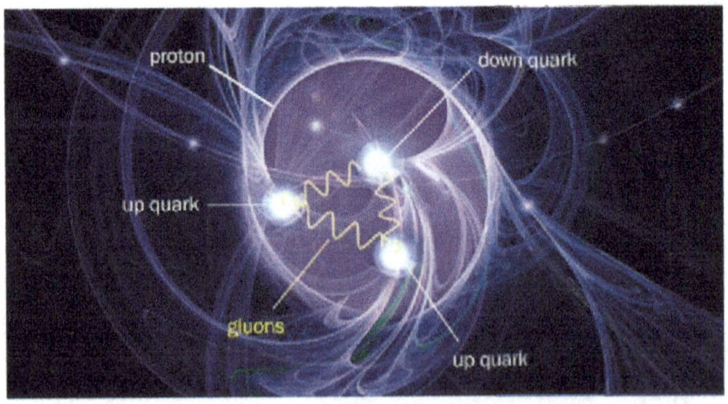

Protons are thought to be made up of a combination of 'up' and 'down' quarks which are held together by a force called gluons.

Protons and neutrons fuse together in the cores stars to form around ten other heavy elements. The largest stars become so dense in their super-hot cores that they explode in only a few million years. These events are known as supernovas. When the supergaint stars exploded, they fused the existing elements together to form another 80 odd heavier elements. Each element has very different properties. And when elements are at different temperatures and pressures, they have different states with different properties.

The four main states of matter are plasma, gas, liquid, and solid.

Many elements form chemical bonds with one or many other elements to become **molecules**. Molecules have very different properties as well. When these elements combine, their combinations create almost infinite combinations. This incredible diversity of matter combined with an enormous diversity of conditions creates almost infinite opportunities for matter to have any state imaginable.

Stars that have different masses have different properties. After enormous numbers of stars formed in the early universe, gravity drew them

together to form large conglomerations known as galaxies.

THE EVOLUTION OF LIFE

On planets with the right conditions, complex arrangements of molecules formed complex organic compounds **that could replicate** - like crystals do - but in more complex ways. These chemical structures formed **amino acids** - **RNA and DNA.** RNA formed much more complex **viruses,** which were not quite life. Eventually, more complex DNA became the basis for **single-cell organisms** - the first life forms to exist. These cells became the basis of more **complex multi-cellular life forms, which** evolved into much more complex forms of life. Over a couple of billion years, life developed a **broad diversity of ever-improving senses, ways of moving around, means of digesting different types of foods, and developing numerous types of biological weapons and defenses.**

The most advanced life forms evolved varying levels of **intelligence and increasingly greater memory,** both of which became ever more complex. Ultimately, over another couple of billion years, the most advanced **intelligences evolved, and they invented high technology.** This caused the evolution of these life forms to speed up dramatically. They quickly learned how to manage genes to make themselves more advanced, and to build other life forms through **genetic engineering.** When they discovered how to expand their minds beyond their brain, they invented **computers.** This caused their evolution to explode exponentially! These intelligences evolved very quickly into **non-biological, super-intelligences** with unbounded potential. This is the level of evolution that we humans are now experiencing.

If life is common throughout the universe, as we think it probably is, then there are possibly innumerable intelligent life forms in the universe, many of which evolved billions of years before life on Earth even started to evolve. By now, they would have evolved immeasurably far beyond our relatively primitive state of consciousness.

A BROADBRUSH EVOLUTION OF THE UNIVERSE

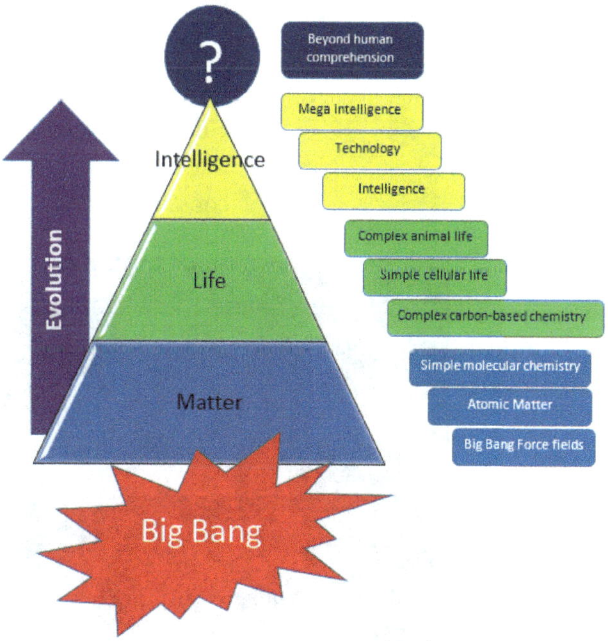

The evolution of the universe - as far as humans can determine to date.

MAJOR MILESTONES IN THE EVOLUTION OF THE UNIVERSE

Elapsed Billions of Years Since Big Bang	Billions of Years Ago	Comments here assume life evolves across the universe
0	13.8	Big Bang
0.4	13.3	Matter & light decouple
0.6	13.1	First stars formed
0.8	12.9	First supernovas occurred
1.2	12.7	First galaxies formed
1.5	12.3	First planets formed
1.8	12.0	Simple molecules form then complex organic chemistry commences
2.1	11.7	Building blocks of life evolved
3.6	10.2	First cells evolve on suitable planets
6.1	7.7	First advanced multi-cellular life equivalent to dinosaurs evolved
6.3	7.5	First intelligent biological life forms create non-biological computer-based super minds
9.1	4.7	Our solar system formed
13.8	**Now**	Humans on Earth create the first non-biological, computer-based mega-intelligence. Humanity starts to become aware of the role that intelligence and consciousness plays in the evolution of the universe.

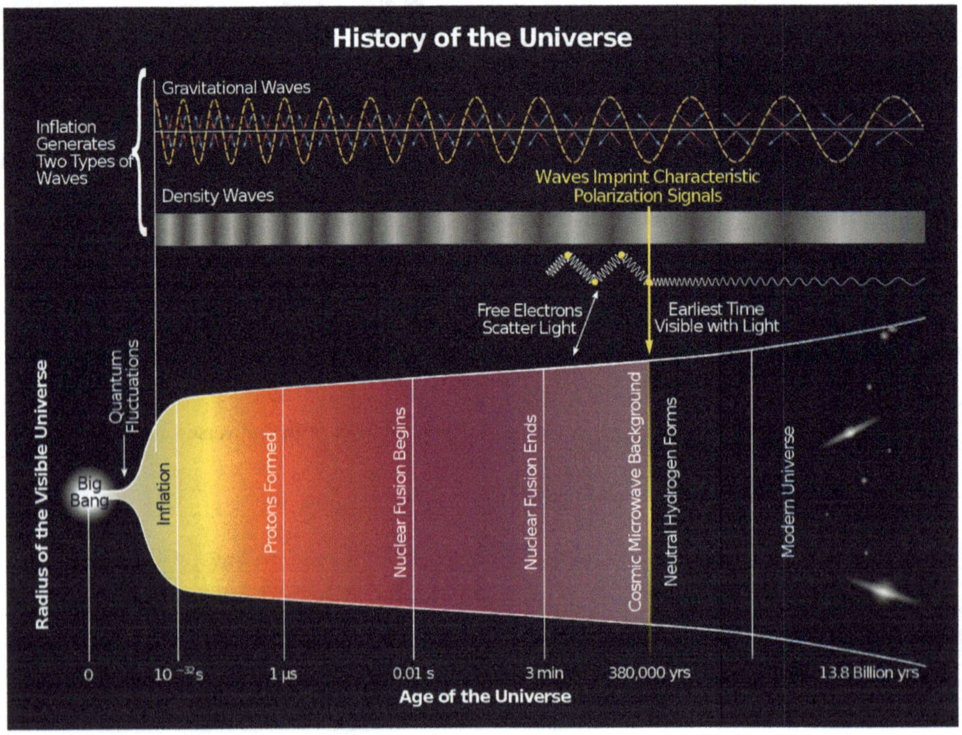

This diagram defines important stages in the evolution of the quantum universe.

WHAT IS THE STRUCTURE OF MATTER?

The following sequence illustrates how matter evolved. Because it is unknown what matter is really like, the following diagrams are purely conceptual to help visualize how matter may have formed. (All diagrams by Gregg Thompson and Nicole Brooke.)

Four force fields are thought to have 'froze' out of the background unified field after the Big Bang occurred and started to cool as the universe expanded.

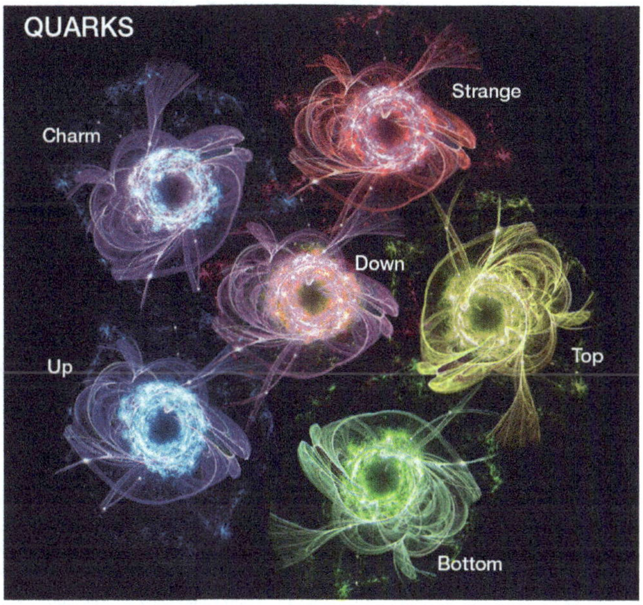

Three of the force fields cooled, allowing the smallest particles of matter - six quarks - up, down, top, bottom, charm, and strange - to 'freeze' out.

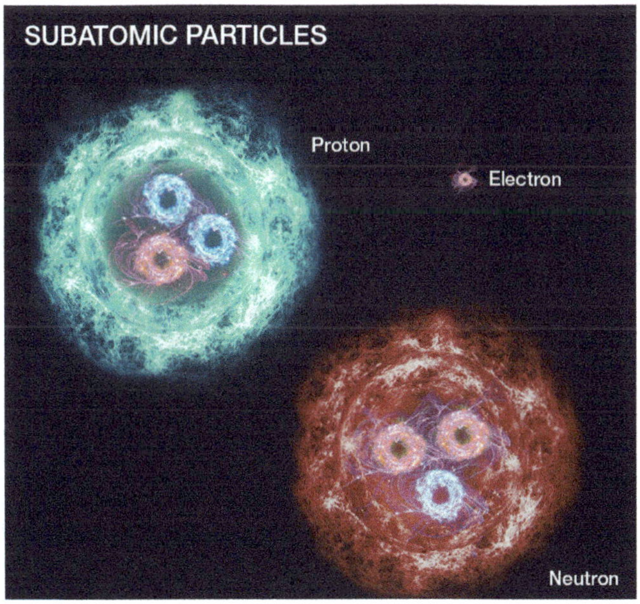

Subatomic Particles - Quarks combined to form nucleons, protons and neutrons. Protons (the turquoise outer haze) have two Up quarks (light blue) and one Down quark (light pink). Neutrons (the red outer haze) have one Up quark and two down quarks. Electrons remained free at this stage.

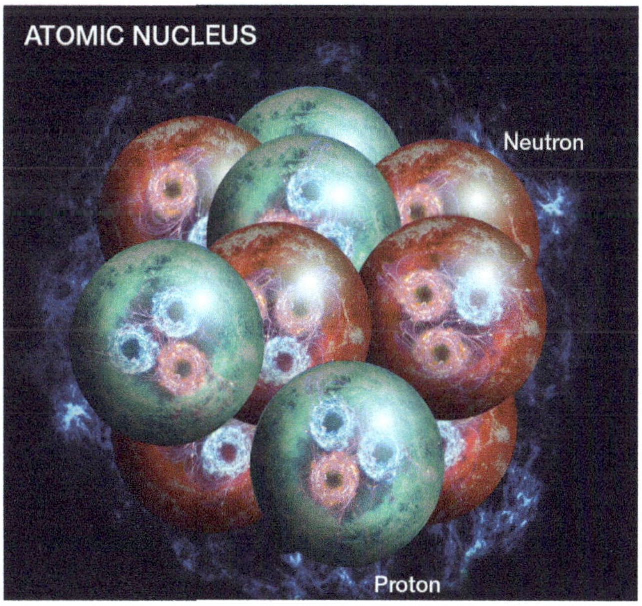

Protons and neutrons combine to become the nuclei of atoms. When the temperature dropped, they were able to capture electrons to form atoms. Atoms are typically represented as spheres for simplicity, but they are probably nothing like that in reality.

SUBATOMIC PARTICLES, ATOMIC NUCLEI, ATOMS, GLUCOSE MOLECULES, ORGANIC MOLECULES, CARBON POLYMERS

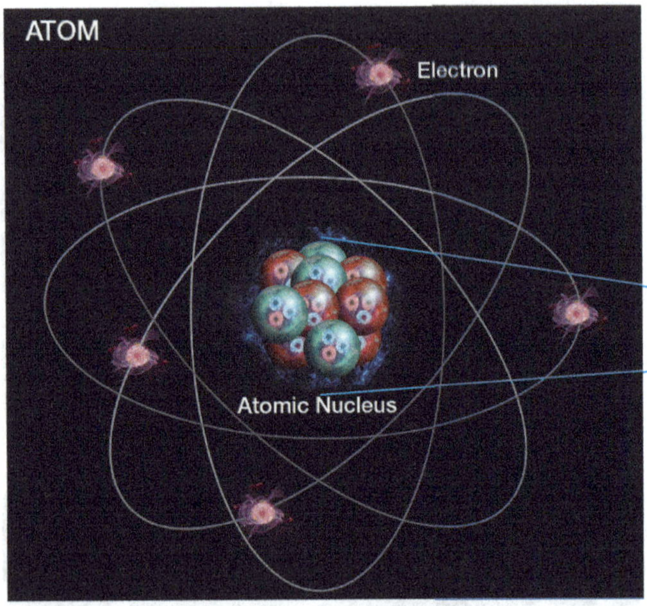

Electrons 'orbit' atomic nuclei to form atoms. Electrons could whiz around more than one atomic nucleus to bind atoms together to form molecules. In reality, electrons do not 'orbit' a nucleus like a satellite does around the Earth. They are more like very fuzzy clouds that can be in many places at once. Atoms are so small that 10 million of them would be required to sit end to end to span the full stop here!

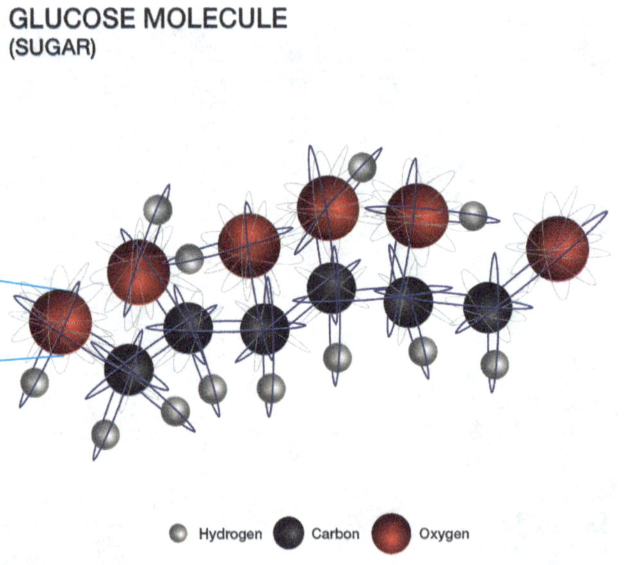

Atoms of carbon (black), hydrogen (gray), and oxygen (red) have combined here to form this molecule of glucose (sugar). Each atom shares its outer electrons (blue ellipses) to hold the molecule together. The more energy atoms have, the more they vibrate. If they have too much energy, the molecule 'shakes' apart into its basic atoms.

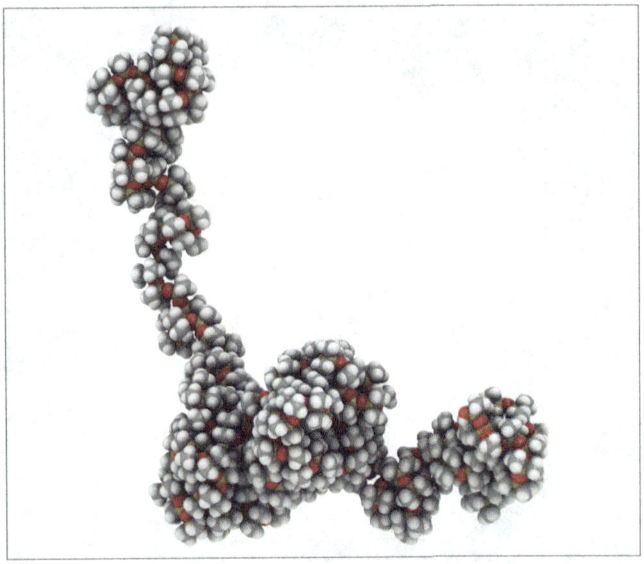

Numerous molecules combine with carbon to form giant chains of organic molecules knowns as polymers. These become the foundation for constructing life.

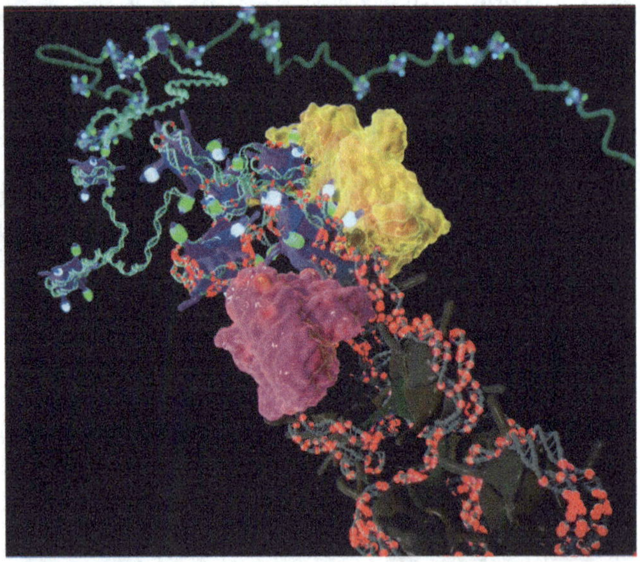

Nature uses long carbon polymer chains to build very complex machinery inside cells to allow them to move, create the molecular structures they need to have senses to find food and to process it to get energy, and to eject waste, to defend themselves, and to replicate. Credit: Ricken Research

In the Big Bang, subatomic particles known as quarks combined to form nucleons, protons, and neutrons. Protons (shown with a turquoise outer haze in the diagram above) have two Up quarks (light blue) and one Down quark (shown in pink). Neutrons (the red outer haze) have one Up quark and two Down quarks. Electrons remain free at high temperatures and densities.

Protons and **neutrons** combine to become the nuclei of atoms. When the Big Bang's temperature dropped enough, nucleic were able to capture **electrons** to 'orbit' them to form atoms. Atoms are typically represented as spheres for simplicity, but they are not like that in reality.

Electrons are able to whiz around more than one atomic nucleus to bind atoms together to form molecules. In reality, electrons do not 'orbit' a nucleus like a satellite does around the Earth: they are thought to be more like fuzzy clouds that can be in many places at once.

An **atom** is so small that 10 million of them would be required to sit end to end to span the full stop here!

In the diagram of a glucose **molecule** above, atoms consisting of carbon (black), hydrogen (gray), and oxygen (red) have joined together to form this molecule of sugar. Each atom shares its outer electrons (blue ellipses) to hold the molecule together. The more energy atoms have, the more they vibrate. If they have too much energy via heat, the molecule will vibrate apart into its basic atoms.

Numerous molecules combine with carbon to form giant chains of **organic molecules** knowns as **polymers**. These become the foundation for constructing life.

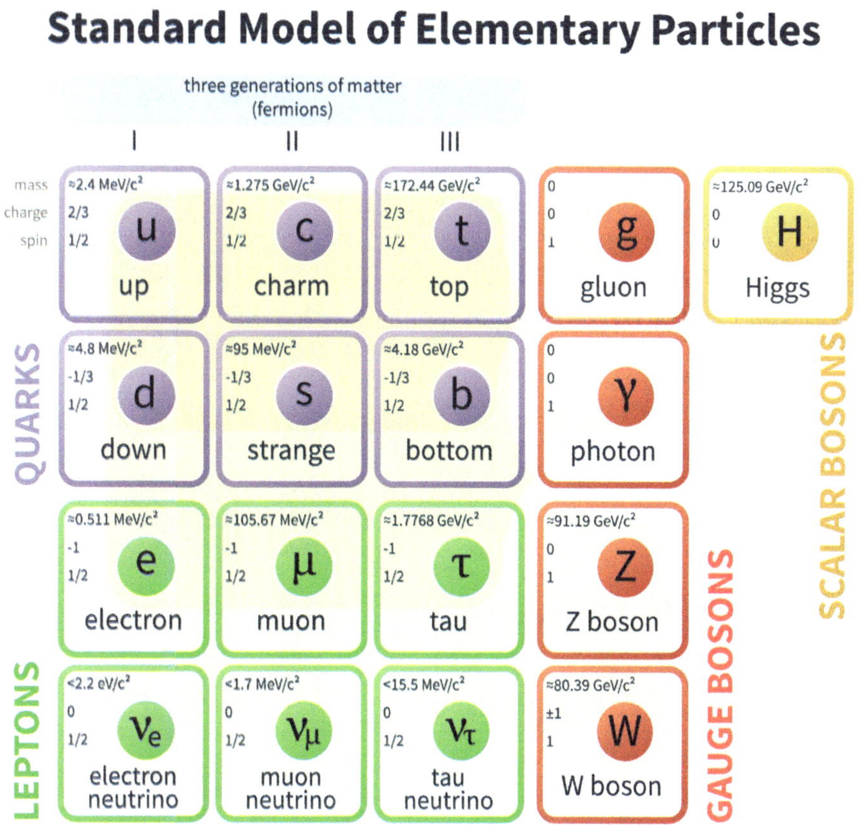

HIGH ENERGY QUARKS
Top and Bottom

MEDIUM QUARKS
Charm and Strange

LOW ENERGY QUARKS
Up and Down

This chart shows the energy levels, the spin, and the charge (in small black type) for each type of known particle. For each quark and lepton, there is an antiparticle.

Quarks can't be isolated. They must always exist in pairs. The more they are pulled part, the greater their energy grows because they absorb the energy used to pull them apart. If they are pulled a long way apart, which is extremely difficult, new quarks form in the space between them! There are three versions of quarks and leptons with each having more massive versions of the same particle for reasons we do not understand. The heavy versions only exist in high energy events like the Big Bang.

Neutrinos are almost massless particles that move at almost the speed of light. They are formed in the cores of stars in huge quantities. They do not interact with matter. It would require a light year thickness of lead to stop them! Sixty-five billion pass through our body every second!

Because a **photon** of light travels at the speed of light, it does not experience time, so to a photon, the moment they are emitted, they are absorbed, whereas we see some photons taking up to billions of years to traverse the universe.

"There is nothing in the universe that does not have a reason for it being the way it is. To know a reason for something, we have to be smart enough to work it out. Of course, there will be reasons that are far beyond the intelligence and knowledge of all of humanity, so we can never expect to know those reasons."

Gregg Thompson

Chapter 3

WHAT IS THE NATURE OF THE UNIVERSE?

WHAT IS THE 'OBSERVABLE' UNIVERSE

The term 'observable' universe is often mentioned in astronomy, and some chapters in this book, so it's important to understand what this term means. **The 'observable' universe is the portion of the Greater Universe that we can observe.** This is similar to the amount of ocean we can see from a ship at sea, or the area of land we can see from a mountain. The observable universe is expected to be a small part of the entire Greater Universe that is thought to extend enormously far beyond it. The observable universe has a radius of around 13 billion light years from any location. We can see nothing beyond this distance.

The distance to the horizon on Earth is limited by our planet's curvature and the observer's altitude above sea level. However, **the distance to the universe's horizon is limited by the speed of light**. If the speed of light was infinite, we would see the entire Greater Universe. But because light travels very slowly with respect to the size of the universe, this means that the deeper we look into space, the longer it takes light from distant galaxies to reach us. This means that **the deeper we look into space, the further we are looking back in time. When we look back as far as we can, we see almost to the birth of the universe in every direction. Incredibly, it surrounds us!** Because this is where time began, we cannot see beyond the beginning of the universe. Understandably, this is the boundary to our 'observable' universe. Nowhere do we see the universe as it is today because **everything we see is as it was sometime in the past** - even our nearby Moon and the planets.

When we look into the universe, we are not seeing galaxies as they are now; we see them as they were when their light left them millions or billions of years ago. Intelligences that live in galaxies near the edge of our observable universe would see our galaxy near the edge of their observable universe, so our Milky Way would appear to them as being new ly born. In the opposite direction to us, they can see galaxies for another 13 billion light years beyond our horizon - just as we can see galaxies 13 billion light years beyond their horizon in the opposite direction to them. (See the diagram below.)

The galaxies that we see at the edge of our observable universe **are now the same age as our galaxy, but we see them as they were around 13 billion years ago when they were forming. Because the universe is expanding, we would expect that over 13 billion years, they would have moved much further away. They are now 46 billion light years away.** Because galaxies beyond our horizon are receding from us faster than the speed of light, their light can never reach us, and our light can never reach them. Therefore, no information can ever be exchanged between galaxies that are beyond each other's observable universe. It's as if they do not exist.

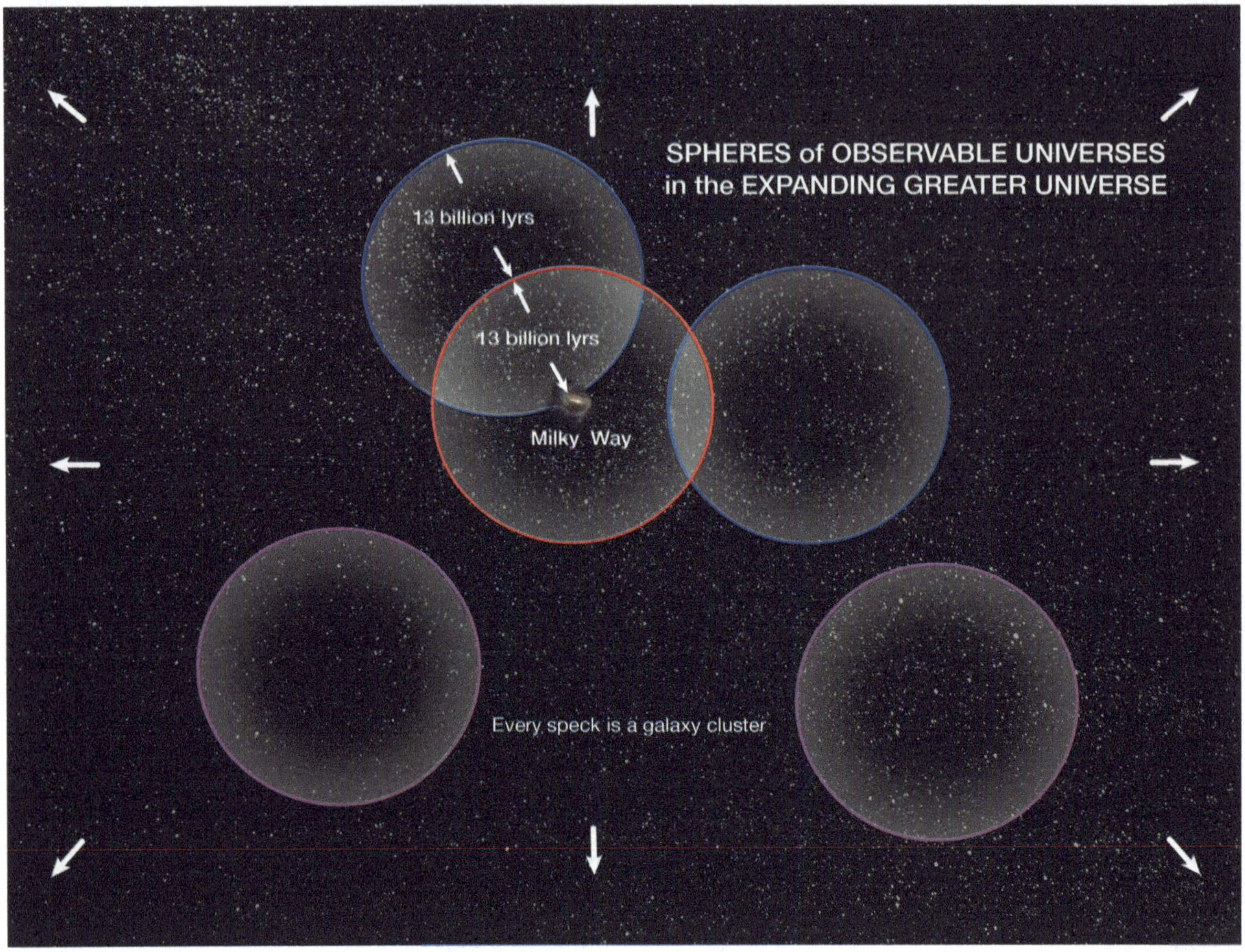

Here we see a part of the Greater Universe in which the circles represent spheres of observable universes within it. The red circle represents the edge of our observable universe, which is centered on our Milky Way galaxy. Blue spheres represent other observable universes that overlap ours. Those with violet edges are completely separated from our universe because they do not intersect ours. They can never interact with ours because light (i.e. information) cannot be exchanged between them. Other galaxies can only see our galaxy if they lie within our observable universe. It encompasses galaxies within 13 billion light years. It is expected that all galaxies in the Greater Universe have an observable universe the same size as ours, presuming that the speed of light is the same everywhere. The arrows represent the expansion of space, which is pushing galaxy clusters apart increasingly faster. Due to this, as time goes on, each observable universe will have fewer and fewer galaxies in it. Ultimately, in the very distant future, each observable universe will contain only one huge mega-galaxy, which is the result of galaxy clusters merging.

DOES TIME AFFECT HOW WE SEE OUR UNIVERSE?

Our universe's horizon is both a distance horizon, and a time horizon. This stops us from ever seeing, or being able to interact with, the Greater Universe that lies beyond our observable universe. Galaxies that are beyond our horizon may as well be in another universe because no matter what happens in space-time beyond our observable universe, it will never affect us. Space across the universe is expanding faster than the speed of light, so information from the Greater Universe beyond our observable universe can never reach us because it would have to travel faster than light to do so, and that is impossible.

When we see a distant horizon on Earth beyond which we cannot see, we know that the Earth extends far beyond that. Similarly, astronomers believe that our universe extends far beyond the horizon of our observable universe, possibly for an unimaginable distance. We know this must be so because when the universe first formed, there were far more galaxies within our observable universe than there are today. This is because the universe was much more compact then, and the rate of its expansion was less. But since then, galaxies that were far away from us when the universe was young have moved beyond our horizon due to space expanding faster than the speed of light. (Matter traveling through space cannot travel faster than the speed of light, but space can expand faster than light.) This can be likened to ships at sea near an observer's horizon sailing beyond it and becoming invisible.

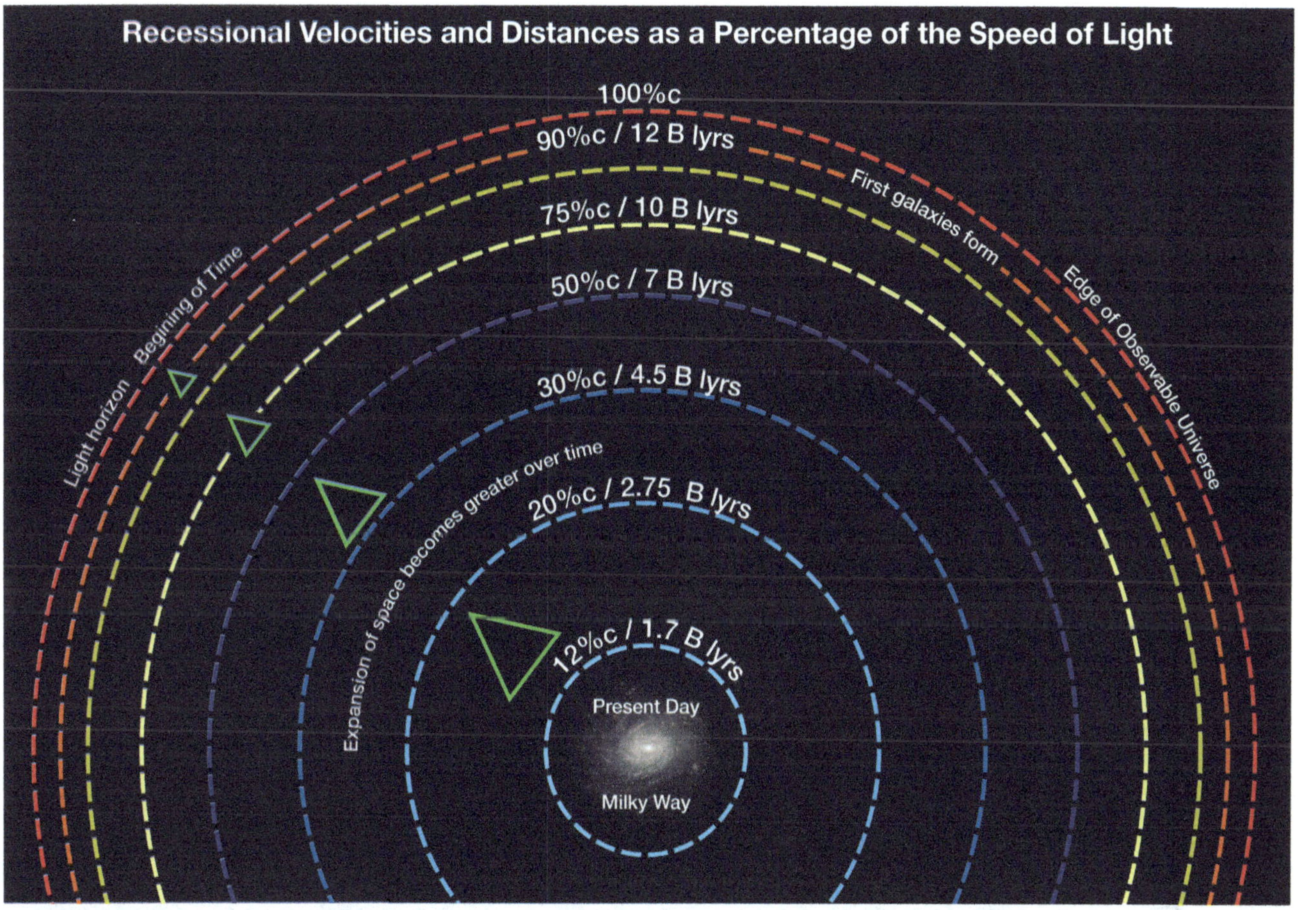

The distance to the edge of our observable universe is 13 billion light years. The dimensions of space are represented as triangles. They appear more contracted the further back in time we go because the universe was much smaller and more compact then. The dashed circles represent the percentage of the speed of light (c) that the universe is receding from us at those distances. The closest circle is 1.7 billion light years away and it is receding at 12% of c. Galaxies at 7 billion years are receding at 50% of c. The most distant galaxies we can observe are 12 billion light years away. They are receding at 90% of the speed of light!

Gravity in galaxy clusters holds them together. Most galaxies are in clusters of galaxies. These clusters merge with other clusters to form supermassive clusters of galaxies. At their centers are the most massive galaxies. These eventually merge into supermassive galaxies.

The concentration of gravity in galaxy clusters causes space inside them to contract. This causes the galaxies to become increasingly more concentrated towards their centers. Over time, they merge with other galaxy clusters to form massive superclusters of tens of thousands of galaxies. Over billions of years, they will eventually merge with other superclusters to form supermassive clusters. The centers of these have so much gravity that all their galaxies eventually become one enormous mega-galaxy.

However, the space between the mega-galaxies is expanding due to the expansion of the universe. This causes each mega-galaxy to move further apart. In the very distant future, these mega-galaxies will move beyond each other's observable universe. If intelligences still exist in them, they will not know that other mega-galaxies exist. Nor will they know how the universe formed, or how it evolved. All they will see is their own mega-galaxy, which they would think is the entire universe.

Over hundreds of billions of years into the future, many supermassive clusters will merge to form a single, gigantic, mega-massive elliptical 'galaxy' consisting of hundreds of thousands of galaxies and millions of giant globular clusters. Their cores will be blindingly bright and full of supermassive black holes. As the universe ages, all galaxies will eventually become a part of one of these monsters.
Illustration by NASA

IS THE GREATER UNIVERSE THE SAME AS IT IS IN OUR REGION?

We assume that the Greater Universe beyond our observable universe looks the same as it does in our region, but does it? At least the nearby Greater Universe should look the same because the parts outside our observable universe were inside it after the birth of the universe.

Is it possible that the Greater Universe could have many different environmental regions like the Earth does e.g. oceans, ice caps, deserts of rock and sand dunes, rainforests, grasslands, wetlands, plains, and mountain ranges etc, but there is no indication to suggest this is the case.

Could the density of galaxies vary across the Greater Universe? Could gravity have different strengths in different locations as it does on Earth from the bottom of the ocean to the top of mountain ranges, and where there are dense masses inside the Earth? Could time tick at different rates in some places in the Greater Universe? Could space be more compressed or expanded in different directions? Does the Greater Universe expand evenly in all directions? Is it possible that the Greater Universe is rotating so that its velocity in certain parts might vary from one region to another? Could the geometry of the Greater Universe vary like the surface of the Earth does from a flat ocean surface and flat plains to deep ocean trenches and high mountains? How would we ever know?

We now know how large our observable universe is. As far as we can tell, the geometry of its space appears to be flat. But that may be an illusion because the curvature of the Greater Universe could be so slight due to it being so large that we cannot detect it – just as an ant would not detect the curvature of our Earth by walking across a salt lake a kilometer wide. If astronomers could determine the geometry of space, this might help us determine how large the Greater Universe might be. But how could that be done?

Almost as soon as the universe formed, cosmologists postulate that it experienced an extremely fast but an incredibly short period of extreme inflation to explain how the universe is so similar in every direction. **Alan Guth** says that the universe's size before inflation took place, was its age multiplied by the speed of light. If so, then the Greater Universe would be in excess of 250 times the size of our observable universe today. There are other estimates that give its size as being close to infinite. On the other hand, if the universe is finite but unbounded (like Earth's surface is), then its boundary might be smaller than our observable universe (less than 13 billion lyrs). This is because the distant galaxies we see could be those that are nearby, but we are seeing them from the opposite direction. Their light may have gone around the universe and be returning to us from the opposite direction. One of these galaxies could be our own! This can't be tested today because, like people, galaxies look different at different ages.

With future technology, we should be able to see slightly beyond the cosmic microwave background caused by the Big Bang. If so, we might be able to detect the relic of the neutrino background that occurred a little earlier than the microwave background of the Big Bang. If we can build highly sensitive gravity wave telescopes in the future, we may even be able to see gravitational waves emanating from the Big Bang.

For astronomers living more than a century ago, the amazing concepts in this section would have been impossible for them to conceive. These concepts still remain inconceivable to well over 99% of people alive today! You are one of a very small percentage of humans who may understand the concept of the observable universe and how time stops at its edge because this is where the universe was born.

THE UNEXPECTED FILAMENTARY STRUCTURE OF THE UNIVERSE

When astronomers recently mapped the locations and distances of hundreds of thousands of individual galaxies and galaxy clusters, they found that they were *not* evenly spaced as expected. They formed long chains in every direction. This was an amazing, unforeseen discovery that was unimaginable even as little as 25 years ago. Major new discoveries like this never cease to amaze us.

This HST deep-field image shows chains of galaxy clusters at different distances across the universe. These form the filamentary structure discovered throughout the universe.

The filamentary structure of galaxies across the universe and the voids between them appear somewhat similar in structure to a close-up view of froth, which has bubbles of many sizes that are joined together.

From a distance, these chains of galaxies look like filaments but they are thought to be the edges of bubble-like voids. Where filaments meet, clusters of galaxies form because this is where gravity is more concentrated. Where many filaments come together, gravity is at its greatest, so this is where superclusters are found.

The web-like filamentary structure is similar to froth because the filaments of galaxies form structures like the membranes of bubbles in froth. Inside the filaments are bubble-like voids that have different sizes. Like bubbles in a froth, the filamentary membranes are connected to those around them. But the 'bubbles' in the universe's filamentary structure are not round. The voids have few galaxies in them and none at their centers. Why is the universe structured like this?

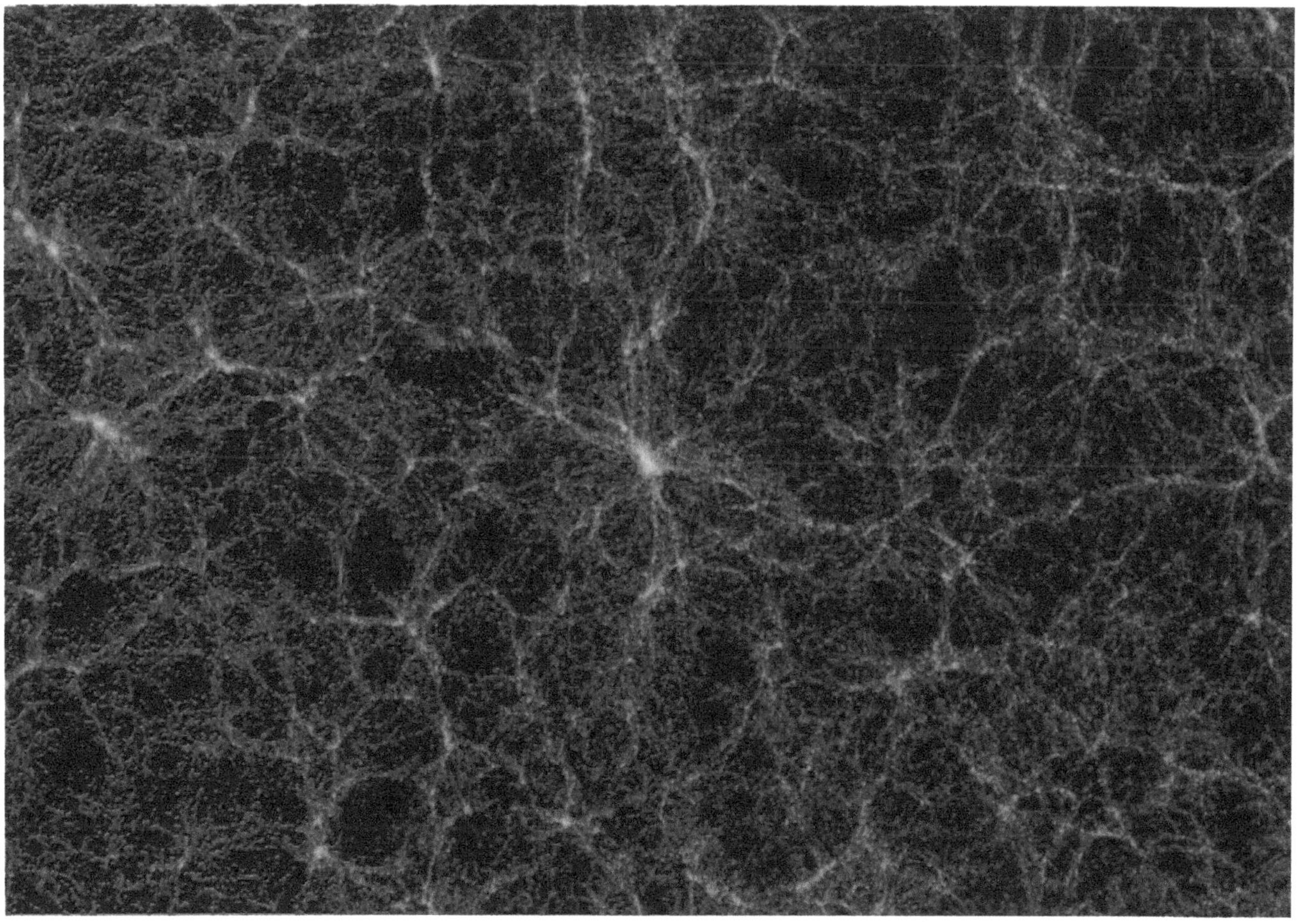

This is a 3D computer plot of the location of hundreds of thousands of galaxies. It reveals a totally unexpected, three-dimensional, froth-like structure that extends throughout the universe in every direction. This was an amazing discovery.

It has been suggested that the expansion of the universe is occurring because an antigravity force, possibly in the form of dark energy, is growing in the voids causing them to expand more quickly with time. While space is expanding inside the voids, it does not appear to be expanding where the galaxies lie along their borders because gravity is holding space together there.

When seen at a close range, the filamentary structure does not appear to be homogenous; however, when seen on a large scale, it is the same in all directions - just as the Big Bang Theory predicted.

Slight fluctuations in the uniformity of the background energy level at the very beginning of the Big Bang is what is thought to have caused this structure. It is seen in the microwave background, which occurred soon after the Big Bang.

This filamentary structure is the largest known structure in the universe. Because it has taken billions of years for the light from galaxies in these distant filaments to reach us, we are seeing them as they were when the universe was only around a third to a quarter of its current age. How might it have changed since then?

THE SPOOKY ALIGNMENT OF THE POLES OF SUPERMASSIVE BLACK HOLES

Astronomers have measured the orientation of the poles of supermassive black holes that reside at the centers of 93 highly luminous galaxies. **Most surprisingly, the axes of these giant black holes are not randomly orientated! Nearly all align with the filaments in which the black holes in each galaxy lie.** The probability of this being through random chance is less than 1%! What force could possibly align the spin axes of supermassive black holes across the universe when all galaxies seem to be randomly orientated? Could the gravitational fields along the filaments somehow align the black holes' spin axes? The few that don't look aligned could be due to a filament running at right angles to other filaments in front of or behind it. Incredibly, the alignments extend for over billions of light years.

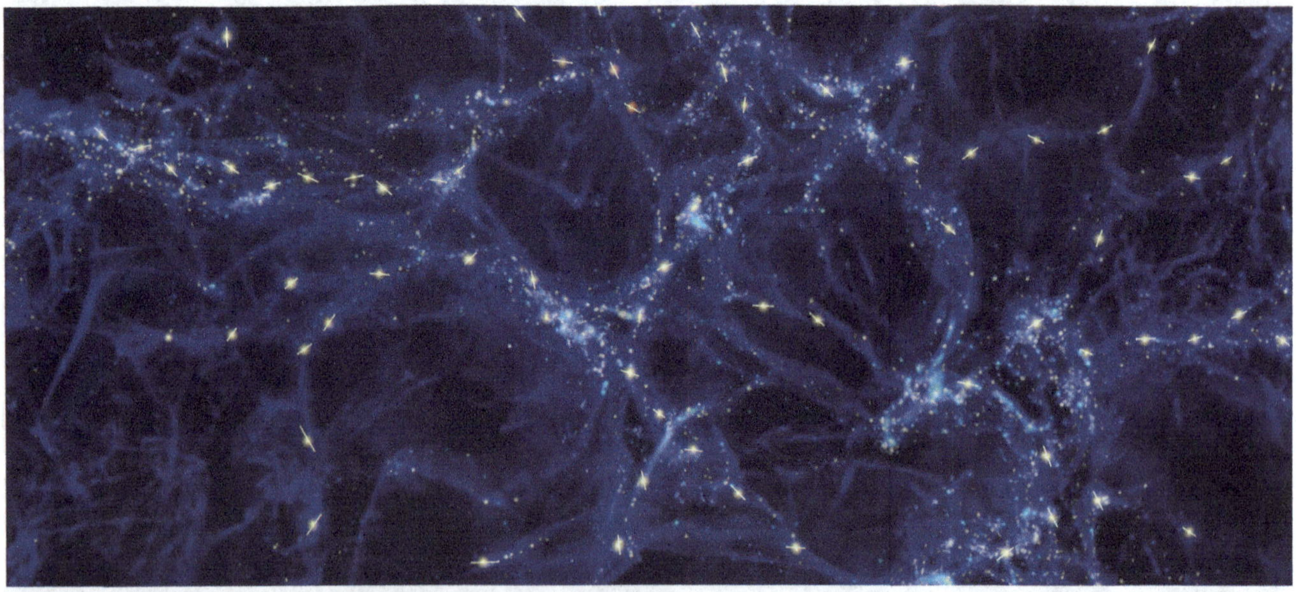

This is an illustration of the inexplicable alignment of the spin axes of supermassive black holes that lie at the centers of very active quasar galaxies that emit huge amounts of electromagnetic energy, particularly in light and radio waves. The rotation axes of the black holes are shown by a small line extending through each quasar. Illustration: ESO VLA telescope, Kommesser.

THE EXPANSION OF SPACE AND TIME

It's very important to understand that the Big Bang was an explosion *of* space, *not* an explosion occurring in some pre-existing space. Contrary to what illustrations and computer-art videos depict, there was no central point from which the universe expanded. It is thought that space right across the Greater Universe expanded at every point. Because this is impossible to illustrate, and very difficult to visualize, the old myth of the expansion starting from one point, will continue to remain a misleading part of the Big Bang theory.

The science of the Big Bang does not say where space, energy, and time came from, or what made the universe come into being. It is often assumed that there was nothing before the universe was born, i.e. there was no space, and that time did not exist. However, the Big Bang theory does not make this claim one way or the other. It assumes that space, energy, and time may have already existed, perhaps in another form as a single compressed force. It suggests that the universe was highly compressed in a singularity until something caused the Big Bang to occur. *If* this is so, it can be thought of as a spring that is wound so tightly that it becomes imperceptibly small. Then something happens that allows it to unwind at an incredible speed for a very brief time (inflation) due to it having infinitely compressed energy.

But if space was infinitely compressed, then the three dimensions of space would be compressed into a one-dimensional point, and matter would not exist in that form. It would be compressed into pure energy with no dimensions. Because energy is equivalent to mass, this infinitely small point should contain infinite mass and therefore, infinite energy. The dimension of time would also be infinitely compressed, so time would not 'tick'. If that is so, how could space expand when there was no time in which it could do so? What could have made space expand and time start ticking? **How the Big Bang occurred is mankind's biggest mystery.**

I suspect that because we know so little about time and quantum physics, and almost nothing about the conditions created by infinite compression, we cannot conceive of what might have caused the Big Bang to occur. In time, with the help of mega-computers, we should discover the answer - as long as it is comprehensible by our limited intelligence.

Space is able to *expand* far faster than the speed of light, but matter cannot *move* faster than just under the speed of light, which is 300,000 km/sec in a vacuum i.e. the universal speed limit. Why this particular speed is the limit for the speed of light is another enigma for science to solve.

At the instant of the Big Bang, it is thought that the density of matter and the gravitational field that matter creates must have been infinite (or very close to it). All the matter that is in our observable universe *and that in the Greater Universe beyond* was compressed into an almost infinitely small volume. This would have made gravity, pressure, and temperature virtually infinite. What ramifications this has is totally unknown.

The Big Bang theory explains very well the expansion of the universe and its evolution because there is much evidence for this. But it does not explain where the mater/energy for the Big Bang came from. **It is often assumed that there was nothing before the universe was born, i.e. there was no space and time did not exist. However, the Big Bang theory does not make this claim one way or the other. It assumes that space, energy, and time**

may have already existed but in another form as a single infinitely compressed force. That being the case, the universe would be a singularity with no dimensions or time. But if its energy was in a singularity like a black hole, time would not tick, so how could it have exploded when an explosion requires the dimension of time to exist? Black holes do not explode, so the Big Bang must have been different in some way. At the outset of the universe, the energy levels and the temperature were infinite or very close to it. Our present theories of gravity and quantum mechanics break down under these conditions, so we don't know what is possible.

What could have made space expand and time start ticking is a huge mystery.

The Cosmic Microwave Background radiation (CMB) recorded by satellites showed that the observable universe's temperature was remarkably evenly distributed around 400,000 years after the Big Bang occurred. But at this time, there had not been enough time for light from the most distant part of the universe in one direction to the most distant part in the opposite direction for light to travel that far to even out the heat. This has been explained by the Inflation Theory, which suggests that almost instantly after the Big Bang occurred, the universe underwent an inconceivably fast expansion of space, which the theory of General Relativity does not permit. It suggests that the universe started at a subatomic size and suddenly expanded exponentially fast by a factor of at last 10^{26} (a hundred trillion, trillion) times to the size of a grain of sand, after which it slowed down virtually instantly to its current expansion rate. This sounds like a good way out of a bind to explain the evenness of the universe. It seems like an extremely long stretch. A new theory says that the CMB is not from the Big Bang but from a vacuum plasma that fills the vacuum of space. The theory suggests that this plasma is caused by high energy particles popping in and out of existence in quantum states. It may interact with photons from all the stars in the universe and thereby make it look like it is coming from the distant universe. But surely, this would cause changes in the CMB over time.

A problem with the Big Bang theory is that it predicts that there should be enormous numbers of **magnetic monopoles** (magnets with only one pole) created in the Big Bang, but no monopoles have ever been discovered.

Some of the conditions for the Big Bang appear to be the same as those for a singularity in a black hole. However, there are differences. The Big Bang caused space to expand from what was a singularity (a one-dimensional point), whereas the infinite gravity of a black hole does the opposite: it causes space to *contract* into a singularity. If the universe was initially a singularity like a black hole, then how did it manage to explode? Black holes cannot explode, so what is different about the Big Bang? For one, black holes exist in pre-existing space, whereas with the Big Bang, there was no pre-existing space and no time - that we know of. Cosmologists need to solve what makes the Big Bang and black holes so different because this may open new doorways into other aspects of cosmology and quantum physics. Some physicists have come up with some possible reasons to explain this, but they are very long shots that are too complex and too difficult to explain here. These concepts can be found on the Internet for those who are interested.

HOW DOES THE EXPANSION OF SPACE WORK?

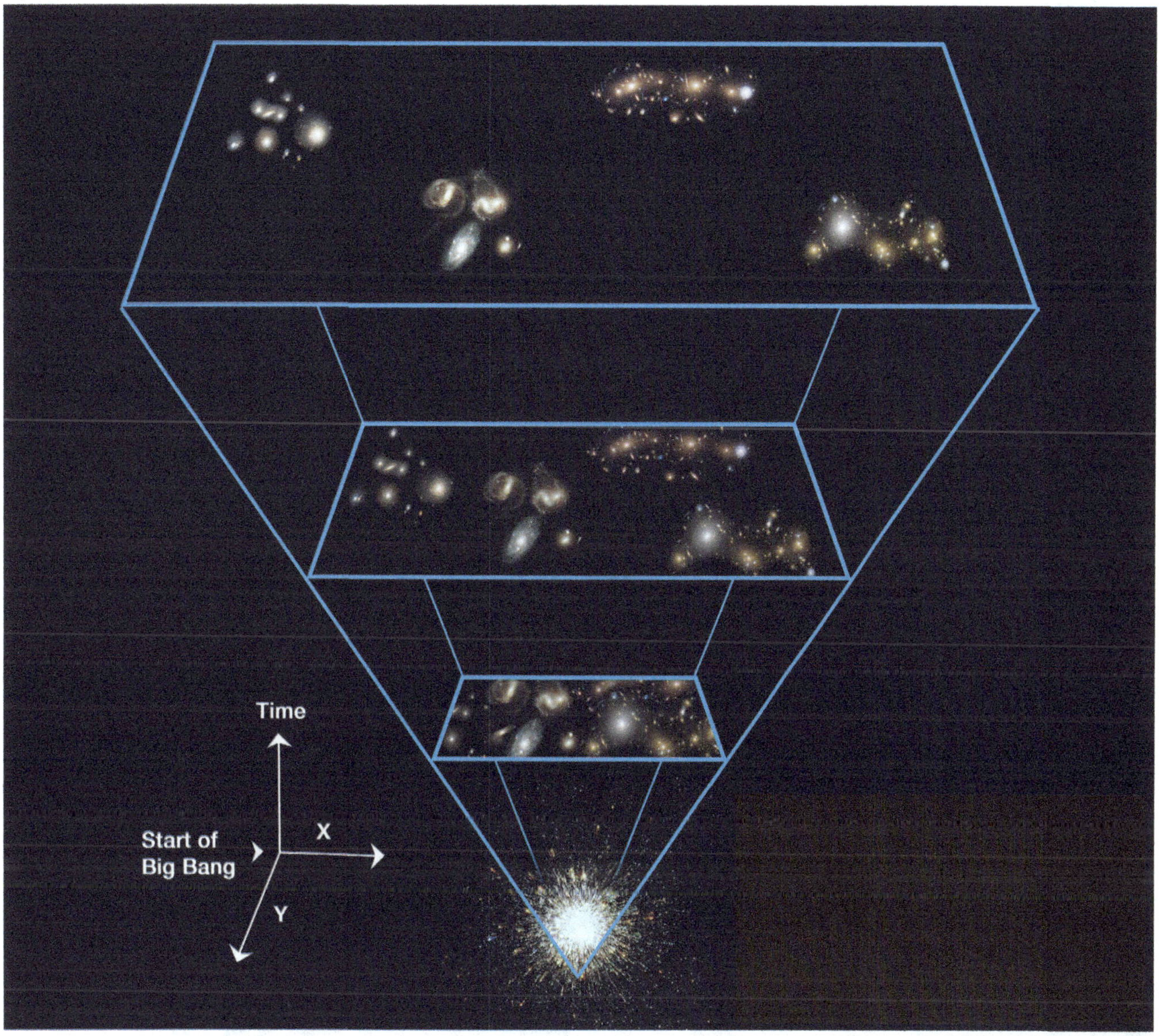

The expansion of space in the universe made galaxy clusters move apart, but not the galaxies themselves. The expansion of the universe is not yet making galaxies or galaxy clusters increase in size. They are staying much the same because their gravity stops the space within them from expanding. But ultimately, space is expected to expand so much that even mega-galaxies will be stretched apart. Credit: Gregg Thompson, Nicole Brooke.

Leading cosmologist **Alan Guth** believes that our Greater Universe may have come from another universe where dark energy caused space to contract to a singularity. *Somehow,* this may have caused our Big Bang to occur in a new 'space-time' and this became our universe. (See 'Dark Energy' in the following chapter.) Astrophysicists **Lawrence Krauss** likes this idea despite there being no evidence for it and nor can there be because there is no way of knowing anything about what can happen beyond our universe. If a new universe can be born from old ones, one has to ask, *"Where did the original universe and its space-time come from"*? The concept of universes being born from parent universes in the same way that life reproduces is a cute idea, but it is just passing the buck to another universe to explain its origin. Such explanations have as much credibility as believing that some god created the universe, a concept that Lawrence Krauss abhors. Other physicists are finding startling evidence that supports the possibility that our universe may be a simulation. Just as worlds in a virtual reality game are controlled by a computer program that exists outside the virtual world, so might there be some control program that exists outside the dimensions of our universe. If this were the case, then where did the energy come from in this other universe? But to prove another universe exists will be very hard, if not impossible – unless we find the power cable inlet! But even if our universe was born from a parent universe, we would then be back to asking, "*Where did the parent universe come from*"? I suspect the answer to where our universe came from is so far beyond a human mind that we cannot imagine the answer any more than an ant could conceive of the universe we have discovered.

THE ILLUSTRIS SIMULATION OF HOW THE UNIVERSE EVOLVED

The diameter of this region of this computer-generated universe is about 300 million light years. It was created as part of the Illustris supercomputer simulation. The distribution of dark matter is shown as dark blue filaments. Individual galaxies are shown as pink star-like dots. Clusters of galaxies appear white. Orange depicts the distribution of gas around the hubs of galaxy clusters. A supermassive galaxy cluster lies at the center. Supermassive black holes eject gas from their poles back into intergalactic space. This simulation is for the current age of the universe. Credit: ESO

This is a single frame of an incredibly detailed video simulation of the formation of the filamentary structure of galaxies. At the center, we see a supermassive supercluster.

The stunning video of the ***Illustris simulation*** is available on YouTube. It was produced using the world's fastest supercomputers available at the time. It is the most detailed simulation of the evolution of the universe ever attempted. It starts just after the Big Bang. The camera moves around the universe, allowing us to see how the universe evolves. It depicts the first galaxies forming and the filamentary fabric of the universe emerging. A useful timeline of the universe's evolution is given. Enormous gaseous bubbles explode from around the centers of large clusters of galaxies. These bubbles are due to the poles of supermassive black holes ejecting gas into intergalactic space. The vision also shows a comparison between dark matter and gas temperature in a cube that measures 35 million light years across. The video then zooms into the centers of galaxy clusters to reveal amazing detail. For anyone who wants to know what the big picture of the universe looks like, this video is a must. It is one of the greatest astronomical simulations you will see, especially when seen on a large, high-resolution monitor or better still, in an OMNIMAX theatre.

THE MICROWAVE BACKGROUND RADIATION

The filamentary structure of galaxies appears to be evenly distributed across the universe with no imbalance in any direction. This is a consequence of the uniformity revealed in the **Cosmic Microwave Background Radiation** map of the universe. It shows the minutest variations in the leftover heat from the Big Bang. This background heat is a frigid -270.15°C – a mere 3°Kelvin above absolute zero. The radiation we are seeing today has taken 13.7 billion years to arrive at our location in the universe. It has traveled from the edge of the observable universe where the Big Bang occurred at the beginning of time.

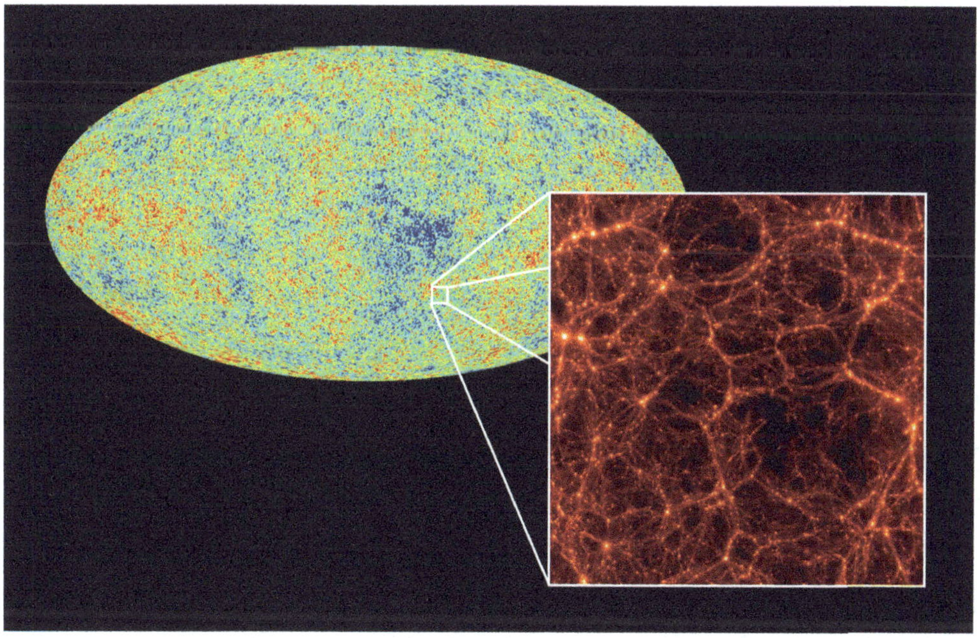

This 3°K Cosmic Microwave Background (CMB) radiation occurred one billionth of a billionth of a second after the Big Bang occurred. Because this image is so clear, it literally brought tears to the eyes of astronomers because it told them so much about the birth of the universe. The WMAP satellite that later mapped the entire sky used even more ultra-sensitive infrared (heat) detectors, and it confirmed the same result in more detail. Credit: NASA, CMB

This oval map is a full-sky picture that shows extremely small temperature variations of only 0.0002°K! These appear as variations in color. The slightly warmer regions are shown as red. This is where dark matter and normal matter became concentrated. This is the reason why superclusters of galaxies formed here. Yellow-green regions are where most galaxies originally formed before they were drawn into clusters. The slightly cooler regions are dark blue. These are voids, presumably of antigravity dark energy where very few galaxies exist. The square at right illustrates an expanded view of a small part of the CMB as seen in visible light. It shows the locations of the galaxies we see today, and it shows the filamentary structure that arose out of the minute temperature variations at the commencement of the universe.

DO WE LIVE AT A SPECIAL TIME?

Because the universe is expected to continue to expand as it now is, **intelligences evolving a hundred billion years in the future will not see numerous individual galaxies and galaxy clusters strewn across the universe and receding from one another as we do. Galaxies in the very distant future will all be drawn into supercluster that will merge to form mega-clusters. Given enough time, gravity will cause all the galaxies in each mega-cluster to merge to become one mega-galaxy full of very old, dim dying stars. No new stars would be born because all the hydrogen will have been consumed long ago by its existing stars.**

Because the universe will keep expanding, all these mega-galaxies will move beyond each other's observable universe. Intelligences in these mega-galaxies will only see their own galaxy, so they will not know that other mega-galaxies exist. They will see no evidence of the Cosmic Microwave Background Radiation because it will be far beyond the horizon of their observable universe. Intelligences in these galaxies will have no way of knowing that the Greater Universe exists. They would not know where the universe came from, and they would not know that the Big Bang occurred, so they would not know how the universe evolved, and they would not know how old it is. They would see no evidence of stellar evolution or the birth of stars. And nor would they know that space beyond their mega-galaxy is expanding rapidly or that there is a universe beyond their galaxy.

On the other hand, **intelligences that evolved say around 10 billion years ago in the early universe** would only be aware of a much younger universe where galaxy clusters are just starting to form, and galaxies are much closer to one another than they are today. They would see far more galaxies than we can see because all the distant galaxies that we can't see because their horizon would include numerous galaxies that have now moved beyond our horizon due to the universe expanding since then. They would see only small groups of galaxies starting to form because large clusters would not have had time to come together. It would be too early in the universe's evolution for them to know how long most stars will live or how stars and planets will evolve in the future.

When you think about this, it is quite remarkable that we are lucky enough to exist at a time where we can not only see back to the birth of the universe, but we can also look forward to see how the universe will most likely evolve in the distant future. We appear to be living in a very special time of the universe's lifespan. Is this luck or by design? Or, are we deluding ourselves because there is so much that we do not know?

IS THE UNIVERSE EXPANDING EXPONENTIALLY?

To the great surprise of astronomers, observations made around two decades ago of supernovas occurring in distant galaxies indicated that the deeper we look into the universe and, therefore, the further back in time we are seeing, the slower the universe was expanding. This meant that the expansion of the universe is increasing at an exponential rate. This conclusion was deduced from observations of supernovas. But it is now becoming known that those observations had significant unrealized biases in the selection criteria for the supernovas. This now throws doubt over an antigravity force and therefore the existence of dark energy. (See Dark Energy in Chapter 4.)

WILL THE UNIVERSE DIE?

It appears that the universe will die, as all things in it do.

It seems that **it will suffer a Heat Death and a Disintegration**. In around 50 billion years time, there will be no hydrogen left to form new stars. Over the following trillion odd years or so, all stars will stop giving off heat and light. The residual heat will be spread evenly over the whole universe, so it will become so cold that it will only be a few degrees above absolute zero. It will become totally dark, consisting of black cinders of stars and black holes.

The increasing expansion of space will ultimately tear the space-time continuum apart, causing galaxies, stars, and ultimately all matter to disintegrate.

Like everything on Earth, elements in the present universe are recycled to create new stars and new life, but will this happen to our universe?

Of course, we may make discoveries in the future that could change both of these scenarios. If intelligences evolve to become enormously more conscious than we are, the state of the universe may have no bearing on them.

Chapter 4

COULD DARK ENERGY, THE MULTIVERSE, AND VIRTUAL UNIVERSES EXIST?

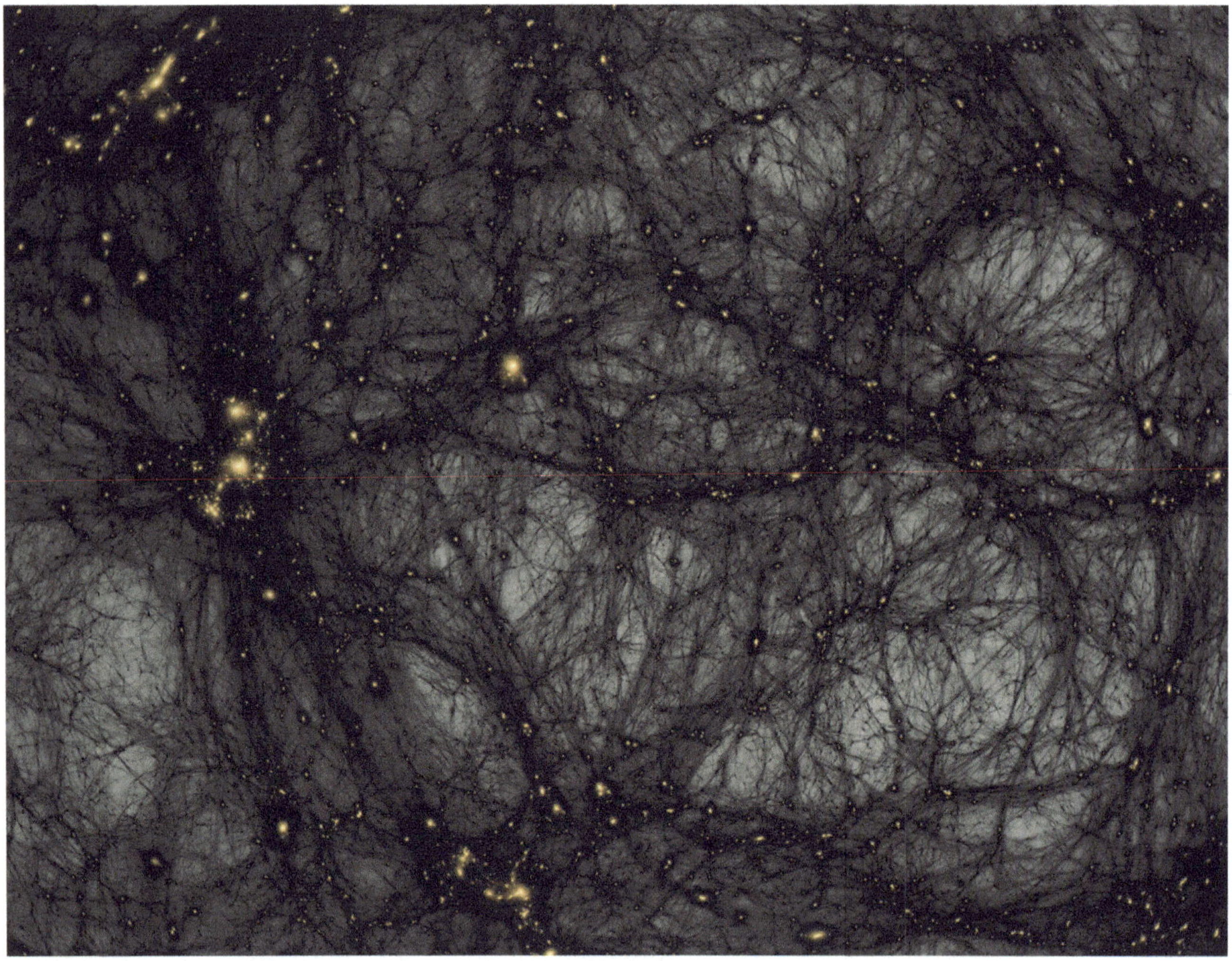

Dark Energy is thought to be what causes galaxies (yellow spots) to form along the filamentary structure that exists at the edges of the bubble-like voids. This is where galaxies reside. They collect into clusters at the nodes where the filaments of adjoining voids intersect. Dark matter is envisaged as being an invisible, anti-gravity force that pushes normal matter away. It is thought to be driving galaxies to the outer edges of each void. This image is from the IMAX movie 'Dark Universe'. Credit: Kavli Institute of Particle Astrophysics & Cosmology at Stanford University and SLAC National Accelerator Laboratory.

WHAT IS DARK MATTER & DARK ENERGY?

There has been a lot of discussion for two decades on the mystery of dark matter and dark energy, but it now seems that there may be errors in observations that have made it appear that dark energy was causing the expansion of the universe to increase. I will discuss the apparent reasons why both dark energy and dark matter have been thought to exist.

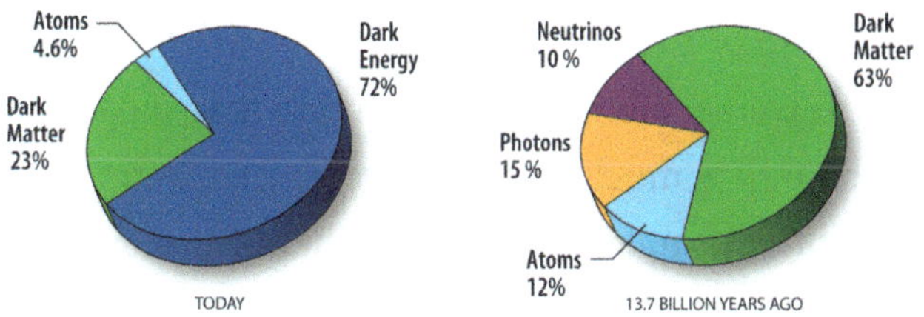

THE DISTRIBUTION OF MATTER AND ENERGY IN THE UNIVERSE

All the visible matter we see in the universe is in the form of stars, planets, gas, and dust in galaxies. It *appears* to constitute just a mere 5% of the total mass of the universe! In an attempt to explain how galaxies and clusters of galaxies rotate without flying apart, astrophysicists thought that there must be invisible *dark matter* involved to hold them together. If so, it would constitute around only another 27% of the universe's mass. Its properties do not fit those of any known particles. Dark matter is referred to as 'dark' because we can't see it. It does not emit light, and it does not absorb light, which is curious. It must be transparent because it does not hide matter that lies behind it. No one has determined what dark matter actually is! It seems to be associated with bright, massive stars, galaxies, and particularly with massive galaxy clusters. It acts like normal matter collecting together over time. Dark matter predicts that there should be density peaks in the cores of galaxies, but observations have not found this to be the case. If it exists, it may not be matter but something we know nothing about.

To explain the balance of the universe's missing matter, astrophysicists also postulated that there must also be *dark energy* that constitutes the other 68% of the universe's missing mass. (Energy and mass are interchangeable because one can be converted into the other.) Some scientists think dark energy may be a new force. Some think it does not exist because there may be an error in Einstein's equations for gravity. (The exact percentages of dark matter and dark energy vary slightly from one source to another). Some astronomers think dark energy may be the anti-gravity force that Einstein's calculations predicted. Some think it might turn out to be the fifth force of nature.

If dark energy's anti-gravity force turns out to be real, many astronomers believe this is what is driving the *apparent* exponential expansion of the universe. The physics of dark energy predicts that it does not dilute with the expansion of the universe, as matter does but rather, it grows stronger. This is because it is thought that it may be continually being created in the centers of each void in the filamentary structure of the universe. How this highly contentious, mysterious process could occur is based on *speculation*, so it is to be taken with a grain of salt.

But is the universe really expanding ever faster? A team of astronomers led by Perlmutter, Schmidt, and Riess made observations of distant exploding stars known as Type 1a supernovas that explode in a very regular way compared to other supernovas. By measuring their brightness, they could determine how far away they were because, the fainter they are, the further away they are. And the further away a supernova explosion is, the more its light is red-shifted. This is due to the expansion of the universe causing the Doppler effect, which stretches a star's light into the red end of its spectrum. And the further away a supernova is, the further back in time we are observing it. The brightness and the redshift of a supernova is directly related to its distance from us. But the Perlmutter team found a difference between these results that they could not explain, so they went and checked all their results for an error, but there were none. They interpreted this as being due to the expansion of the universe increasing exponentially around 6-7 billion years ago. They put this down to dark energy becoming increasingly stronger. In 2011, they won a Noble Prize for their observations. *However, their explanation is now in doubt.*

A 2019 paper by astrophysicists Colin, Mohayaee, Rameez, and Darkar, explains that Perlmutter's team's observations were made well over 20 years ago and that they used a relatively small sample of only 110 supernovas, which were mostly from one direction. They did not have thousands from across the whole sky, as many assumed. Since then, data has been collected from over a thousand Type 1a supernovas from across the whole sky. This included the rate at which they were moving away from our galaxy – data that was not available to Perlmutter's team back then. When all this new data was analyzed, **it produced no major difference between the redshifts of supernovas and their distance**. Perlmutter's team did not think to take into account that our galaxy is moving through space in one direction at a particular rate, while many others are moving away in different directions at different velocities. Light from supernovas that are traveling away from us at high speed are more redshifted than those traveling more slowly at right angles to us. **When this is taken into account, the increasing expansion of the universe disappears!**

David Wiltshire and many other astronomers are now starting to think that **dark energy may not exist**.

COULD DARK ENERGY CREATE A MULTIVERSE?

In my mid-teens, when I had learned about the basic structure of the universe, I conceived the idea that our universe might be nothing more than an atomic particle in a much larger universe than ours. And possibly, this mega-universe might be no more than a speck in an even larger universe, and so on. Some decades later, I discovered that cosmologists and others had come up with much the same idea. The reason for this might be that **the more we learn, the larger and more complex our perception of everything becomes - including the universe**. When we look into a microscope or a telescope, we discover seemingly endless levels of detail, with one level being a small part of the one before it. So, it's reasonable to extrapolate that there might be one cosmos inside another and so on – like Russian dolls, possibly going on forever. But could this be realistic?

Throughout human history, our perception of our 'universe' has changed remarkably. For cave dwellers, their universe was the land that they roamed across to as far as they could see. To them, the sky was like a mysterious ceiling. Ancient Asian cultures proposed that our flat Earth floated in an endless, fluid-like space.

The Hindu culture thought that what little they knew of our planet was supported by giant elephants, but someone asked what supported them? The answer was that they stood on the back of a giant turtle swimming through some universal ether. Our perceptions of a multiverse are most likely just as naive.

Millennia later, when people traveled much further afield, the known area of land and sea became much larger. The boundaries of this part of the Earth became their universe. No cultures had any conception of the Earth being round and there being inconceivable volumes of space beyond it. Are we in the same situation not able to imagine what lies beyond our observable universe?

Around 2,300 years ago, **Eratosthenes** calculated that the Earth was a sphere. Incredibly, it took until the 1500s for this concept to be accepted! The Earth and the other planets were then known to orbit the Sun. With this knowledge, the universe became our solar system. But it was not known then what stars were. In the 1900s, it was confirmed that our Sun was just one star in a huge body of other very distant stars. By the 1920s, astronomers became aware that our Sun was just one of billions of other stars in our enormous spiral galaxy. By the middle of the 20th C, it was apparent that our galaxy was just one of what appeared to be an infinite ocean of galaxies! By the beginning of the 21st C, we could see to the boundary of our universe and back to the beginning of time!! As we kept building more powerful microscopes, the scale of things in the micro-universe became ever smaller. Technological advances are rapidly expanding our awareness of our macro and micro-universe. At every scale, be it large or small, there appears to be another scale beyond it.

Since the beginning of the rise of science, the more we have learned, the more our universe has grown ever larger and more complex. This trend will surely continue and become ever faster. But it could soon become beyond our ability to comprehend it.

WHAT IS THE MULTIVERSE?

With the concept of ever-larger universes, some cosmologists are now wondering if our universe might exist in a super-space full of other universes. This concept is referred to as the 'multiverse'. Some cosmologists speculate that the multiverse may be giving birth to ever more universes all the time. Despite this concept becoming popular, there is no evidence whatsoever for it. And nor is there ever likely to be because, it is beyond our universe, so we are unlikely to know anything about it.

At the very end of the first '***Men in Black***' movie, the camera pulls back from the Earth to see it and our Moon disappearing into the distance. Our whole solar system then recedes into a profusion of distant stars in our Milky Way galaxy. The camera then leaves our galaxy behind to see innumerable other galaxies in our universe receding away with it. The camera continues to pull back until we see our universe become a tiny sphere amongst innumerable other spherical universes. As the camera pulls back further still, these universes are then seen as being specks inside marbles, which three-fingered, super-alien children of an incomprehensible size are playing with! The audience is left wondering what sort of universe do these super-sized aliens exist in. One wonders if there might be no end to this progression of ever-larger universes. I'm sure numerous people across the world have thought of this concept because it is a natural progression from what we have learned to date; however, there are good reasons following why this concept is probably wrong. If there is anything beyond our Greater Universe, then it will be far beyond what even the most imaginative humans could ever comprehend – just as our solar system and our galaxy were far beyond the ability of all cultures in previous millennia to conceive of.

As a multiverse can only exist outside our universe, there is no way to prove or disprove it. In science, the inability to disprove a thesis is just as big a problem as not being able to prove it.

Some current concepts suggest that there is a **multiverse** in which a universe exists for every possibility. However, the rules of logic state that when there is a concept like a multiverse where anything is possible, then one possibility must be that it is impossible! This problem exists in String Theory.

Cosmologist **Alan Guth**, who brilliantly conceived the concept of inflation of the universe, went on to find observational evidence that strongly supported his theory. He claims that his recent calculations show that half the presumed dark energy in the universe becomes pockets of space-time that ultimately contract exponentially into point source singularities, while the other half of the dark energy causes space to expand ever faster than the regions that are decaying. Guth is exceptionally creative but he has no possible way of explaining what would cause this to happen. Guth uses this idea to ensure that the increasingly faster expansion of the universe will not end. To date, there is no evidence for or against this concept. It sounds like an each-way bet.

This illustration attempts to simplistically visualize a very small portion of the hypothetical multiverse. In it, there is an infinite number of bubble universes that breed infinite numbers of new bubble universes. These, in turn, do the same thing forever! Where would all the energy for this concept come from? And what would be the reason for a multiverse? When infinities are introduced, they typically mean there is an error.

Guth speculates that at high-energy states - such as during the Big Bang or when some dark energy is compressed into a singularity - gravity can act *repulsively*, making a singularity explode. However, Guth does not explain why this does not occur with singularities in black holes. Guth speculates that our universe could have been created from a region of space in a parent universe that decayed into a singularity and then exploded to form a new universe in a new space-time. *If* this is possible in a mega-universe, then infinite numbers of embryonic universes would be being born into whatever super-universe our universe came from. If our universe could give birth to other universes, then those universes might do the same. Where would it end, if ever? There would be infinite numbers of universes to the power of infinity! This would be mathematically and physically impossible.

If there are other universes, some String theorists speculate that each 'pocket' universe could be built around a different vacuum energy level. This would make each universe different.

Our universe requires an extreme level of fine-tuning to exist, so the chance of other random universes having such conditions by *pure chance* should be impossible. If other universes do not have the right conditions, they would be very short-lived, extremely simple, or stillborn. However, physical laws in other universes may be quite different, so this might make some universes viable.

If the multiverse is possible, one would have to ask, *"Was there an original super-universe from which the multiverse was born, and if so, where did that super-universe come from, and how was it created?"* This has the same problem that the concept of gods existing has. A multiverse does not seem to get us anywhere. It just pushes the answer as to where our universe came from back one huge step.

The '*13th Floor*' movie speculated about what could happen when virtual worlds are invented that are just like real worlds. This is a graphic illustrating where the main character drives away from his virtual city and through a barrier to find he had reached the edge of the city's simulation and discovered it is not real. Credit: Centropolis Entertainment

An analogy for the **multiverse concept** could be an infinite forest. Each tree produces large numbers of seeds. The seeds that are lucky enough to land in a rich growth medium grow into slightly different trees to their parent tree. Most do not germinate at all, and many of those that do, die soon afterward due to unsuitable initial conditions. New trees that survive produce multitudes of seeds before they die, and the cycle repeats. Like this forest analogy, if a multiverse exists, it might be that most universes are not viable because their starting conditions are unsuitable. Some may survive for a short time, but only a small fraction of them will grow into long-term sustainable universes. But remember, all this is extreme speculation with not a skerrick of evidence to support it. And because all these concepts lie beyond our universe, it would be impossible to find any evidence to test if even one other universe exists, let alone a multiverse.

COULD THERE BE PARALLEL UNIVERSES?

Some cosmologists have their own personal form of a multiverse - just like most religious people have their own version of their religion: they keep the bits they like, or can live with, and reject the rest. Some cosmologists think there is a multiverse that allows for every possibility that could ever exist – an infinity of so-called 'parallel universes'. But none are parallel: they just branch off every part of some original universe like branches of a tree trunk.

Parallel universes would require a universe for every position of every particle in every universe for every nanosecond throughout all of time! **At every nanosecond, zillions of new 'parallel universe' to the power of infinity would branch off from every universe into new universes that would also do the same every nanosecond because every particle**

would have multiple possibilities. The number of possibilities in every universe would grow exponentially every nanosecond!! This is incomprehensible. Such a concept amounts to infinity multiplied by infinity, multiplied by infinity, an infinite number of times!!! Those who think of these concepts should be writing the most bizarre sci-fi movie scripts. But many people love this idea because there would be an infinite number of themselves existing in every possible way, but you would not know what your other selves were experiencing. That might be good because most could be very bad. To those who support this idea, it's like a new version of heaven where some of themselves will live on forever.

The parallel universe conjecture is the most preposterous and naïve of all concepts. It would be far worse than the simplistic examples its believers give, such as Elvis and trillions of other people who do not die early in one universe, but instead, in an infinite number of other universes they could live on! There has to be every conceivable possibility for when, how, and where every person and all other lifeforms live and die.

This would also be true for every person that has ever lived throughout all of time, and similarly for every life form, and every possible future of our planet, and every other solar system, and galaxy, and so on. This would not be limited to our universe, but to every other universe! Clearly, this concept is insane at every level. Parallel universes and the multiverse are concepts that are nothing more than feel-good ideas for people who want more options in their lives. It amounts to sheer nonsense to the power of infinity!

If there could be a near-infinite number of universes, some cosmologists think that our extremely finely-tuned universe would have had to eventually come into being by pure chance. However, mathematicians have calculated that the chances of this occurring are actually zero, even given infinite time for this to occur. This is because our universe required an incredible degree of fine-tuning across at least 30 of its parameters. The chances of this occurring by chance would be zero. It would be infinitely more unlikely than expecting an ape to type up 'War and Peace' by pure chance. Even if the ape was given infinite time, probability theory makes it clear that the ape would not even type up several sentences by chance, let alone the entire book.

The multiverse concept has been around for many decades, and there are a number of prominent physicists and cosmologists who cautiously support this crazy idea that they childishly think is 'cool'. They all have their own ideas about the types of universes that might exist, and how the multiverse would work. I doubt that any of them have really thought through how infinitely impossible it would be. Without *any* evidence to support the multiverse concept, it is more akin to a religious belief or a philosophy than it is to being a scientific theory. A leading cosmologist, **Paul Steinhardt** sums it up by saying, *"A theory that predicts everything predicts nothing."*

COULD VIRTUAL UNIVERSES EXIST?

Humans are now creating increasingly more realistic 'virtual' or 'simulated' worlds in computer games. In the not-too-distant future, super-intelligent computers that are being developed at present will have unimaginable knowledge, intelligence, and creative abilities. They should be able to create incredibly realistic virtual worlds. Given that there are likely to be numerous other super-intelligences in the universe that have evolved up to billions of years before us, they would have already done this. And in each virtual world, numerous intelligences could evolve in them as well. They might even create virtual galaxies, so there could be an infinite number of them also. This would require infinite energy to run them. Where would this energy come from? There seems to be an underlying flaw in this scenario because, when infinities are introduced, it typically means there is a mistake that has been overlooked. This concept, like those before it, is in the realm of fantasy rather than any possible reality.

Physicists from Oxford University believe they have shown that entire universes cannot be simulated by mega-intelligent computers.

Could force fields permeate higher dimensions in our universe, or possibly connect ours to other universes in ways that today's science cannot imagine?

They discovered a link between gravitational anomalies in the underlying space-time geometry and computational complexity that makes it impossible to construct a computer-simulated universe due to a quantum phenomenon that occurs in metals that exhibit strong magnetic fields very low temperatures. They found that the complexity of the simulation increases exponentially with the number of quantum particles being simulated, as occurs with quantum computers. The amount of computing power required for a simulated universe would double with each particle added, so the task would quickly become impossible – even for a quantum computer. They calculated that just storing a couple of hundred electrons would require a computer memory that would physically require more atoms than exist in the universe! Their work provides convincing evidence that entire universes could not be simulated because the quantum universe would not allow it.

If it were possible to somehow create a universe, then there would be little, if any difference between a virtual universe and a real one, so it wouldn't matter to us one way or the other which one we lived in. What would define a 'real' universe? The concept of creating a small virtual world was the basis for the movie '***The 13th Floor***'. Worlds and possibly solar systems may be simulated, but not entire universes.

Chapter 5

WHAT ARE THE ESSENTIAL PROPERTIES OF OUR UNIVERSE?

HOW COULD OUR UNIVERSE BE SO FINELY-TUNED?

Extraordinary degrees of fine-tuning have been discovered in at least ten physical constants in cosmology and another 20 in quantum physics. This inexplicable degree of fine-tuning in 30 constants is infinitely beyond chance, and it's impossible to explain! This is simply mind-boggling to the nth degree!! And there may be more finely-tuned constraints that we have not yet discovered.

This fine-tuning is known as the **Goldilocks enigma**. Without this extraordinary level of fine-tuning, our universe would not have evolved at all. Over the last century, many leading cosmologists and quantum physicists have written numerous scientific papers and books on this subject, all trying to unravel how this could have come about, but so far, there are no convincing answers to this enigma. But this is not a reason to resort to a belief that the universe was designed by a creator.

There are many conditions in the universe that are precisely fine-tuned, not only for the evolution of stars and planets but also for life to evolve. The following are just a few relatively-easy-to-grasp examples of this incomprehensible level of fine-tuning of some of the universe's constants.

1. **The four forces of nature - gravity, electromagnetism, and the strong and weak nuclear forces - are extremely finely tuned.** Even the minutest change to any one of these fundamental forces would have stopped the universe evolving as it has, or at all!
2. **The expansion of the Big Bang was impeccably finely-tuned.** If the expansion had been just a little slower, the universe would have quickly collapsed into a big crunch and disappeared before life could have arisen. Similarly, had it exploded even the tiniest bit faster, there would not have been time for stars, planets, and life to evolve before matter had spread out too thinly. Inexplicably, it has the right rate of expansion *precisely!*
3. **If there was ever so *slightly* more dark energy, space-time would have expanded too fast for galaxies to form.** To have the universe we have today, the degree of dark energy had to be fantastically finely-tuned to more than 120 decimal places!
4. **If the mass and other properties of neutrons, protons, and electrons were even *infinitesimally* different, there would be no atoms in the universe.** Neutrons were created to be ever so slightly heavier than a proton so that they could decay into protons, electrons, and neutrinos. If

neutrons were even the *slightest fraction* lighter, they would not decay. Alternatively, if they were ever so slightly heavier, there would not be an abundance of hydrogen in the universe to form stars, and there would be far too much helium.

Similarly, if the charge of an electron was just the tiniest bit different, the universe would have no complex chemistry and, therefore no life.

5. **If the minor quantum fluctuations during the period of inflation at the outset of the Big Bang had been just minutely less, then stars and galaxies would not have formed. Had the fluctuations been just a little stronger, then giant black holes would have formed and there would be few stars.**

The leading cosmologist **Paul Davies**, author of '***The Goldilocks Enigma***' said, "*In all these parameters, the universe is balanced on an unbelievably fine knife edge.*" Many leading physicists such as **Robert Dicke, Fred Hoyle, John Gribbin** and **Martin Rees** believe that this degree of fine-tuning is far beyond the laws of chance. **Stephen Hawking** stated that "*the values of these numbers seem to have been very finely-tuned to make possible the development of life.*" These scientists are not suggesting that this was a deliberate intent by a creator god; however, it does raise the question as to how could this incredible degree of fine-tuning has come about? It looks like the parameters for the program that controls the universe's evolution were designed to have these precise values.

Physicist **Victor Stenger** thinks that the fine-tuning enigma may be solved if there is a unified field in which there may be connections between all the physical constants. He surmises that if a change were to occur in one constant, then the 'connections' might cause a change in one or more of the other constants, thereby 'adjusting' the parameters of the universe so it can exist. No explanation is given for what the connections might be or how they would work. This is pure speculation. There is no evidence to indicate that Stenger's explanation is possible, although it is an interesting concept.

Some people have attempted to explain the fine-tuning by suggesting that a supernatural god or some advanced alien from some super-universe outside of ours created our universe. It could equally be said that some super-mega-intelligent computer might have calculated what the conditions would need to be to create a universe like ours. It then programmed this level of fine-tuning before it turned on the Go switch. As I have mentioned before, there is a problem with introducing any type of supernatural entity from outside our universe because there is no evidence to support such a conjecture, and furthermore, it raises the question as to what created the supernatural entity and its universe.

There have been many dilemmas similar to the fine-tuning enigma throughout the history of science, but they were eventually solved when new data was gained and new insights into nature were discovered. Once quantum physicists gain a much deeper understanding of the forces that underlie the quantum universe, it is possible that this may solve the fine-tuning quandary.

Leading computer technologists are now in the process of creating highly advanced, mega-intelligent, self-learning computers that will have intellects immeasurably greater than the entire human race.

They may be able to gather unimaginable amounts of big data to allow them to discover how the universe became so finely tuned. If a solution to the Goldilocks' enigma is discovered, it will surely have far-reaching implications.

"The more we learn, the more complex everything in the universe becomes. The universe is not illogical or haphazard. It appears to be an extremely finely-tuned program".

– Gregg Thompson

THE UNIVERSE'S TWO OPPOSING PROPERTIES - ENTROPY & ORDER

Just as there are many opposing forces in nature, such as the positive and negative forces of electricity, there are two important forces that few people know about, and without which the universe would not evolve. One is **disorder**, known as entropy, and the other is **order,** which drives increasing complexity.

Disorder results from the natural process for everything in the universe, including the universe as a whole, to ultimately move towards disorganization and decay. For instance, entropy causes all stars to eventually burn out and die - just as it causes all life forms to die and decay.

Order grows in pockets amongst disorder to form such things as stars, planets, and life forms. These states of order only exist for a certain period of time before they decay back into disorder. Their elements are then recycled to create new forms of order, which form new structures and increasing complexity. The elements in stars and planets become more complex over time, and life forms evolve to have increasing levels of order in the form of improving senses and greater intelligence and memory. This process of pockets of order opposing disorder is what drives evolution. But in time, everything ages and entropy wins out. For this reason, like all life and non-life, it appears that the universe will eventually die like everything in it. But by then, very high levels of highly intelligent order may change this outcome. It may be that highly intelligent consciousness evolves to a state that is above requiring a physical universe.

The **Second Law of Thermodynamics** simply states that in a closed system (i.e. gas or liquid in a container, a life form, or the universe), heat can only flow in one direction - from a hot region to a colder one - and never the other way around (unless energy is expended to make this happen). Due to this law, over trillions of years, disorder will cause the heat concentrated in stars and planets to eventually spread out across the vastness of the universe. There will be no remaining regions with differences in heat energy, so heat will not be able to be transferred from a hot region to cold one. As a result, matter will not be able to interact with other matter, so no new stars will form, and no life will be able to exist.

If there was no order to counteract disorder, then the universe would not have evolved. For the universe to evolve, pockets of order must occur amongst the disorder. **As order evolves, it becomes more complex. This occurs in molecular structures, astronomical bodies, cells, intelligence, consciousness, engineering, technology, and so on. Because of this, I contend that there is an underlying universal law, which stated simply says, '*the universe shall become more complex*'.**

In a closed system, order leads to increasing complexity. When the complexity of molecular structures reaches the level of cells, we call this closed system 'biological life'. Life is a state that creates order within itself at the expense of *creating disorder (entropy) outside itself.* It does this by expending energy and excreting waste. **Life is a form of organized energy that incorporates highly complex molecular systems.** It appears that nearly all star systems throughout the universe will form planets that have complex molecular structures. On some worlds where conditions are suitable, very complex structures like DNA, cells, and advanced life forms will evolve. But every star, planet, life form, and all cells will have a limited lifespan. This is why there are systems for them to reproduce.

It now appears that life will soon be proven to be a common natural evolutionary process that creates pockets of order throughout the universe.

In the Greater Universe outside our observable universe, there may be pockets of different types of disorder and order, which we cannot imagine. In our observable universe, it appears that we will have pockets of life that evolve on suitable planets, but in some parts of the Greater Universe, this might be rare. An analogy for this on Earth could be that rainforests have an abundance of different life forms, whereas dry, hot deserts and cold Antarctic wastelands do not. We have no way of ever knowing what conditions lay beyond our observable universe because we cannot receive any information from it.

HOW CHAOS DRIVES THE UNIVERSE TOWARDS INCREASING COMPLEXITY

In this section, I will suggest one factor that could be causing the universe to evolve and to have continually increasing complexity.

Edward Lorenz discovered that chaos was actually an unsuspected *higher* level of order to what we were aware of.

THE DISCOVERY THAT CHAOS IS A HIGHER LEVEL OF UNSUSPECTED ORDER!

The brilliant mathematician and meteorologist **Edward Lorenz** is famous for developing **Chaos Theory**. Before Lorenz discovered that the laws of chaos caused many systems to appear chaotic, chaos represented a mathematical abyss to both physicists and mathematicians because it appeared to have no order to it. But **Lorenz discovered that chaos was actually an unsuspected *higher level of order that produced greater complexity*.**

Lorenz found that **the smallest changes in a system could multiply and lead to major changes.** This became known as the **'Butterfly Effect'. This is the universe's way of driving continual change in every aspect of its existence, thereby causing it to continually evolve towards ever greater complexity.**

In the late 1950s, Lorenz was working on trying to improve weather forecasting. In 1961, he used a primitive computer to predict the outcome of the weather. His program for predicting the weather was a 12-component model that measured temperature, wind speed, air pressure, humidity at different altitudes, ocean temperature, and so on for different altitudes in the atmosphere over a region covering Britain, France, Scandinavia, and the North Sea. He calculated predictions for what the weather would do over the coming days using initial conditions to six digits. On one occasion, he ran the program a second time, but this time, to reduce computing time, he decided to round off the initial conditions to three digits instead of six. He assumed that it would make no difference to ignore such minuscule differences in the initial conditions. But to his amazement, this produced a completely different outcome! The graphs matched for the first few days, but as the days went on, they drifted apart. Instead of it being a fine day over Paris in 10 days' time on his first prediction using 6 places, the second prediction using only 3 places forecasted cyclonic conditions! Most people would have shrugged this off as some quirk in the program, but Lorenz investigated what the reason for this could be. He discovered that very small differences in the first six decimal run kept multiplying to cause major changes that did not appear in the three-decimal run. This became a major discovery that led Lorenz to develop **Chaos Theory**. His theory became the key to understanding why numerous chaotic systems in the universe work the way they do.

Chaos Theory showed that uncertainty in chaotic systems increases exponentially with time. However, chaotic systems do have limitations that cannot be exceeded. For example, while the temperature changes all the time across the globe, the average global temperature will never reach 60°C or drop to cause a major ice age under the present astronomical conditions for the Sun and the Earth. Current conditions should exist for at least another thousand years, barring an asteroid impact or a major eruption of a supervolcano such as Yellowstone.

Lorenz had discovered that all chaotic systems (which is nearly everything) are affected by the Butterfly Effect. It got its name from the possibility that under the right conditions, the mere flapping of a butterfly's wings in one region of the Earth could potentially cause a hurricane over another region in the weeks ahead due to Chaos Theory.

Lorenz had discovered that local weather cannot be predicted with any accuracy for any more than 5 days because too many minute changes are always occurring, and these keep multiplying to change the outcome. Beyond 5 days, the chances of a prediction being right diminish rapidly, so much so that by 10 days, it was down to pure chance for most regions on the planet.

Despite computers becoming thousands of times bigger and faster than Lorenz' primitive computer and there being many weather satellites monitoring the globe, the Butterfly Effect is the reason that climate forecasters cannot predict the Earth's climate 10 years ahead, let alone 100. Some errors still occur even in a 5 day window.

All 50 of the computer models used for predicting global warming have always turned out to produce false outcomes due to the effects of Chaos and its Butterfly Effect as well as using false data and flawed computer programs. There are literally thousands of things in forecasting that can cause Butterfly Effects, none of which are taken into account in climate prediction programs.

The biggest advance for predicting global climate will come from programs that integrate the effect of the Sun's weather on the Earth. This would include the output of solar radiation and the number and intensity of the Sun's storms when coronal mass ejections are directed towards our planet, as well as the intensity of the solar wind. Also, the intensity of cosmic ray showers from deep space must be taken into account because they affect the degree of precipitation that occurs in Earth's atmosphere. Incredibly, none of these crucial conditions are used in the programs!

Computer programmer **Michael Chermside** wisely states that with regard to predicting changes in the climate "*too much trust in computer programs can lead us astray. Invalid input, invalid assumptions, invalid equations, and mathematical chaos, all conspire to cause many pitfalls.*"

He also says that "*an experiment that fails may, in the end, be even more useful than one that succeeds (because it) allows us to discover things about nature that we do not understand at present.*"

A SIMPLE FORMULA DRIVES CHAOS

The Butterfly Effect can be described simply in algebraic terms by this extremely basic formula:

$$X^2 + C \Rightarrow X$$

In this equation, 'X' represents anything, and 'C' is a change of any kind on any scale. When X is squared and then changed by C, the outcome value of X is affected. This new value for X is fed back into the equation and squared again, and a new change is added to produce a new value of X. This is repeated ad infinitum. This process produces never-ending change. The rate of change may be extremely fast or very slow, depending on the value of C as well as how fast the feedback system works. If C represents a large change, this will cause a rapidly changing effect; however, over time, very small changes can also multiply and cause large changes to occur.

AN EXPERIMENT THAT VISUALLY DEMONSTRATES THE BUTTERFLY EFFECT

To see how dramatically the Butterfly Effect works, you may like to perform the following captivating experiment. In a dark room, connect a video camera to your TV or computer monitor screen. The camera should see just the blank screen. If the room lights are on, you will see an image of the screen repeating to infinity.

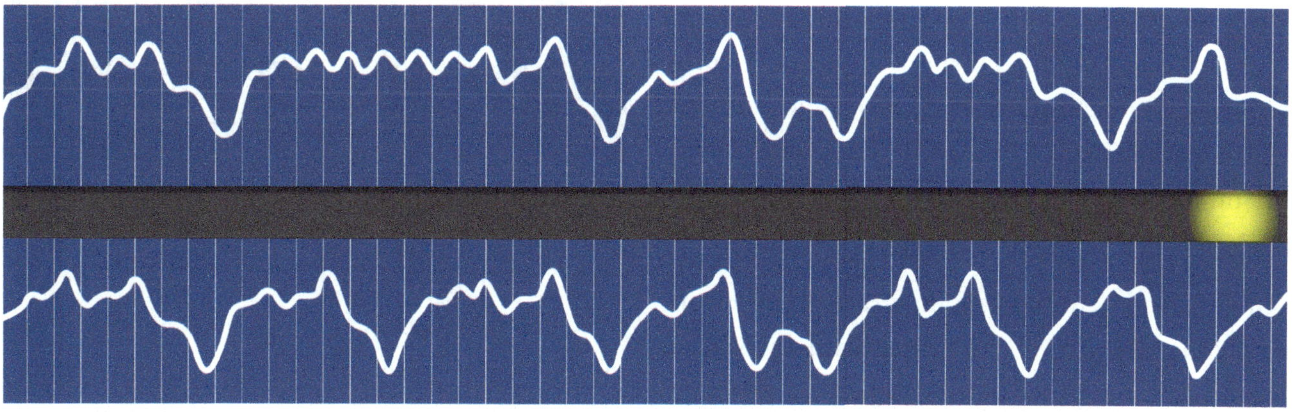

Lorenz's top graph plotted the weather over a month using initial conditions rounded off to six places, whereas the bottom graph used initial conditions rounded off to three places. The outcome stays in sync for about 5 days, but after this, it changes dramatically due to very small Butterfly Effects multiplying. Halfway through from day 14 to day 21, both graphs become fairly synchronized again, but they then drift apart once more to produce very different outcomes. If 12 or 20 places of accuracy were used in calculating the predictions, this would produce a different outcome again.

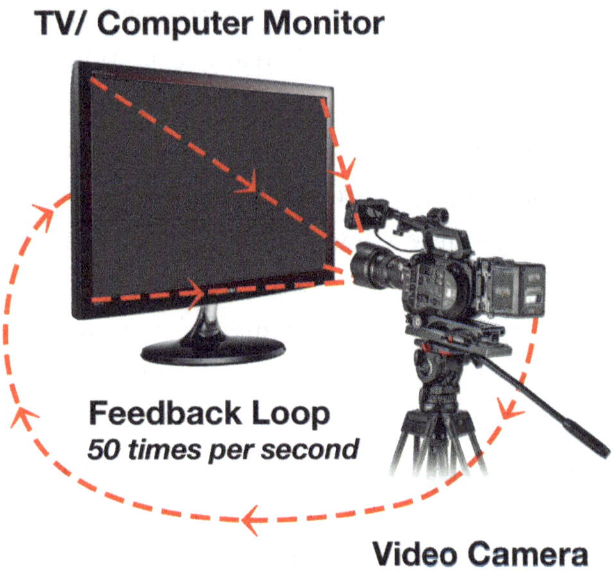

At night, turn the room lights off and hold a lit match briefly in front of the screen. The video vision will show the flame. Now, draw it away quickly, so it is out of the camera's view and extinguish it. You'll notice that the image of the flame will remain despite the match no longer being there! This is because the initial image is continually being fed back into the circuit. But the image will look nothing like the flame! It will have quickly changed in a remarkable way. Due to the feedback loop in this system, you will see a constantly evolving, complex pattern radiating out from the center of the screen. It will be reminiscent of computer-art video vision that simulates traveling through a 'space warp' in a science fiction movie!

The reason the image keeps changing is due to minute changes in the temperature of the electronics in the circuitry of the camera and the screen. Distortions are caused by tiny imperfections in the camera's optics system, small temperature changes in air currents and in the electrical system, and from particles in the air moving between the camera and the screen. Each small thing slightly changes the image 50 times a second. The process is repeated over and over causing a cascade of rapid changes from one frame to the next. This makes it look like a movie. Changing the camera angle very slightly will create a new set of initial conditions that will cause a different image to evolve.

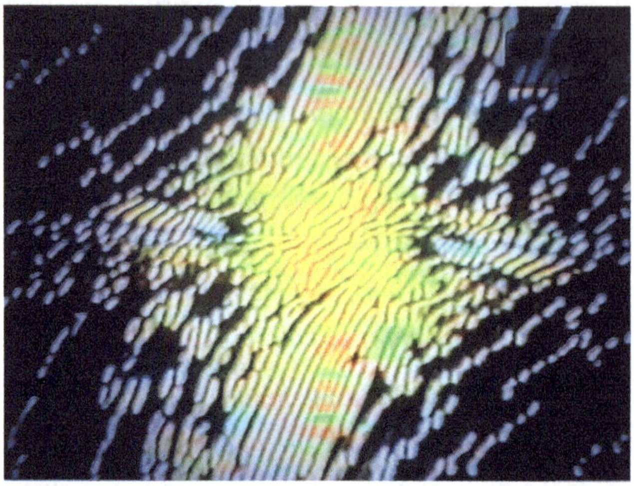

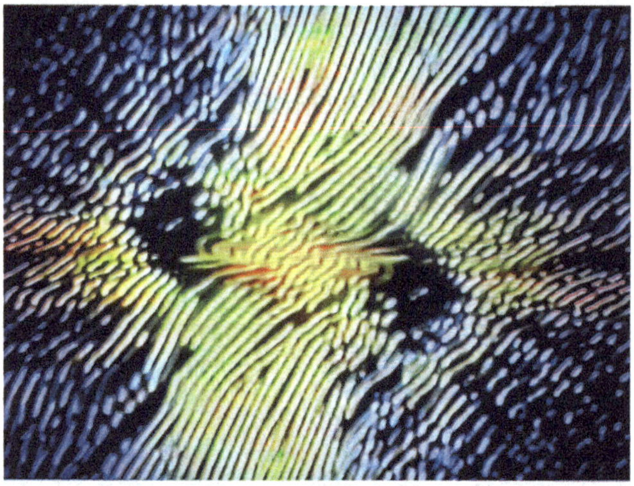

The Butterfly Effect in a video loop causes the feedback video image to continually evolve every fraction of a second, making the image continually change. It appears as if you are flying into it.

Once the effect is running, you can save it as a video. I used a video projector to project it onto a wall at parties to create a great special effect! While the screen image changes endlessly, you will note that there is a limit to its degree of change, as is the case with all chaotic systems. This experiment is a visual demonstration of how dynamic chaotic systems continually evolve and how their changes are limited to the boundaries that the system has. Butterfly Effects like this are occurring everywhere across our world and throughout the universe at every level - from molecules to superclusters of galaxies.

Smoke rising from a cigarette in a still room follows classic chaotic behavior. The smoke rises in an orderly vertical fashion at first, but quickly, minute butterfly effects such as slight variations in the rate of the cigarette burning and, therefore its temperature, cause airflow turbulence. This causes the rising smoke to become increasingly more chaotic the higher it rises. The same occurs with the flow of water and the flow of gas from the poles of supermassive black holes. Seemingly insignificant changes in the initial conditions of chaotic systems multiply to produce unpredictable outcomes.

BUTTERFLY EFFECTS AT WORK IN NATURE

Butterfly effects in one system can interact with those in other systems to produce more complex chaotic effects. As an example, the Earth has numerous complex, interacting, chaotic systems in which butterfly effects are continually occurring. Major ones happen in the Earth's inner and outer cores where very hot electrically charged plumes of molten rock and different metals rise into the mantle. This in turn affects the Earth's crust and this then causes continental drift and volcanic eruptions to occur. More Butterfly Effects occur due to changes in ocean currents, variations in the saltiness of ocean water and changes in the air temperature and humidity at different altitudes. The reflectivity of the Earth's surface, cloud cover, and precipitation also affect our planet's climate as does the Moon's changing gravity effects and the constantly changing angle and direction of the Earth's axis. In the biosphere, there are many changes that occur due to changes in vegetation caused by changes in the weather, insect infestation, and human development.

Earth's climate is always changing gradually due to Butterfly Effects building up. It changes catastrophically very quickly when large asteroid impacts occur. Over long time spans, changes in our planet's orbit and our solar system's orbit through our galaxy cause huge Butterfly Effects. Due to these variables, Earth has devolved an incredible diversity of very complex chemistries and environments. Earth's chemistry has produced numerous types of rocks, gems, and sediments. There are also Butterfly Effects occurring from plate tectonics, ever-changing coastlines, ocean currents, volcanism, river erosion, changing habitats, lightning, volcanic eruptions, fires, and human development, amongst numerous other things. It is no wonder climate programs are unable to predict long term climate changes when almost none of these variables are taken into account!

Chaos rules supreme throughout the universe.

Chaos' remarkably simple formula for the Butterfly Effect is a driving force behind the evolution of all chaotic systems in the universe. It is incredible to think that such a simple equation could be responsible for the universe's infinite complexity.

Evolution driven by the Butterfly Effect not only impacts physical systems, but also affect what most people would think of as being non-physical systems such as the sparking of new ideas and how these can affect one's beliefs and changes in cultures. Random events such as a joke taken the wrong way, a misinterpretation, an unexpected death, a change in politics, a war, new inventions, or new discoveries can all cause butterfly effects that change people's lives, their beliefs, or the economy and much more. This is important because people continually come up with new ideas some of which radically change the future. Butterfly Effects can lead to numerous positive outcomes, such as new scientific breakthroughs, wonderful new inventions, and new forms of art etc. They can also lead to negative outcomes such as conflict, declining standards of living and education, or a collapse in law and order.

When sand grains pour chaotically into a pile, they cause random micro- avalanches in the pile. But these chaotic avalanches even out to maintain a very ordered slope of 45°.

While most systems act chaotically, the overall end result is one of order. Chaotic events on one scale can create order on a larger scale. This process occurs across the universe at all levels, from atomic nuclei, in cells, in large biological systems, in civilizations, in economies, in solar systems, and in galaxies as well as the universe at large. **As one part of a large chaotic system decays, another part emerges to take its place, thereby maintaining overall order within the system.** This is demonstrated through endless wars. Millions of people may be killed leading up to the rise of a victor, however, sooner or later, a new war will commence, and that victor will eventually succumb to a new victor, thereby allowing the chaotic cycle to continue. We have seen this with the rise and fall of many great empires throughout history. This chaotic cycle occurs with all species. Changes in environmental conditions can cause certain species to become extinct, whilst others thrive, and new species evolve, thereby maintaining order overall. Chaos also affects solar systems. Planets can have very chaotic orbits at the outset, with some colliding and some being ejected out of their solar system after close encounters with larger planets. The orbits of planets are continually changing in line with Chaos Theory. The tilt of their axes can flip chaotically. Over time, chaos diminishes only to increase at a later stage.

It is impossible to make accurate, long-term predictions because everything in the universe at every scale can ultimately affect everything else at every scale. We can predict however that everything living, or non-living like a star, will eventually decay. Attempting to pinpoint exactly when this will happen is unpredictable due to the Butterfly Effect.

Since Isaac Newton's time, it was thought that we had a clockwork universe. It was once thought that if we could get enough detailed information about a system it would be possible to predict its future - that was until quantum mechanics was discovered in the early 20th C. It showed that there was a fundamental, inbuilt unpredictability at the sub-atomic level that makes it impossible to determine anything precisely at that level. But quantum uncertainties do not appear to be transferred to our macro universe, for reasons we have not yet identified. In the next 2-3 decades, there will be huge butterfly effects that will affect us and other life on Earth. This will be due to the exponential advances that are now occurring in intelligent-computing, robotization, and genetic engineering - to mention but a few of the many rapidly progressing modern-day technologies discussed later in this book.

Albert Einstein believed that the universe has law and order, but the great quantum physicist **Neils Bohr** thought that Einstein's deterministic laws were incompatible with the laws of chance at the quantum level. When Chaos Theory's Butterfly Effect was discovered, it became apparent that **the universe has rigid physical laws, but they work hand in hand with endless change produced by chaos. These two things make our universe evolve in an ordered way at large scales, but chaotically at small scales.**

It now appears that the universe works like a gigantic blender that keeps changing environmental conditions right across the universe to force it to evolve. Without chaos, the universe would almost certainly have been stillborn.

THE PECULIAR PROPERTIES OF SPACE AND TIME

Einstein's genius was to realize that space and time are 'elastic' in that they both contract in a gravitational field. When the field becomes infinite such as at the event horizon of a black hole (its black sphere), both space and time contract to zero! Einstein also showed that this contraction also occurs under acceleration in the direction of motion. The faster a body goes, the more it contracts in length, and the more time slows. At the speed of light, the length of a body in the direction of its acceleration contracts to zero, and time stops ticking. There is much observational evidence for this. But only light which has no mass can reach the speed of light. Einstein showed that space and time are two sides of the same coin. He called their union 'space-time'. Anything that exists in any place in space only does so at a particular timeframe. And similarly, anything that exists in time must exist at a particular place in space.

An electrical field and a magnetic field were found to be expressions of the same force. Similarly, Einstein discovered that mass and energy are equivalent, and so are gravity and acceleration. Light and other particles act like solid particles under certain conditions, yet under other conditions, they act as if they are waves. We live in a universe of equivalents, but these equivalent properties appear quite different to us. Are each of these 'equivalents' manifestations of some underlying force that we see from different perspectives? These equivalents are telling us something very important about how the universe is structured, but we have not yet grasped its significance.

We need to integrate the theory of quantum mechanics with the theory of relativity, but to do so, we need to understand the true nature of the quantum universe. However, the complexity of this could be like expecting our Stone Age ancestors to have conceptualized that their world was not flat as it seemed, but huge and round, and that it is suspended in a medium we call space. And on top of that, they would need to understand that everything is made of these incredibly small things we call atoms. No part of this concept would make any sense to Stone Age people. We may very well be as far away from knowing the true nature of the workings of the sub-atomic universe and how our universe came into being as Stone Age man was about what we know.

THE PROFOUND EFFECTS OF RELATIVITY

The following aspects of physics are most curious.

There is no place in the universe that is not in motion, and there is no place that is not experiencing some gravitational attraction, so space and time are always contracting to some extent. Individual galaxies and clusters of galaxies are accelerating towards one another due to the pull of each other's large gravitational fields. Because the expansion of the universe is expected to keep accelerating, the effects of relativity should ultimately make space shrink to some degree in the direction in which the acceleration is occurring if it approaches the speed of light, due to Einstein's Theory of Relativity.

Einstein showed that there is no absolute frame of reference in the universe because there is no center to the universe that can be considered at rest. This means there is no way to measure an object's velocity in relation to a stationary point. The velocity of an object can only be measured in relation to other objects that are also in motion. To get around this, astronomers consider our galaxy to be at rest. Einstein said, *"Everything is relative to the observer"*.

The speed of light is always the same in the vacuum of space. It remains the same *no matter how fast the source of the light is traveling*. Light can never travel faster than 300 million km/sec (186,000 mls/sec) in a vacuum because there cannot be the same event occurring at different distances. Therefore, information of any kind, photons, particles, or gravity, can only travel at a maximum finite speed, which is the speed of light. A light beam could be emitted from a fictional spaceship traveling at 90% of the speed of light, yet the light beam will not travel at 190% the speed of light as you would expect: it will always travel at the speed of light and never exceed it - regardless of the speed of the source of the light due to Relativity. This seems impossible, but it is a well-proven fact.

The faster a body moves, the more time slows down. At the speed of light, time stops. If a body is not moving or in a gravitational field, time will tick at its fastest rate. But even between galaxies there are weak gravitational fields and a body is always moving relative to other distant bodies to some extent. Why the universe has these rules, no one knows - yet.

Einstein also proved that the rate that time 'ticks' is relative to the observer. It ticks at a rate that is relative to the observer's acceleration and/or the strength of the gravitational field that the observer is in. So, *if* it was possible (which it isn't) for a person to travel at the speed of light, they would experience the same slowing of time to zero as a person who is in a near-infinite gravitational field such as that at the event horizon of a black hole. Because space and time can contract and stretch like elastic, the speed of light stays the same for all observers, regardless of their acceleration or the strength of the gravity field they are in. If this were not the case, the same event could be witnessed at different distances at the same time.

Imagine that a fictional person has spiraled into the inner accretion disc of a black hole. He is spinning around it at say, 99.9% of the speed of light. To him, his watch would appear to be ticking as usual, but observers outside the effects of the black hole would see his watch almost stopped! It would take billions of years on his watch for one second on our watch to pass. As he spirals into the black hole's event horizon where time stops ticking altogether, he feels like he is stationary. But due to the effects of relativity, when he looks outward, he sees everything in the outside universe moving at virtually the speed of light! The outside universe is evolving so fast that he could see to the end of time! These two opposing occurrences happen because, as Einstein proved, everything is relative to an observer's location and therefore, his conditions i.e. the speed at which he is traveling and the strength of the gravitational field he is in. This determines what an observer perceives to be 'real'.

THE PHOTON PARADOX

The speed of light creates the most curious paradox for photons.

Because we experience a four-dimensional universe (i.e. three physical dimensions and one of time), we observe photons (particles of light) traveling through space and through time. However, **if we could sit on a photon, the universe would appear to be a dimensionless point where neither time nor space exist!** To a photon, its entire universe is a point, known as the **zero dimension**. This dimension is what the universe would have been just before space and time were created during the universe's birth.

We don't really know much about time other than the arrow of time always travels from the past to the future, but in the mathematics of physics, it can go either way. Time is associated with a transition from order to disorder e.g. from life to the decay, or from a new building to a ruin.

As stated earlier, Einstein proved that the faster an object goes, the more the passage of time slows until, at the speed of light, time stops ticking altogether. Because photons always travel at the speed of light, the dimension of time does not exist for them. Therefore, **in the photon universe, nothing can change because there is no time in which change can occur.** With his Theory of Relativity, **Einstein proved that length shrinks to zero at the speed of light in the direction of motion. This means that there is also no distance in the photon universe** because, for photons, every part of the universe in their direction of motion would be at the same point, the zero dimension. Space has no length at the speed of light. Because distance does not exist for photons, this is another reason why time does not exist - it takes no time to go no distance. So, **to anyone sitting on a photon, they would appear not to be moving, space would have no dimensions and time would not exist. Along a photon's path, a photon would seem like it was at all points in the universe at the same time.** If we could travel at the speed of light, it would take no time to go to what people in the below-light-speed, 4-dimensional universe call infinity!

In summary, photons exist in the zero dimension that has no spatial dimensions, and no time. Because we exist in 4 dimensions, we can see photons traveling across space and through time. This demonstrates what a difference extra dimensions make.

In considering the photon paradox, also remember that photons can appear like a particle, while at other times, they act like a wave. (See Chapter 28). They are most likely neither, but rather something far more exciting but which we are so far unable to imagine.

IS THE GREATER UNIVERSE THE SAME IN EVERY DIRECTION?

We assume that the Greater Universe beyond our observable universe looks the same as it does in our region, but does it? At least the nearby Greater Universe should look the same because the parts outside our observable universe were inside it after the birth of the universe.

Could the Greater Universe have many different environmental regions like the Earth does? It has oceans, ice caps, deserts of rock and sand dunes, rainforests, grasslands, wetlands, plains and mountain ranges etc. Could the Greater Universe have different physical laws from one part to another?

Could the density of galaxies vary across the Greater Universe? Could gravity have different strengths in different locations, as it does on Earth from the bottom of the ocean to the top of mountain ranges? Could time tick at different rates in some places? Could space be more compressed or expanded in different directions? Does the universe expand evenly all over? Is it possible that the Greater Universe could rotate so that its velocity varies from one region to another? Could the geometry of the Greater Universe vary like the surface of the Earth does from a flat ocean surface and flat plains to deep ocean trenches and high mountains? How would we ever know?

We now know how large the observable universe is. As far as we can tell, the geometry of space in the observable universe appears to be flat. But that may be an illusion because the curvature of the Greater Universe could be so slight due to it being so large that we cannot detect any curvature – just as an ant would not detect the curvature of our Earth by walking across a salt lake 10 kilometers wide. If astronomers could determine the geometry of space, this might help us determine how large the Greater Universe might be. But how could that be done?

Chapter 6

DID A GOD CREATE THE UNIVERSE?

(As this chapter challenges religious beliefs about the existence of gods, those who are strongly religious are advised to jump this question.)

It's highly unlikely that a god created the universe because all the evidence points to gods being created by men. There is simply no scientifically valid evidence for the existence of any god. **If gods were real, we would be swamped with evidence proving their existence, but we are not.** Mere belief in a god is not evidence.

If a god did create the universe, then we would ask, "In what universe did this god exist and who created it?" It is impossible to answer this. We would then have to ask, "What created the god and its universe?" This would go on forever!

Supposing a god did create the universe, which god of the thousands that humankind has worshiped would it be? Evolutionary biologist **Richard Dawkins** correctly points out in his very popular, best-selling book ***'The God Delusion"*** that **those who believe in any one of the world's 4,500 religions and just as many cults are atheists about every god except their own!** Dawkins says that religious people only have to be atheistic about one more god (their own) to be realistic. Thinking logically, **religions are based on illogical stories invented by persuasive men over many tens of thousands of years from tribal clans right through to recent times.** There are many books with evidence that that supports this.

HOW DID RELIGION EVOLVE?

The first god concepts in different regions were probably conceived around 60,000 years ago or maybe earlier. This had to occur well after language became well established. At that time, humans knew almost nothing about the universe, nature, or themselves, so the brightest amongst them came up with gods to explain how nature was created and why it works the way it does. They became the first shamans. They invented gods to give meaning to human existence and crucially, why people should obey the shamen because they were delivering god's will. The most creative talkers became shaman by making up god stories for their tribe while sitting around a fire at night.

By observing how modern-day religions and cults spring up, it is not hard to visualize how ancient religions involving all-powerful, ethereal gods came into being. The god story most likely started something like this. A clever shaman in a primitive tribe came up with the original god concept. Their god was typically a supreme being living in the sky. This was an attempt to explain things like life and death, disease, war, the Sun, the Moon and the stars, natural disasters, and the weather. A single shaman may have been the first philosopher to develop the god concept to explain these things. In doing so, he found that most of his tribe thought he had great insight. They believed him when he said his knowledge was given to him by the all-powerful god that only he could communicate with it. It is impressive for someone to think of such an all-encompassing concept at that time. The first shaman found that he gained much power over his tribal members including the chief, by being in direct communication with his 'god' that created the world.

This god story spread like wildfire to other tribes. Other smart opportunists in other regions realized what a big opportunity the god story was to allow them to have power over their tribe, so they promoted it amongst their own people. The god story was a big hit when it was promoted by a good salesman who was a convincing actor. Shamans became oracles who used their god stories to gain much power and wealth. When the first tribes to farm animals and crops settled down in one place, they were able to create the first form of civilization. They developed agriculture, the farming of animals, and permanent home structures, so they no longer had to be nomadic. (See Chapter 15). This created great wealth for those times. Shamans were given great privileges, a life of comfort and much power by leaders because their followers believed their stories to be true.

As the god story spread, it was embellished, and varied by each shaman/priest to suit himself and his tribe's culture. Biblical historians have shown that nearly all Catholic Popes changed biblical scripture to suit themselves. They certainly did not follow the laws they preached because secretly, they did not believe in a god, as they knew it was a made-up story, so they didn't have to please their false god. Overwhelming evidence shows that this process is just as true today for those who start modern religions and cults, as well as those in the most established churches. This is why they have so much pedophilia and corruption in them.

To manipulate their believers, **shamans used showmanship**. In tribal times, they would dress in scary costumes, speak in tongues, and go into trances to look like they were in contact with a supreme being who was giving them directives for their believers to follow – directives that typically benefited the shaman of course. When civilization began, churches for a particular religion were built. Their priests consolidated their power and influence through their traditions and theatrics, which they performed in large, formidable churches built out of the earnings and hard work of their followers. In their grand edifices, priests wore rich, flamboyant costumes and performed symbolic rituals to impress their believers. They made out they were in direct contact with their god, and most ordinary people believed this.

The simple trickery that the god preachers used still works well today on those who are typically unable to think for themselves and those who are desperate to find something they can believe in to give them meaning to their existence. Believing in a Christian god gave billions of people hope that their savior loved them, often when nobody else did. It also gave them a belief that there was an **afterlife** that is far better than the hard, retched life they had on Earth. An afterlife in heaven gave many people hope that they would be reunited after death with their dead loved ones. They did not realize that there would also be those who they did not get on with.

There are those who pretended to believe in their religion so they would not be excluded from their tribe.

Priests use their power to direct the lives of their gullible followers in order to influence political outcomes, often to benefit themselves. Politicians then and now give church leaders big incentives to influence their followers to support their needs.

When large churches became established millennia ago, these powerful institutions took maximum advantage of controlling those under them. The hierarchy of the Church became the government, so they had ultimate power. This made the high priests very rich and authoritative. Their wealth and power assured the survival of themselves and their god myths. Anyone challenging them and threatening their power was likely to be murdered or imprisoned.

Over time, some shaman developed the concept that the Sun, the Moon, and the planets, as well as some groups of stars that the Sun and the Moon travel through during a year (the Zodiac), had the power to determine people's future. This myth spread far and wide and became known as **astrology**. Like so many mythical beliefs, it became a form of religion - just as socialism and global warming are today. Religions of any kind require no *valid*, provable scientific evidence to support their myths - just blind faith. All religions are promoted by a small group at the top so that they can gain wealth and power from the many simple people who believe that their myth is real.

In the 16th C, the huge and powerful Ottoman or Turkish Empire, which stretched from the Persian Gulf to Hungary, reached its peak, becoming the most advanced civilization in the world. At that time, it was well ahead of all other empires in science, math, geometry, architecture, shipbuilding, strategies and weapons for war, and all aspects of the arts. It flourished for centuries until the Islamic religious leaders took total control around 700 years ago. They realized that the practical explanations underpinning the rapidly advancing science at that time would undermine their religious doctrines, and therefore their hold on power. So, they suppressed scientific development. Those that tried to continue doing science were typically put to death. Through the fear of death, they abolished scientific progress, so few scientific discoveries and engineering breakthroughs occurred. This eroded their civilization drastically, causing it to eventually become much less powerful. Religious control has remained a barrier to Islamic nations being able to progress scientifically and technologically. They have had to employ the best minds from Western nations to use their oil wealth to develop their countries. This has stopped their integration into the developed world. While most developed western countries made enormous progress scientifically, Islamic nations did not. Their religion stopped them from remaining a leader in the world through not inventing new technologies. If it were not for their luck of having western companies finding oil on their lands in the Middle East in recent times, they would be unimportant third world nations, struggling to survive in an inhospitable landscape.

Today, **charismatic preachers use sing-alongs, hypnotism, and simple special effects and illusions** to give their simple-minded followers a false sense that their preacher has supernatural powers bestowed upon him by their god. Their followers want so much to believe what these preachers say is true that they make no attempt to question if it is all a con. People flock to the charlatans who start their 'churches' in warehouses, old halls, and large tents. The preacher brainwashes his brethren and makes them feel that they are special and loved. Because they are with like-minded people, they feel like they are in a caring family – as long as they contribute their funds of course. Once these charismatic preachers gain a healthy stash of cash from their believers, it is common for them to disappear with the loot.

ALL POWERFUL GODS ARE UNCARING

If an omnipotent god existed, as preached by most religions, then it would have the power to make everything on Earth perfect. It would not have the need to make so many innocent people and animals suffer so much. **People who worship such a god do not realise that it does not look after those who worship it.** In fact, it allows the likes of mass-murdering despot dictators, drug lords, Satan worshipers, thieving bankers, corrupt politicians, and numerous paedophiles etc. to get away with their crimes, most of them for their entire lives, yet billions of hardworking good people are seldom rewarded for worshipping their god. **Their god did not stop any of the wars throughout all of history. This is evidence that gods do not exist.** As the fable goes, **the Christian god even sent his only son to Earth to teach humans some of his 'rules' for life and let him die a painful death. Yet he was totally unsuccessful because 2,000 years later, nothing about humanity has changed.** There are numerous things like this that prove beyond all doubt that the Christian god does not exist. It is the same for all religions.

In which universe would a god exist? And how would he observe everything on Earth? Given that the universe is very likely filled with innumerable intelligences, would a god be able to oversee all life in the universe? Why would he want to? Would his rules for other intelligences be as insane as they are for humans?

When one takes the time to think through **how unrealistic all religions are**, it becomes clear that their stories don't make any logical sense! And they all have contradictory aspects to them! For example, **the Christian religion states that their god is all-powerful, yet he is unable to get humans to do what the priests say he asks.** If he was truly all-powerful, he could make people do what he wants in an instant. As the unbelievable Christian story goes, in order to 'save' Christians, God sent his only 'son' Jesus to Earth to preach his will to the people, but the 'evil' Romans killed him for causing trouble. As the fable goes, God allowed his son to be painfully murdered. What an unempathetic god he must be. Being all-knowing, he would have known the outcome, but of course, this story was made up by humans. As the story goes, Christians are told that Jesus miraculously rose into heaven when he was dead! What a cracker of a story that is.

Throughout both the Old Testament and New Testaments of the Bible, there are numerous contradictions and evil commandments. If the Christian god is all-knowing, why wouldn't he have told us at the outset what the meaning of life is, how he designed the universe to evolve from a primeval fireball, how and when the Earth came into being, and how life evolved. The reason all gods in all religions do not teach us these things is because there are no real gods - just make-believe ones. Instead, unknowledgeable Christian priests tell us an unbelievable creation story that the Earth was created in just 6 days? The Bible is full of these crazy stories which were made up by men. Modern-day religious stories are not much different than those of the most primitive tribal cultures.

There is **no evidence at all for any religious story** being true. All religious believers choose what bits of their religion they want to believe and they ignore the parts that don't suit them. Many non-believers make out they believe in their religion due to cultural pressure. They do want to be banished from their herd.

When I was in preschool and in grades one to three, I asked my priests many questions like why god causes good people to die of cancer, or in car accidents, and why God does not show himself to us, I also asked why he allows wars to kill millions of people, and why he doesn't use his power to make the world a better place. Not one priest has ever been able to give any reasonable answer to such reasonable questions. They just duck them by saying "god works in mysterious ways". I asked, "Why does he do that?" They do not answer and walk off. I knew even as a young child, that the god story must be false.

BLIND FAITH

When people want to believe in something that makes them feel good, they dispel all realities that go against their belief in order to retain their sense of fulfillment. This is understandable for those who do not have self-worth. Unfortunately, there are many people who have so little love, respect, or joy in their life, that they need to believe in something greater than themselves. To survive daily life, sadly, they have to hang onto any belief that deadens the physical and mental pain that reality causes them to suffer. These people are grist for the mill for the opportunists running churches. They milk the purses of their followers for their own wealth and power.

There are, however, some religious groups that do not exploit the weak. They offer true pastoral care - meaningful love, companionship, advice, and practical ways of dealing with illness or being dealt a bad hand in life. The principles that underpin some religions are intended to provide a moral compass, such as the ten commandments of the Christian faith, many of which are unacceptable today. Local churches try to promote social cohesion and order, but most religious leaders, be they of large or small dominions, do not follow what they preach, and many are exposed for spending the Church's money on themselves.

Many of the hierarchy in the mainstream churches worldwide are now being exposed as hypocrites, frauds, unempathetic psychopaths, grand scale thieves, pedophiles, and smugglers. Many of the banking Elite have been exposed as worshipers of Lucifer! The magnitude of the corruption that has been exposed amongst the hierarchy of the major churches is so great that it is very hard for both religious and non-religious people to come to grips with it. This is because it is beyond what normal people can relate to. Astonishingly, the churches seem to survive endless major scandals because the bulk of their followers do not want to give up their simplistic faiths.

The best religion of all is to believe in is oneself, but this is largely controlled by one's genes, as you will learn in Chapter 25. **Our genes give us little say in what we do or what we believe.** We have no ability to make our own logical decisions if we fail to analyze bogus religious teachings for what they really are.

THE ROLE OF SCIENCE AND ITS IMPACT ON HUMANITY

It is looking increasingly more likely that the universe is a set of mathematical formulas designed to create ever-higher levels of complexity. (See Chapter 30.)

It appears that the reason for the universe's formation turns out to be the need for it to become increasingly more complex so it can evolve ever-higher intelligence and consciousness. If this is so, then we must ask, "Why does the universe seek this?" If the universe is the ultimate program, what is its end goal? Does it have one? I suspect that any answers we might come up with would be as naive as those of a 5-year-old child trying to work out what their life is about and how their world came into being. Most young children do not question how Santa Claus can visit billions of homes across the world and carry tens of billions of presents for children in his small sleigh - and do it all in one night! They just like the belief and don't question it. Aren't adult religious believers the same?

Science has provided innumerable benefits in the form of miraculous engineering solutions in the form of great architecture, better transportation, amazing advances in communication, extraordinary entertainment, irrigation, piped water, sanitation, medical care, brain surgery, the Internet, and an abundance of food and energy to mention just a few. But because these benefits are nearly all developed in western nations, Islamic nations have to buy in their technology from the West. They have to send their scientists and engineers to be educated in western universities. Most Islamic scientists have been confined to mostly weapons development. This prevents them from progressing in the sciences, which would challenge their religion.

As science progressed, the most advanced scientific thinkers worked out that the Sun and the Moon were not gods, but bodies out in space that moved around. By the 16th C, astronomers had calculated that the planets orbited the Sun. In the 17th C, the first cells were discovered by **Robert Hooke** using a primitive microscope. It became apparent that cells played an enormous role in the evolution of life and people's health. No religion would ever have made these discoveries and numerous others like them.

When the **Renaissance** occurred in Europe in the 17th and 18th centuries, it introduced a wave of scientific investigation known as the **'Age of Enlightenment'**. This allowed science and progressive thinking to flourish. Across the western world, the Church no longer had a stranglehold on scientific inquiry and new knowledge, so the rate of scientific progress exploded. Many topics came off the taboo list. But today, the religion of the Elites and their leftist communism, together with the Catholic Church is trying to destroy science and all rational learning, to make unquestioning adults believe that the fraud of global warming will destroy the world.

Until recently, Christian cosmology has preached that its god was the creator of our world and all life. But the Bible never says how all the immense detail in the world was achieved. We are expected to believe it was the greatest of all magic tricks! But recently, the Catholic Church finally accepted the insurmountable evidence that our world, the universe, and life was created by natural evolution commencing with the Big Bang nearly 14 billion years ago - not in six days as they preached for millennia. This means that their Christian god deceived them about how the Earth was created and how long it took to form? Or is it more honest to say that the hierarchy of the Church knew all along that there was no god and that they did not know how the world was created? The pretense that they were in touch with a higher universal power was clearly a ruse. In the late 20th C, the Catholic Church finally admitted that they were wrong to have killed, imprisoned, and punished some of the world's greatest scientists hundreds of years ago because they proved that the Earth went around the Sun and that Earth was *not* at the center of the universe, as the Church preached. The Church also had to accept that the Sun was not god as they had preached. They retained it symbolically as a halo.

By the 19th C, **Charles Darwin** showed that all life forms evolved from simple cells and that the diversity of life was accomplished by natural selection. The realization that evolution acts without any need for godly intervention to create all forms of life drove a stake into the heart of religions that preached that their god created all life. Back then, Christian-based religions illogically believed that the beginning of time was only 4,000 years ago, as Newton had calculated. Before this, it was thought that the universe and all life on Earth was created in a matter of six days! It was believed that all life was the same as it is today despite numerous fossils being discovered that showed that different life forms existed at different times in geological history extending back up to billions of years.

Charles Darwin and his wife were very devout Christians, and his son a preacher. When Darwin realized that life evolved and that it was not created by a god, he faced a serious personal dilemma and a crisis of faith because his thesis of evolution dispelled the most fundamental teachings of his church – that God created all life forms at the same time. Darwin held back publishing his book on evolution to near his death because he agonized over the impact it would have on his family's association with the church and many of their friends. His deeply religious wife ultimately supported the publication of his work. When his **Theory of Evolution** was published, Darwin suffered enormous personal consequences because he was virtually shunned by the religious society in which he lived.

Despite the fact that the stories that religions preach have been soundly disproven, few religions die out because they offer their believers simple, feel-good concepts to make them feel like they are secure and loved. This need is understandable because understanding the real world is far too complex for most people.

WILL RELIGION BE A COMMON STAGE OF DEVELOPMENT FOR ALL INTELLIGENCES?

In the context of the evolution of intelligent life in the universe, one has to wonder whether religion might be a fundamental evolutionary phase for all intelligent life at the stage of human development.

At the outset of tribalism and the beginning of civilization, religion aided in slowly developing an organized society. It provided a sense of community for many people and a safety net for the survival of some by way of charity. It also produced a value system for conducting one's life and an explanation for one's existence. The power of the Church was able to create the first forms of government. The Church's system of forced donations evolved into government taxes. It seems like everything has a purpose for a period of time – even religion.

The rise of religion as a unifying force for social order may be common for most emerging organic intelligences in the universe.

IS THERE AN AFTERLIFE?

Many people who believe in a god do so because they want to believe in an **afterlife**. A comical origin to this belief was cleverly illustrated in the 2009 movie '***The Invention of Lying***' where the son of his dying mother tells her that instead of an eternity of nothingness after she dies, death will bring her a joyful afterlife in a wonderful place. By telling her this lie, her son thereby introduced the concept of heaven, so his mother would die happy. When his conversation was overheard by nursing staff, word of this concept spread like wildfire. He became the new messiah. A belief in heaven is still held today by many people despite there being not an iota of evidence for it. But there is much evidence that proves it does not exist.

If heaven existed, it would be a very undesirable place to go to. In fact, it would be hell. Some people imagine it to be like sitting on a cloud listening to harp music. How attractive would that be for eternity? Most people would like hard rock music while others would want Indian, Chinese or tribal chants etc. Others imagine heaven to be like a English garden. Imagine being stuck in heaven for an eternity with relatives and people you detest. There would be fights over everything just as there is on Earth. You would have no body so you would not be able to enjoy the pleasures you liked on Earth. What would you do for an eternity without a physical body? All 'good' people from the caveman days would be there. Everyone would have different value systems so they would not get on. All diseased, and mutilated people would be there as they were when they died? Would babies that died be there forever as they were when they died? Would people all be as they are when they died - old and decrepit or disfigured after being in a war, in a car crash, or from horrific disease etc? That would be terrible. Would there only be humans, or would there also be all good animals as well? Could God or Saint Peter be trusted to decide who gets into heaven? It would be a very, very crowded place. What would dead people do in heaven? If they were there for an eternity, they would become insanely bored, wishing that they had just died and there was no heaven. Would there be any way out? Heaven would be as bad as being sent to hell. It too would have as many insane realities about it, if it existed.

Being reincarnated in another body would be a lottery. You could forever get a bad deal each time. No matter what religion one subscribes to, they all have similar insanities involved. The people who want to believe in them are not thinking people, so they believe in the most simplistic concepts that bear no relation to reality. No human has ever invented a god that would be workable or rational in our universe. They are all make-believe gods with as much reality to them as Santa Claus or the Easter Bunny. It is truly incredible that there are some people who are intelligent and knowledgeable about certain subjects that they have studied but they are completely naive about the god they believe in. Studies of genes show that religious beliefs are so genetically imbedded in the psyche of those who believe in gods that they seldom question how real their belief is.

Religious followers want a feel-good belief system that gives them a feeling of security and meaning to their life. They also want the promise of an afterlife for themselves and their loved ones when they die. In contrast, most scientists are not afraid to ask difficult questions to discover things that change their belief systems. Scientific beliefs have often had to change when new evidence comes to light. Over the last century, we have learned so much about nature and the universe, as well as ourselves. Before this, we would not have thought these discoveries could be possible. That is the way of science. People who believe in science will change their mind when solid evidence demands they do so, but religious people cannot do this, they stick to their feel-good beliefs, no matter how unrealistic they are. None of the answers in this book require the existence of a god.

Some religious people think that because science has not discovered every answer to every little aspect of nature, this is a reason not to believe in God. They think that their "*god must exist to fill in the gaps*" in scientific discoveries. But there are very few 'gaps' that remain in present scientific knowledge that can be used to support the existence of a god. The 'gaps' have been filled many discoveries. The extraordinary claims made by religious fanatics require extraordinary evidence: not supposition and speculation, or wishful thinking.

Had scientific thinking not evolved, we would know very little because religion has never discovered anything other than how to manipulate people. In contrast, science has greatly bettered the lives of humans and many animals. By contrast, religion has led to the most unjust barbaric punishments, and it has caused innumerable wars driven by religious megalomaniacs in order that the leaders of a church-run government may consolidate their power over those who cannot think and learn for themselves.

"Science promotes questions, some of which may take a long time to be answered, whereas religion promotes answers that can never be questioned."

Author Unknown

IS RELIGION USED TO CONTROL PEOPLE?

The control of large populations by way of religion has been so successful that it has survived through the millennia in nearly all countries despite the advancement of science using analytical deduction to discover reality. But many religions are now starting to diminish due to their corruption being exposed. In just the last couple of decades in western countries, the Elite bankers and the governments they control have taken much power away from religious leaders to erode their ability to influence their believers. This has allowed Leftist governments to more easily control their populations and to preach socialism as their simplistic religion- but it never delivers what they preach.

DO RELIGIOUS PREACHERS PRACTICE WHAT THEY PREACH?

The ethical and moral base of the Christian religion that teaches us not to kill, steal, lie, or engage in pedophilia is good; however, hypocritically, the hierarchy of many churches and their preachers have adeptly practiced these things, proving that they know the god they preach about is a false one. Natural moral values don't require a religion: they are common to most societies, including those that do not believe in a god. Churches and cults are largely unethical criminal businesses that pay no tax and are often involved in the highest financial and moral crimes. Despite numerous crimes being exposed in all types of religious organizations over many decades, most of their believers ignore the evidence of their criminality. They have blind faith that requires no evidence to justify it.

GODS ARE UNEMPATHIC

If gods exist, they are unempathetic to create so much physical pain in humans (and all life forms) and to create so much psychological anguish for people. This would make them psychopaths. Most gods, including Yahweh and Allah, say they are jealous of all other gods. The Koran and the First Testament of the Bible state that those who worship another god must be severely punished or killed. Having them jealous of other gods was an admission that they are not the only god. These books clearly reveal that Yahweh and Allah had human emotions because the texts for these religions, like all others, were written by men. Jealousy and murder are not the actions that one would expect from an all-knowing, omnipotent god: these are human emotions that were infused into the gods that humans invented. The concept of emotive gods wanting to kill those who do not believe in them was developed by early shamans in an attempt to stop other shamans from stealing their followers.

DOES INTELLECT, KNOWLEDGE OR GENES AFFECT ONE'S BELIEF IN A GOD?

Many psychological tests have shown that the more educated one becomes, especially in science, and the more intelligent people are, the less likely they are to believe in religion, occult beliefs, or what governments tell them without proof. This is because science gives them numerous logical options from which to choose what is real and what is not. Poorly educated people have few options to consider.

Neuroscience studies have shown that people who have genes that predispose them towards having a dominant right brain are more likely to have religious beliefs that do not require evidence. Those with a dominant left brain are more likely to be rational and not religious. It has now become clear that it is our genes that determine whether we will be religious or not. (See Chapter 27.) A person's genes can dictate that they will believe in a higher intelligence that will create order in the world and that there will be no random, unpredictable evolution as long as people do as their god preaches. This has never happened because there have been continuous wars that create near total chaos and the extermination of many hundreds of millions of religious people every century so that megalomaniac dictators can try to gain more power.

Even when believers in a religion are provided with strong evidence that a god does not exist and that their god is not all-knowing, their genes that drive their emotive decision-making will not let them change their mind. They have **Confirmation Bias** (See Chapter 25, page 340) that stops them from being able to rationally question what they are told. Until genetic engineering advances, there is nothing that can be done about this.

It's amazing that most highly religious people are unable to rationalize any of the logical points in this section, so they can't see how religious leaders have manipulated them for their own gain. It is so sad that so many people have so little in their lives that they have to hold onto the vaguest hope that there is someone like a Pope, or an ethereal god, that will rescue them from their misery, fear, or insecurity. Religious people who are not badly off just want a simple belief to give them some direction to live their lives. There are many followers of religions who know they are unknowledgeable, so they want their preacher to do their thinking for them. There are also those who feel like they are not accepted in most social groups, so by going to religious gatherings, they feel a sense of friendship and belonging.

The faiths of the Muslims, Judaists, and the Christians and many of the world's other major faiths arose from one original ancient religion. This is why they have so many aspects that are very similar. In most if the major religions thousands of years ago, the Sun and the constellations were similar. Many religious cultures personified the constellations and their god's relationships to nature. The Sun was the giver of life and it represented goodness, while darkness represented evil. Many gods were born by a virgin when there was a star in the East. The child god was adorned by three kings. When 30, the god had 12 disciples. He could perform miracles. He was betrayed by a disciple and crucified for 3 days and then resurrected. Other concepts that were very similar were the Great Flood, Moses the Law Giver, The 10 Commandments, The Afterlife, Heaven and Hell.

It seems that the more we learn, the stranger and more complex the universe becomes. This is immensely exciting for thinking people, but for most people, it is quite challenging to think beyond their encrusted old beliefs, most of which are based on simplistic thinking and a severe lack of knowledge. We are now able to learn so much about all aspects of the universe that we cannot help but wonder if there will ever be an end to the seemingly endless discoveries that are being made. We cannot possibly imagine what we will learn even in another decade, let alone in the next 50 years.

"Learn from yesterday, live for today, and hope for tomorrow. The most important thing is to not stop questioning."

Albert Einstein

LIFE IN THE UNIVERSE

Chapter 7

IS THERE LIFE OUT THERE?

When we look into our universe with the knowledge of modern astronomy and biology, we cannot help but wonder how much life there might be out there. And we can't help but muse about how many other intelligences could be wondering the same thing and looking back at our solar system. Credit: Gregg Thompson, Joe & Rebecca Gjerek

The most common question people ask me as an astronomer is, "*Do you think there is life out there?*" In this section, we will endeavor to evaluate the chances of both simple and intelligent life existing elsewhere in the universe. Given the size of the cosmos, and the number and diversity of planets that are now being discovered, it does not seem reasonable to think that our Earth is the only planet in the entire universe teeming with a huge diversity of life. Life here exists in almost every environment - on land, at the poles, in the oceans at all depths, in the atmosphere, and inside the Earth's crust.

As has been expected, astronomers are now finding numerous planets orbiting other stars. Given the way in which solar systems form, nearly all stars will have planets, but only a small proportion are likely to be suitable for advanced life.

Now that biologists have a good handle on the conditions that life requires to evolve, we are now gathering enough knowledge to make an educated estimate of how many life-bearing planets there might be in the part of the universe that we can observe (the observable universe). Of this total number of planets, we can make a reasonable estimate of how many planets might evolve simple life, and how many might have an intelligence emerge that can develop technology, as we have.

To make these estimates, we need to know how many planets there are likely to be in our observable universe. (The term 'observable universe' is explained in Chapter 3, page 36.)

HOW MANY PLANETS MIGHT EXIST IN THE OBSERVABLE UNIVERSE?

To estimate how many planets there are likely to be in our observable universe, we need to estimate:

1. how many average-sized galaxies there are in the observable universe,
2. the average number of stars there are in each galaxy,
3. the average number of planets that each average star may have,
4. the number of planets that might have conditions suitable for simple life, and,
5. the percentage of those planets that might evolve intelligent life evolve that develops technology.

Let's start from the top and see what numbers come out of conservative estimates in each case.

This is a very small region of the sky that shows numerous large and small galaxies. Background galaxies are so far away that they appear star-like. Every speck is a galaxy: there are no stars in this image! The smallest dwarf galaxies in the universe contain, on average, around ten billion stars while the very largest may have over ten trillion. Credit: Marco Lorenzi

HOW MANY GALAXIES MIGHT EXIST IN THE OBSERVABLE UNIVERSE?

As we look ever deeper into our universe's cosmic ocean, we see endless galaxies in every direction and in every orientation. Most are in clusters, many of which are in the process of merging with one another to form superclusters. Across the sky, galaxies appear as numerous as microscopic plankton in our oceans! But if they were at the same scale, galaxies would be immeasurably further apart. Using the HST, we have been able to count trillions of galaxies in our observable universe. But this number does not include the most numerous galaxies of all – small, faint, dwarf galaxies, most of which are too faint for even the HST to record them. In the Greater Universe beyond, galaxies could be almost infinite, but we do not know.

The most recent estimates of **the number of *average-sized* galaxies in our observable universe is around 2 to 3 trillion.** To be conservative, I have used 2 trillion as the number for the estimates following.

The Large Magellanic Cloud shown here is a dwarf galaxy. It orbits our much larger Milky Way galaxy. Dwarf galaxies are much smaller than average-sized galaxies. They contain only several billion stars. Credit: AAT David Malin

DWARF GALAXIES

Dwarf galaxies like the Magellanic Clouds are by far the most numerous of all galaxy types, but they are the faintest and smallest because they contain much fewer stars than an average-sized galaxy like our Milky Way. In nearby galaxy clusters, we see the largest and brightest elliptical galaxies surrounded by the brightest spirals and dwarf galaxies, which appear as specks. They are like bees around a disturbed hive. The vast majority of dwarfs are too faint to be recorded. It is therefore, unknown just how numerous they are. Estimates of **the number of stars in the largest dwarf galaxies can be as many as a hundred billion**. **Although dwarf galaxies are numerous in number, to be conservative, they are not included in my calculations for the total number of galaxies in the observable universe** because they have comparatively small populations of stars compared to average-sized galaxies. However, due to the large number of dwarfs that exist, they could account for a considerable percentage of the total number of stars in the universe, and therefore a reasonable percentage of planets. So my estimates for the likelihood of life existing beyond Earth are very conservative.

Globular star clusters contain hundreds of thousands to many millions of stars, but the stars in globulars that formed planets

may have had their outer planets stripped away during close encounters with other stars. It is thought that stars in globulars may only retain planets that orbit relatively close to them, but they could also gain planets that are roaming freely. Some planets will be thrown into their star, while others might be ejected from the globular altogether during close encounters. This makes it hard to estimate how many life-bearing planets globulars may have. Credit: Misty Mountain Observatory

GLOBULAR STAR CLUSTERS

Globular star clusters are like mini dwarf galaxies. They swarm around average-sized galaxies in numbers varying from scores to hundreds. But in giant super-massive elliptical galaxies, which form from mergers of hundreds of galaxies, their globulars number in the tens of thousands to perhaps millions. Globular star clusters typically have stellar populations ranging from half a billion stars for average-sized globulars up to 10 billion for the very largest. Those that are larger than this are classified as dwarf galaxies.

The densely populated inner regions of globular clusters are where most of the stars exist. Planetary systems have already been found around some stars in globulars. It's presently unknown what percentage of a globular's stars have planets and what the average number might be.

Many stars in globulars have orbits that take them into the dense inner regions of the globular, where they have gravitational interactions with other stars. Close interactions pull their outer planets of stars out of their orbits, so there would be many planets in globulars that are unattached. With no star to warm them, these planets would be bitterly cold and unlikely to sustain life - unless they have substantial inner heat from their cores, or from tidal heating. To be conservative, **I have not included the number of stars in globulars or their planets in my calculations.** Even if the stars and planets in dwarf galaxies and globulars were included, the total number of stars and planets in the universe *may* not increase significantly.

Here we see the barred spiral galaxy NGC 1073 showing innumerable, brilliant, blue giant stars in its spiral arms. For every blue giant visible, there are hundreds of thousands of the more common red dwarf stars that are too faint to be seen. Old yellow stars populate the galaxy's central bar. Credit: HST, NASA

HOW MANY STARS ARE THERE IN AN AVERAGE-SIZED GALAXY?

An average galaxy like our Milky Way has recently been estimated to contain at least a trillion stars. The estimated number has gone up considerably since astronomers discovered how numerous small, faint dwarf stars are. This should have been obvious earlier because the smaller stars are, the more numerous they are, and the fainter they are, and therefore harder to detect.

HOW MANY PLANETS ARE STARS LIKELY TO HAVE?

How many planets are there in other solar systems is a question that we have long wanted to know. Rapidly advancing technology is now allowing us to answer this question.

Supergiant stars may have up to hundreds of planets in their solar systems when they form.

Planets form in the accretion discs around stars. The discs around giant stars would have far more gas and dust than dwarf stars, and their discs would be far larger, so they would have far more planets than dwarf stars. (Our Sun is a large dwarf star.) Giants and supergiants could have over 20 large gas giant planets, some of which are likely to be much larger than Jupiter, so they could have many earth-sized moons. Some of these could support life. And there could be a dozen or more large, Earth-sized rocky planets. But giant and supergiant stars are only a small percentage of all stars, so their large numbers of planets would not increase the average number of planets around an average star by much. Our Sun is an average-sized star. But more importantly, supergiants are very hot, so they burn their hydrogen fuel very quickly. This causes them to have very short lives lasting only a few million years before they explode. When a supergiant goes supernova, their planets are destroyed. In view of this, we must ignore the number of planets around supergiant stars because life would not have time to evolve on them.

This is an illustration of the solar system Kepler 11. Illustration by Kepler Telescope, NASA

Dwarf stars are by far the most dominant in number, but most are as small as the Earth. It appears that, on average, the number of planets that they have could range from a few to a dozen or so. Their planets are expected to be mostly small rocky ones the size of Mercury and our Moon, or smaller like Pluto, Triton, Titan, and the Galilean moons. These solar systems are very small, being around the size of Mercury's orbit.

Small red dwarfs are much smaller and cooler than our Sun. They have large solar flares that emit extremely high levels of deadly X-rays and UV light, so their solar winds are lethal to life. In view of this, their planets are not likely to support life on their surfaces unless some of these planets are large enough to have a very dense, protective atmosphere like Venus has. Life could exist in deep oceans and underground in cavernous crusts. When all of this is taken into account, the number of planets around red dwarfs that could support simple life could be as low as 1% of their total number. But that could still be a significant number considering how numerous red dwarfs are.

For our calculations, I have estimated that **the number of planetary bodies (including large moons) around most stars will range between 5 to 100. I've assumed 10 to be the average number of planets per star.** This is likely to be very conservative.

DO PLANETS EXIST OUTSIDE OF SOLAR SYSTEMS?

Astronomers are now becoming aware that during the formation of solar systems, the orbits of planets, asteroids, and comets are chaotic because there are so many bodies that have frequent close encounters with one another, especially when solar systems are forming. Gravitational forces whiplash many bodies into their star or out of their solar system altogether. **Astronomers estimate that for every planetary body in orbit around a star, there may be a similar number, or even more, ejected at such high velocities that they would go into interstellar space.** Many may end up in their galaxy's halo. This is a huge spherical volume of space about ten times larger than the obvious part of a galaxy seen in photographs. Despite these planets not having a star, they could have hot interiors that might allow life to evolve inside aquifers or cave systems in their crusts. Life could evolve on their surface if they have a dense atmosphere that retains heat from the planet's interior. For this reason, a small proportion of planets in the cold, dark, outer regions of solar systems may also harbor life. However, these planets have not been included in the total number of planets that might support life, as their number is unknown. Again, another conservative approach.

In summary, to be very conservative, planets that have been excluded are those that exist in:

- dwarf galaxies,
- globular clusters,
- the halos surrounding galaxies,
- interstellar space, and
- the cold, dark outer regions of solar systems.

While these groups may account for a majority of all planets in the universe, to be conservative, they have not been included because it is unknown how plentiful they might be, and how many could support life.

Stars abound in the billions in the central hubs of galaxies. Galaxies with a double hub like this one form when two galaxies merge into one. Illustration by ESO

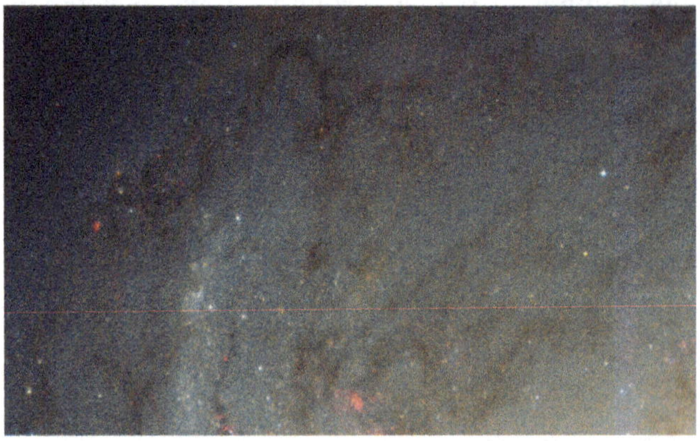

In the outer arms of galaxies, there is not enough gas (pink nebulas) to produce large numbers of supergiant stars, so astronomers assume that there should be few supernova explosions that produce heavy elements, which are necessary for life to evolve. But supernovas are frequently seen to occur in these outer regions of galaxies, so our theories in this regard may not be correct. Credit: HST, NASA

HOW MANY PLANETS ARE THERE LIKELY TO BE IN THE UNIVERSE?

By multiplying the estimated average number of planets (10) around each star by the estimated number of stars in an average-sized galaxy (1 trillion), equates to **10 trillion planets per average galaxy!** If we multiply this number by **2 trillion average-sized galaxies in the observable universe, this gives us the minimum number of planets as being around 2,000 trillion, trillion.** That's a lot of planets! But few of these are likely to have conditions to support life.

Einstein's planet', known as Kepler 76b (black sphere), is a "hot Jupiter" that is so close to its star that it orbits it in only 1.5 days! Its diameter is about 25% larger than Jupiter, and it weighs twice as much. To have a large planet this close to its star was thought impossible, however we now know that there are plenty of stars that have large planets orbiting very close to them. This artist's conception shows how the planet's gravity tidally distorts its star into a football shape by pulling its atmosphere towards it. Illustration by ESO

FINDING EXOPLANETS

Using the Kepler Space Telescope and land-based telescopes, astronomers are discovering tens of thousands of planets orbiting the small selection of stars used for this study. Two methods are used to detect planets around stars. **One is the transit method. It measures the way a star wobbles due to the gravitational tug of its planets as they move around it, and the second method measures the slight drop in a star's light that occurs when a planet passes in front of it.** This is known as a planet *transiting* its star. Transits only occur in the rare occasions when the plane of the star's planets lie almost exactly along our line of sight to the star. If this is not the case, then the star's planets will pass above or below the star, so no transits can occur, and no planets will be able to be detected. This is the case for most stars.

At present, both methods can only find planets that are orbiting close to their star and which have orbits with short periods under 3 years.

The transit method has to ensure that any drop in a star's light is caused by a planet transiting its star. Planets need to orbit their star at least a few times within the few years that the star is being monitored to be sure they are planets. Planets that are close to their star have short orbital periods, so they pass in front of their star frequently and predictably. This proves that the slight decrease in the star's brightness is caused by a planet and not from some random variation in the star's light output caused by dark sunspots.

The **gravitational method** is also limited to finding planets that orbit close to their star. The planet must be close enough and massive enough to make the star wobble back and forth as it counterbalances the planet's gravitational tug on it as it orbits it. Presently, both methods are limited to stars within 5,000 light years of Earth that have sizeable planets.

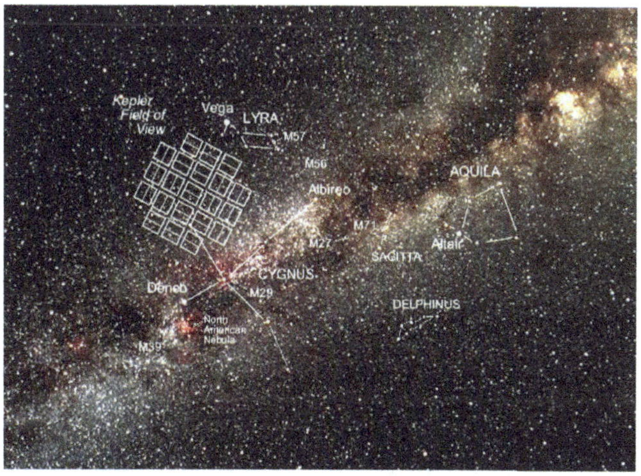

The Herschel Space Telescope searches for planets within the squares shown on this small area of the Milky Way in the constellation of Cygnus. Illustration by Kepler Mission

The Herschel Telescope's search for exo-planets is restricted to a small area of the Milky Way. The Large Synoptic Survey Telescope (LSST) can monitor all stars in the sky that lie within 10,000 light years, as well as some bright stars that lie far beyond this.

The planet-hunting Herschel Space Telescope observes the transit of a planet across the face of its star. Illustration by W. Stenzel, Kepler Space

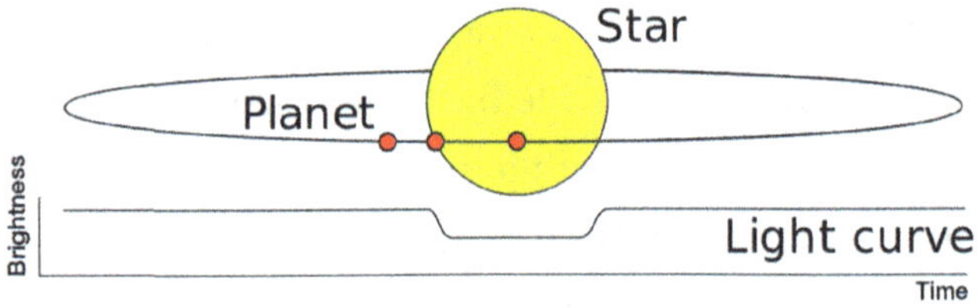

A star dims minutely when a planet passes in front of it. By this method, the planet's size and its orbital period can be determined. Diagram: Wikipedia

WHAT ARE EXOPLANETS LIKE?

Astronomers have been amazed to discover large, Jupiter-sized planets orbiting very close to their star at distances of only a few million kilometers! That's around 15 times closer than Mercury orbits our Sun. Planets that orbit their star this close whiz around it so quickly that their year is measured in mere hours! Until this discovery was made, it was thought that large gas-giant planets like Jupiter or Saturn could not form this close to their star. These exoplanets are known as '**hot Jupiters**'.

Even a newly formed planet as large and as voluminous as Jupiter, which has a dense atmosphere, would start to lose it once its star starts to emit light when nuclear fusion occurs in its core. Over hundreds of millions of years, the star's solar wind and radiation pressure would blow much of the planet's atmosphere away, but the planet's very dense metallic hydrogen mantle might remain, and certainly, its molten rock and metal core would. The cores of these stars would be much larger than the Earth and extremely hot. These types of planets are around 2-3 times the mass of the Earth. They are known as **super-earths**, but of course, they are nothing like our Earth. They would be like small versions of Uranus.

Many large rocky super-earths are also being discovered. They are by far the most common of all exo-planets discovered to date because small Earth-sized planets are hard to detect unless they are in relatively nearby solar systems and close to their star.

Some super-earths appear to have surfaces covered in lava, and some have diamond crusts! Others have 'hot ices' made of metal-like compounds that are in a liquid to semi-solid state due to their temperatures being so high. Many super-earths are thought to have raindrops of iron falling from dense clouds! Others appear to be covered in water. Many have their rotation locked to their star, so their day equals their year. This causes one side to always face their star, thereby making the dayside boiling hot while the nightside is frozen. Some may rotate so slowly that it could take a century for one day to occur.

Many hot gas giant planets that orbit very close to their star have been discovered. Their atmospheres boil away, creating a long comet-like tail behind them. Illustration by NASA

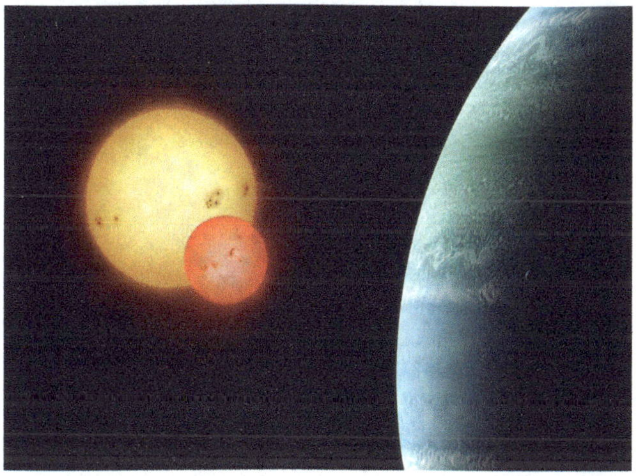

A gas giant planet (blue) has been discovered orbiting close to the binary star system Kepler 453. Illustration by Mark Garlick

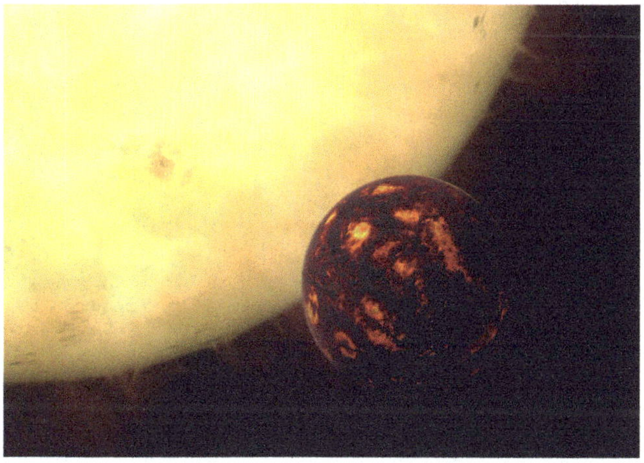

This illustration depicts the super-Earth planet 55 Cancri, which orbits so close to its star that its year is only 18 hours long! On the sunward side, surface temperatures would be around 2,000°C.

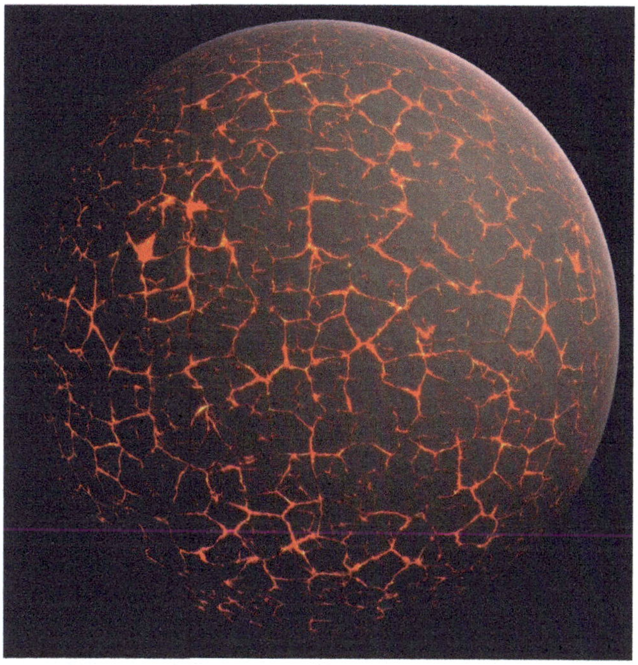

This is a **rocky super-earth planet that is covered in lava.** Illustration by NASA

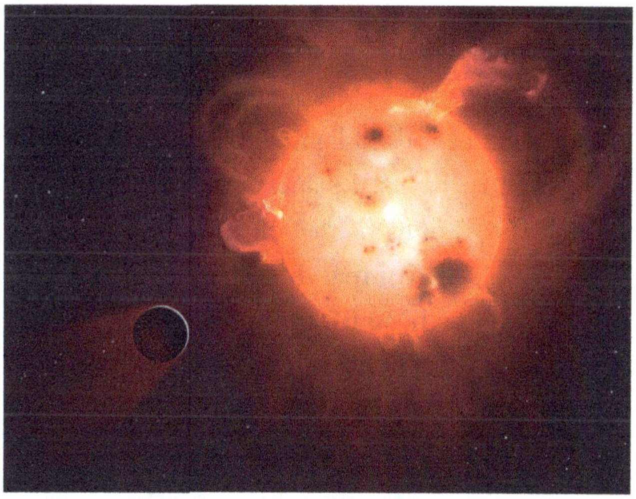

Red dwarf stars like Pegasi 374 regularly have huge flares. When they are directed towards a planet, it is engulfed in lethal X-rays and UV radiation, thereby stopping life from evolving on the planet's surface. Illustration by Mark Garlick

Some super-earths are the right distance away from their star to receive a similar amount of energy as Earth does from our Sun. Many of these planets are likely to have a deep, dense atmosphere of hydrogen and helium like our gas giant planets have. Deep below their cloud tops, they would have an atmospheric pressure that is hundreds to thousands of times greater than Earth's atmosphere at sea level. Such a planet would be a mix between a large rocky planet and a small gas giant. Life could only evolve in the mid-level of the atmosphere.

The large populations of small, cool red dwarf stars are expected to have planets. Astronomers are now finding planetary systems around red dwarfs that lie only tens of light years away. **The sizes of these systems are very small, being similar to Jupiter and its Galilean moons.** In many systems, some planets will lie within the dwarf's habitable zone, so they may have liquid water and life. One such system is only 40 light years from us. It has at least seven Earth-sized planets orbiting a red dwarf with a mass of only 8% that of the Sun. **At least six of its planets are comparable in size and temperature to our Earth, so life might exist there** – that's as long as explosive flares from the red dwarf do not kill life living on the surface of these planets or in their lower atmospheres. Life might survive under the surface of these worlds in giant aquifers or caves, or in oceans.

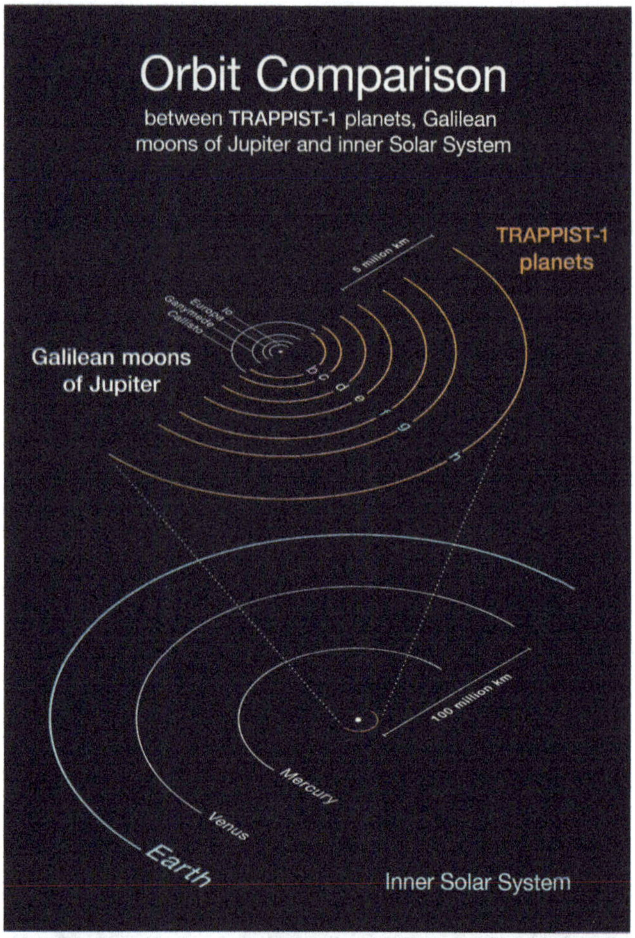

This diagram illustrates just how small solar systems around red dwarf stars are. They would fit well inside the orbit of Mercury.

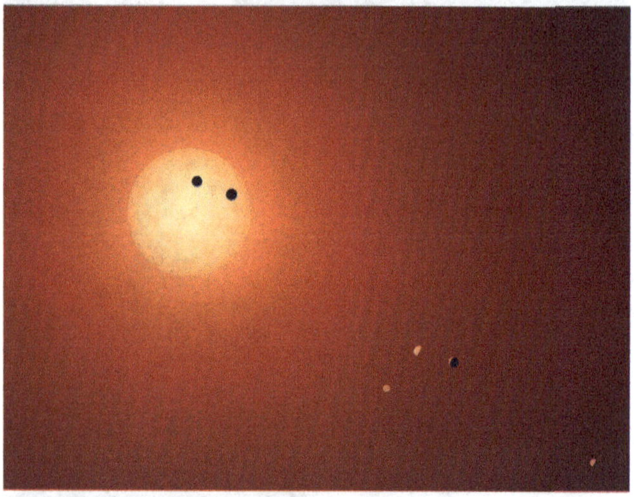

The small Trappist 1 solar system has at least seven rocky planets orbiting a red dwarf star. Most of these planets are expected to have conditions suitable for life, possibly under the surface or deep in an ocean. Illustration by NASA/JPL

Some life forms on hostile worlds might evolve a means to shield themselves during a flare to allow them to live on the planet's surface. The orbits of these planets are close enough to one another that when they are at their closest to one another, they would appear larger in their sky than our Moon does to us. Geological details and clouds could be seen with the naked eye when looking from one planet to another.

Planets around dwarf stars that are in its habitable zone where life could evolve will orbit very close to their star because the temperature of a dwarf star is comparatively cool compared to our Sun. But being so close to their star makes such planets tidily 'locked' to their star, just as the Moon is to the Earth. This means that the planet's star will always be in the same place in the sky on the sunward facing side. At the center of the dayside where it is the hottest, their star will be overhead. If such planets have enough water, there may be a huge hurricane at the center of the dayside. It would permanently rage, causing continuous torrential rain, except in the hurricane's huge central eye. Outside the hurricane, temperatures could be temperate and suitable for life. Rain would pool in lakes and oceans. The dark side would always be dark and permanently frozen.

Because red dwarf stars burn slowly, they live for an extraordinarily long time, so there is a possibility that if it was possible for advanced life to evolve, **intelligent life might arise on some of these worlds many times over.**

Most planets will not have homogeneous surfaces, so many will have different environments in which some might be able to support life and others not.

In 2016, a planet named **Proxima B** was discovered in orbit around the small, cool, dull red dwarf star named **Proxima Centauri.** It orbits the double star **Alpha Centauri**, which is the closest star system to us. Proxima is only 1/8th the mass of the Sun. Because it is a very small star, its planet orbits eight times closer to it than Mercury does to our Sun. It has a year that is only 11 Earth days long.

Because Proxima Centauri is a dull star, its planet, Proxima B, receives only 2% of the light that Earth does - even though Proxima B's is very close to Proxima Centauri. However, due to its closeness, it receives 400 times as much X-ray radiation as our Earth does, and Proxima B's solar wind would be 2,000 times as strong as what we receive on Earth from our Sun! This means the radiation levels would be lethal to life and Proxima B would not be able to retain an atmosphere - if it started out with one. So, the chances of life evolving even under the crust on that type of planet is likely to be low.

Proxima B may look like this. It orbits the red dwarf star Proxima Centauri. It may have small moons in orbit around it, and there may be other small planets in the system. Illustration by NASA

Already, it's becoming possible for astronomers to roughly determine the chemical composition of a planet's crust and its atmosphere, as well as any lakes or oceans it may have. Our knowledge of life and exobiology (biology beyond the Earth) will expand so quickly in the near future that **it will be possible for advanced, intelligent computers to accurately calculate the likelihood of life existing on these exo-planets, and even what type of life it might be!**

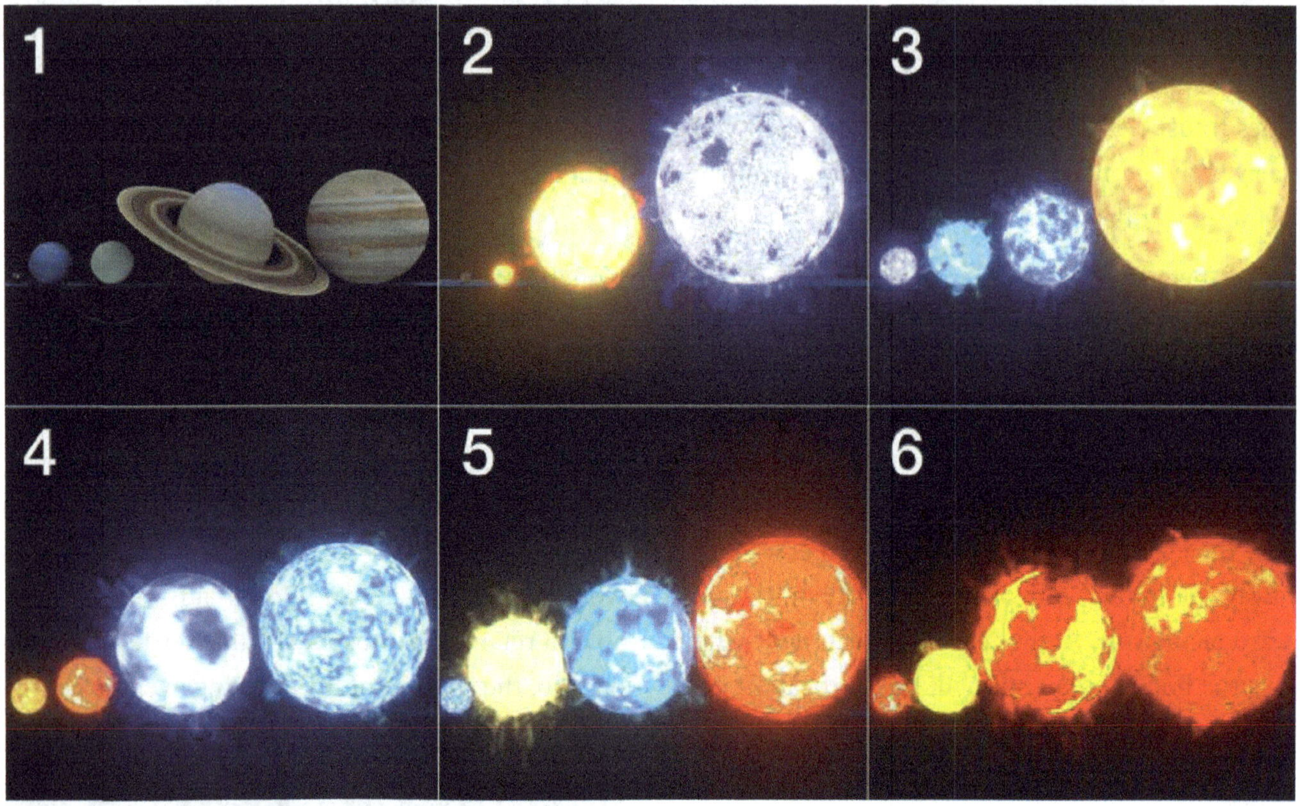

Types of stars. The real differences in sizes are enormously reduced to fit them into a manageable diagram. In each square, the largest object in the previous square is the smallest in the following one. The smaller the star, the more likely it is to have planets. Illustration by Dave Jarvis

1 - Our gas giant planets. 2 - Jupiter, Red Dwarf, Sun, Sirius 3 - Sirius, Alderamin, Maia, Kraz 4 - Kraz Arcturus, Canopus, Rigel 5 -Rigel, Sadr, Pistol Star, Mira 6 - Mira X Carinae, Mu Cephii, VY Canis Majoris

HOW MANY PLANETS MIGHT HAVE INTELLIGENT LIFE?

To estimate the number of planets that have intelligent organic life forms, I have limited this to those planets that might have beings that are at least as intelligent as Stone Age humans, which are on the path of evolution to super-high intelligence.

With regard to estimating the number of organic intelligences in the observable universe, I will be very conservative and estimate that **only 1 in 10,000 life-bearing worlds will have conditions that allow a technological intelligence to evolve. If that is the case, there would be around 2 million, trillion planets in the universe where an organic intelligence has evolved at least once. In this scenario, our Milky Way galaxy would be host to at least 1 million organic intelligences. These intelligences are almost certain to develop non-organic computer intelligence. Over billions of years, several intelligences might evolve on a single planet that has a stable star.** For instance, other organic intelligences could evolve on Earth in tens of millions, or billions of years' time. We humans evolved in only tens of millions of years after the demise of the dinosaurs, so more intelligences could easily evolve in the two billion years that Earth will be around in its present form before our Sun starts to become an unstable red giant.

If a significant number of planets can evolve more than one organic intelligence, this would considerably increase my estimated number of intelligences in our galaxy, and in the universe. (I qualify intelligence as being 'organic' because once an intelligence invents computers, they will create numerous non-organic intelligences that will be far superior to them.

QUALIFIER

There is a remote possibility that our planet might have so many rare qualities (See Chapter 34) that life bearing planets may be much rarer than we might think.

ESTIMATED NUMBERS OF GALAXIES, STARS, AND PLANETS SUPPORTING LIFE, AND THOSE WITH HIGH TECHNOLOGY INTELLIGENCE

Number of stars in Av. galaxy	Av. number of planets around stars	Av. number of planets in Av. galaxy	Number of Av. galaxies in the Observable Universe	Total number of stars in the Observable Universe	Total number of planets in the Observable Universe	Total number of planets in the Observable Universe that could support life	Total number of planets in the Observable Universe where a Technological intelligence evolves	Number of planets in the Milky Way where life evolves	Number of planets in the Milky Way where high intelligence evolves at least once
Includes all types of stars from supergiants to red dwarfs	Assumption: Number of planets range from 5-100 per star. Av. estimated to be 10	Assumption: 10 planets x number of stars in an average galaxy = 1 trillion		Number of galaxies 20^{12} x 10^{12} stars	200^{24} stars x 10	Assumption: Total no of planets = $2{,}000^{24}$ x 1 in 1,000 planets	Assumption: 1 planet in every 10,000 planets with life	Planets with life in the Milky Way 1 trillion x 1 in 1,000 planets suitable for life	Assumption: of 10^9 planets in the Milky Way with life. intelligent life may only evolve on 1 in 10,000
10^{12}	**10**	**10^{13}**	**20^{12}**	**200^{24}**	**$2{,}000^{24}$**	**2^{21}**	**20^{17}**	**10^9**	**10^5**
10,000,000,000,000	10	100,000,000,000,000	2,000,000,000,000	200,000,000,000,000,000,000,000,000,000	2,000,000,000,000,000,000,000,000,000,000	2,000,000,000,000,000,000,000	200,000,000,000,000,000,000	10,000,000,000	100,000
1 trillion stars when all the red dwarfs are included	**Ten planets per star**	**10 trillion planets per av. galaxy**	**2 trillion galaxies**	**200 trillion, trillion stars**	**2,000 trillion, trillion planets in the universe**	**2 trillion, trillion planets that could support life in the universe**	**200, trillion planets with an intelligence that could develop technology**	**10 billion planets with life in the Milky Way**	**1 million planets with intelligent life in the Milky Way**

The above table attempts to conservatively estimate the number of planets that there might be in the universe. It also estimates the number that might have simple life evolve on them. Of those types, it estimates how many might have intelligent life evolve that is capable of producing high technology. It shows that intelligent life might be rarer than we would expect, given how many planets there are likely to be. However, many solar systems may have planets that evolve intelligent life several times over. This would significantly increase the number estimated.

It may be that planets with conditions suitable for the evolution of life maybe 10 to 100 times less common than I have estimated, but given how resourceful life is in adapting to harsh conditions, it may be 10 to 100 times more common than I have estimated. I expect that life will be prolific throughout the universe: it is just a matter of how far separated each life-bearing planet will be from its nearest neighbor.

If we take the area of the Milky Way where most stars exist, which is thought to be around 1,000 square light years, and divide this by the estimated 1 million planets in the Milky Way where I estimate that intelligent life evolved at some time or other, this would mean that, on average, **planets with intelligent life would be separated by about 10,000 light years.** That's a long way apart. But organic intelligences would have evolved at different times over the last 8 billion years or so. At any one time, they could be much more sparsely separated than 10,000 light years in any one galaxy. On the other hand, there would be planets where more than one intelligence arose over the planet's life and there may well be solar systems with a number of planets that evolve life at the same time with more than one intelligence.

Given that intelligences will evolve exponentially fast once they develop high technology, they will very quickly become so advanced that they will become undetectable to an intelligence at our level. (See Chapters 15, 16 & 17.) In this case, there could be masses of super-intelligences that we would not be aware of – just as we are unaware of 99.99999% of cicadas because they lie dormant underground for around 16 years before they come to the surface briefly to mate. Most people have probably never seen a cicada even though there are countless trillions of them. And 99.9999% of ants in the world have probably never seen a human, and they would not recognize us as being a super-intelligence if they did.

Most intelligences are likely to be far ahead of us because most would have evolved billions of years before us. After intelligences develop technology, they will very quickly develop genetic engineering, which will allow them to create innumerable other organic life forms. They will develop computers, which will quickly become mega-intelligences that could replicate in vast numbers. These computers would become highly intelligent, extremely knowledgeable, and highly conscious. Mega-intelligences could be as numerous in the universe as ants are on Earth. Mega-intelligences will have capabilities that are incomprehensible to us. We will have no idea where their capabilities will lead to. For this reason, I have not included the impact that they will have on the numbers of intelligences in the universe.

Chapter 8

WHAT DO WE KNOW ABOUT LIFE?

In this chapter, we will investigate what we now know about life and the possibility of it existing beyond Earth.

Over recent decades, there have been scientific debates over whether life is unique to Earth, or whether it might abound throughout the universe. Recent discoveries are now weighing the scales heavily towards life being a fundamental aspect of the evolution of the universe.

HOW DID LIFE EVOLVE?

This chapter is important because what we have learned about the evolution of life on Earth should be able to be applied to life across the universe. You will see that there are convincing reasons why this is the case.

When **Charles Darwin** published his landmark book '***On the Origin of Species***' in 1859, his **theory of evolution** beautifully explained how life evolved from the simplest life forms. It is now thought that all life on Earth evolved from one common ancestor cell. Because Darwin was able to show that each species no longer had to be created individually by a god, his brilliant concept drew ferocious antagonism from those who believed in the Judeo-Christian religion's view of creation that said that their god created all life and the universe in only six days. The Catholic Church has now abandoned its teaching of creation because of the overwhelming evidence for evolution.

Life appears to exist along a continuum of increasing complexity that started when the first sub-atomic particles appeared shortly after the universe was created.

Following are the major developments in the evolution of organic chemistry that biologists believe led to life evolving. These developments are actually very complex chemically, but I have tried to keep the explanations simple enough for non-biologists to find them easy to understand. This is a rough guide as to what scientists have discovered about the chemical processes that they think may have led to the evolution of life. There are divergent views on exactly how these steps may have occurred and in what order. Now that scientists are rapidly discovering much detail about life's chemistry, these evolutionary milestones will become much more refined in the next decade or two. Some of the assumptions will no doubt be amended, and some may be overturned when new, unexpected discoveries are made. With rapid advances occurring with genetic engineering, it will not be long before we know how to create life ourselves.

THE MAJOR STEPS IN THE EVOLUTION OF LIFE

Simple Chemical Reactions: When atoms started combining and reacting with one another, chemical complexity commenced. Atoms formed many diverse combinations known as molecules. Each different molecule (and there are huge numbers of them) has different properties.

In the early 1800's, scientists showed that there were a number of chemicals that are unique to life, such as **urea**. It cannot be made from simpler chemicals: it can only be made by life forms. It is common in the urine of mammals and fish. This discovery led to a slow realization over 200 years that life was the outcome of increasing chemical complexity. This crushed the idea that life had some special quality called ***vitalism*** that made living organisms special and different to non-living matter. It also shattered the Bible's explanation that God created all life instantaneously.

Crystals That Replicate: Life is thought to have occurred after a number of stages of chemistry evolved in hot, mineral-laden water that percolated through innumerable fissures and aquifers in the Earth's crust. This water was rich in both hydrogen and carbon dioxide, as well as sulfur, iron, and compounds of carbon. This mixture allowed the first crystals to form. It is thought that crystals are important in the story of life because they were the first molecules that could replicate their structure.

Amino Acids and DNA: It is postulated that when alkaline underground water passed through hydrothermal vents on the sea floor where the continental plates were pulling apart, it mixed with acidic seawater containing amino acids. This probably produced the chemistry for building the long chains of **nucleotides** that formed **ribonucleic acid (RNA)**. This long molecule contained information for how molecules could be replicated. The first simple viruses used this to replicate. When RNA and amino acids interacted, they created the very long double helix molecule **deoxyribonucleic acid (DNA)**. This could also replicate. And because it could hold much more information for replication than RNA could, it could produce more complex instructions to build more complex life forms. It became the molecule of life that held the genetic codes for all life forms. It may be common to all life in the universe.

Long Polymer Chains of Carbon: Carbon atoms are the only element that can combine with many other elements to create a wide range of very long, complex, **carbon polymer chain molecules**, many of which perform a wide range of specialized tasks. These produce **proteins** and **enzymes which** can also replicate. These are essential to the workings of all cells.

Cell Formation: **Chemical complexity continued through many steps to a stage where the first cells evolved. Science defines cells as the beginning of life.** About 4 billion years ago, it is postulated that fatty, bubble-like molecules that were coated in iron-sulfur molecules became the first **cell membranes**. These encased all the workings of a cell for protection. The very first, simple, **prokaryotic cells** to evolve were bacteria. This is when life is considered to have commenced. Prokaryote cells evolved small chemical 'machines' that we call **enzymes**. These could extract chemical energy from seawater that existed outside the cell's membrane. This provided the energy that cells needed to perform their functions, and to move around. These first cells evolved into two branches known as **bacteria** and **archaea**, neither of which have a nucleus. At this time, a great diversity of cells evolved. Nature's survival-of-the-fittest weeded out those that did not have the best chances of survival. This left the survivors to replicate and to evolve into more complex and superior life forms.

It is thought that the mixing of underground fresh water with sea water produced chemical reactions that **prokaryote cells** used as a weak source of energy to grow. Later on, larger, more advanced **eukaryote cells** evolved that could generate their own electrochemical energy within their cell. This allowed them to move away from the hydrothermal vents into the wider world. These cells have a number of smaller bubble-like membranes inside them, known as **organelles**. They have specialized functions. They are essentially *organized* 'machinery' systems that perform specific functions in a cell. They give cells the name 'organisms'.

Increasing complexity continued to create a broad range of cells, each of which had special features. By chance, some cells found that there were advantages in combining to form **multi-cellular organisms**, which were much more complex, and which had more capabilities and options. These exploded into the huge diversity of increasingly complex forms of simple life, many of which still exist today.

Over time, life developed more **senses**, which evolved to become more sophisticated. As an example, over millions of years, simple light-sensitive cells curved to form balls filled with fluid. Through trial and error, they developed a gel lens behind a hole in the eyeball. Light passed through this to produce a sharp, focused image. After this, many eyes developed the muscles to form an iris to control the amount of light entering the eye. Mono-color eyes evolved into two colors, then three colors to provide more information. Similar improvements occurred in all senses. Senses permitted life forms to make more informed decisions to improve their chances of survival.

It has recently been discovered that life adapts to a far wider range of environments than we had previously thought possible. Because life is programmed to want to survive, it evolves ingenious ways to do so through trial and error so it can adapt to harsh conditions. The cells that adapted to the harshest conditions are known as **extremophiles**. This process of adaption is known as '**survival of the fittest**'. This will be common on all planets with life.

As **intelligence** evolved in advancing life forms, this increased their chances of survival. They learned how to manage threats to ever greater degrees. Brains grew to have increased computing power and to have more memory. This considerably improved life forms' chances of survival. When the most intelligent life forms evolved over billions of years to the stage of creating technology, intelligent life quickly evolved a means of expanding its intelligence beyond its brain. It did this by developing non-organic intelligence with greatly enhanced memory in the form of computers. As a result, intelligence was no longer constrained by the physical limitations of organic brains. This stage of evolution marks a major milestone in the evolution of life.

Understanding how the 'miracle of life' became more complex through numerous small steps makes it easy to comprehend how evolution works. From this perspective, life should be an inevitable outcome of the universe's evolution.

What is truly remarkable is that nature was designed with programs from the outset that make atoms grow into increasingly more complex molecular structures that lead to high intelligence, then to increasing consciousness, and whatever lies beyond that stage.

Like everything in the universe, **all life abides by the laws of physics.** **It converts chemical energy into order** - until the life form dies. At this time, the matter in a life form returns to disorder until it is recycled into another life form. Overall, life creates more order than disorder. The disorder created, is discarded as heat and waste, thereby abiding by the Second Law of Thermodynamics. **Incredibly, a human body is extraordinarily efficient in converting food into energy. An average weight person has to walk for 40 minutes at 4km/hr (3mph) just to burn off the energy in an average-sized apple. The human body generates 6,000 times more heat energy per kilo than the Sun does!** This is because each cell is burning energy, whereas in the core of a star, only a very small percentage of its atomic nuclei fuse with other nuclei to release energy. The nuclei must impact one another at exactly the right angle and with the right amount of energy to fuse - and this is very rare. Stars radiate much energy because there is an enormous density of matter in their cores, so only a minute number of atoms need to fuse with one another to create the energy that stars release.

In essence, life is about manipulating chemical processes and organizing information. The more complex a life form becomes, the more flexible it is likely to be.

Death is a program written into life to cause lifeforms to have a rapid turnover so that new ones are being created all the time. These have slight variations, some of which are helpful to their species survival and evolutionary advancement. Without death, evolution would not occur.

Some negatives against life evolving on certain planets are:

1. Planets orbiting red dwarf stars are thought to produce deadly flares that preclude life evolving on their surface or in their atmospheres unless they are very dense,
2. Supergiant stars explode within millions of years - not long enough for life to evolve on any planets they have,
3. Many planets will be so cold that chemistry would be far too slow and limited to permit life to evolve,
4. Gas giant planets are unlikely to have conditions suitable for life in their clouds,
5. Life is likely to only evolve on planets relatively close to a star so it receives adequate heating. This greatly limits the number of planets suitable for life to evolve.

There may be few Earth-like planets with Earth's characteristics i.e. a stable star, the right strength of gravity and temperature range, plenty of oxygen, liquid water, nitrogen, CO^2, a rich mix of elements for life to evolve into complex forms, a protective magnetic field, and so on.

Some positives for life evolving where it was thought it could not:

1. Advanced life may evolve under the surface of planets that have stars that emit deadly radiation,
2. There might be much larger numbers of yellow dwarf stars like the Sun in the cores of galaxies, in their spiral arms, and in globular clusters that can support advanced life,
3. There may be significant numbers of rogue planets not attached to stars in the spiral arms and haloes of galaxies where life may exist due to them having internal heat from a hot core, radioactivity, or tidal friction.

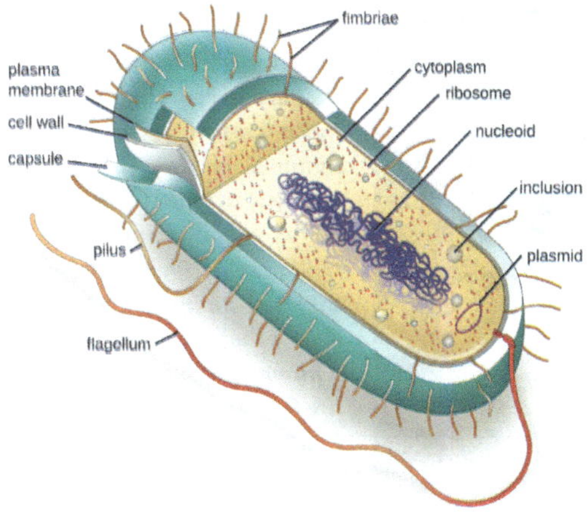

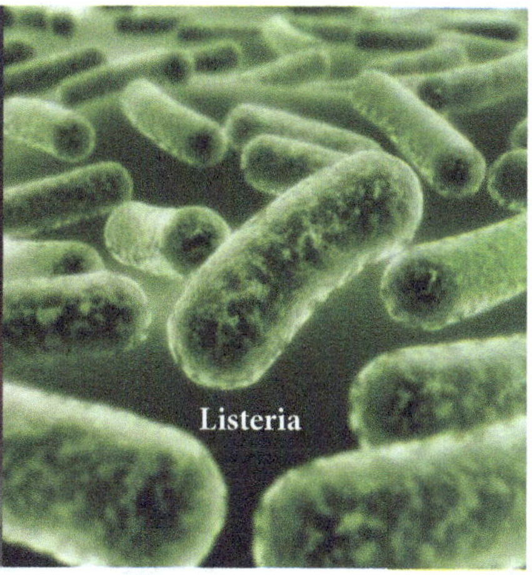

This diagram shows the structure of a typical prokaryotic bacterial cell. It is enclosed in a three-layered membrane. Curled up in the center are strands of chromosomal DNA (purple) known as nucleoids. They are surrounded by a fluid known as the cytoplasm. Floating around in this fluid are small ribosome 'machines' that make proteins. On the outside of the membrane are hair-like pilli fibers or fimbriae that are used by the cell to move around, to hold onto surfaces and onto other cells and to sense its environment. Some bacteria have a tail called a flagellum, which is used for propulsion. Bacteria, which are also known as microbes, are the most basic form of life. In the center image, we see the Escherichia Coli bacteria. Most strains are harmless. They live in the lower intestine, but strains like the one above can cause deadly food poisoning. Listeria on the right, is a very hardy bacteria that is found in soil and water. It can kill humans when it gets into unpasteurized milk, cheeses, salads, raw vegetables, cold diced chicken, and pre-cut fruit. Bacteria are highly likely to be the most prolific and most basic life form throughout the universe.

This diagram of the internal structure of a eukaryote cell illustrates just some of the complex structures (greatly simplified here) that make up the basic components of a cell. In reality, a cell is millions of times more complex than this illustration suggests! If a protein were the size of a human, a cell would be like a hugely complex city the size of the Earth! It would be full of chemical factories that have billions of molecular machines moving around inside it performing their specialties in organic chemistry.

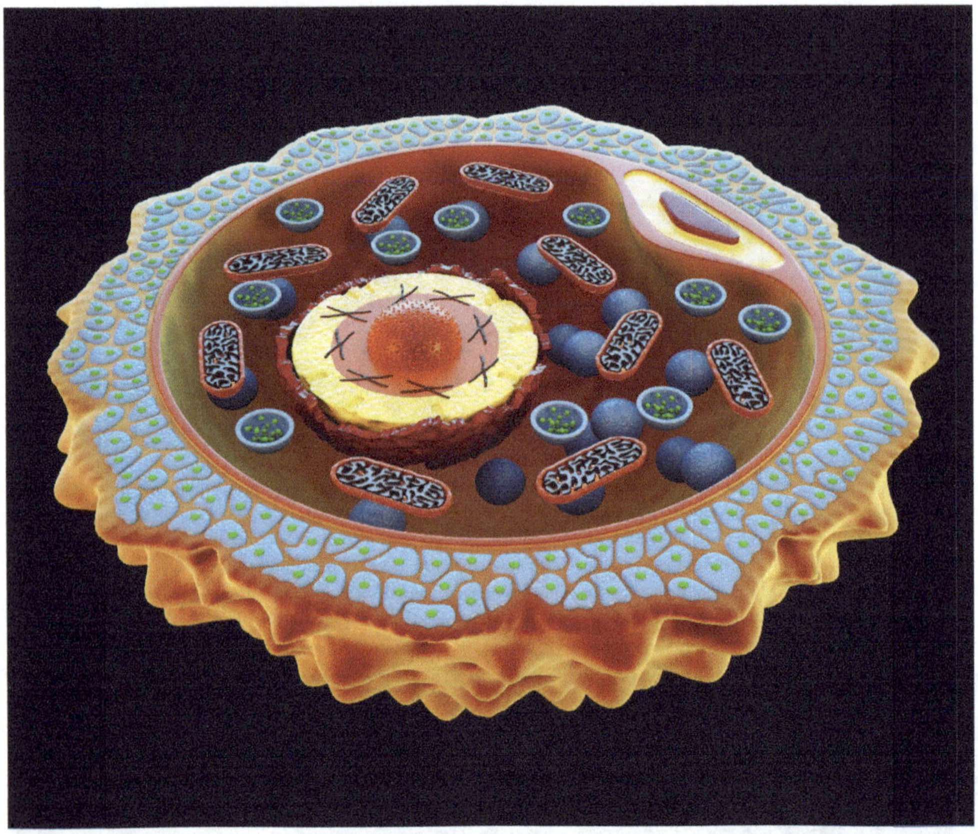

This is a cut-through of a human egg – the ovum. It's hard to conceive that this microscopic egg has innumerable programs that enable it to build all the parts of a human body composed of 100 trillion cells. Try to imagine that! It is unknown by most people that the bulk of cells in a human, or any other animal, are foreign bacteria that are in a symbiotic relationship with the body's own cells. The following is also amazing. When a sperm cell fertilizes a human egg, it is incredible that the fetus begins its growth symbiotically in tune with its foreign bacteria that become a part of the baby. How does it manage to do that?

A human female releases about 400 eggs in her life up until around age 50. Incredibly, a female's eggs are in her ovaries when she is an embryo!

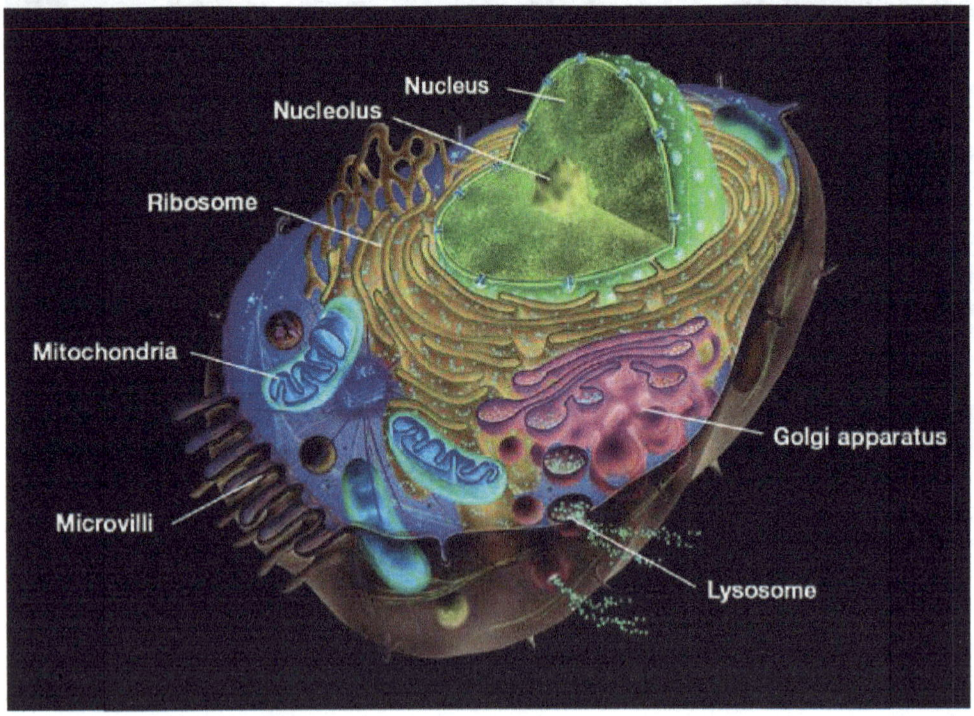

An illustration of the structure of a eukaryote cell.

Only carbon atoms can produce the huge variety of long polymer molecules required to build the structures necessary for cells. These include 20 amino acids, proteins, enzymes, carbohydrates, sugars, vitamins, RNA, DNA, viruses, chromosomes, hormones, membranes, cell nuclei, flagella, and pilli etc. Organisms take their name from being highly 'organized' structures. Credit: Person Education Inc.

The earliest life on Earth that is known to have formed colonies were stromatolites. Some life on other planets may start in a similar way.

The earliest extremophile cells on Earth evolved in hot mud springs. They are seen here as green, yellow and brown material.

HOW IS LIFE DEFINED?

In general terms, life is self-organizing, dissipative bodies of matter and energy. Life converts energy and particles into organized atomic structures that can grow and replicate. When a lifeform dies, its energy is released back into the environment in the form of its component states.

Life is a chemical system that is able to consume energy to grow and reproduce. But a wildfire could also be said to do this, yet it is not considered to be alive. We cannot add 'be active' to the list because life forms such as rock lichen or fungi do not move. We all think we know what life is, but after 200 years of amassing much knowledge in biological science, **biologists still have difficulty coming up with a definition for life that covers all species. The definition at the start of this section is the most appropriate.**

When we can work out exactly how life develops, then we will be able to work out what types of life forms are likely to evolve on other planets, providing we know the chemical composition of the planet, its conditions, and the type of star it orbits.

This real, but alien-looking life form has 4 eyes, a fake face, 8 legs, and two fangs.

WHAT DOES LIFE NEED TO EVOLVE?

Life has evolved to use the most common elements in the universe. We know that it requires at least the following:

A planetary body – **Life cannot exist in empty space, in a nebula, or on a star: it can only evolve on a planetary body.** These can range in size from large asteroids to planets – but probably not gas giant planets unless life is able to exist in their upper atmospheres or where their atmospheres start to become liquified at lower levels.

Life must have the following basic requirements as well as the right conditions:

- **Carbon** - **It is the only element that can produce the staggering array of long, stable polymer chains, which are *essential* for life.**
- It is the only element that can make cell structures, and also perform metabolism.
- It will bond readily with all the elements that are necessary for the complex chemistry of life. Carbon combines readily with - hydrogen, oxygen, nitrogen, phosphorus, and sulfur, as well as metals such as iron, magnesium, and zinc. Carbon also bonds with many other elements that can exist in all three states of matter - solid, liquid, and gaseous.
- Astronomers have already found approximately 100 complex carbon-based molecules in the interstellar medium in nebulas, in comets and asteroids, and in the atmospheres and crusts of planets and moons.
- **Because carbon is the only element that has so many unique properties and such extraordinary flexibility, all life throughout the universe will be carbon-based. It is the only**

element that can produce the chemical complexity for cell structures that are required for life.

There are about 30 elements that are necessary for simple life. Higher life forms require trace amounts of most of the other heavier elements in the Periodic Table to perform specific functions. A wide variety of elements are required for cell growth, metabolism, reproduction, and to ward off foreign bodies. These elements must be in the right proportions, as too much or too little of any element has the potential to kill a life form. Astronomers have found that **the elements necessary for life are plentiful throughout the universe.**

- **Hydrogen** is the most abundant element comprising 50% of the universe. Its electrons form strong nuclear bonds with many elements to build the molecules necessary for life. Hydrogen is the major constituent of water, which is a major part of all life forms.
- Cells contain around 24% **oxygen**, which is the third most abundant element. In most life forms, oxygen is required to convert food into energy.
- Most cells have around 18% **carbon**, which is the universe's 6th most abundant element.
- Cells also have 4% **nitrogen** - the 7th most abundant element. It is necessary for the construction of genes and proteins.
- The rare element **boron** together with a number of other elements are also necessary for life because it is a critical element for ribonucleic acid (RNA), which is an important component of life. Boron is a nutrient for all plants.

Liquid water is essential to life because it has many unique properties.

- Water is common in the universe but most of it is in its solid form as ice, or in its gaseous state as vapor. Neither of these states are useful to life: **life must have *liquid* water**. However, water does not have to be in its liquid form all the time. As long as it is in a liquid state some of the time, then life will grow and replicate at those times.
- Liquid water permits numerous chemicals to dissolve in it, **far more so than any other solvent.** Liquid water transports chemicals around inside a life form, as well as throughout the biosphere of a planet.
- On average, **most cells consist of around 70% water.**
- **Water appears to be very common on most worlds in our solar system**, but even though it may only be in its liquid form on a small percentage of worlds, this would still constitute an extremely large number of worlds in the universe. It appears to have been found on at least several worlds in our solar system. Celebrity scientist **Brian Cox**'s excellent two-part documentary '***Wonders of Life***' explains a number of these facts very well, although he does close this documentary with the naive belief that life on Earth is the only life in the universe, simply because we have not yet seen evidence of life beyond the Earth.

An acceptable temperature range – **This is a 'must' for liquid water to exist. Many chemical reactions do not occur without suitable temperatures, and the few that do react far too slowly.** If the conditions are too hot or too cold, molecules cannot react or assemble, so life cannot evolve. If conditions are too hostile on a planet's surface, then suitable conditions could exist under its surface. (When the term 'planet' is used, this includes large moons, and dwarf planets/large asteroids.)

A suitable source of energy – It was once thought that all life in the universe would get its energy from a star in the form of heat and/or light. A planet would therefore have to be in the star's **habitable zone**, which is also known as the ***Goldilocks zone***. Here, there is just the right amount of heat to permit some water to be in a liquid form. However, we now know that **heat energy can also come from inside a planetary body** by way of convection from a hot core, volcanism, tidal heating, chemical reactions, and radioactive decay. All these forms of energy are relatively plentiful in the universe.

Scientists have discovered a means for life to obtain energy from an ocean by combining hydrogen with carbon dioxide. This reaction produces methane as a by-product. Seasonal releases of methane have been discovered on Mars. The only way known for this to occur is from volcanism or microbes in the soil.

Saturn's moon Enceladus appears to have a huge internal ocean that should have the elements necessary for life, so there is a possibility that it could provide chemical energy for simple cells, but the water is extremely cold, being far below 0^{C}, so life is unlikely there. Enceladus' water is in a liquid state because of the extreme pressure it is under due to the weight of the thick overlying crust. This is also the case for the Galilean moons of Jupiter.

A magnetic field – This *may* be necessary for life to exist because a magnetic field blocks solar wind radiation from the planet's star. However, some simple bacteria on Earth have evolved to survive in a nuclear reactor, so a magnetic field may not be necessary for some simple life forms. Whether these forms of life could evolve into higher life forms is unknown. If a planetary body does not have a protective magnetic field or a dense atmosphere to block a star's radiation, then life may be able to live underground, or in a deep ocean.

An atmosphere – This is a requirement for advanced life forms that live on the surface of a planet. An atmosphere that is too thin may not protect life from deadly forms of radiation from the planet's star and from deep space. And life may need the atmosphere's gases to evolve.

An **oxygenated atmosphere** *may* be necessary to provide enough energy for advanced animal life. When oxygen levels were much higher on Earth during the Jurassic Period, life forms like the huge dinosaurs grew far larger than the largest animals today. As an example, dragonflies grew up to a meter (3') across, and the tallest trees were three times taller that they are now. High oxygen levels provide more energy to allow life forms to grow larger and expend more energy.

A planet's star needs to be relatively stable - If a star is unstable, its planets will be plunged into cycles of life-destroying overheating and strong radiation followed by deadly cold. Some simple life forms would adapt to this, but advanced life forms may not.

Plate tectonics – This might be an essential requirement of life because it enriches a planet's crustal chemistry. Tectonics dynamically transports minerals, gases, and liquids from the planet's interior to the surface and back again. It acts as a fertilizing process, which advanced life may need for healthy growth.

The electric field of a planet – This may be beneficial but not essential. We do not know enough about this to be sure. It is thought that if the charge is very high, it is likely to stop life from evolving.

Suitable environmental conditions – If conditions are too harsh, then advanced life is less likely to evolve. Examples of major environmental catastrophes that kill life forms on the surface of a planet are as follows:

- Frequent meteoric bombardment,
- Powerful solar winds, with high levels of UV radiation,
- Strong cosmic ray showers,
- High energy electromagnetic radiation emitted from supernova explosions, merging neutron stars, and black holes. Depending on the strength of the source, if these events occurred within 10 to 100 light years, they would be lethal to life that we are aware of.

As long as environmental change is not so devastating that it causes a total extinction of *all* life, it will drive diversity by forcing life forms to evolve in order to adapt to change.

On Earth, severe environmental change has extinguished numerous species when it occurred too quickly for them to adapt. This occurred due to:

- **The Earth's orbit changing shape** going from being almost circular to becoming elliptical. Over long periods of time, this causes large, regular, periodic changes in climate. Relatively short, warm periods like the present are interspersed with long deadly ice ages.
- **Large changes in the tilt of the axis,** which causes extreme changes, in the seasons.
- **Periods of massive volcanism** have poisoned the air and the water and also blacked out the Sun, thereby collapsing the food chain.
- **Large meteor strikes have made nearly all life on Earth extinct several times.**
- **The Sun's orbit around the galaxy** takes it above and below the galactic plane. When it passes through the galactic disc every 60,000 years, this can cause a drop in the Sun's output and therefore Ice Ages. It can also cause our solar system to have close encounters with other solar systems and stars causing the orbits of some planets and smaller bodies to change causing meteor impacts and mass extinctions.
- Many species on Earth have also become extinct due to lesser environmental changes such as:
- continental drift,
- large variations in sea level,
- changing ocean currents,
- reversals of the Earth's magnetic field and,
- large solar flares, known as mass coronal ejections

Major environmental changes have been responsible for at least five mass extinctions of life that combined, killed an estimated 95% of all life forms that ever existed. Fortunately, enough life survived each time to allow new life forms to repopulate the planet within hundreds of thousands of years to tens of millions of years depending on how devastating each extinction event was. This recovery process occurs in a way that is similar to how new flora and fauna grow back after a devastating wildfire. Environmental change will be common to all worlds. It drives evolution.

We must remember that there is much we do not yet know about the evolution of life and the evolution of planets. And we do not know how diverse and resilient life may be on other worlds, so the dot points above may be too limiting - or there may be other conditions that are required for life that we are not yet aware of.

CAN LIFE ONLY EXIST IN THE 'HABITABLE ZONE' OF A STAR?

For a planet to support life, astronomers have often said for a planet to support life, it needs to be in the 'Goldilocks Zone' where the temperature is 'just right'- like Goldilocks' porridge. This zone is the right distance from the star for a planet's surface to have liquid water. Inside this zone, closer to the star, water on a planet's surface would only exist as steam, and outside it, it would exist only as ice. Earth sits in this Goldilocks zone, but Venus does not: it is too close to the Sun. Mars may once have been in the Goldilocks' Zone when the Sun was more active billions of years ago, but today Mars is outside of it, so it is bitterly cold. Some leftover bacterial life may exist under its surface if there is internal heat remaining in some regions where liquid water might exist. There is some evidence for this.

Life could exist well beyond the Goldilocks Zone in extreme cold if there is a source of heat coming from inside a planet. This is the case for many planets. At least two or three of Jupiter's Galilean moons and one of Saturn's moons show strong evidence for enough heat and pressure being produced inside these moons to allow liquid water to exist. This heat is presumably due to gravitational tidal heating, high pressures, or from residual heat from the core.

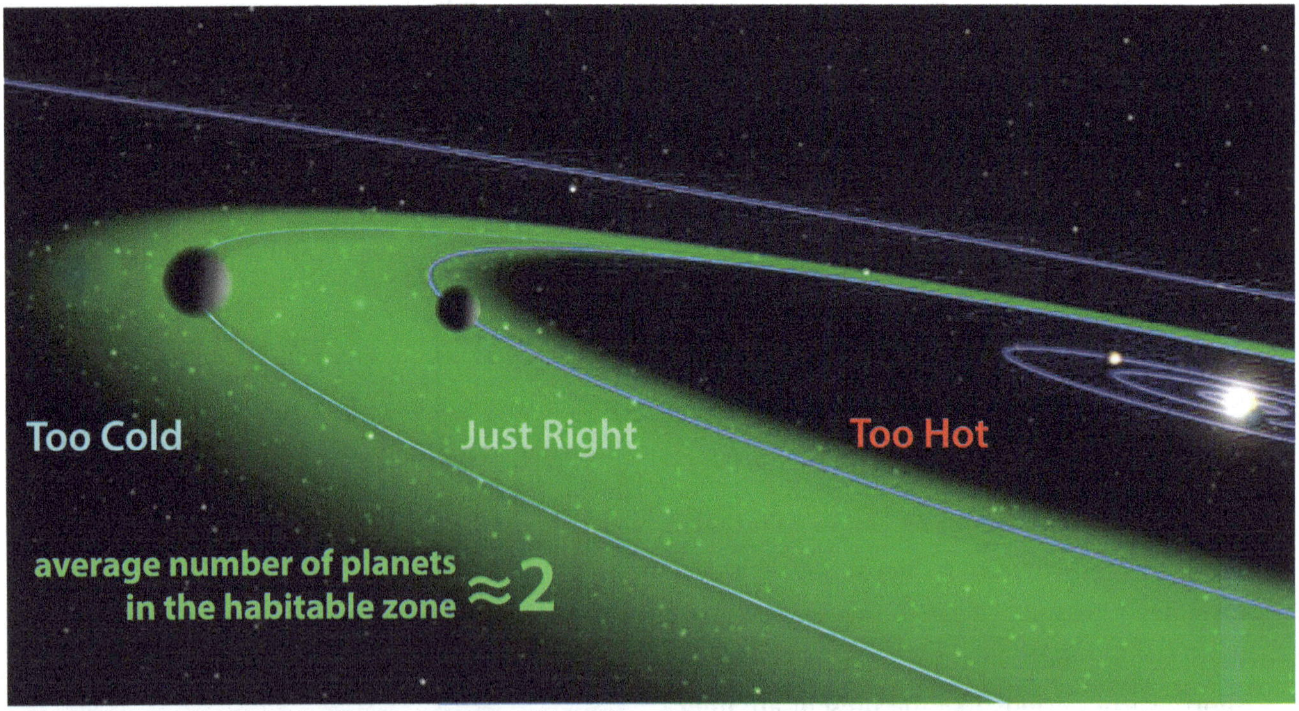

The habitable zone around a star is the region where a planet's temperature is just right for liquid water to exist on the surface. Liquid water is required for life to evolve. We now know that life can live outside this zone if there is heat coming from within a planet to allow liquid water to exist under the surface.

On Earth, biologists are discovering many forms of bacteria living inside rocks in very deep mines kilometers underground where the temperatures are in the hundreds of degrees Centigrade. This heat rises from the Earth's core, which is as hot as the surface of the Sun! These basic forms of life also thrive in aquifers, caves, and between layers of rock in the Earth's crust. The bacteria feed on sulfides, minerals, and other chemicals, and some on each other. Biologists are amazed at the volume of bacteria they are finding under the Earth's surface. They estimate this biomass to be between 3 to 20 times greater than all other life living on the land and in the oceans! This is beyond all expectations!!

Given how much simple life exists inside the Earth's crust, it is likely that numerous other worlds that have water and warmth within their crusts will also have simple life existing there. Perhaps advanced life could also evolve inside large caves, aquifers, and lava tubes as well as in sub-surface oceans deep under the crust of many planets.

If you watch **James Cameron**'s documentary '***Aliens of the Deep***' you will see extraordinary vision of alien-looking life forms that live in the deepest parts of the Earth's oceans. At the end of the film, he provides excellent CGI vision that imagines what life could look like in Jupiter's moon Europa's sub-surface ocean. We may discover such life when NASA sends a probe there in the not too distant future to go below Europa's icy crust.

It is also very possible that life might exist in internal oceans on some other moons in our solar system and on some newly discovered exo-planets if there is some heat source. The Galilean moons are expected to contain more saltwater than all of Earth's oceans! Such oceans may be common across the universe. If so, simple life might be very common.

Most likely, it would seem to be impossible for life to survive on the surface of worlds that do not have a reasonable atmosphere - unless life there evolves some form of protection from the vacuum of space, and from its star's radiation.

CAN LIFE ADAPT TO EXTREME CONDITIONS?

Many extremophile life forms (those that live in extreme environments) have recently been discovered on Earth, thriving in conditions that were once thought to be far too hostile to support any form of life. Scientists have discovered extremophiles living in boiling thermal pools in volcanic regions, deep in ancient ice laid down millions of years ago in Antarctica, and also in the driest desert on Earth in Chile where it never rains (but there is morning mist). As stated earlier, extremophile microbes live deep inside the Earth's crust under very high temperatures and pressures. They have also adapted to living in very high levels of salt, and in lakes that are highly acidic, and also in those that are very alkaline. They even live where there are high levels of radiation in nuclear reactors as well as high in Earth's atmosphere at the edge of space where there is strong radiation from the Sun and deep space.

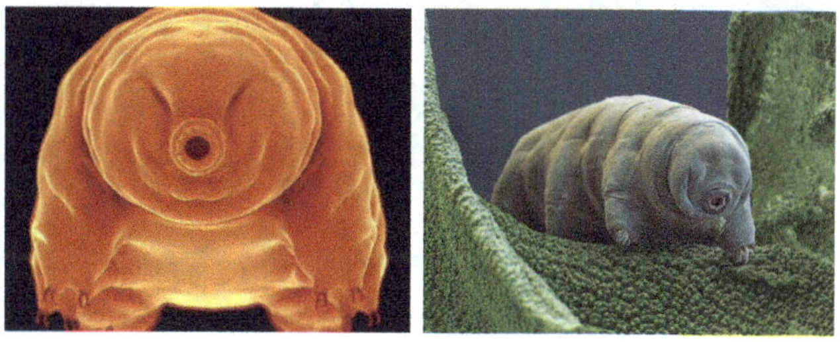

This is an electron microscope image of a tardigrade (water bear) extremophile. It lives in water for 60 years and can survive not eating for 30 years. (It looks like it's smiling about its ability to survive!)

The most resilient extremophile so far discovered is the eight-legged tardigrade, which is 1 mm long. It can stand being boiled, frozen, squashed under extreme pressure, exposed to the vacuum of space and to radiation, with no ill effect! Other planets are bound to have extremophiles that cope with extremely harsh conditions. **Extremophiles tell us that life is very likely to exist in numerous places that appear to be very hostile to life.**

Cells in plant leaves evolved to convert the Sun's light into energy through photosynthesis. Plants became the base of the food chain for more complex, plant-eating animals, known as herbivores. Carnivores evolved to eat the herbivores and other carnivores to obtain the energy they need.

Until recent decades, it was not thought possible that there could be any other form of energy other than sunlight to provide the base energy required for life. But this long-held assumption turned out to be wrong. To the amazement of biologists, ecosystems were discovered that were thriving in total darkness in the depths of the ocean around thermal vents coming from submarine volcanoes. Here, there is extreme pressure and extreme cold, as well as superheated subterranean water flowing from the vents at temperatures of hundreds of degrees! At these extreme depths, the water's high pressure stops it from boiling. The thermal vents pump out dissolved sulfur-based chemical nutrients from deep within the Earth's crust. These nutrients are poisonous to most life forms, but to the extremophile bacteria that have evolved there, they are food. These bacteria evolved to extract what they need from the water coming from the vents. By converting chemicals into energy, the bacteria could grow and replicate. The bacteria became the base of the food chain for higher life forms such as tubeworms. More complex life, such as crabs, eels, and shrimp, also exist there because they eat lower life forms.

Life on other planets will adapt in similar ways so it can survive and evolve in what we consider to be extreme conditions. The conditions we like may seem extreme to them. **This being the case, the chances for life existing elsewhere in the universe may be much higher than my estimates in Chapter 6.**

Left: **Deep ocean hydrothermal vents that are often referred to as 'black smokers', release superheated water saturated with sulfides. The water coming out of them provides chemical energy for bacteria to survive. These bacteria are food for other simple life, which in turn, is food for higher life forms.**

Right: **Thermal submarine vents create conditions suitable for supporting ecosystems containing tubeworms, eels, crabs, shrimp, fish, jellyfish, and other life forms that do not require sunlight to survive.**

Most rocky planets will have a great diversity of caves and aquifers within their crusts. Life might exist in these in greater abundance even to that which biologists are finding inside Earth's crust.

This is a cave in an underground river where large square crystals formed when lava cooled quickly in running water. Credit: Niacaa

Son Doong cave in Vietnam is the largest and most incredible cave in the world because it has so many diverse environments and features. For scale, zoom in to see the campers on the sandy beach. It is a good example of what mammoth cavities might be like inside the crusts of other rocky planets. Some would be lava tubes. On low gravity worlds, lava tubes are kilometers in height and hundreds of kilometers long - big enough to contain huge cities. There would also be enormous caves cut out by fast-flowing rivers. Huge underground lakes and oceans would also exist. Such environments would allow advanced life forms to evolve and flourish without the need to be on the surface. Underground habitats would be away from the deadly conditions that exist on the surfaces of many planets. Certain life forms have evolved in total darkness in caves on Earth. Credit: Kieu TramBuiThi

Inside the crusts of rocky planetary bodies, advanced life may evolve in large lava tube cave systems, similar to this Phong Nha Cave in the Phong Nha-Ke Bang National Park, in Vietnam.

Amazing aquifers like this one in Mexico exist deep within the Earth's crust. It contains giant crystals that took tens of thousands of years to grow to this size. Many forms of bacteria have been discovered in such caves. The water in this one has been pumped up to the surface so scientists can study the crystals and bacteria more easily. The air temperature this deep underground is so hot that scientists have to wear thermally cooled suits and have a refrigerated work station. They can only stay down there for short periods due to the heat and the high air pressure. Credit: www.naica.com.mx

This is one of many large, open spaces in spectacular limestone caves in the Rock of Gibraltar. Caves like this are very likely to exist on other rocky planets and life could exist in them. Credit: Gregg Thompson

HOW VULNERABLE IS LIFE TO CHANGING CONDITIONS?

Most life on Earth exists in a somewhat delicate state of balance. Changes caused by the Earth's varying distance from the Sun, the eruption of super-volcanos, or significant changes in the levels of oxygen and nitrogen in the air, will kill most life. Even a relatively small reduction in the current level of trace gas carbon dioxide (CO^2) would kill most plant life. This would collapse the food chain, thereby starving all animal life.

At present, Earth has one of the lowest levels of CO_2 in the planet's history. Contrary to many false news reports, even a significant increase in this trace gas would not be dangerous. In fact, it would be very beneficial because it would increase plant growth substantially and therefore food production. The very slight increase that has occurred in CO^2 over the last several decades has increased the greening of our planet by an area twice the size of the USA! And contrary to popular belief, it has caused no unnatural warming. When CO^2 was ten times higher in the geological past, the fossil record shows that plant and animal life thrived. When high levels of CO^2 existed, temperatures did *not* rise, contrary to popular belief. Our media does not tell people that CO^2 in the atmosphere is a mere trace being only 1/400^{ths} of a percent! And 97% of that is produced by nature.

When people discover that there is only one CO^2 molecule in 100,000 air molecules, they do not believe that this could heat the planet - and they are right. The CO^2 myth is the second biggest con ever perpetrated in history. For over 50 years, the mainstream media that the super-rich Elite bankers and industrialists control have promoted this false fear. They also use socialist government leaders globally to push this agenda. They are frightening people across the world that our planet is going to overheat catastrophically, even though there is no real scientific evidence for this. They want people to be so fearful of this killing them that they will agree to the Elites forming a world government to stop this happening. Their One World Order global government that they will control will subjugate everyone in a feudal system. Their published Agenda 21, and now Agenda 2030 papers that detail this real conspiracy have been presented at climate conferences. The Elites plan to exterminate 90% of world's population and make most of those left their slaves. This sounds unbelievable I know, but this is their insane megalomaniacal objective. Many people are now becoming aware of this. There has been numerous books and documentaries about this prepared over the last century by very responsible, knowledgeable, and highly respected authors of history, government, economics, and science.

A change in the Earth's rotation, its atmospheric pressure, or a reversal (or reduction) of its magnetic field, would drastically affect many life forms. But when life forms become extinct, new ones evolve that can adapt to and exploit the new conditions.

MAGICAL DNA AND GENES

It's almost magical that inanimate matter can arrange itself into complex molecules such as DNA in such a way that it can act as computer code that holds numerous formulas for cell construction and instructions for how genes should fold. Genes carry instructions to construct a multitude of complex cells that form the working parts of all organs and other parts of a life form's body.

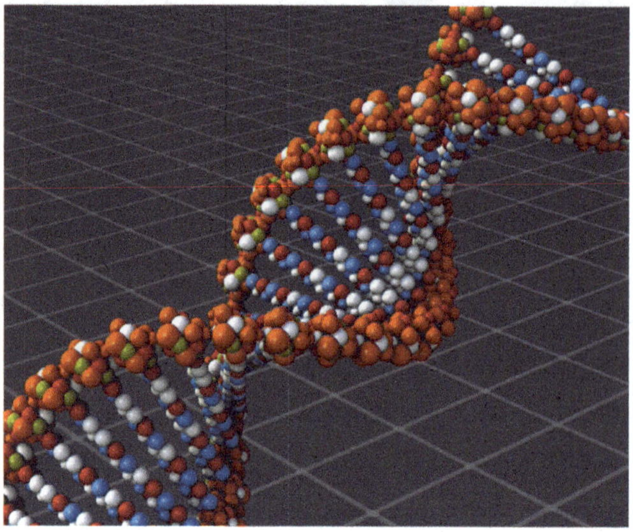

The DNA molecule carries codes for the construction of life. Could all organic life in the universe be based on the information system of DNA? That is very likely.

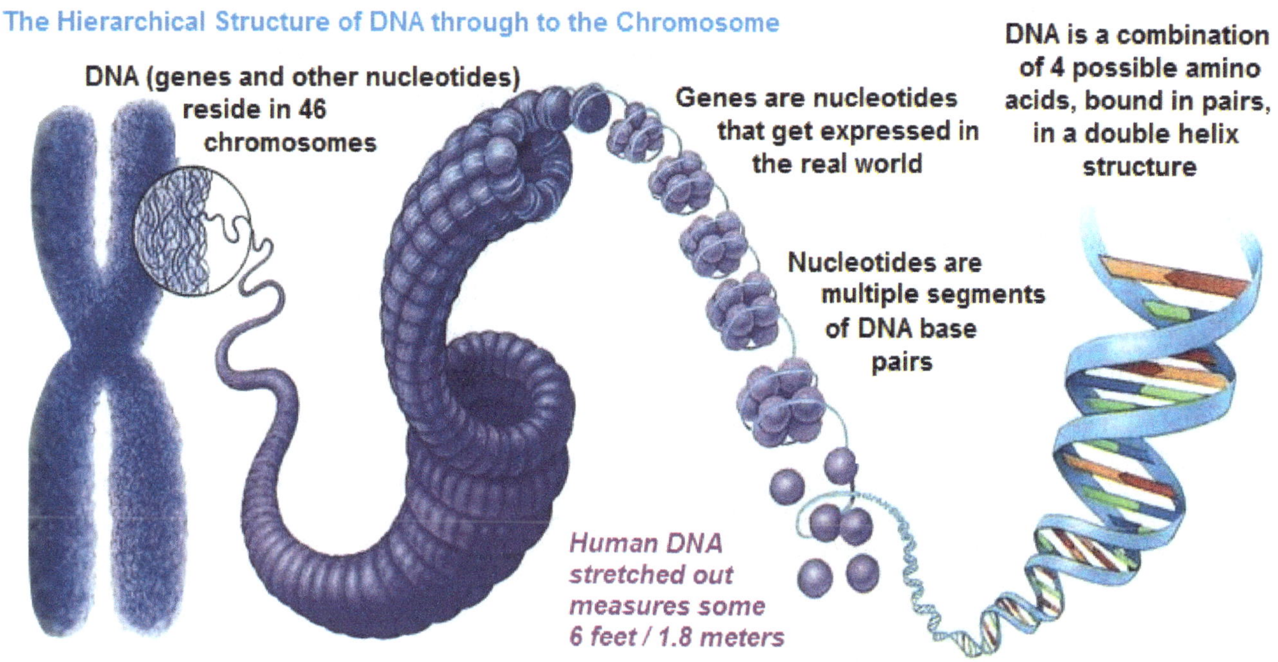

It is simply amazing that molecules work out how to curl and twist in ingenious ways to maximize their information content in a minimum of space. DNA curls up to form genes, which in turn form spirals within larger spirals. These twist up into X and Y chromosomes. There are only two sexes that exist for all types of sexually replicating animals - males and females.

Genes have the ability to create an almost unlimited variety of cells to create an endless number of life forms. However, not all variations of life are viable or stable. Biologists suspect that there are only certain states for life (known as **attractor states**) that produce viable life forms. Biologists are trying to define what these states are and to understand the process (the program) for moving a life form from one state to the next more advanced state. Because there are still numerous simple life forms on Earth that have existed for billions of years, not all life advances to become a higher life form. But why not?

COULD LIFE ON EARTH BE UNIQUE?

There appears to be nothing about the Earth that would suggest it is unique in the universe in its ability to harbor life. Even in our solar system, there appears to be a number of worlds that may have conditions suitable for harboring at least simple life forms.

Mars may still have regions below its surface with remaining thermal activity, particularly near its large extinct volcanoes, that may still have enough residual heat that could allow liquid water to exist. During Mars' summer, there appears to be salty water leaking out of crater walls and channels, as well as from some rounded mountains and cliffs. (But these runs could be sand slides.) Salty water on Mars would not evaporate as quickly as fresh water would, due to Mars' low atmospheric pressure. NASA now believes that finding salty water could indicate a probability of microbial life still surviving under the surface of Mars. There are microbes on Earth that can survive using CO^2 instead of oxygen, so these would be likely to evolve on other suitable worlds. The first life that evolved on Earth did so when our atmosphere was almost all CO^2.

Jupiter's volcanic moon Io could provide conditions for bacteria that live off sulfurous compounds underground like bacteria on Earth do - as long as there is some liquid water there. If simple life is found on other worlds in our solar system, it will provide compelling evidence that at least simple life could exist in innumerable planetary systems across the universe. Where there is simple life, there is a chance that conditions on some of these worlds would be suitable for advanced life.

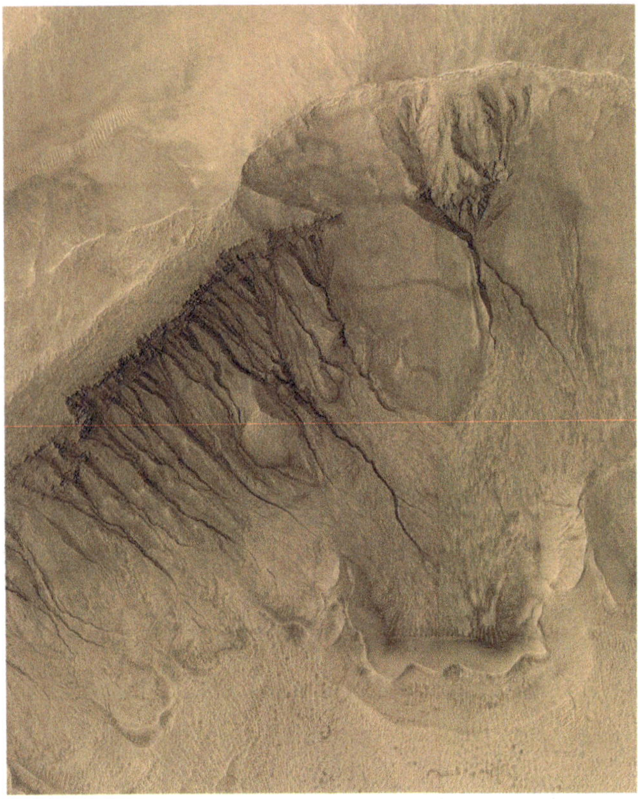

The dark lines appear to be water seepage from a cliff on Mars. Note the dam-like formations at the bottom that suggest larger quantities of water may have existed long ago. However, the dark runs may be small sand slides. Credit: NASA/ ESA

Evidence is rapidly building to support the likelihood of life being common throughout the universe for the following reasons:

1. **Life appears to be a natural part of the evolution of the universe** due to the universe having increasing complexity. Complexity increases through chemistry and cells combining in a near-infinite number of ways.
2. **Organic molecules that are the building blocks for life are being found in nebulas, comets, asteroids, and on some moons and planets.** These molecules will probably be found on nearly all planets across the universe. Where conditions are suitable for life, it will evolve from these molecules.
3. **It now appears very likely that solar systems form around almost every star. This means that there will be an incredibly large number of planets and moons in the universe.** Even if only a small fraction of these worlds support life, there will be a huge number of life-supporting worlds.
4. **Earth has evolved an enormous diversity of life that has adapted to a very broad range of conditions.** Biologists believe that wherever possible, life will evolve to suit the conditions (within certain extremes). Extremophile life forms strongly indicate that some simple life can adapt to very harsh conditions. This means that **life is very likely to be common across the universe**, even where conditions may seem far more extreme and inhospitable than they are on Earth.

Intelligent life may only be able to evolve where the conditions are not extreme.

There does not appear to be any reason to think that life on Earth is unique.

From top left to bottom right - Mars, Enceladus, Europa, Ganymede, Callisto, and Io. All these worlds have internal heat, and all except Io have large volumes of water ice on their surfaces, so there should be a strong likelihood of sub-surface liquid water existing inside them due to internal heat. There is a possibility that internal seas could harbor at least simple microbial life. Credit: All images by NASA

HOW QUICKLY DID LIFE EVOLVE ON EARTH?

The very first cells appear to have evolved over 4 billion (4 thousand million) years ago. **Fossils of single cells have recently been found in sedimentary rocks that date back to 3.8 billion years**. That's only a billion years after Earth formed! Some scientists think that the first cells may have started to evolve around 200 million years after the period of heavy meteoric bombardment ceased. If life evolved this quickly on Earth, it would probably evolve that fast on all worlds across the universe where the conditions permit it to do so. That means that **the first life in the universe might have evolved soon after the first planets formed between 12 and 13 billion years ago.**

On Earth, the cells of every life form contain DNA and genes. **All life appears to have evolved from a single common ancestor cell known as LUCA** (the Last Universal Common Ancestor Cell). One would think that our planet might have evolved many trees of life from a variety of initial types of cells, but surprisingly, **there is no evidence for multiple trees of life**. Could it be that the molecular structures that produced the LUCA cells are the only ones that produce life anywhere in the universe? Or is it that other original cells did not survive very long.

It is likely that all organic life in the universe will use DNA and RNA (or something very similar, if there is such a molecule) as a means of coding information to construct life forms. Carbon-based chemistry of all life in the universe is expected to be essentially the same as it is in life on Earth because it does not look like there are any other viable chemical options. This is due to the laws of physics and chemistry that exist throughout the universe.

If we discover life on other worlds in our solar system, or around other stars where there is no chance of an exchange of cells between planets, we would know that life originates on its own accord. There would be no need for it to spread from one world to another. To *physically* find and analyze life that should exist on some exo-planets around other stars is going to be impossible, even for the nearest stars, because it would take an eternity for us to send probes there. After conducting highly detailed analyses of the chemistry and conditions on other planets, mega-intelligent computers will probably be able to work out if other life forms could exist beyond Earth and what form they may have. Once mega-computers discover a great deal about how life evolves and develops under different conditions, they should be able to determine what life will be like on other worlds.

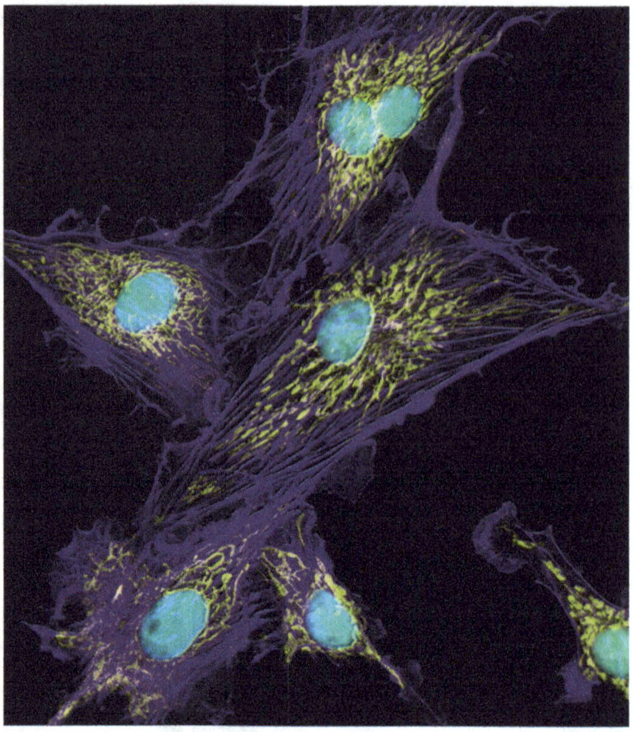

Phosphorescent proteins highlight complex cellular structures in this micrograph.

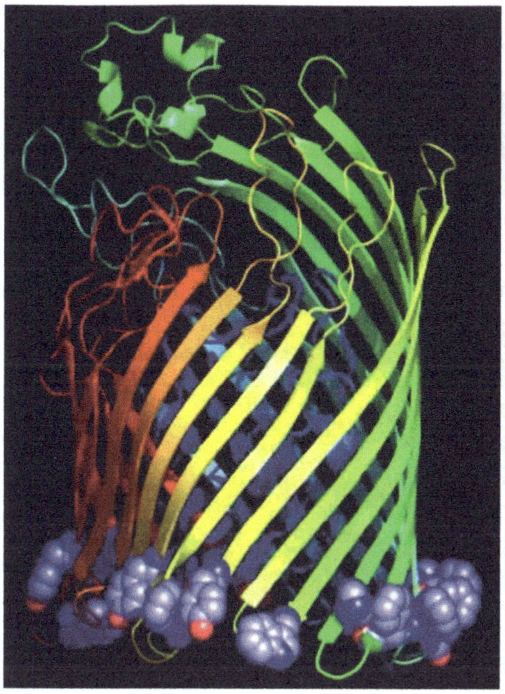

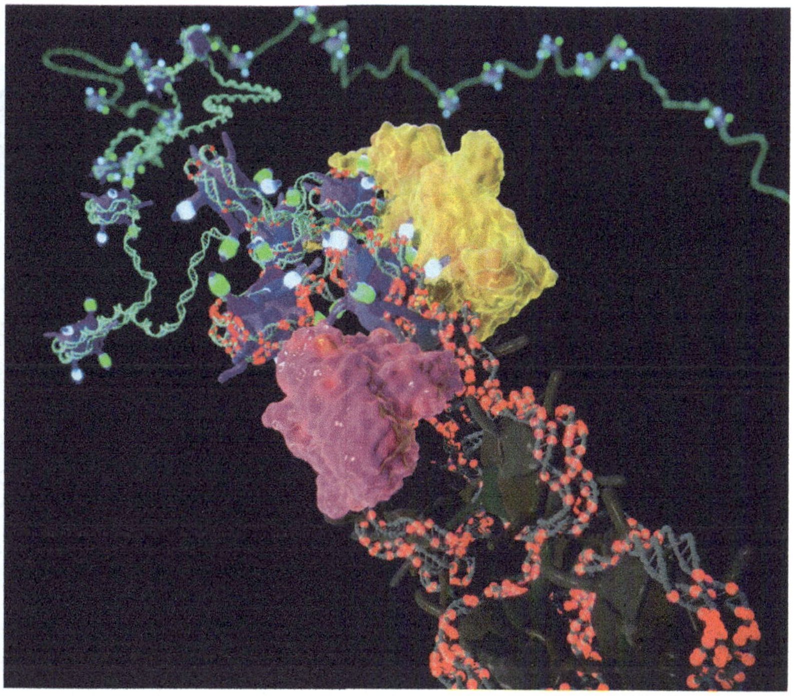

Even the simplest multicellular life forms had to be built from complex molecular structures like these. This will be the case on all life-bearing planets.

Left: *An illustration of a protein under construction.*

Right: *Here we see the structure of a multi-cellular organism being constructed from organic molecular structures. Illustration by Ricken Research*

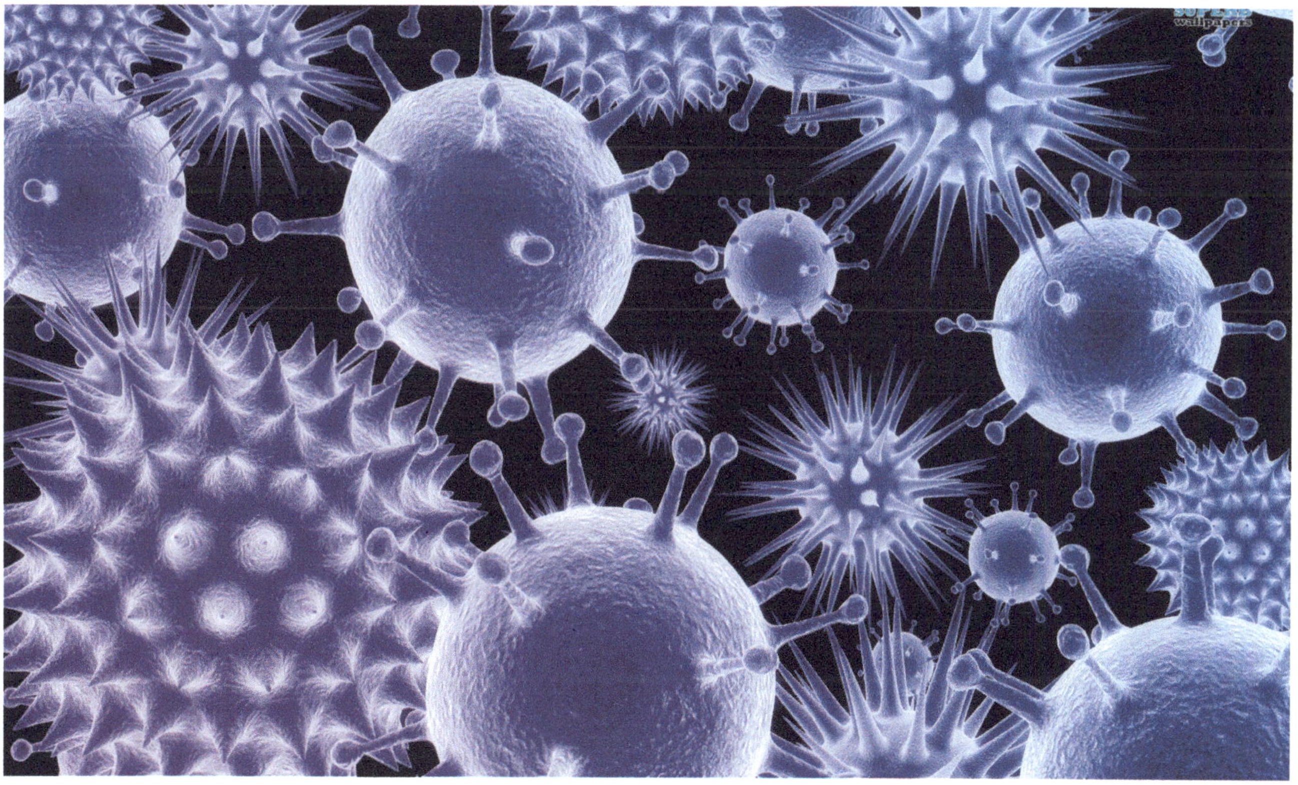

These semi-life forms are known as viruses. They may be pre-life that will arise on all life-bearing planets along with the first cells when they start to evolve.

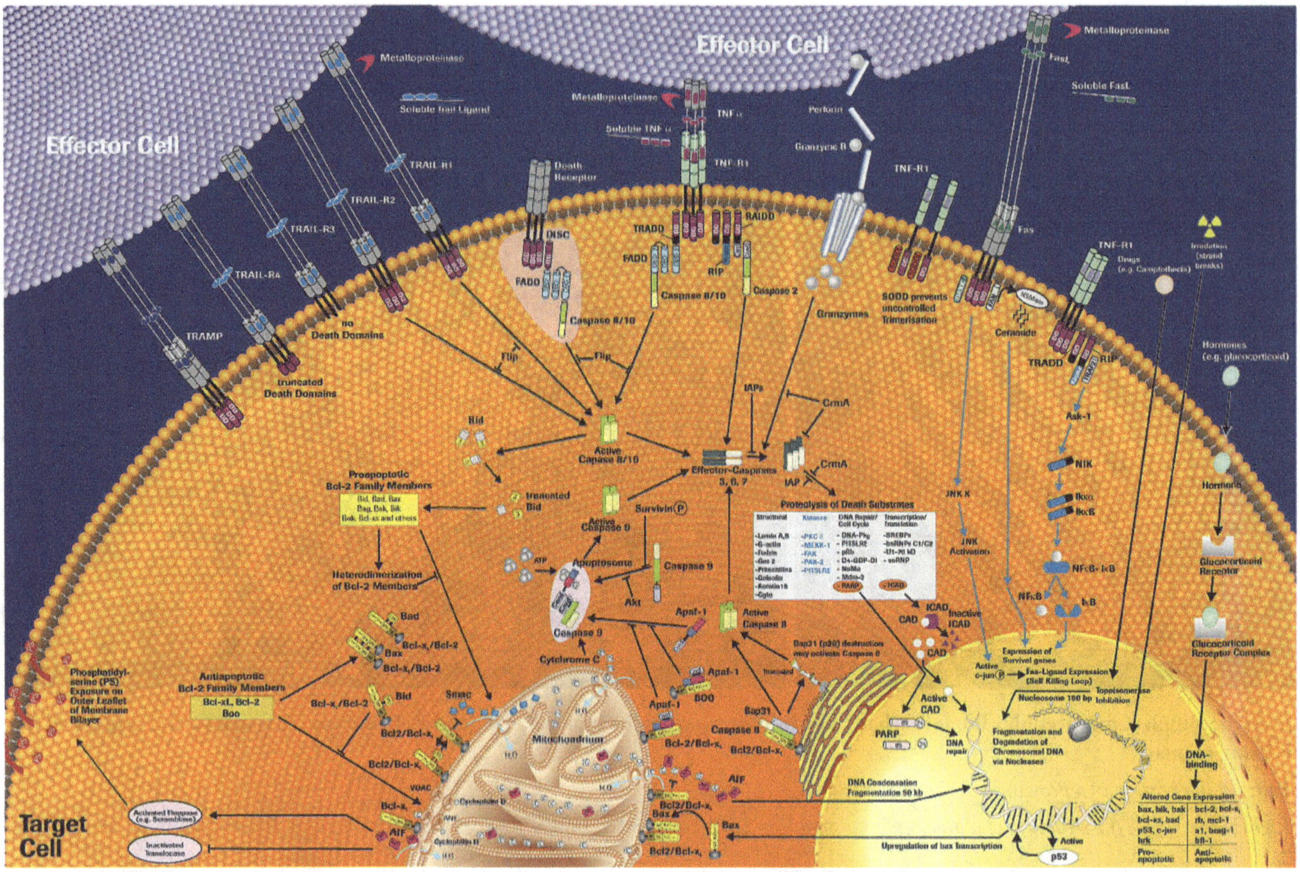

Most people assume that a cell is a very simple organism, but that is not the case. It is extremely complex! This diagram is too scientific for most readers to understand what the chemical structures shown here are. The real molecular structures are not like the symbols shown here. They are very complex. (Look on YouTube for incredible computer animation of how complex cellular structures are and how they operate.) This diagram is included to assist you in conceptualizing the incredibly complex activity that occurs inside cells. The biological 'machinery' represented here is just a small part of a cell's defense mechanisms. Cells also have complex chemical factories for building the machinery required for energy production and the manufacture of proteins, enzymes, amino acids, carbohydrates, sugars, and vitamins, etc. These are chemical machines needed to build membranes, locomotion structures, waste removal apparatus, and replication systems. If a cell was enlarged to the size of the Earth it would appear like the most incredibly complex enormous space station full of billions of very busy chemical factories, transporters, security functions, walls, long DNA structures, and numerous other molecules. It has everything that is needed for this gigantic cellular structure to function, survive, and ward of invaders. All cells throughout the universe are likely to be designed very similar to this.

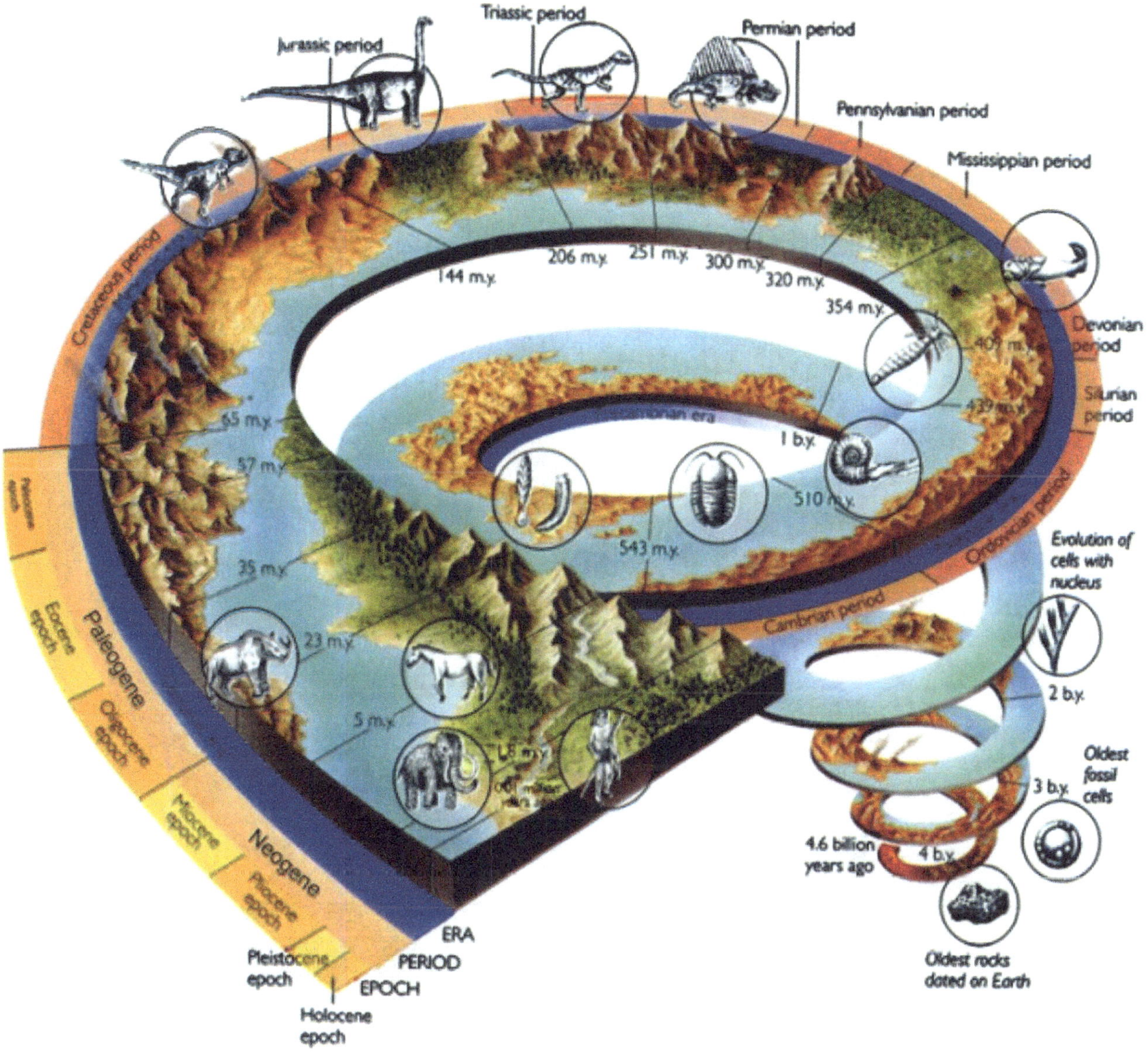

This diagram illustrates the timeframes in which life emerged on Earth. Take some time to inspect it.

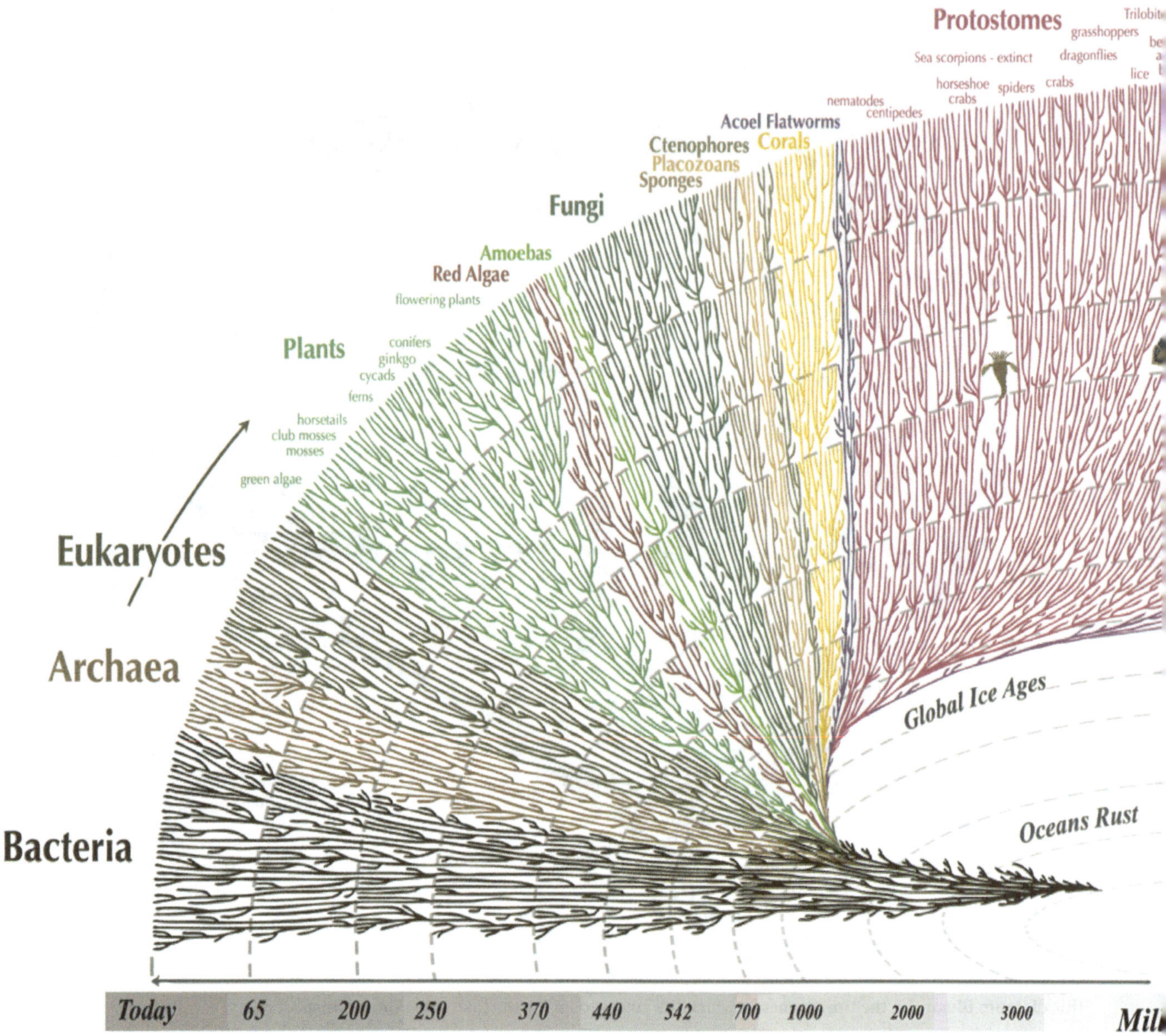

All the major and many of the minor living branches of life are shown

This chart illustrates all the variations of life on Earth that evolved from one original cell over 3,000 million years ago. The lower right side of the evolutionary paths show six mass extinctions. Gaps in the branches indicate the extinction of particular groups of life forms. Only a small number of these are shown to keep the diagram's size manageable. Mankind is the new twig at the far lower right above 'Today'.

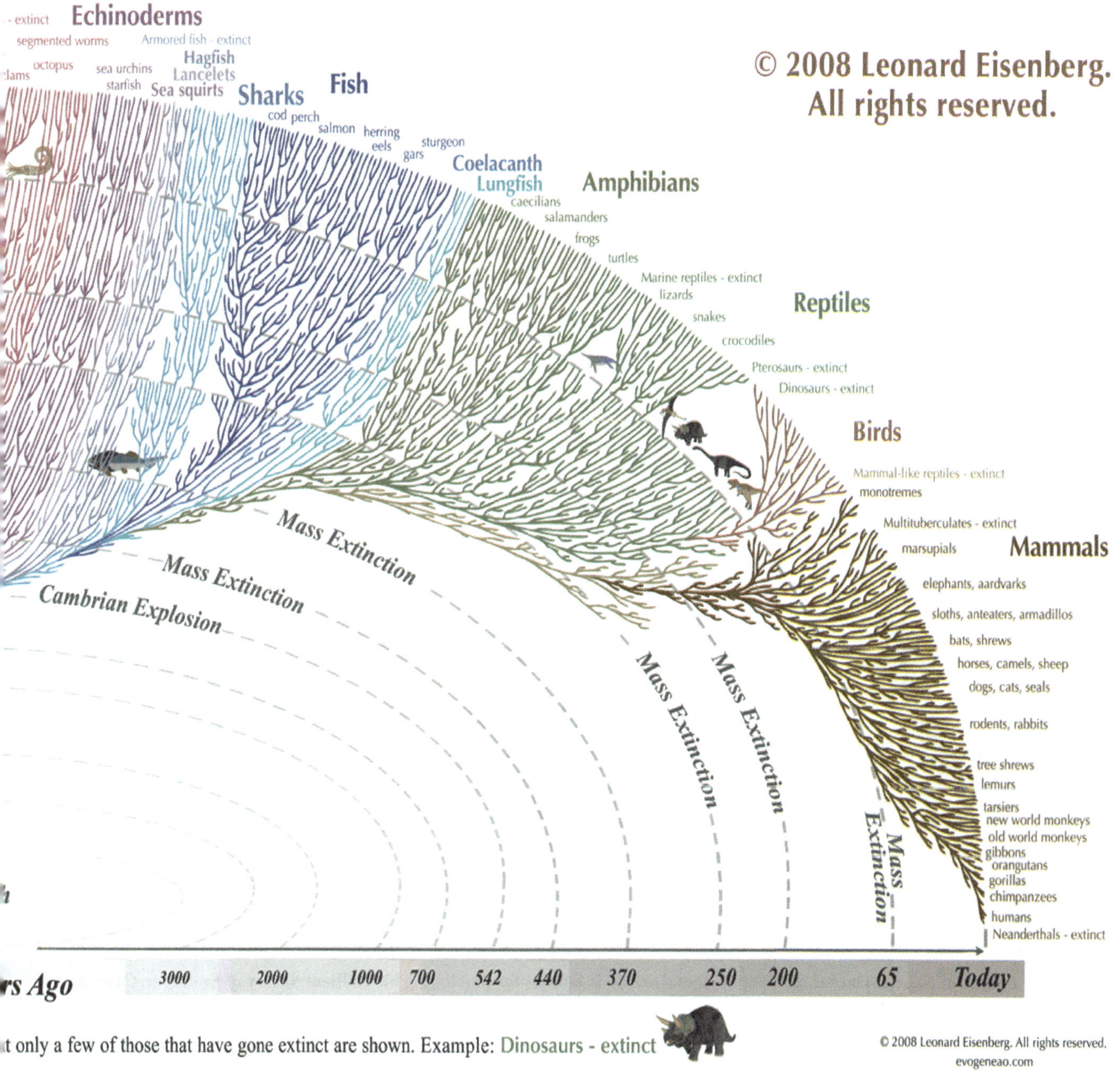
- extinct
Echinoderms
segmented worms
Armored fish - extinct
clams
octopus
sea urchins
Hagfish
Lancelets
starfish
Sea squirts
Sharks
Fish
cod
perch
salmon
herring
eels
gars
sturgeon
Coelacanth
Lungfish
caecilians
Amphibians
salamanders
frogs
turtles
Marine reptiles - extinct
lizards
snakes
Reptiles
crocodiles
Pterosaurs - extinct
Dinosaurs - extinct
Birds
Mammal-like reptiles - extinct
monotremes
Multituberculates - extinct
marsupials
Mammals
elephants, aardvarks
sloths, anteaters, armadillos
bats, shrews
horses, camels, sheep
dogs, cats, seals
rodents, rabbits
tree shrews
lemurs
tarsiers
new world monkeys
old world monkeys
gibbons
orangutans
gorillas
chimpanzees
humans
Neanderthals - extinct
© 2008 Leonard Eisenberg.
All rights reserved.
Mass Extinction
Mass Extinction
Cambrian Explosion
Mass Extinction
Mass Extinction
Mass Extinction
rs Ago
3000
2000
1000
700
542
440
370
250
200
65
Today
t only a few of those that have gone extinct are shown. Example: Dinosaurs - extinct
© 2008 Leonard Eisenberg. All rights reserved.
evogeneao.com

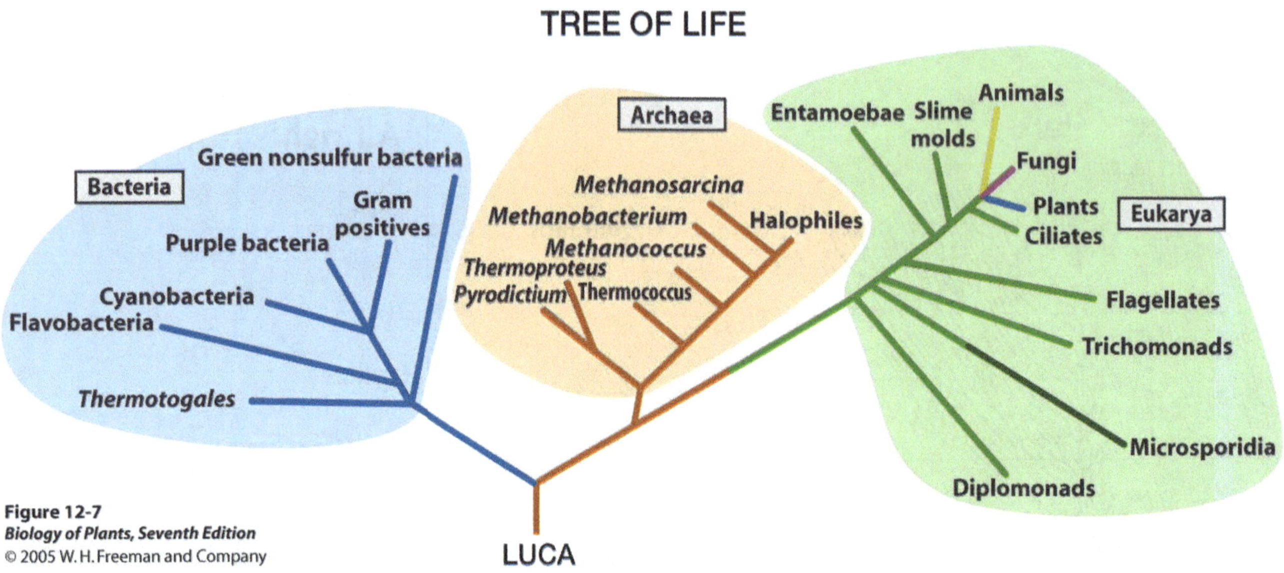

This diagram shows how the three early branches of the tree of life evolved from just one first ancestor cell known as LUCA.

Here are three images of Charles Darwin at different ages. Darwin is famous for developing his brilliant *Theory of Evolution*. Over his life and through his travels, he found many proofs for his theory, as did the English biologist Alfred Wallace who had similar beliefs. But Wallace's science was not as well developed as Darwin's was. Darwin waited until near the end of his life to publish his work because he knew it would cause a furor in the Church, and this would create vicious opposition to his theory. Darwin's ground-breaking theory explained how all life originated. Evolution made sense of the fossils that geologists were finding at different depths in the Earth's crust.

WHY IS THERE A PREFERENCE FOR LEFT AND RIGHT-HANDEDNESS IN THE MOLECULES FOR LIFE?

Most complex organic molecules such as sugars and amino acids have molecular structures that are both left and right-handed. The construction of these molecules is chemically identical, except that they are mirror images of each other. **Inexplicably, life only uses right-handed sugar molecules for energy and left-handed amino acids to make proteins.** Why this is so is unknown.

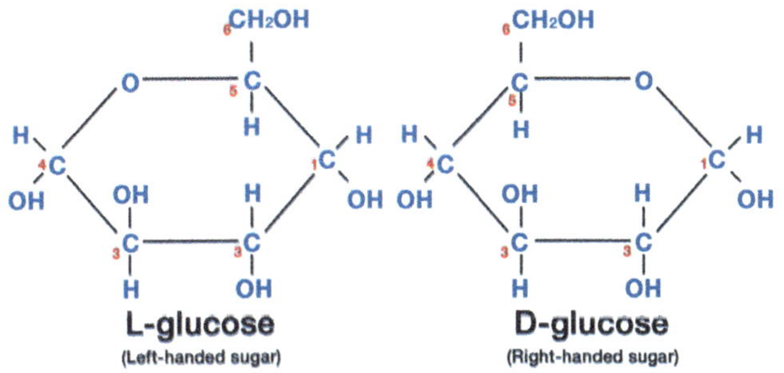

The structures of left and right-handed glucose (sugar) molecules are a mirror image of one another yet inexplicably, they have different properties, but they taste the same! For unknown reasons, life requires right-handed sugars.

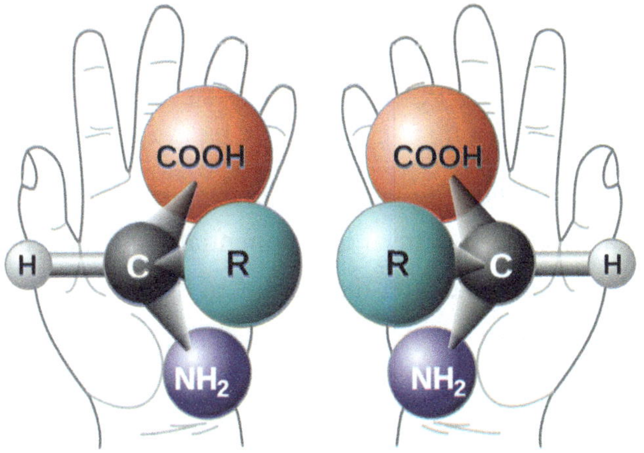

Life only uses left-handed amino acids to manufacture proteins. Amongst many things, proteins make hair and nails and they regulate chemical reactions in cells. Left-handed amino acids are found in meteorites but not right-handed ones. Why? Credit: NASA

When it comes to **nucleotide bases that comprise RNA and DNA, these are all right-handed.** DNA always twists like a right-handed screw. Why this is the case is a mystery. Life on Earth absorbs energy from naturally produced right-handed sugar molecules, but inexplicably, life cannot absorb energy from artificially produced left-handed sugars, despite their molecular structure being exactly the same other than the atoms being mirror-reversed. Even though left-handed sugar tastes the same as right-handed sugar, the body cannot absorb energy from it! Why not? If we could use left-handed sugars in many foods, this would solve many health problems caused by excessive sugar intake. Unfortunately, left-handed sugar is not available because it is very expensive to manufacture. The reason why nature does not use it might be something to do with the electron bonds.

Asymmetry is common in the universe. Symmetry is the state of two things being equal or the same, whereas asymmetry is the lack of equivalence between parts of something. For example, our eyes and ears are very similar, (but not perfect replicas of one another). **Asymmetry goes back to the universe's formation. If there was no asymmetry between the quantity of matter and antimatter when the universe formed, then the universe would not exist because matter and antimatter would have annihilated one another, leaving no matter at all.** Fortunately, there was an imbalance for unknown reasons. This created a little more matter than antimatter. That 'little bit' is what constitutes our universe today.

CAN LIFE FORMS EVOLVE QUICKLY?

Until recently, the evolution of advanced life has been thought to be a slow process, but scientists are now finding many cases where advanced life can evolve very quickly. It is well known that new strains of bacteria can evolve within as little as hours, but although advanced life forms cannot, they can evolve surprisingly fast within only decades or less.

When **Charles Darwin** went ashore on the **Galapagos Islands**, he studied how lizards, birds, and plants there had evolved into different species to those on the mainland, from where they originated. These species had drifted out to sea from the mainland on driftwood and grass mats after major floods occurred in rivers in Ecuador. Some birds may have flown the whole way. They came ashore and took root. Because the conditions on each island were different to one another and to those on the mainland, only those that could adapt to each islands' different landscapes, food sources, and predators were able to survive. These survivors were different in some important way. Their survival traits were reproduced in their offspring. This process continued so that each species evolved until they suited their conditions well. Most species on the Galapagos Islands diverged into sub-species as they adapted to the different conditions that existed in different regions on each island, and from one island to another. These evolutionary changes had to occur relatively quickly for each species to survive and multiply.

Some decades ago, biologists relocated a threatened species of lizard to a remote Indonesian island where it had no predators. To the surprise of the biologists, in only 20 years, the lizards had grown rapidly in number and spread across the island. In doing so, they diversified into 20 different sub-species! Each species evolved differences in their body features such as their size, skin coloring, and markings. The diets of each group changed so they could survive off what was in the environment that each colonized. Those that survived evolved characteristics to suit their environment. Their environments ranged from sandy to rocky coastlines, coral beaches, savannah land, rainforests, and cool highlands. This showed that significant evolutionary change can occur quite quickly, even in advanced life forms.

When humans migrated out of Africa and across the globe, those who inhabited different environments evolved different skin color, types of hair, facial features, body shapes, and diets.

It is now known that the shorter the lifespan of a life form, the shorter its reproduction cycle is, so the quicker it can evolve to suit a new environment. This is why viruses and microbes can mutate quickly within hours. When humans expanded out of Africa, their average life spans were typically only 25-35 years - that is for the few that survived childhood. Due to this, they had children in their early teens. This promoted quicker evolutionary change than we have today. Due to migration, their changing environments also speed up their evolution.

The following graph of the evolution of life on Earth shows a pattern of exponential growth occurring before and after mass extinctions.

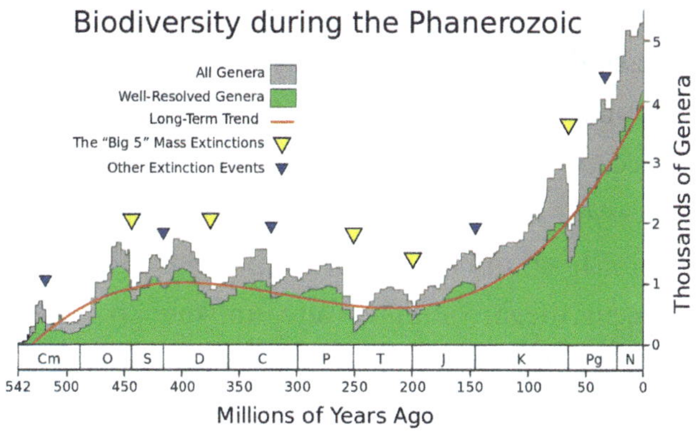

The evolution of life on Earth demonstrates an increase in the diversity of life forms and an increase in the intelligence of the most advanced ones. This is despite several mass-extinction events, which appear as troughs in the graph. They have yellow arrows above them. Our present time is on the right of the graph at 0. As can be seen, biodiversity has risen exponentially over the last 150 million years. It has not decreased as many believe. The application of genetic engineering in the coming decades will make biodiversity sky-rocket even faster.

EMOTIONS

Scientists are now starting to discover the role that emotions play in the animal kingdom. Emotions are distinguished from rational thought by being feelings derived from a life forms' circumstances such as - experiencing threats, moods swings, difficulties raising young, feelings for a partner, a parent's love for a child, beliefs, and relationships with a group or a leader. **Basic emotions appear to extend back to well before the age of the dinosaurs. They are instinctive drivers that prompt immediate decisions on such things as whether to fight or take flight to protect oneself and one's offspring. They are strongly present when seeking a suitable sexual partner. In quite a number of life forms, emotions appear to be fundamental in developing a tightly knit group. This usually enhances a group's chances of survival.**

Emotions may turn out to be a basic element in the evolution and survival of organic life forms across the universe. Even simple ants experience the basic emotion of panic and act confused when they sense fellow ants a meter or more away being killed. Do they scream out in fear or emit chemical odors?

Emotion is a dominant force in the way nearly all humans make decisions. When we are threatened and do not have time to logically think through possible solutions, we respond quickly by using primitive brain circuits that provide an immediate flight or fight reaction. These responses were programmed into our genes billions of years ago, yet they are still automatically activated today in most humans when they are under stress. Emotion is an automatic response that uses a very small part of our brain, which evolved well before the dinosaurs did. Primitive animals had a quick emotional reaction programmed into their genes to help them survive. Even today, for many people, this system takes control when there is a hint of a threat but not necessarily any immediate danger. If people today were to use the power of their huge neo-cortex (super-computer) part of their brain, rather than have their primitive instincts control their decisions, they would be able to think through more options to threats so they could choose the best outcome. But to use their neocortex option takes longer for the brain to analyze a range of solutions, so it is not frequently used by most people. Of course, it cannot be used when there is an immediate impending danger. Generally speaking, most humans tend to make quick and easy, impulsive, emotional decisions rather than rational ones that take more time to think through to find the best solution. **Psychological tests have shown that decisions that are made emotionally are nearly always wrong.**

Technological advances in the decades ahead will make the future more different than we could ever imagine.

In the very near future, **human intelligence is expected to rise exponentially due to advances in computer/brain technology.** Presumably, this will be accompanied by much quicker and wiser thought processes and higher consciousness. This should diminish emotional decision-making. Highly intelligent mega-computers are unlikely to require emotions for their thinking because high intelligence should lead to thinking logically and quickly without the need for primitive emotional responses. This does not mean that high intelligence will not have empathy and morals. To the contrary, it may have enhanced levels. When human intelligence increases, emotions may not be required.

Chapter 9

WHAT MIGHT ALIEN LIFE BE LIKE?

Extra-terrestrial life forms would look at least as different to us as most other life forms on Earth are different to us. But some alien creatures are likely to be much stranger than any life forms on Earth. Consider how different the creatures that inhabit our deep ocean depths are to anything on land, especially the many types of bioluminescent jellyfish species. Many aliens creatures may be far stranger still. A huge bioluminescent underwater advanced intelligence from another world was the basis for the excellent movie '***The Abyss***'.

In many science fiction movies and books, intelligent beings are typically depicted as having a humanoid form, but in all probability, real aliens will bear little resemblance to us. In most films made before computer-generated art was developed, humanoid forms were used for aliens so that an actor could be inside the costume to make it move. Most aliens in movies typically have one or two eyes, a mouth, and two arms and two legs like us, but this is unlikely to be the case for many aliens. With the use of computer-generated vision, this trend is slowly changing in modern sci-fi movies. A very innovative documentary '***Alien Planet IV***' (available on YouTube) cleverly visualizes many new forms of life on another planet. These life forms are quite different to those on Earth but also somewhat similar. The documentary also considers different types of meteorology that can occur on other planets.

If a more advanced intelligence did come to Earth, it may have a form and powers similar to the intelligence in the movie '***Annihilation***'. Hopefully, it would be far more intelligent than that! But intelligent alien life forms will not visit Earth for the reasons explained in Chapters 11 & 13.

Survival Tactics: It is inherent in all life to survive by using whatever strategies it has to enable it to do so. Life on Earth uses a wide range of means to survive including: camouflage, illusion, exaggerating one's size, fake eyes, flight, hiding, armor, stingers, electric shocks, poison, thorny exteriors, wide-angle and telescopic vision, an acute sense of smell, overwhelming physical strength, speed, stealth, darts, acid spray, suffocation, strangulation, inflating body parts, sharp claws and teeth, threatening colors, playing dead, foul smells, jumping, rolling, distraction, secret vibration communications, hunting in packs, laying their eggs in another life form's nest to raise them, or laying eggs in another life form's paralyzed body, copying another life form's appearance, creating fake bait, protection in numbers, and many more. These strategies are likely to be common across the universe.

Senses: Each life form will perceive its environment based on the type of senses it has. These evolve to suit the environment in which it lives. Many life forms will therefore have very different senses to ours and therefore different perceptions about nature.

Most extra-terrestrial life forms will have a similar diversity of senses to Earthly life forms. Where there is light, they will evolve eyes because eyes deliver large amounts of information quickly over a wide range of distances. On Earth, eyes have evolved independently at least seven times, so they are sure to be common where light is available. Where sunlight is not available, bioluminescence will evolve. Some alien's eyes will not only see visible light, but they will also see ultraviolet and infrared light, just as many insect's eyes can. Hearing will also be common because sensing vibrations through gases, liquids, and solids is an easy way to collect information, so there will be alien life that can sense ground vibrations like elephants, ants, and some beetles do. Senses that detect variations in temperature and pressure will be common because there will be many planets that have sudden and dangerous changes in their weather that life forms will need to be able to identify quickly so they can to seek shelter. Life forms will use many types of senses that their skin or exoskeleton can accommodate. There are sure to be some life forms that have senses that are unlike anything that life on Earth has evolved.

Physical Forms: Certain physical characteristics of life forms will be dictated by the conditions in which they live. For instance, land-based life forms living on a planet with much higher gravity than Earth would most likely have small, squat, strong bodies, whereas those living on worlds with low gravity would typically be tall and spindly, like insects. They might be jelly-like or have thin inflated membranes so they could float like a balloon. In low gravity, they could have lightweight, fragile bodies that would not be damaged in a fall because they would fall slowly incurring no damage when they land – as occurs with insects on Earth. There would be many life forms that would be essentially weightless due to them living in liquids or in a dense atmosphere. Such life might be similar to life in our oceans. Those that exist in the atmospheres of giant planets could be like gigantic inflatable jellyfish measuring kilometers across. Such lifeforms may exist in the clouds of Jupiter and the other gas giants.

There will be species that have multiple forms. In the movie '***Starship Troopers***', Earthlings battle a colony of intelligent, giant insect-like alien creatures. Some alien life may be like gigantic versions of ant colonies on Earth that have different types of individuals that specialize in certain activities. If it were possible for a tiny ant-like life form to be intelligent, it would see humans as an enormous collection of specialized life forms (cells and bacteria), each with very different functions living as one entity. In all animals, almost all cells work together in a symbiotic relationship - just as most people in a civilization do. In cases where some cells don't work for the common good, the animal gets sick and typically dies.

It may be that some aliens are so different to life on Earth that they could appear to us to be ethereal entities that communicate through forming geometric patterns in the air by manipulating dust particles and water vapor. We would not be aware of their presence unless they tried to communicate with us in this way. Credit: Annihilation movie by Paramount Pictures

In the movie '*Starship Troopers*', intelligent alien creatures were based on large lifeforms that had colonies similar to ants on Earth.

There are enormously large life forms on Earth that few people know about. One is **mushroom mycelium**. It is a single life form that can extend for well over a square kilometer under the surface of grassy fields or forests! Mushrooms are its fruit. Only recently has it been discovered that the Redwood forests in northern California are actually one huge single life form! Each tree shoots from one gigantic root system that extends under the entire forest! There are sure to be many examples of life like this on other planets. Some may cover much of an entire planet, yet they will permit other life forms to exist within them, just as trees on Earth do.

There are only a few ways to walk, run, hop, roll, slither, swim, use jet propulsion, and fly efficiently, so life on alien planets will evolve these types of movements, as has many life forms on Earth have.

Alien life forms will evolve numerous ways to replicate, although many will be similar to the ways life on Earth reproduces. Some aliens might learn to hunt by copying the DNA of their prey, so they look like them. This way, they would not be seen as a threat when they are about to attack them. This was the basis of the terrifying creature in the second and best sci-fi movie version of '***The Thing***' that starred Kurt Russell.

Alien Ways of Thinking: It's likely that all life forms will have their own ways of thinking - just as cats have certain aspects of their thinking that are different to dogs and to us.

But many aliens may think very differently to any life form on Earth. We assume that there is only one logic, but is it? Could there be other forms of logic to ours? If so, it would be impossible for us to comprehend a logic that is alien to ours.

Space artist Mark Garlick envisages large bird-like life forms darting between the clouds of an alien planet that has two suns. Enormous inflated jellyfish-like creatures float high in its atmosphere.

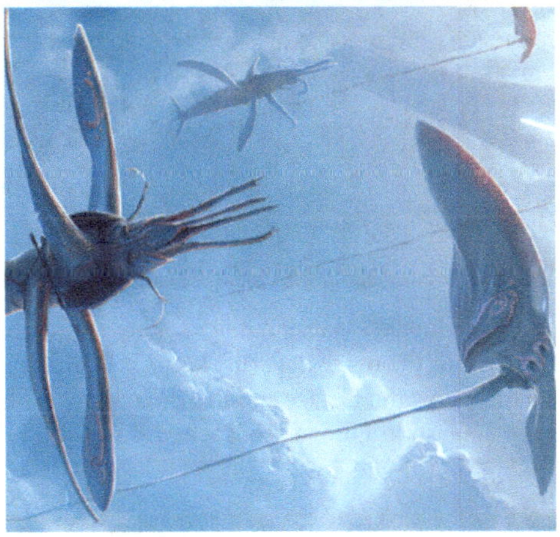

Above: **Some advanced alien life could look something like these creatures.**

Left: **Many planets will have atmospheres as dense as our oceans, just as Venus has. Huge life forms as large as whales, or even far bigger, could swim or fly through the air – just as whales, giant squid, and jellyfish do in our oceans.**

In the near future, NASA will search for simple life that may exist in oceans under the icy crusts of Jupiter's Galilean moons, Europa, Calisto, and Ganymede. Heat from the cores of these worlds is expected to melt the ice. The pressure from the overlying crust keeps it liquid even though its temperature would be far below 0°C. Europa's ocean is expected to commence 60 km below the surface and it is thought to be just as deep. Just as life developed around volcanic vents deep in Earth's oceans, it may also have done so in Europa's ocean, and those of other Galilean moons.

Some life will develop exotic forms of display and dance routines for mating that might be similar to those of birds of paradise in the forests of New Guinea. Others will use exaggerated display features for defense or camouflage, as some insects and sea creatures on Earth do.

The earliest life forms on Earth may have evolved and flourished in thermal ponds on the surface before it evolved to be able to move onto the land.

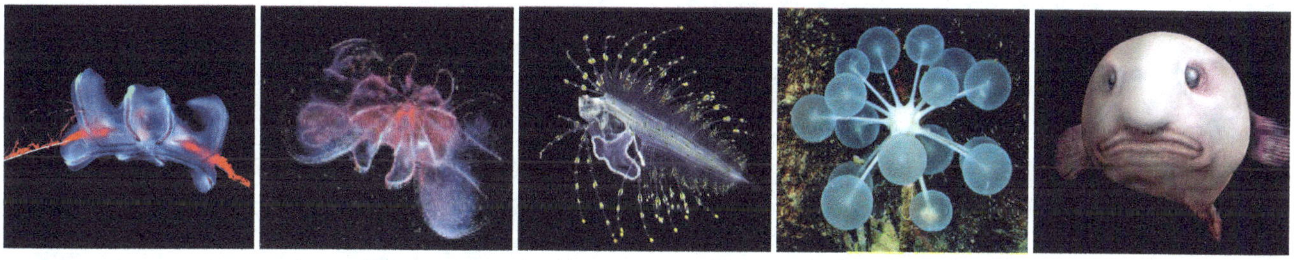

Life forms on other worlds that live in low gravity, or in dense gases or fluids, are more likely to be fragile like the Earthly creatures shown here.

Many life forms will evolve so that they can cloak themselves in an illusion for protection and to confuse predators. This squid (left) and this seahorse (right) are examples.

Redwood trees in forests are not individual life forms: they sprout from one huge underground network of roots. Other planets are likely to have life forms similar to mycelium and redwood trees that spread over large areas of their planet.

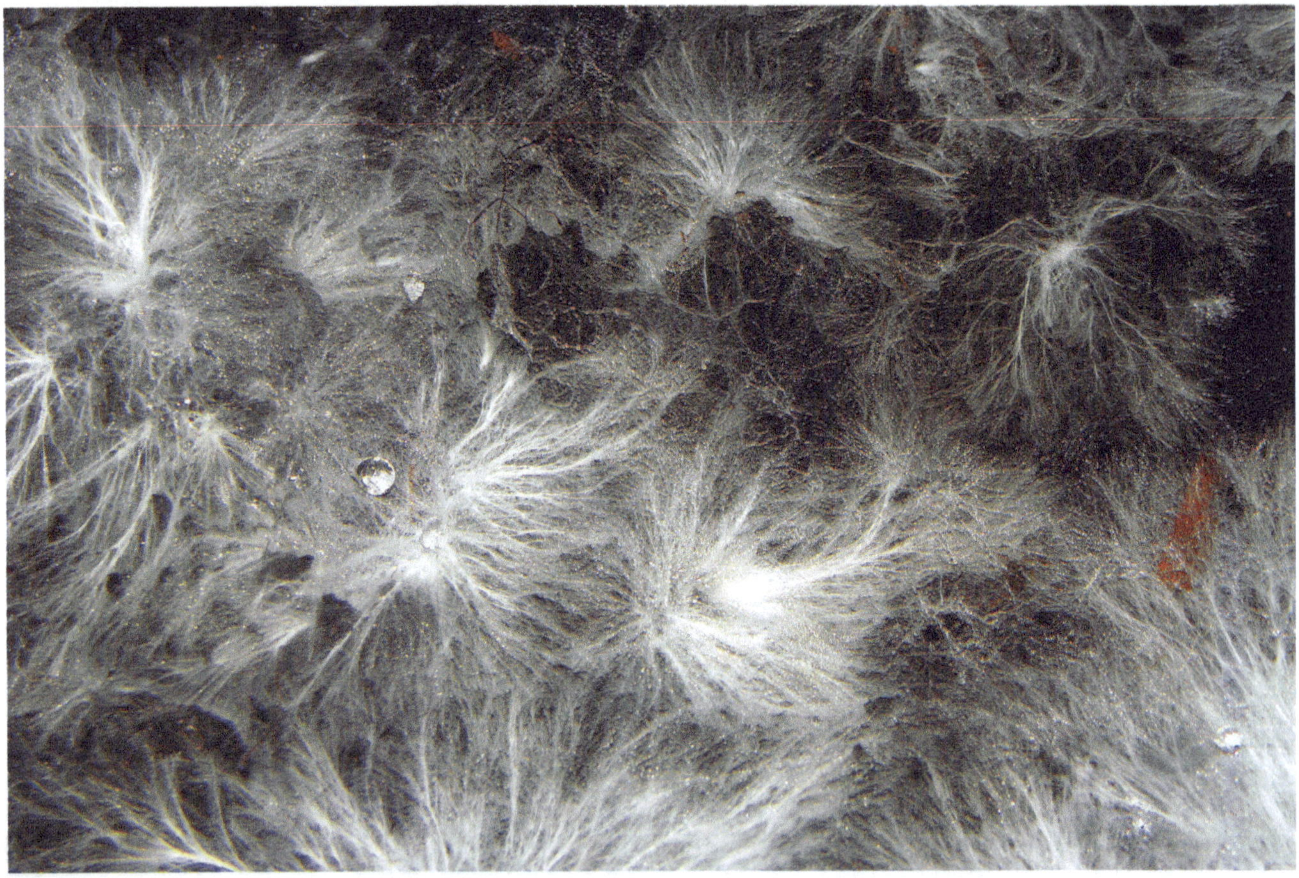

Mycelium is a life form that can be more than a square kilometer in size. When conditions are right, it fruits by sprouting mushrooms, which drop spores to continue the growth cycle.

GENETIC DIFFERENCES AND INTELLECTUAL SEPARATION

The facts about genes in this section are fascinating and unexpected.

The differences in the genes of humans and chimpanzees are said to range from 4% to as little as 0.1%. For there to be such a large difference from one source to another by the most prestigious scientific research organizations studying genetics means that there is a lot about genes that we are very uncertain about. Until recently, it was thought that the relatively small difference in the genes of chimps and humans is what is responsible for the enormous difference in the intelligence and consciousness between a human and a chimp. It was thought that this small percentage of different genes is what allows us to do such things as - talk, farm food, be artistic, develop written language, do math, problem-solve, develop science and technology, and build innovative cities. These are just some major things that we can do that chimps cannot. We humans have developed all manner of tools, yet chimps have only managed to use hand-held rocks as a hammer and use sticks and grass strands as a tool to draw out termites from their nests. Could the small percentage of differences in our genes compared to those of chimps be the reason for our much greater intellect? It's very unlikely.

Despite there being only this small difference in the genes between humans and apes, chimps have not evolved over the last 2 million years to work out how to start a fire or maintain one lit by lightning or lava. No other life form on Earth has learned how to build anything other than a nest, and only a few have developed simple traps like a spider's web to catch food. Yet once humans developed science, they learned how to build numerous things such as very comfortable, safe homes and furniture, sophisticated skyscrapers, cars, ships, aircraft, and space stations. Science also enabled us to invent electricity, microscopes, telescopes, atomic energy, and computers etc. Could such a small percentage of genetic differences be the reason for us developing science when no other animal has?

It would be very useful to know what has driven our extraordinarily rapid growth in intelligence because this would help us evaluate the likelihood of alien life on other planets developing technology. It would also help us to know how far ahead of us aliens may be if they have just small improvements in their genes. But are genes the whole answer?

If the difference in mental capability between humans and chimps is solely driven by genes, then an advanced alien species that has genes that are just a mere 0.1% more advanced than ours would be unimaginably more advanced than we are.

The number of genes a species has cannot be the only cause of greater intelligence for the following reasons. The number of genes a fruit fly has is around 13,000, whereas a mouse has approximately 23,000, and a dog has around 25,000. On that basis, one would assume humans might have around 40,000 genes or more, but surprisingly, humans only have around 20,000 – that's less than a mouse! Astoundingly, rice has in the order of 50,000 genes!

Clearly, this tells us that **the number of genes a life form has does not determine its intelligence. The determining factors are more likely to be**

1. **how many genes are actually useful,**
2. **how well do they operate,**
3. **how well do they interact with other genes,**

4. how well do they organize brain activity and reasoning, and

5. how well do they adapt to changing circumstances? At this stage, geneticists do not know what drives large differences in intelligence between species, although it seems that **the number and type of proteins created by a lifeform's DNA seem to play some role in how mentally capable a life form is.** How this works is unknown.

If genes are not the reason for our superior intelligence on Earth, what is? Genetic science is very complex and little-understood at this stage, so it is not clear what role it plays, but the following may be involved.

The human genome-mapping project has only been recently completed. Its results are not yet well understood, but this will change dramatically in the near future with intelligent computers being able to solve problems quickly that humans might otherwise take decades or longer to solve.

There are likely to be numerous computer-based intelligences in the universe that are unimaginably more intelligent than organic beings like us, but computer-based life forms do not have organic genes. They have computer code as their equivalent.

It is common to hear many scientists, science fiction writers, and people generally imagining advanced aliens that might come here as being very similar in intelligence to us, but the chances of that being so are actually zero! If they were able to get here, they would have technology beyond our wildest dreams. They would be so far ahead of us intellectually that they would see us as we see insects. With the exponential development of technology, they would only need to be slightly ahead of us for that to make an incalculable difference. (See Chapter 14.)

Chapter 10

WHERE ARE THE ALIENS?

This section looks at how realistic it would be for advanced extraterrestrial intelligences to visit the Earth.

In 1950, the renowned quantum physicist **Enrico Fermi** made a pertinent statement when he said, *"Our galaxy should be teeming with civilizations. Where are they?"*

Some scientists have come up with the following reasons why we do not see advanced aliens:

Answer 1 assumes that *all* intelligences reach a level of technology that they cannot go beyond, so this stops them from developing interstellar space travel or communication. But this is at odds with the way life keeps evolving towards ever greater complexity and greater intelligence to enable it to solve ever more difficult problems. Highly advanced transhumans and mega-intelligence will start to evolve on Earth within only a couple of decades. There is no reason why either would stop evolving to become exponentially more intelligent and technologically capable. This will allow them to work out ways of studying other planets and alien life from afar without physically visiting other worlds.

Answer 2 assumes that all civilizations will destroy themselves before they develop a means of traveling or communicating across interstellar distances. There is no reason to believe that *every* civilization that has ever existed in the universe over billions of years will all be so *un*intelligent that they will destroy themselves. This is an ultra-pessimistic explanation that has no evidence to support it. Why would the universe develop life to ensure that it destroys itself? This statement says more about its proponent's depressive psychological state and lack of love of life than it says about reality. When technology advances exponentially, it causes intelligence, knowledge, wisdom, and consciousness to also advance exponentially. This produces a greater ability to solve problems to avoid total destruction. Even if some intelligences manage to totally destroy themselves, most would not. Most intelligences will keep evolving at an exponential rate by adapting to serious threats along the way.

Answer 3 assumes that we humans are the only intelligent life in the universe. This naïve, egotistical answer goes against nearly all the scientific evidence. Very few scientists, or even non-scientists think this is a realistic answer. This is an ancient Christian religious concept that goes against all that we are learning about the evolution of life, the enormous number of planets there are in the universe, and the number that are likely to support life. There are likely to be numerous planets that evolve a high intelligence, just as Earth has. It is very unlikely that Earth is unique - but there's a very slim chance it may be for reasons we do not know about.

Very few scientists are yet aware of the dramatic effect that the exponential evolution of technology has, and how rapidly this occurs for all intelligences that reach our level of development. This opens up incredible possibilities that our most advanced brains are just starting to glimpse. Most people, including most scientists, are locked into the extremely naive idea that space travel will be like early humans exploring new lands by sailing across Earth's oceans. They do not appreciate that physical interstellar space travel and radio communication is *completely impossible* due to the incredibly vast distances and the extremely long timeframes that would be involved. And nor are they aware of the physical constraints that limit how fast spaceships can travel, regardless of how advanced an intelligence may become. (See Chapters 11 and 12.)

In the BBC's documentary series ***'Human Universe'***, it is surprising that science presenter **Brian Cox** did not think of any of the following reasons as to why we have not seen evidence for other intelligent life in the universe. Because of this, he believed that Answer 3 is the reason, but he now seems to have changed his mind.

The following reasons – Why we do not see evidence of alien civilizations or why they are not visiting us should be fairly obvious to any intelligent scientific thinker who is knowledgeable about the leading edge of science and technology.

Popular science fiction culture has romanticized the myth of advanced extraterrestrials invading Earth. In movies, despite aliens being extremely superior to us, incredulously, we always manage to destroy them! That would be like Stone Age man using spears to defeat a USA aircraft carrier fleet. There are many reasons why aliens would not bother to travel across the gulfs of space for eons to visit a common planet like Earth merely to have physical contact with us lowly humans whom they would view, according to the first '*Men in Black*' movie, as 'barely conscious pond scum'. Illustration: '*Independence Day*' movie

WHY DO WE NOT SEE EVIDENCE OF ALIEN INTELLIGENT LIFE?

1. **We humans are at a very early stage of technological development, so we currently have no means by which we can detect other intelligences in other solar systems.**

2. **Alien intelligences would not use radio transmissions to communicate with us, or other intelligences, because radio takes far too long (centuries to millennia) to traverse even relatively short interstellar distances.** As well, radio signals start to become distorted and very weak very quickly over just a light year or two. This makes radio communication infeasible. (See Chapter 14.)

3. **No super-intelligence will travel physically across interstellar space because it is impossible to do so in any practical timeframe. Furthermore, it would require an impossibly large amount of energy. This makes interstellar space travel impossible, no matter how advanced an intelligence is.** (See Chapter 12.)

   Highly advanced intelligences will invent other means of exploring distant worlds that will not require physical transportation e.g. highly advanced virtual reality and programs to accurately predict evolutionary trends of alien intelligences and their planets when enough data is collected.

4. **Advanced intelligences would know all there is to know about us** through means beyond our comprehension, so they would have no need to come here physically. **Biological life on Earth is unlikely to be any more special than biological life elsewhere, so aliens would not try to visit us. They have probably been monitoring us for millions of billions of years,**

5. **Our resources are readily available in most solar systems, so aliens would not come here to plunder us.** Furthermore, advanced intelligences would synthesize their own resources,

6. **Once advanced biological aliens develop technology, they will quickly invent mega-intelligence. It will become so advanced so exponentially fast that we would not recognize it, or be aware of its existence if it did visit Earth.** (See Chapter 12.)

7. **Advanced intelligences would know that the universe is filled with creatures like us.** They would know that we are so naïve that we cannot work out the most obvious reasons why they do not bother to communicate with us.

   It would be as pointless for an advanced interstellar intelligence to visit us as it would be for us to take ages to walk to a common ant's nest amongst millions of others in a distant remote desert. Just as it would be impossible for us to tell the ants what we know about the universe because they would not understand anything we tried to tell them and because they are so unknowledgeable. **Advanced intelligences would not be able to explain to us what they know, as we would not understand a thing they were trying to communicate to us.** It would be far too complicated for our level of intelligence and knowledge.

Advanced mega-intelligence may have no obvious physical form as they might exist everywhere in all things at some atomic level. It would be far beyond our ability to conceive of such an intelligence – just as insects cannot conceive of the level of existence that we have. Once mega-intelligences advance past the need for what we arrogantly call 'high technology', they might be able to achieve a level of consciousness that we might consider to be ethereal. Mega-intelligences will be so advanced that they would not only surround their planet but also their entire solar system with a nearly infinite number of nano-cell sensors. This would be their nervous system. If a mega-intelligence exists in this form, we would not be aware of their existence.

WOULD WE RECOGNIZE ADVANCED INTELLIGENCES?

Any intelligence that is experiencing exponential technological development will be unrecognizable to us even if it is only a mere century ahead of us technologically and intellectually. This is because it would be trillions of times more advanced. Its being could be infused into every part of our planet and our entire solar system in a way that we could not detect. It may not even require a physical form the way we do, so it would be invisible. Its presence would be so subtle that if we saw any aspect of it, we would think it is just a part of nature. Would simple life forms like worms recognize us as at all, let alone as superbeings? No, we would be in the same position relative to a super intelligence. A super-intelligence and a super-consciousness would see us as we see worms.

I once saw a cartoon that had two blue-green algae cells in a puddle in a gutter in the middle of a city where numerous people were walking by. One algae cell said to the other *"If there are other advanced life forms out there, then they would be visiting our water world and trying to communicate with us. As this is not occurring, there must be no other life forms beyond our puddle."* This sums up the explanation to Answer 3 above. In the near future, even the mega-intelligences that we are incubating now will become so advanced that we will not be able to recognize them even in a few decades. Because **we do not see other intelligences does not mean they do not exist.** A tribe of Stone Age people in a remote jungle who have never had contact with other people would not know that us technologically-advanced humans exist because they have never seen us, yet there are billions of us across the world.

Even during the mid-20th century, there were many examples of primitive tribes in the jungles of New Guinea, the Amazon, Borneo, and the Congo, as well as on remote mid-ocean atolls that did not know that modern humans existed. These people did not recognize our technology for what it was when they saw it. They interpreted things like airliners flying over them, satellites in the night sky, and ships at sea as being respectively, peculiar but natural things such as strange birds, moving stars, or unusual floating islands. They had no reason to suspect that these things could be technologies of a much more advanced civilization than their own. Even today, most people, including most scientists, have a similar inability to conceptualize things beyond their limited imagination and experience when it comes to subjects that they have not thought much about.

Melanesians thought that US pilots landing on their island in WWII were sky gods who gave them magical presents. In an attempt to lure them back after they left, they built life sized models of their aircraft.

A good example of primitive people not recognizing technologically advanced, modern-day people for who they are, occurred during the Second World War. A primitive Melanesian tribe living on a mid- Pacific Ocean island thought that American air force pilots that came out of the sky in their planes to land on their islands were gods. The fact that they had white skin and wore strange clothes added to this belief. These 'gods' also had magical things that defied explanation such as mirrors, guns, flashlights, radios, cigarette lighters, watches, sweets, and so on. Modern man's advanced technology was so foreign to these Stone Age people that they could not be expected to recognize these things as being advanced people like themselves who had technology far more advanced than their spears and grass huts. This is understandable, as they had not even invented the wheel.

WISHFUL BELIEFS

People who believe in **unidentified flying objects (UFOs)** are entranced by the fantasy of aliens in flying saucers coming from outer space because this is exciting to them. Some want the aliens to 'sort us out' about war, and others hope they will educate us to live together in blissful harmony. This was the theme of the original '***The Day The Earth Stood Still***' movie. Some UFO believers hope that aliens will get rid of anyone who is not as enlightened as they mistakenly think they are. There are those who think of UFO occupants as being all-knowing equivalents of gods who will give us great wisdom, while others think they will wipe us out for being so destructive. There are those who believe they will come here and take our resources. Some theorize that they may come here to mate with humans to populate Earth with their kind before their planet dies. All these reasons for believing in UFOs visiting us are naïve in the extreme, for the reasons given in this chapter and others.

WHAT ARE UFOS?

The term 'flying saucer' became popular to describe photographs of what were said to be alien spacecraft. These pictures have been shown to be faked. They were created by photographers who stuck the likes of teacup saucers together with teapot lids or half balls on top. They threw them into the air, or they suspended them on invisible fishing line and then photographed them over a landscape to simulate an alien spacecraft flying through the air.

In 1952, the US Airforce commenced **Project Blue Book** to scientifically analyze flying saucer reports to see if those that looked real might be a threat to national security. The Airforce was concerned that some UFOs might be Russian surveillance aircraft. Astronomer **Dr. J. Allen Hynek** was enlisted as the scientific consultant for the project. He used the more scientific term '**unidentified flying object**' to replace 'flying saucer'. He said the term UFO should not mean more than it says i.e. an object that was not (quickly) identified by the observer(s). The media liked this term, so it stuck. Despite this sensible terminology, the term UFO became synonymous with 'an extraterrestrial spacecraft controlled by intelligent aliens', so to most people, it made little difference changing the name from 'flying saucer'. The military that were experimenting with new aircraft and missiles were happy to see their top-secret experiments reported as flying saucers so enemy powers would not try to discover what they had invented.

Almost all people who report UFOs have no experience in recognizing unusual examples of natural phenomena that are meteorological, astronomical, or biological in nature. And they often do not recognize unusual manmade objects, or deliberate fakes. Typically, after being investigated, nearly all UFOs are *identified* as natural phenomena or manmade objects.

In the decades before mobile phones, when the bright planets Venus, Jupiter, and Mars were at their very brightest, they were often mistaken for UFOs because they were so much brighter then than other stars. But this seldom happens today due to the public being much more educated about the night sky and due to today's light pollution making the planets appear less bright than they did when the night sky was much darker decades ago. Importantly, people today use Apps like 'Starmap' and other similar Apps to identify stars instantly. Brilliant fireball meteors and uncommon cloud formations were routinely reported as UFOs until the 1980s, but people are much more aware of these natural phenomena today because many people photographically record them on their smartphones and they are regularly caught on security cameras, so they are quickly identified. Manmade objects seen at night such as helicopters with their searchlights on, rocket launches, satellite reentries, small candle-lit hot air balloons, fireworks, special effects searchlights on clouds, and flares on parachutes were often reported as UFOs, but again, people are more able to identify these things today because they have become common and familiar. Daylight UFOs often turned out to be aluminized weather balloons, clusters of party balloons, secret experimental aircraft, and hoaxes.

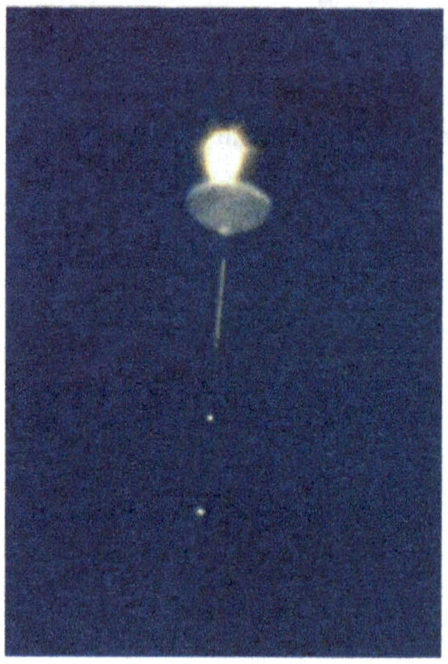

This high-altitude weather balloon was photographed in the 1980's over Argentina and thought to be a UFO. Sunlight is reflecting off its upper aluminized balloon skin that holds helium. Instrument packages hang below it. These balloons travel around the globe on jet streams in the stratosphere.

In past decades, when UFOs were identified, most of the media did not want to report what they really were because UFOs made sensational reports that increased public interest, and this increased their sales. Fortunately, there is much less of this today because most readers are much less interested in fraudulent and exaggerated reports.

I have had parties in the past, where at midnight on special occasions, I gave guests giant balloons 800 mm (30") in diameter filled with helium. Attached to the balloons were three 1 m (40") long sparklers with their ends wired together so they would burn for over 20 minutes. When each person lit the bottom sparkler, the balloon was released into the air. The night sky filled with brilliant golden stars moving across the sky. The public then (and even now) had not seen this before so on two occasions years apart, this spectacle was reported in the newspapers and on television news as being UFOs. The somewhat random movement of the balloons on air currents at different altitudes made some of them travel at different speeds. They appeared to be steering themselves between one another. This caused them to be reported as moving under 'intelligent control' - even by a high school 'science master'! He videoed them and was interviewed on TV. Some students at the party deflated his great observation when they explained what his UFOs were the next day at the school parade. Other 'experts' that were interviewed said, *"There was no way to explain such a phenomenon"*.

UFO HOAXES

There have been many excellent books written that demonstrate just how easy it is to fool even educated people with hoaxes, fake photographs, and uncommon natural phenomena to make them think they have seen an extraterrestrial spacecraft.

It is easy to deceive UFO believers into believing they have seen a UFO with hoaxes because they want to believe they are real aliens. As an example, decades ago, engineering students at a US university built a somewhat elaborate, fairly large, 5 m (200") diameter, lightweight, wireframe 'spacecraft' in the shape of a typical flying saucer. They covered the frame in thin, translucent tissue paper and backlit this with lightweight special effects bud lighting that was made to chase and change its colors and patterns. This lighting was also installed around the perimeter and on the bottom of the fake flying saucer. A series of lightweight battery-powered speakers sent out UFO sounds recorded from sci-fi movies. An intermittent voice repeated, *"Attention, attention, people of planet Earth......"* This model saucer was suspended by fishing line far below several invisible, dark helium weather balloons. They were given very little positive buoyancy so that it would not rise too high into the air. This allowed the model UFO to float on air currents at a low altitude around 100 meters (300') above the ground over the UFO believers. There was no way of determining its size in an open sky because the separation of our eyes is too small for us to tell how large or distant such an object is if it is more than 25 m (75') away. Only when something goes in front of a structure of known dimensions can its distance and size be estimated. In an open sky, this is not possible.

The students released the model UFO upwind of a convention of UFO believers who had gathered in the desert near Roswell, New Mexico, for their yearly meeting. When they saw the fake UFO in the night sky, they were ecstatic because they believed that they had finally seen a real extraterrestrial spacecraft – and what a beauty it was – just like those in the movies! These UFO observers assumed it to be enormous in size and at a high altitude when the reverse was true.

The next day, the convention organizers informed the media of their 'amazing' observation and displayed photographs of the UFO they had witnessed. This seemed like undeniable proof that extraterrestrials existed. Soon after however, the student hoaxers revealed that their 'UFO' was actually a fake. They displayed video of them building it and releasing it. Despite this, about 20% of the UFO 'researchers' refused to believe that what they witnessed was a hoax. (See Confirmation Bias, Chapter 23, page 327.) Today, programmed drones can be used to create a flying saucer that could change altitude and direction. It could even release small fireworks. At night, this would be very convincing.

Believers in UFOs are all too keen to jump to the assumption that any observation they make that isn't quickly identified must be an alien spacecraft. Very few of them attempt to try to explain their observation rationally. Lights in the sky at night are more likely to be reported as a UFO when observers are alone and frightened of the dark and the unknown.

SCIENTIFIC INVESTIGATIONS OF UFO REPORTS

I confess that in my naïve youth, I thought that maybe 1% of UFO sightings might turn out to be real extraterrestrial spacecraft so, in my twenties and early thirties, I had a camera ready at most times when observing the night sky with my telescope in case I saw one. But I have never seen anything that I did not manage to identify quickly. Over my life, I have investigated numerous UFO sightings in the hope that one might be the real thing, but alas, none were! All turned out to be natural phenomena or manmade objects. Many other UFO investigators who used scientific methods to research UFO sightings also found this to be the case. There are many books on this. Discovering that there was not a shred of evidence for any real alien craft was initially depressing until I realized that the logical reasons were for UFOs not visiting us were much more thought-provoking, as you will see in Chapters 12, 13, & 14. These reasons turned out to be much more interesting than the naive wish that little green men in aerodynamic, tin-can spacecraft were visiting us.

Many UFO observers jump to fanciful explanations to explain their sightings because this excites them, and it draws attention to them to make them feel special.

I find it amusing that those who claim to have been contacted by aliens 'to take a message to our leader' are typically middle-aged, unemployed men living in old rusted campers in remote locations. Wouldn't it be more logical for the aliens to land on the White House lawn if they wanted to communicate directly with our leader, or hover over the Pentagon if they wanted to disarm the planet? They wouldn't even have to do that because they could transmit their message across all TVs, radio stations, and the Internet worldwide from space. They would have the technology to make our defenses useless within seconds.

Each time UFO observers recall their sighting, they typically exaggerate it a little more. Their exaggerated story then becomes their memory of it for the next time they talk about it. Over time, their observation takes on epic proportions. I remember a man who saw what others observed as a brilliant star-like meteorite. He said it was "*the size of a bus.*" Some people interpret a very bright light as being large in size, while others exaggerate their sighting to get their message across.

When credible explanations are provided to those who love the idea of UFOs, they can become angry and reject a rational explanation. For those who passionately believe in UFOs, their belief becomes a religion whereby factual evidence that undermines their belief is not tolerated, let alone considered.

It's easy to get carried away with explanations for things you have not experienced before. The following is a classic example where no explanation seemed plausible.

I was asked one night to investigate hundreds of reports of thousands of star-like 'UFOs' up in space, forming geometric patterns. Observers on the Gold Coast in Australia thought that aliens were trying to communicate something to us. I went outside my observatory and looked with my large binoculars over the Gold Coast some 70 km south of me where the reports were coming from. I could see nothing unusual in the sky so, I deduced that it must be a local phenomenon. I phoned an amateur astronomer friend of mine on a small mountain overlooking the area, but he could not see anything even in his telescope. This made it a very local event. I called a physics PhD student who was an astronomer, to ride over to the area on his bike, which he did. When he got there, he used a local's home phone to call me. He described the phenomena as bewildering. There were thousands of yellowish-green, star-like objects in the sky. People were laying in their yards, on grassy footpaths, and on the nearby beach watching this inexplicable event. He said they were forming geometric patterns. I asked him to ask people if they had a pair of binoculars he could borrow, as he was an investigator. He soon had a pair, and was able to see that they were insects flying around only 16m to 20m or 50' to 70' above the ground. It turned out that they were fireflies from a nearby rainforest swarming in a mating activity. Had I not gone through the steps I did, then this observation would have been recorded as being a much greater mystery than it was, and it would have been one of those highly exaggerated, unexplained UFO phenomena.

There are however, reports on the Internet from seemingly respectable US military personnel such as Navy pilots who have filed reports of their observations of UFOs. Some of those in the military say they were told by high ranking military personal to never mention their observations. Top secret observations are very difficult to investigate. One can only listen to the stories of some of the observers that have come forward, but there is no way of validating them. Some claim radar reports support their first-hand observations made of saucer-shaped, cigar-like and sphere-like craft being seen at military bases and then traveling away much faster than the speed of sound with no noise!

Navy pilots who claim to have filmed a small tic-tac, pill-shaped 'UFO' flying at phenomenal speeds and into the ocean and out again without slowing seem unconvincing when you see the bland footage that is hard to interpret. There are claims that these UFOs are visible on radar. It's possible that these 'observations' are faked by the Pentagon to make China or Russia think the US is way ahead of them technologically. Without any solid evidence or logical explanations to support these 'classified' sightings, it is hard to place any credible reliance on them being accurate, especially as they defy the laws of physics! Aliens advanced enough to traverse the immense gulfs of space should be immeasurably more advanced than these sightings suggest. Why would the military be so keen to want to keep such observations secret after all this time?

It should be noted that the 'observers' often receive speaking fees and travel around the country giving talks. There are sensationalized UFO documentaries with clips of well-known politicians and even some presidents, saying aliens exist with no evidence whatsoever to support this claim. They say "the public should be told the truth". But they themselves lie about all sorts of things to their advantage. These politicians would not know even the most basic laws of physics that clearly demonstrate why UFOs could not be extraterrestrial spacecraft with aliens inside them. This is their childish wishful thinking. They should read Chapters 12, 13 & 14 to realize how unrealistic physical interstellar space travel is - even for advanced intelligences.

There is now a fad claiming that aliens are walking amongst us, yet there has never been a blood test or a DNA test that reveals aliens exist. One has to wonder why an advanced intelligence would only make themselves visible to a relatively small number of unimportant people who have no high standing in the community and who are typically scientifically illiterate.

There have been numerous tall tales told about UFOs by persons, who at first seem quite believable, but later, it is discovered that they deliberately deceived those who would listen to them. They did this in order to make money out of telling their story and writing books and making videos.

THE MARTIAN INVASION!

In the late 19th and early 20th C, it was thought that there might be intelligent beings on other worlds. This was an exciting possibility for many people. In the previous two centuries, it was commonly believed that because Earth had life, other planets would also have beings living on them. In those days, the Moon was considered by many to be a world somewhat like Earth. Some scientists thought that there were probably Moon beings living there.

Astronomer **Percival Lowell's** belief that intelligent beings existed on Mars took hold in the popular press in the early 20th C.

Due to the rise of dictators like Hitler and Stalin in the years before World War II's commencement in 1939, there was a lot of uncertainty and anxiety about another world war occurring, especially given how devastating World War I had been. At this time, engineers built fighter and bomber aircraft and talked about building rockets for space flight. This made people think that if there were more advanced aliens in outer space, they might already have rockets and spacecraft and be able to attack our Earth. Science fiction movies, magazines, and novels fueled this fear, which became firmly implanted in many people's subconscious.

In 1938, **Orson Welles** produced his famous US radio play '***War of the Worlds***' based on **HG Well's** book of the same title. It was a play about Martians invading Earth and destroying cities with their death rays. Listen to the radio broadcast of this play on the Internet in the dark and see how surprisingly realistic it seems. With no outside distractions, you will think they will be outside your home any time soon!

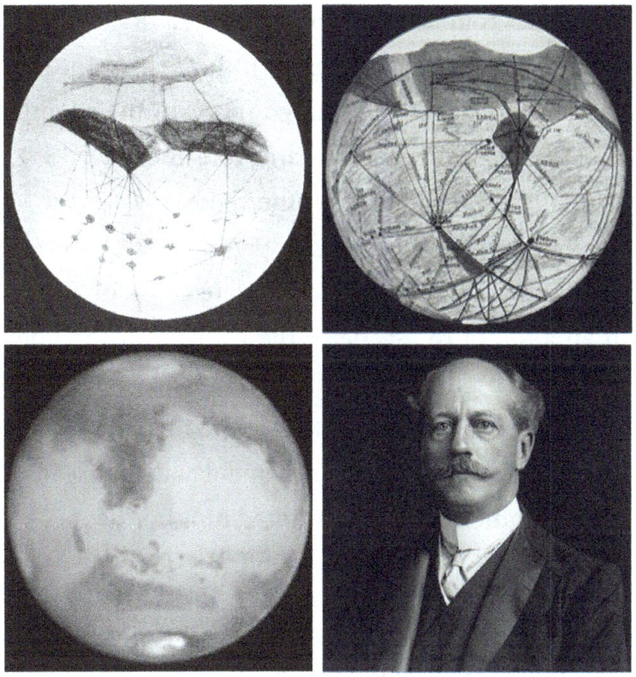

Drawings of visual observations of Mars made by astronomer Percival Lowell around 1900 (top left) and 1895 (top right) show fine lines that Lowell (lower right) thought he could see. He proposed that they were canals that intelligent Martians had built to feed water from Mars' polar ice caps to cities and farmlands in Mars' temperate and equatorial regions. Compare the large dark marking in Lowell's 1895 drawing to the high-resolution HST photograph of Mars (lower left) taken around a century later. Both show a fairly similar view of Mars, with the angle being slightly different. The black and white HST photograph shows much more detail, but it reveals no canal lines. If you compare Lowell's drawing with the photograph, you will see that Lowell's proportions are inaccurate and that he used a large degree of artistic license. Other visual observers could not see Lowell's fanciful lines because they were far too fine for the eye to detect them - even in the largest telescopes of the day. Lowell had a very fertile imagination to make his drawings look more detailed than what could be seen in a telescope. However, it must be said that making drawings at the eyepiece of a telescope is difficult due to air currents making the small image of Mars wriggle around.

Planetary photography did not exist in Lowell's day. The power of suggestion, artistic license, and the effects of poor atmospheric viewing conditions often resulted in an observer seeing what they wanted to see, rather than what they could see. (The angle from which we observe Mars is always different and because the planet is also rotating, images and drawings made at different times are seldom ever orientated in exactly the same way.)

H G Welles' novel 'War of the Worlds' became a classic. This is the cover of Jeff Wayne's outstanding musical version of this story. It is beautifully narrated by Richard Burton. His marvelous, deep, authoritative voice lends much gravity to the story. The music in this album is brilliant, unique, and timeless.

Many people across America listened intently to radio presenter Orson Welles' very convincing portrayal of Earth's invasion from outer space. He presented what were assumed by many of his audience to be actual live news broadcasts from places across the country and overseas where Martian spacecraft had landed. Welles cleverly made it sound like real reporters and eyewitnesses were describing spacecraft coming out of the sky like meteors. The studio's voice actors made out they were reporters interviewing 'astronomers'. Listen to this on the Internet in the dark and see how real it sounds.

Welles told listeners that this was a fictional play at the outset, but most listeners tuned in after the play had started. Many only tuned in when their family and friends phoned them and told them to do so. Many people believed that the reports were real, so this caused many listeners to panic. Hysteria spread rapidly across the US. People left cities in droves in an attempt to escape to the country in the hope that they might survive there. For weeks afterward, Orson Welles was castigated by the media for causing such panic despite his statement at the outset that it was a radio play. Many wanted him imprisoned!

EARTH TO BECOME UNLIVABLE!

I remember a similar cause for concern after the BBC broadcast on television a so-called documentary titled '***Alternative 3***'. It went to air across several nations in 1977. In the week before it was broadcast, it was promoted with saturation advertising as a serious documentary. Prior to this mockumentary being televised, the promotional advertising stated that every citizen needed to know what this top-secret documentary would reveal. It was said it was being sent to television stations worldwide under high security. This guaranteed large audiences.

It described a top-secret 'brain drain' program in the UK and other countries to transfer humanity's most creative, intelligent, and knowledgeable people to the Moon and Mars because manmade climate change would soon make Earth's climate deteriorate to such an extent that it would become unlivable. This was around the time that the Global Warming deception began in earnest. The mockumentary had interviews with fake astronauts, and so-called NASA insiders, and whistleblower politicians who said they were prepared to expose the 'truth' at the risk of losing their lives. It claimed that 'top scientists' had disappeared as a part of the brain drain while others were murdered so they could not reveal the truth.

Being familiar with the space program, I noted that the astronauts interviewed were not real astronauts, but the public would not have known that. After the program finished, I was surprised that some astronomer friends phoned me, due to them believing what this 'documentary' revealed. It seemed that many of those I spoke to in business the following day who watched this hoax were equally concerned. Those who believed it did not think about how ridiculous it would be to go to the Moon or Mars when neither have any life, no breathable atmosphere, no water, no food, and unbearable temperatures! No matter how bad Earth's climate might become, it would never be remotely as inhospitable as the Moon or Mars. But back then, very few people were aware of what the Moon or Mars were like. The number of people who should have been able to rationalize that this program was an obvious fraud was a frighteningly small percentage. Would it be much different today given that scientific literacy is so low?

As with Welles' play, at the outset of '***Alternative 3***', an announcement told viewers to watch the special message at the end of the broadcast. But very few watched it to the end to see in the credits that the program was an April Fool's Day joke, and that the astronauts and scientists interviewed were all actors. Most people were too shocked by what they had heard that they didn't watch the message or the credits at the end. They immediately phoned friends to discuss how they could get on the passenger lists for the Moon or Mars so they could survive! This program is available on the Internet. It demonstrates well how gullible most people are when they have little knowledge of reality. Today's Climate Change and the benefits of Alternative Energy (that all rely on coal-fired power stations) is a real-life, modern-day version of this.

These media programs are just two simple examples that illustrate why scientific illiteracy is so dangerous. If most people were scientifically literate, such programs would not gain an audience other than for light entertainment.

Across the world, scientific illiteracy has become increasingly worse over the last two generations because, in most democracies, students have not been taught scientific methodology. Instead, they are taught untruthful histories to suit a political narrative. They are not taught how to consider realistic alternatives to problems, and they are no longer taught how to question what they are told. On top of this, the media has strongly encouraged the population to only be interested in mind-numbing social entertainment, sport, and *un*Reality shows at the expense of learning about intelligent issues. Those who know some science typically only know about it in their field of interest or due to the type of work they do, so they too are easily hood-winked on subjects outside what they know.

This lack of general scientific knowledge combined with people thinking that they can believe what the mainstream media tells them has led many of the public to believe in the following myths.

RECENT MYTHS ABOUT WORLD-THREATENING CATASTROPHES

1. In the 1960s, **a new Ice Age was predicted to destroy the world.**

   The Real-World Result: From the 1970s to the mid-2000s the reverse happened! The climate became warmer and drier due to the Sun becoming very active. Due to this, the elite globalists bankers who developed this fake fear, changed the threat of a new ice age to 'Global Warming' caused by manmade CO_2 rising by a minute amount. This became a new catastrophe. But by the mid-2000s, the warming fear was also shown to be false because the climate became increasingly cooler again with huge snow storms for over 20 years due to the Sun becoming dormant again - like it was from the 1940 to the 1970s.

2. **In the 1970s ultra-green organizations and the elite globalists promoted the fear that acid rain caused by industry would kill all forests worldwide.**

The Real-World Result: No forests have died from acid rain. New technology together with Clean Air and Water legislation has greatly reduced air and water pollution generally.

3. **Air and water pollution would become extreme due to pollution from burning fossil fuels and the bulldozing of forests.**

   The Real-World Result: Our cities and much of the world environment in developed countries has become much cleaner due to anti-pollution laws and new anti-pollution technology that has been introduced in most western countries. Satellites show that Earth has become much greener, and forests have enlarged in total area.

4. **Most species would become extinct due to human development.**

   The Real-World Result: Overall, more species exist today because humans have created many more habitats in which they thrive. Many laws have been introduced to protect most endangered species.

5. **Overpopulation of the world would cause hundreds of millions of people to starve, and pollution to run rampant.**

   The Real-World Result: With new farming methods, fertilizers, irrigation, genetically developed crops, and new farming technologies, food production has skyrocketed, thereby eliminating mass starvation. This is despite world population having doubled when this scare was raised. Due to a greening of the planet through slightly increased CO^2 from technology, forests are now growing more quickly and more healthily. This is because slightly higher CO^2 helps plants grow faster and with less water. Despite claims by the Elite in the 1970s that millions would die from starvation due to overpopulation, there has been a massive *over*production of food. Food waste globally now runs into hundreds of billions of dollars annually. In the 70s, the world population was around 3 billion. Today it is almost 8 billion, yet we have far more food produced on less land, and poverty has been drastically reduced! We have better health, longer lives, and far less pollution. With more people, we discover more ways to make the world better.

The real threat is not overpopulation, but ***depopulation*, especially of Western countries. This is** due to too many self-centered young people not trusting marriage, males being fearful of losing any wealth they have to a female partner in a divorce, and increasing numbers of females enjoying their freedom. They do not want to be tied down to children or a permanent partner like their mothers were in marriages that often become unloving relationships. Also, women are now more educated, so they can support themselves, and they have social security in times of need, so they do not need a male for support. Many have chosen not to take on the responsibility of children. Young generations in the West are less likely to commit to marriage, and therefore children. These reasons are causing western nations to have reproduction rates well below 2.1 children per family, which is the minimum to have a stable population. However, Muslim families are having between 3 to 7 children, so they will quickly outbreed

Western nations. They are already starting to take over some western governments by having Muslim politicians elected due to them have ever-growing numbers of voters in certain electorates.

6. **In the 70s and 80s, the Club of Rome globalists predicted that the world would run out of resources such as fossil fuels and metals if the world's population was not drastically reduced quickly.**

   The Real-World Result: New technologies discovered numerous new deposits of coal, oil, and gas, as well as minerals of all types. New mining methods and technology made them easier and cheaper to extract.

7. **A growing ozone 'hole' would make living outdoors unlivable.**

   The Real-World Result: As the Sun became less active, the slight thinning of ozone over the south pole returned to normal.

8. **The 2000 (2YK) Bug would destroy our computer age** unless a trillion dollars was spent by industry, ordinary people, and governments to avoid this disaster.

   The Real-World Result: The money was spent was wasted because nothing at all happened!

9. **Manmade Global Warming will cause the ice caps to melt, resulting in drastic sea level rise, which will destroy coastal cities around the world by 2000.**

   The Real-World Result: There has been no sea level rise for over a century despite false claims to the contrary by climate alarmists. Even places just 1m above sea level have not experienced any sea level rise – only occasionally do short-duration storm surges, which are natural, cause king tides to rise but only for hours or a day. Some island nations such as the Maldives have mined their natural coral reefs to make cheap cement rather than import expensive supplies from other countries. They now have no storm surge protection from large waves produced by cyclones. To obtain large foreign aid donations from other countries, these island nations' councils create the misleading appearance that the sea level is rising. It's a fiction that serves them well because western nations pay them unjustified compensation to build more billion-dollar condominiums. Western nations have woken up to this ruse. The icecaps are not melting: the depth of ice at both poles has increased! Glaciers are not retreating: most are surging due to increasingly colder winters. Sea level is not rising since 2008.

10. **There would be water wars by the mid-2000s due to a lack of water caused by Global Warming drying out the planet.**

    The Real-World Result: There has been an overabundance of water in most countries most of the time due to there being numerous floods since the Sun became dormant. This allows more cosmic rays to enter the atmosphere, and they produce more cloud, rain and snow. Most dams have been close to full or overflowing since 2008. But there are always places that go against the general global trend, and they can have droughts for a while.

WHAT ABOUT ROSWELL'S CRASHED SPACECRAFT AND ALIEN BODIES?

After the end of WWII in 1945, the US military developed large hydrogen-filled weather balloons to stay aloft at high altitudes of 30,000 m (100,000') for weeks on end. This was so they could monitor sudden changes in electric and magnetic fields in the atmosphere to detect nuclear bomb tests by a foreign power. To have balloons stay aloft for that long, they experimented with developing balloon skins made of thin, lightweight paper with a coating of ultra-thin aluminum sprayed onto the paper to slow the leakage of hydrogen through the balloon's membrane. This coating was the forerunner to the aluminized birthday message balloons sold today.

When sunlight glinted off these balloons, they were often misidentified as UFOs. Their highly reflective skin, radio antennae, and instrument packages suspended below them were unlike anything people had seen before. When one of them came down outside Roswell, New Mexico, USA in 1947, a local farmer who found a part of the skin jumped to the conclusion that the material was part of some advanced technology from a crashed alien spacecraft. This was enhanced by the US Air Force conducting secret test flights of experimental aircraft from Area 51. By coincidence, in the same area soon after this, some newly developed crash dummies that were used by the Air Force for experimental ejections from aircraft were recovered by the military. A local small-town newspaper reporter who was desperate for news combined the two events and drew the conclusion that a flying saucer had crashed and that the dummies must have been dead alien bodies from the spacecraft. The reporter did not get to see the dummies because they were taken away under tight security. This small incident started the Roswell incident of a government cover-up. Over a decade later some people who worked in movie special effects created the cheap humanoid figures shown below, with one having an injured leg. They were claimed to be alien bodies recovered from the so-called crashed alien spacecraft that the military was keeping secret. A short, very poor-quality video of the dead alien on the right was made. It can be found on the Internet as well as in some poor-quality documentaries about UFOs and aliens. For decades, the 'Roswell incident' became a religion for conspiracy-loving UFO enthusiasts who believed that aliens visit our planet. The Roswell myth has largely withered away.

A part of the remains of an aluminized onion-skin paper weather balloon was believed by some to be evidence of high technology from an advanced alien spacecraft that crashed near Roswell, New Mexico.

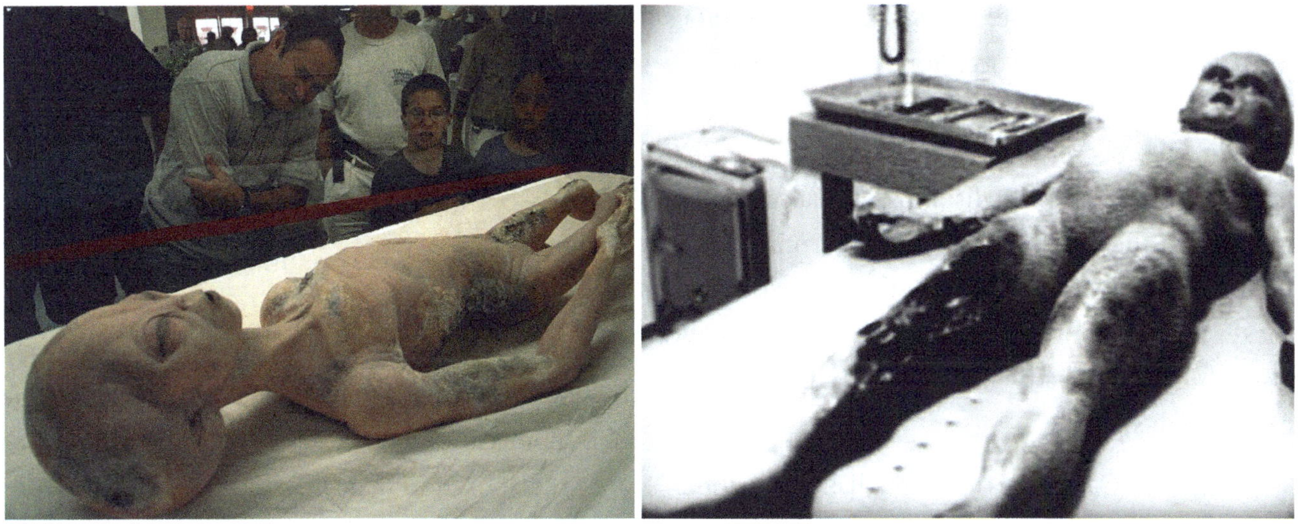

Note how different the claimed alien bodies are that supposedly came from a crashed spacecraft in Roswell, New Mexico in 1947. The color version is in a display case. Imagine the one on the left having to give birth to an offspring that had a skull that large and a lower body that small!

Had the recovery of an alien spacecraft and alien bodies from Roswell been real, this would have been the biggest story of all time, and it would have frequently been front-page news around the world for years, but it wasn't. Major nations at the UN would have demanded that the US give their scientists access to the alien bodies and spacecraft technology because there could potentially be a huge risk of alien microbes affecting all people and animals around the globe. As well, the US would have sole access to the spacecraft technology, and this could potentially allow the US to take over the world.

If there really was a crashed alien spacecraft, tens of thousands of experts in many fields from many nations would have been seconded to investigate this discovery and plan a course of action in case more craft were on their way. This would have resulted in numerous leaks and pillow talk that would have been leaked to the media. But the story didn't rate in any respectable newspaper or television news anywhere in the world. It was given no credence by any government because it was obvious to all informed, intelligent people that it was a beat-up with no evidence to support it.

SO, WHAT ARE THE UFOS THAT PEOPLE SEE?

Between the mid-1950s and the early 1960s, many UFO sightings were reported as being cigar shaped flying discs, which were often sighted during daylight. These were commercial aircraft seen edge-on in the distance. In those days, commercial airliners were new and few in number, so most of the public in country areas were not familiar with them. Air travel was expensive, and flights were infrequent, making commercial aircraft uncommon outside major cities. When these planes were seen side-on in the distance, and often under hazy conditions, their thin wings became invisible against the cigar-shaped fuselage. And because the tails of the planes were smaller back then and not brightly painted with the company logo as they are today, they were difficult to see in the distance. The cigar-shaped fuselages were interpreted by some to be flying saucers seen edgewise.

For a couple of decades after the late 1960s during the Cold War, occasional secret nighttime military rocket launches were observed in the US and Europe. Secret US experimental aircraft that looked like UFOs from a distance, were also seen. In the time of the Cold War, several governments and their military wanted to keep their experimental rocket launches and new aircraft secret, so they did not debunk the idea of alien UFOs being responsible for such sightings.

Faked UFO images were published by indiscriminate UFO magazines and by a media that pushed sensational news items to increase their sales to gullible readers.

Today, using Photoshop, UFO images are easily created by unscrupulous hoaxers who sell their pictures and stories to UFO magazines. Self-deceiving UFO enthusiasts swallow these fake images without having any need to verify if they are real because they want so much to believe that they are.

In recent times, UFO sightings have become considerably less in number despite there being far larger populations of people who have smartphones with cameras that could capture such sightings. Nevertheless, a small number of fake images and videos are still being produced by some computer artists to satisfy greatly diminishing audiences who want to see 'evidence' of alien spacecraft.

Dedicated sky watchers such as experienced amateur astronomers, pilots, and weather-watchers observe the sky far more than any other group of people, yet they almost never report UFOs because they can quickly identify uncommon sightings. (I have not included professional astronomers because, over the last few decades, many now work indoors on computers connected to remote telescopes, so they do not have a lot of experience in directly observing the night sky or identifying unusual natural night sky phenomena.)

During my adult lifetime of investigating UFO sightings, I have found that most people who think they have seen a UFO are unknowledgeable about science and most aspects of nature, so they are prone to jump to irrational explanations. They have little ability to discriminate when they see lights or objects that they cannot easily identify. Those who are predisposed to believing in extraterrestrials are not objective observers, and they are very prone to exaggeration.

A number of saucer-shaped flying wing experimental aircraft were designed and built by the US Air Force. Occasionally, they were spotted in the distance by civilians who thought they were UFOs. These crafts turned out to be too unstable in flight to be practical.

Left: **For a worldwide marketing campaign in the late 1990s, Whitman's Chocolates had an internally illuminated blimp flying over many cities at night. When the blimp was seen in the distance, the signage could not be seen so it looked like a UFO. Many people were fooled into believing it was one. As it turned around, it would change from a cigar shape to spherical. This further added to the perception that it was a UFO. When seen in the distance, I found it hard to realize what it was until it moved closer. Only then could the logo lettering be made out.**

Right: **A typical, fake flying saucer image created on a computer.**

Creating a Photoshopped image of a UFO over a scene is easy for anyone moderately competent in producing computer art. It is a simple matter of dropping a UFO image into a background, as has been done in this illustration. When looking at UFO photos, it becomes obvious that most UFOs have the same general form. This is because those shapes are the easiest to produce. UFO computer artwork is readily available on the Internet to create fake images like this one.

MAJOR PROOF THAT ALIEN SPACECRAFT DO NOT EXIST

Today, there are thousands of satellites in orbit around the globe photographing every part of our planet in many different wavelengths, day and night, and in incredible detail, yet overall the years this has been happening, not once has a UFO ever been recorded! This is undeniable proof that alien spacecraft are not flying around observing us, or flying with or past commercial aircraft, or past US fighters, or over people, or landing in deserted places. As well, many astronomical observatories film the sky every night and nor have they ever recorded a single UFO. THAT SAYS IT ALL!

As I have explained earlier in this chapter, aliens have no reason to come here as Earth has nothing that is not common in the universe. It would take them an eternity to cross space, so they would never waste their time doing so in their 'tin can' spacecraft. And no matter how advanced technology may become, no spacecraft can exceed the speed of light, which is very slow when interstellar distances are taken into account. Even approaching the speed light in something as massive as a small or large spaceship is against the most fundamental laws of physics.

We may as well accept the reality that UFOs don't exist, despite the fact that some people love to dream that they might.

SPACE TRAVEL

Chapter 11

WILL HUMANS EXPLORE MARS AND OTHER WORLDS?

Many people want to see a manned mission to explore Mars. I have since I was a teenager, and especially after I watched the Apollo Moon missions. Those who want to see humans exploring Mars will not want to hear how impossible this will be because humans would die in the process for the multitude of reasons explained in this chapter. The cost alone would be prohibitive. The hurdles that need to be overcome are impossibly high. Robot technology is advancing so fast that before 2035, highly intelligent robots will be far more competent than any human, so they will explore Mars far sooner, far more efficiently, much more effectively, and at a far, far cheaper cost, with very little risk.

Within a decade, the Moon will be mined robotically for water and most of the minerals needed to build facilities there for human habitation, lunar tourism, and exploration. It will also have scientific research centers as well as launching facilities to send probes into the far reaches of the solar system. By then, scenic flights around the Moon and landing on it will be straightforward and safe. It will only take 2-3 days to go to the Moon, and the same time to return. **However, a manned mission to Mars is a very different story because it would take 2-3 years to get there, explore it briefly, and return. Every aspect of a manned mission would be exceedingly more difficult than going to the Moon. As well, there would be an almost certain risk of failure.**

Despite seven decades of fanciful talk about sending a manned mission to Mars, no detailed plan has ever been developed.

Mars is a bitterly cold, desert world, with only 1/100th Earth's atmosphere, no oxygen, no liquid water, no vegetation, and only ¼ Earth's gravity.

MANY MANNED MARS MISSIONS HAVE BEEN PROPOSED FOR DECADES

Since the middle of the 20th Century, many vague plans for a manned mission to Mars have been announced by several nations, and recently by private companies. But not one has a practical means of doing so, and none have come off the drawing board. (See 'Human missions to Mars' on Wikipedia.) Those who push for a Mars mission, including some at NASA and Elon Musk, do so to obtain research and development grants, which are paid for by taxpayers. Serious researchers know that there is no realistic possibility of a manned mission to Mars. The problems highlighted from hereon clearly expose the realities involved that make a manned mission to Mars not possible.

Approximately 99% of the Saturn 5 rocket's payload consisted of fuel to propel the very small lunar module capsule at the very top of the rocket. The Lunar module and the Lunar Lander are shown in the inset. Nearly all of the rocket's fuel was consumed in lifting the fuel itself off the Earth and into space! Imagine how much fuel would be needed to propel a much, much larger payload to go all the way to Mars, land safely, and take off and return to Earth. Any small increase in the payload requires a massive increase in the fuel required. A manned mission to Mars is implausible for numerous practical reasons, and it would be financially unaffordable. Credit: NASA

Proposals by two private companies have been put forward to attract public donations and NASA research funds. The ***Mars One Project*** was a private concern run by a Dutch entrepreneur **Bas Lansdorp** who claimed he would land the first explorers on Mars by 2020! But this project had almost no funds, and they had not developed any hardware or solved any of the problems for such a massive venture. Mars One did not developed any serious plan to achieve such an enormous task. Many people saw Mars One as a publicity-seeking scam to promote the company and to obtain funds from gullible space enthusiasts. The date was pushed back to 2028, but this would be an impossible timeframe to meet. When the company asked for expressions of interest from those wanting to go to Mars, the company stated that they had well over 200,000 people register. The real number was only around 2,000. Those applying were naïve, having no idea how impossibly hard and deadly a trip to Mars would be.

To a public ignorant about the realities of a mission to Mars, a movie like the very popular '***The Martian***' made it seem plausible that an astronaut could survive on Mars – even alone! But nearly every single aspect of that movie was completely unrealistic. Of course, these types of movies are not made to be realistic, but entertaining, and it was. Audiences loved it because it showed a human overcoming impossible odds.

Elon Musk's company **Space X** claimed at the outset that it could send many people to Mars for a total sum of a measly $6 billion! As of 2019, Space X had not even raised a million dollars to do this! They did not secure the expected sponsorship for a *un*'reality' TV program that would show the development of the project over the years. Television companies did not believe the project was real or doable. (See comments about how realistic Musk's early plans are on the Internet.) Musk's plan is to populate Mars with a *million* people over the next 40 to 100 years. Everything will change drastically for mankind over the next 20 years, so it is impossible to make any predictions that far into the future. Space X may not even exist by 2030. Going to Mars is a childhood dream for Musk, like it was for me before I realized the enormous difficulties involved. Initially, Musk planned to sell tickets for $200,000, but they were *one-way* tickets! This is because he has no realistic way of getting them there alive, let alone back to Earth! He is a leader in rockets, communication satellites, artificial intelligence, and driverless cars, so he should be able to see the huge pitfalls in his dream to populate Mars. These problems are exposed in this chapter. Musk is a very intelligent person, a very clever business developer, and he puts everything he has into his dreams, but he must surely be aware of the enormous difficulties involved in building spaceships to send humans to Mars, let alone developing a base there and returning them. He must surely know that such a mission would have an extremely high probability of being fatal for all and prohibitively costly. Intelligent robots will do it far more quickly, inexpensively, and safely. As we will see, humans today, and for some time to come, are just not built for space travel.

NASA says it is developing its Orion capsule to take four astronauts to Mars. They could be saying this to keep research funds coming in from the government. If NASA were to attempt a manned Mars mission, it would take them decades to be ready for such a mission, but in that time, extremely advanced intelligent robots will have done most of the exploration hundreds of thousands of times more effectively than humans could, probably before 2035. The following pages explain the major problems that a manned mission would encounter and why a robotic mission would be far superior and very cost-effective. **Extremely hardy, versatile, highly intelligent, sophisticated robots will probably have explored pretty much all of Mars by 2035.** The extraordinary exponential rate of the advancement of intelligent robots and the dramatic effect they will have on space exploration cannot be overstated.

IS THERE A POLITICAL INCENTIVE FOR HUMANS TO GO TO MARS?

Politically, no government has an imperative to get to Mars despite one American president after another claiming they will send astronauts there. The 1960s and 1970s was the time of the Cold War, so there was a race by the Americans and Russians to be the first to put a man on the Moon to prove who was the most technologically superior nation. They hoped that the winner would deter their opponent from engaging in a pre-emptive nuclear attack on them. This situation does not exist today, so there is no imperative for the US government to go to Mars other than for scientific research.

NASA's first mission to the Moon was in 1969. Five more missions followed in the 1970s, yet after almost 50 years, NASA has not been back to set up even a small human habitation base there, despite how close the Moon is. Most of NASA's technology from the 70s was allowed to deteriorate, and some of it has been destroyed because there was no plan to go back to the Moon for a long time to come because America had proven it was technologically superior to Russia. NASA decided to concentrate on its Space Shuttle program, planetary missions, Earth monitoring satellites, and space telescopes. But President Trump said he wanted a manned base on the Moon quickly because China is planning to do the same.

If NASA, Space X or Blue Origin have to develop the technology to go back to our nearby Moon at very considerable cost, how can they possibly justify an extraordinarily more complicated and extremely expensive, highly dangerous manned mission to distant Mars? NASA's so-called 'Mars' rover, which is now in development, was revealed to the media in 2017. It will more likely be used for a return mission to the Moon, not Mars. It is much larger and heavier than the rovers the Apollo astronauts used, so it will be very expensive to launch and send to Mars. A few rovers would be required for Mars.

IS THE COST OF A MANNED MISSION TO MARS AFFORDABLE?

The total cost of *all* the Apollo missions to the Moon was around $150 billion in today's money. NASA estimates that one manned mission to Mars will cost $700 billion to $1.5 trillion!! But this is a totally unrealistic budget, as you will see. The real cost will likely be over a quadrillion! But NASA cannot divulge that because they will not get any funding for the research and development work they are doing. Due to the US' enormous debt, the budget for a manned Mars mission will never get congressional approval. NASA is already fighting for every penny it can get for much simpler, well tested, and far less expensive probe missions to asteroids, comets, and the inner planets.

If a manned Mars mission could ever have a chance of getting off the ground, it would have to be jointly funded by many countries, and probably also by private enterprise. Assuming the funds could be found, negotiating such a complex arrangement would slow down the development time by another decade. But over that time, technological development would advance greatly, so this should considerably reduce current costs, and it should speed up technological solutions. Space X is rapidly developing its new large rockets and new fuels so this will speed up missions into the solar system, but this will not solve how fragile humans are physically and metally compared to intelligent robots. **By 2030, technological advances in intelligent robot technology will be so sophisticated and so much cheaper than sending humans to Mars that a manned mission will not be viable. Robots will provide a much higher scientific return at a far lower cost while providing a much lower risk of failure than a manned mission. Robots will be self sufficient needing very little backup assistance compared to the enormous amount required for humans.**

HOW MUCH FUEL WOULD BE REQUIRED?

Just to go to the Moon with the Apollo mission's small payload required the gigantic Saturn 5 rocket, the largest rocket ever built! **A rocket to go to Mars would require in the order of hundreds of times as much fuel as that required for each Moon mission. And for such a long duration mission, at least a thousand times as much in the way of supplies would be required. And far more equipment would be needed.** Even if the payload for Mars was lifted out of Earth's gravity with scores of launches and then assembled in Earth's orbit, this would require even more fuel making the total cost enormous.

The Apollo Moon missions required only fuel for a small, lightweight space capsule to carry only three astronauts and just enough supplies for 8 to 12 days, as well as a number of small, scientific experiments, and a rover that could carry two astronauts. However, to explore Mars will require far more of everything. One huge rocket like the Titan used for the Moon missions will not be anywhere near large enough. Mars will require a fleet of them! **Mars is roughly 240 times further away than the Moon,**

so a manned mission would require that much more fuel for starters! But the Mars ships will have far more mass than the Lunar Module and lander because a manned mission to Mars would need 6 – 12 crew, and much more equipment, as well as two or more much stronger and heavier rovers, science equipment, living quarters, and enough food, water, oxygen, nitrogen, and medical supplies to last a minimum of 400 days, or up to three years, depending on the mission route. All this mass will greatly add to the amount of fuel required. It will be hundreds of times as much as the Moon missions.

Mars has a three times higher gravity than the Moon, so it will require far more fuel for landings and take-offs. It also has an atmosphere, so the craft will need heat shields, so they do not burn up upon entry. The landing craft for Mars would have to be far larger, much stronger, and therefore much heavier than the lunar landing module, so this will also require much more fuel. Great quantities of fuel would be required for: **1.** launching from Earth **2.** landing on Mars **3.** blasting off from Mars **4.** the return journey to Earth, and **5.** re-entry into Earth's atmosphere and landing safely. For the return trip, the astronaut's capsule will require far more fuel for lift-off than the Lunar lander required due to its much greater mass and due to Mars' gravity being higher than the Moon's. All the fuel, food, water, oxygen, nitrogen and general supplies will have to be brought all the way from Earth for the outward journey, as well as the time on Mars, and what is needed for the return journey. And it has to be enough for a much larger number of crew and for a much longer duration. This is an extraordinary amount of mass that requires much fuels. But what is forgotten is that the more fuel that is required for these purposes, the more fuel is required to lift the fuel itself. This is like a dog chasing its tail. Around 99% of the fuel needed would be required to propel the fuel!!

Even with technological costs coming down dramatically over the next two decades due to incredible advances in computer intelligence and robotics, trying to take humans to Mars will be extremely expensive.

When the following are taken into account, **the total cost would be unaffordable** - even if all nations contributed to it, which of course they would not. They would see a manned mission to Mars as being a huge waste of their money with an extremely high risk of failure.

A single problem with any one of the numerous critical parts of the mission would jeopardize the entire mission. Despite the enormous cost, there would be no commercial payoff so investors would not support this venture.

ROBOTS VS HUMANS

Sending inexpensive robots would require only a modest amount of fuel. They could stay there, so there would be no fuel required to bring them back. Furthermore, robots could stay there for decades and explore the whole planet, whereas humans could only stay for a couple of months and they could not venture far from their landing base in a rover as it would need to be recharged. Astronauts could not stay out for long because they would have to eat, drink, go to the toilet, and sleep whereas robots could work day and night. And radiation damage would accumulate. The power in their backpacks would not last long, and they would freeze during the night. Robots do not require food, water, sleep, or living quarters, and other equipment. They would be self-contained and able to roam endlessly. Robots would not have psychological problems, which humans would have due to the long journey out and back, and having to live in a very confined space. Robots would not get sick from weightlessness from Mars' poisonous soil and dust.

Each landing module would have to operate flawlessly, because there would be little redundancy to keep weight down and restrain costs. The surface landers would need to carry the crew's living quarters, laboratories, heavy-duty rovers, and all the food, water, air, and medical supplies to last, not just for a week as with the lunar missions, but for a minimum of 400 days for the shortest route! All the equipment for all the experiments would have to be included. The crew's capsules would have to be much larger than those for the lunar missions because they would have to accommodate at least 4-8 crew somewhat comfortably. All modules would have to land very close together. This could be difficult.

WHAT ARE THE REALITIES OF A MANNED MISSION TO MARS?

Once people know the realities of a manned Mars mission, they soon realize that human exploration of Mars would be nothing like the fantasies presented in science fiction movies, or the plans marketed by Mars One, Space X, and even NASA. Why? Consider the following:

TECHNICAL RISKS

Some of the orbiters and landers sent to Mars over the last few decades by the US, USSR, and the EU have suffered unexpected failures. The 2016 NASA mission had a relatively simple parachute failure that caused the total loss of a sophisticated landing probe that took six years to plan, test, construct, and travel to Mars. The failure was due to a simple air pressure sensor causing the parachute to open at the wrong altitude. Even with present-day robotic missions, there are so many complex, crucial steps that it only takes one to fail for the whole mission to be lost. A manned mission would increase the risk of failure more than a *thousand times* compared to a robotic mission.

TRAVEL TIMES TO MARS ARE FAR TOO LONG TO BE PRACTICAL

The duration of a mission depends on the spacecraft's acceleration, its trajectory, and how close Mars comes to Earth at its closest point. A close approach to Mars only occurs every two years. Because Mars and Earth have oval orbits, the closest of all approaches only occurs every 15 years. Missions have to fit into these windows of opportunity because the closer Mars is to Earth, the less fuel is required to get there, and importantly, the less travel time the mission will require.

To travel to Mars along the lowest energy orbit to minimize fuel would require 9 months to get there and 9 months for the return voyage. On top of this, the crew would have to spend a grueling 500 days on the surface waiting for the right configuration of Earth and Mars for their return trip. The total time away would be nearly 3 years! There is no mission plan that would permit astronauts to survive such a long time away from Earth. To have a shorter duration trip of 400 days would require a lot more fuel for acceleration and deceleration. This would make it much more costly. However, this shorter mission would provide little time on the surface for exploration, thereby making the scientific returns much less than what might be achieved by a longer mission.

HIGH-VELOCITY SPACE DRIVE TECHNOLOGY IS UNLIKELY TO BE INVENTED

Space travel enthusiasts who are eager to have humans explore Mars place their faith in future technology greatly reducing the time it takes to go to and from Mars. They hope that technology will become so advanced so quickly that new space drives will be invented that will reduce travel times considerably. They also hope that such drives will be so powerful that they will be able to carry much bigger payloads than is currently possible.

There will be huge advances in many technologies, but there is nothing on the horizon to suggest that there will be any major advances in space drives that will provide much shorter travel times within coming decades. Unfortunately, interplanetary space drives and spacecraft cannot ignore the laws of physics. For the foreseeable future, the only means of creating enough propulsion for human space travel is to convert mass into energy in a controlled manner, as with rocket fuel. The faster a craft can accelerate, the more energy it needs, and the more energy it requires to decelerate at its destination. There is no other viable form of energy than rockets. Ion drives have very slow acceleration, and they are not powerful enough to accelerate a large spaceship fast enough. They are only useful for small, low mass probes that can take their time to reach their destination. Ion drives need large and heavy solar cells for all the equipment that is needed to go to Mars. These cells would take a long time to accelerate. A nuclear reactor engine is also far too slow, and it is impractical because it would be very difficult to protect the crew from radiation.

To overcome the problem of needing so much fuel, a solution would be to somehow extract a lot of energy from space. This would eliminate the need for heavy chemical rocket fuels. But there is very little energy in empty space, so this is not possible. Sunlight is a feeble energy source, and it fades out very quickly beyond Earth, so it too is not a viable energy source.

FOOD

A round trip mission would require very large quantities of food for the crew. **Just to transport 1 kilogram (2.2 lbs) of food to Mars one way would cost $2 million!** So, imagine how enormous the cost would be to transport and store all the food needed for an entire Mars mission lasting for 1.1 years to 3 years for say 6-8 crew. A smaller crew of only four that NASA has considered will be too risky due to the possibility of losing one or more specialized people.

There is **no ability to grow food in the Martin soil because it is toxic**, so all food would have to be brought from Earth for the entire duration of the mission. Plants cannot be grown there in earth soil efficiently because sunlight that far from the Sun is too weak. As plants would need soil and fertilizers brought from Earth, the weight of this would provide no net gain. If the plants did not grow well, the crew would starve. Trying to grow food there is too risky.

Food that astronauts would want to eat, which has to last for at least 400 days and be palatable, has not yet been developed. NASA found that most air-sealed meals start to spoil after a year, no matter how well it has been preserved. This alone would stop a manned mission.

All food has to be sterilized and sealed. **If anything went wrong with even a part of the food, the astronauts would starve to death** because there is nothing on Mars that can be eaten. In a short time, the limited selection of bland food would soon become a major problem psychologically, especially on such a long-duration mission. (A robotic mission would not require any food, or water, or cooking equipment.

Food must be eaten even if a crew member does not like it or does not feel hungry. There would not be enough food to waste. Used food could not be left in a refuse bin because it will quickly stink out the small cabin. In the weightlessness of space, crumbs are a problem as they float around in the cabin, so food has to be prepared that does not have a crumbling topping or texture. When in space, weightlessness means liquids cannot be poured because droplets would float around in the cabin and coat the interior surfaces, and they are difficult to get rid of. Liquids must be kept in sealed containers and sucked out via a tube. Food smells cannot be evacuated by opening a window. It takes time for air filters to remove smells.

The crew will not be allowed to boil water, use a conventional oven, or have a refrigerator in the capsule due to the power these items consume. Food that needs to be kept cold would have to be stored outside the capsule in the cold while on Mars. Keeping the cabin at room temperature would require most of the energy produced from solar cells. A naked flame could not be used due to the fire risk it would pose in the event of an accident.

Toilets are a major inconvenience while in space and while on Mars because, in a very restricted space capsule, it is very hard to provide much privacy and to contain bad odors. (Robots do not require costly sanitation equipment.)

LIFE-THREATENING RISKS DUE TO LOW AND ZERO GRAVITY

A mission to Mars would require astronauts to be away for at least more than a year. For the bulk of that time, they would be in zero gravity in space on the way to and from Mars. And when on Mars, the gravity there is only a quarter of that on Earth. Both zero and low gravity cause serious life-threatening health problems. Astronauts cannot stay in space for too long because their bones become thin and brittle. Astronauts **in space lose up to 20% of their muscle mass in just one week!** This is due to not having to hold the weight of their bodies up against gravity. Their **bones deteriorate ten times more quickly than people with osteoporosis!** When the Mars astronauts return to Earth, their bones can snap, and their weak muscles can tear even when just trying to walk or twist around. Astronauts in the International Space Station that are in zero gravity for much shorter times exercise to try to reduce the rate at which their bones and muscles waste away, but this does not have much effect on overcoming this problem. After being in zero to low gravity for all the time, a Mars mission will cause **the crew's lung diaphragm muscle and heart to be so weak that they will have trouble breathing and pumping blood when they return to Earth's crushing gravity. Even when they land on Mars after several months in zero G, they will feel weak, even in Mars' low gravity. They may not be able to do much exploring for a week or two because they will feel weak and bilious.**

In space, astronauts often do not feel hungry because they use little energy due to not having to fight against gravity. Not eating enough leads to a lack of energy.

Unlike the space station, there will not be enough room for much more than a bike or weights to use for exercise in the Mars capsule, but these are of little benefit. (Robots do not lose their strength in low gravity.)

In space, body fluids are not held down in the belly. This makes astronauts feel like they are upside down. Fluids move up into the sinuses, making them feel like they have a constant cold. This is hard to get used to.

They are also likely to have **distorted vision** due to their eyeballs expanding and going out of shape. In low gravity, CO^2 can pool around the head, decreasing the amount of oxygen in the blood. This causes rapid breathing and a high heart rate, as well as nausea, vomiting, clumsiness, stress, and fatigue.

In space, **the immune system weakens, and very painful kidney stones are more likely. The chance of heart disease increases.**

Due to there being no up or down in weightlessness, this causes **astronauts to suffer nausea**. Until they become accustomed to this over a few days, they can have an attack of vomiting come on quickly before they can grab a sick bag to capture their stomach material. In space, this floats around in the cabin, and it is very hard to get rid of it. It makes the air putrid inducing other crew members to have a similar attack. These problems would be intolerable in a cramped space on a very long journey. (This problem does not affect robots.)

There is no practical way to create artificial gravity on such a small spacecraft. And even on a large

The cosmonauts here returning from a Soyuz mission are so weak that they have to be lifted into chairs, as they cannot walk. They take months to recover. NASA astronauts that have had long missions in the International Space Station also suffer this problem.

space station, the effects of artificial gravity such as dizziness, headaches, nausea, and disorientation are hard to get used to.

WORKING ON MARS

While on the surface of Mars, astronauts would suffer from being in reduced gravity. Their weak muscles and bones will make them unfit to do work. On Mars' surface, they will find it easier to skip along as the Apollo astronauts did on the Moon because trying to walk is more tiresome. And due to Mars' low gravity, they will only have a quarter of the traction on the ground that they have on Earth, so they will find physical work hard to do without slipping. Their very restrictive space suits will make work difficult because they won't be able to bend down without kneeling. Getting up is difficult, as one cannot easily keep one's balance with a heavy backpack. And they cannot twist around in a pressurized suit. Being in Mars' low gravity for months will induce vertigo and dizziness in some of the crew. All these physical problems will mean that astronauts will not be able to get much work done when they don't feel well. (Robots will not have any of these problems.)

During the entire mission, the smell of food, flatulence, toilet use, body odor, sweat, bad breath, sickness, and engine oil lingers in the air.

RADIATION EXPOSURE

Astronauts on long missions will be exposed to high-energy particles coming from solar flares and the solar wind as well as cosmic rays from exploding supernovas all over the universe. Both of these forms of radiation will go straight through the space capsule, their suits, and through the crew's bodies damaging some of their cells, including brain neurons. **This would cause some loss of eyesight as cells in the crew's eyes are increasingly damaged by radiation. When they sleep, they will see flashes of light as cells in the retina are damaged. They could also suffer cataracts from the Sun's unfiltered UV light due to Mars having no substantial atmosphere. Radiation exposure will cause astronauts to have a high probability of getting cancers and having brain damage. This can cause long-term dementia and increased anxiety. Their DNA may be irreparably damaged.** This can cause them to have **cancers**, and it may affect their reproductive organs, and therefore any future offspring. **Cognition** is also affected by radiation exposure, as it kills off brain cells. To reduce this risk, radiation shielding would have to be very thick and heavy. To get it into orbit would be extremely costly, yet it would only be partially effective. **Radiation sickness is a serious problem that could easily destroy a mission.** Radiation doses on a Mars mission would be 10 times that of astronauts on the ISS because it orbits inside Earth's Van Allen radiation belts that absorb much radiation. (Radiation problems will not affect robots, as only a very small area of their susceptible chips would need to be shielded.)

PSYCHOLOGICAL PROBLEMS

Contact with Earth would be important for humans, but due to Earth and Mars being far apart, there would be very frustrating delay times of up to 20 minutes both ways when they are on opposite sides of the Sun. Imagine saying a sentence and having to wait up to 40 minutes to hear a reply. (This would not be a problem for robots.)

When in space, the crew would not have ever-changing views of our beautiful and detailed Earth to look at just below them, as crews on the space station enjoy. Mars crews would see only black space and faint stars through a small window. This would make their isolation difficult to deal with, even for crews that are trained to deal with this. Boredom can become a major problem on both the outbound and return journeys. (Robots will not suffer such psychological problems.)

The size of a space capsule designed to hold four to possibly eight or twelve crew would have to be minimal in size in order to keep the payload as low as possible to try to restrain the cost and the enormous amount of fuel and supplies required. In such a confined space, flatulence frequently makes space travel most unpleasant. This was the case on the Moon missions, and it remains so for space station crews, even though the ISS is much larger.

There is no room in such a small capsule for privacy. Even going to the toilet will not offer much privacy. This would be the case during the entire time they would be away. Over that time, **both male and female astronauts will have sexual needs that will be very difficult to fulfill due to the lack of privacy**. They will need bromide in their drinks to suppress natural sexual urges, otherwise there is likely to be outbreaks of serious infighting. (Robots do not need privacy.)

The crew would not be able to escape one another. Such conditions over such a long period can cause a **high level of stress for the crew. Small annoyances between different personalities are greatly magnified under such conditions. Very serious differences of opinion and serious personality clashes can quickly lead to physical fighting, injury, depression, and even insanity, thereby seriously threatening the mission, even before arriving at Mars**. (Robots will not suffer this.)

Hibernation is often put up as a solution to overcome long space flights, but we know of no way to achieve this. Even if it were possible, the human body would deteriorate in hibernation due to inactivity and no gravity. (Such problems do not apply to robots.)

If a crew managed to successfully land on the surface of Mars, their risk of serious psychological problems would continue to grow. They would not have any more comfortable accommodation on Mars because they would probably have to live in their capsule. It would be enormously expensive to provide larger separate private living quarters, so **psychological problems would be likely to set in.** With no way out of this situation, this would be a recipe for disaster. (This would not be a problem for robots.)

The crew would have to be able to ignore the continual noise of the spacecraft, as do astronauts on the ISS. There would be no separate room for recreation or entertainment, no advanced medical facilities, and probably no research labs. (Robots will not need relaxation or medical facilities. They would have miniature research labs built into them.)

The death or a mental breakdown of a crew member would have devastating impacts on the remaining crew members and place the mission in jeopardy as each crew member would be chosen for their special abilities.

LONG DURATION ISOLATION

Being so far away from Earth and other people for such a long time would be very difficult for nearly all people, no matter how much they trained for this.

The **Biosphere 2 experiment** was conducted in 1991. It consisted of an enormous three-acre complex of glass domes, pyramids, and other shapes. It cost over $200 million to build this multi-story glass biosphere designed to be home to four men and four women to see if they could survive on their own on another world with no help from outside for two years as on a mission to Mars. It was a failure from the beginning! Within two weeks, a female crew member had accidentally cut the end of her finger off and had to be removed against the rules. Two weeks later, she returned with a duffle bag of extra supplies. Staff members made many deliveries to Biosphere 2 sending additional seeds, vitamins, mouse traps, and other supplies twice a month! This was against the rules, because this would not be possible on a space mission.

There was an unexplained loss of oxygen, which would be sure to happen on Mars due to the dust there making seals leak. By 18 months, the oxygen levels dropped so low that the crew could barely function. This was caused due to too much CO^2 in the air. Bacteria in the rainforest and savanna soil environments built up and consumed oxygen and produced CO^2. The plants should have produced oxygen through photosynthesis, but the CO^2 reacted with the concrete wall in the biosphere to form calcium carbonate and water. Once this cause was discovered by engineers and chemists on the outside, the oxygen levels still remained a life-threatening problem. There was also a dangerously high level of di-nitrogen oxide, which causes brain damage due to a reduction in vitamin B12. Outside engineers fitted a CO^2 scrubber, so the CO^2 problem could be artificially managed. All these things destroyed the scientific viability of the experiment,

There was a variety of plant, insect, and animal species, but the extinction rate was far higher than expected. Ants, cockroaches, and katydids ran rampant, and morning glory choked every other plant species.

The water became polluted with too many nutrients.

They could not produce enough food, so the crew was terribly malnourished, being little more than skin and bone over 2 years, despite receiving illegal outside help.

There were also major disputes between the crew and mission control.

Problems such as these and many more would plague a Mars mission, especially as it will be very cramped, with zero to low gravity. Deep space and Mars are very hostile unforgiving environments. The long duration time of a Mars mission significantly adds to the mission's chances being a failure. There will be no ability to say, "I'm out of here!"

The Biosphere 2 facility featured environments such as a 9,000 square-foot ocean, mangrove forest, and a desert environment. Credit: *Madman Films*

MEDICAL THREATS

If small amounts of liquid water exist on Mars during mid-summer (very small chance), this raises the **prospect of viruses and microbes existing in the Martian soil. If so, they could find their way into the capsule to fatally infect humans who would not have antibodies to fight them off. Infection could quickly destroy the entire mission. And perchlorates in the soil could poison the crew.** (Robots would not be affected.)

If **serious medical situations occur**, it would be very difficult to deal with this as there is no hospital backup for Mars astronauts. They would be on their own. Two crew members would have to be trained surgeons. It would be hoped that both of them did not have a serious medical issue at the same time. Serious health failures in crews wintering over in Antarctica have led to the death of two crew members. Winter weather conditions prevented the extraction of a dying scientist for several months, and another who developed cancer died later at home because he had a six months delay in getting treatment. (Robots will not have these problems.)

DUST

Mars is a planet-wide desert covered in fine rust-like dust. There are huge dust storms on Mars. At mid-summer in the Southern Hemisphere, when Mars is at its closest to the Sun, a planet-wide dust storm occurs for some weeks. There are also hundreds of thousands of whirlwinds on Mars that send dust high into Mars' very thin atmosphere. Astronauts spacesuits would be covered in dust each time they went outside. Despite all attempts to stop dust getting into the cabins, it will quickly cover the interior - just as Moondust did inside the Apollo lunar lander modules. Dust will get into spacesuits, and the airlocks in the living quarters, causing air pressure losses. It will stop air seals working, which could easily become fatal. (Robots can be designed to not have problems with dust.)

Spacesuits get very dusty on the outside and sweaty on the inside, but they will not be able to be easily cleaned during the mission! They would have to steam them. Their suits are very bulky, so the astronauts would only be able to have one spare suit. When the astronauts take them off, the dust on them will inevitably be brought inside the living quarters despite high-tech attempts to stop this. The Apollo astronauts found that when the lunar dust on their suits was exposed to air in their landing module, it had a pungent, wet ash smell. Dust will be a big problem with no simple solution. (Dust is not a problem for robots.)

A failure of any one of the life support systems in the crew's capsule, or in their space suits, would be fatal. (This does not apply to robots.)

SPACE SUITS

Future space suits need to be like the one on the left - trim, lightweight, highly flexible and strong so they are not easily torn. But such a space suit is extremely difficult to design and build because a spacesuit needs air pressure inside every part of it. On the right is the best suit that NASA has so far designed. It is relatively safe but, despite Mars' low gravity, it is heavy and very bulky with a heavy backpack and a helmet that restricts vision and movement. This makes it very tiring to move around in. Because air pressure does not make the joints flexible, it is difficult to bend down and twist in. Even moving one's fingers takes a lot of effort due to air pressure in the glove. It also takes much effort and a long time to get into it and out of it. It requires two people to get the suit on. The suit on the left uses tight pressure on the body to overcome the need for air pressure, but it has many problems that are too hard to describe here.

The Apollo astronauts took an hour to suit up and half an hour to get out of their suit. This wastes a lot of valuable time.

On Mars, the temperature outside is colder than Antarctica in winter. **It is only 2°C (35°F) during the day, and it drops to a bitterly cold -70°C (-105°F) during the night).** This will require a lot of power for heating space suits and habitats.

The astronauts would not be able to breathe Mars' very thin, suffocating carbon dioxide air, so the crew would have to wear a spacesuit with oxygen tanks when outside their capsule. (Robots will not need to breathe air, or eat for energy. They would have a small nuclear power supply.)

Existing space suits are far too bulky and heavy, so astronauts use up half of their energy just trying to move around in them. New suits would need to be designed and debugged because astronauts can easily fall over in the current top-heavy suits due to their backpacks' weight. Their backpacks contain their air, heating, and communications. And because the backpack is almost as heavy as the astronaut, this causes them to overbalance, making them more likely to fall over and risk tearing their suit and sudden death. Mars' higher gravity compared to the Moon would make their suits heavier than those that the Apollo astronauts used. (These problems are not applicable to robots.)

We have to accept that we humans are just not made for space exploration.

ROVERS

The first Mars base will need at least two rovers that will only be able to travel up to tens of kilometers from base camp. A second rover would have to be fully charged to bring them back. The rovers would have to be designed to carry four astronauts in an emergency. In case of an emergency, seating for four astronauts would have to be built into the design.

A LACK OF MAINTENANCE FACILITIES

As there will be almost **no major machinery available and no workshop**, astronauts would not be able to make any substantial repairs. (Compared to what robots could do, humans will be very limited in what they can accomplish.)

DEPRESSING LANDSCAPES

It is seldom realized that because Mars orbits at an average distance of 72 million km (43 million mi) further from the Sun than Earth, there is 35% less sunlight. This makes the landscape appear considerably darker than it does in enhanced NASA photos, or in movies where computer vision is created. In reality, the Martin landscape would appear like a dull cloudy day on Earth. **It will appear as being a dim, rusty-colored landscape. While it will be exciting at first, it will quickly look like a relatively colorless desert** because there is little color on Mars - only brownish rocks, yellowish sand, and dusty white ice in the Polar Regions. The sky appears a dull salmon pink.

There are hardly any clouds. There is no rain, no oceans, no lakes, no rivers, no sunset colors on twilight clouds, and none of the beautiful colors created by vegetation and other life that we are accustomed to on Earth. Mars is a bitterly cold, bland desert that would quickly become very boring. After a long hard trip to Mars, this would become depressing for humans who are used to lots of color and changing weather. (A boring landscape would not affect robots. They will have high sensitivity cameras that will enable them to see much better than we could in Mars' dimmer daylight. Robots will even be able to see well at night, so they could explore 24/7.)

We should be much more grateful that our Earth is so colorful. Due to our weather and seasons, most regions on Earth are ever-changing. This makes it much more interesting than Mars. Earth has a range of extremely diverse environments compared to Mars - or any other world in our solar system. Our planet is by far the most beautiful and exciting in the solar system.

While images of the Martian landscape from the surface are not overly exciting, many of those taken from space do look very interesting, especially those of enormous canyons, massive volcanoes, the polar icecaps, great mountains, and a marvelous diversity of craters, but astronauts will not see it like that from the surface.

NASA cannot make it public that there will be no manned mission to Mars because: this would alienate a lot of their supporters in the public and, they would lose funding for some of their research. They cannot admit that a manned mission would be unaffordable and that it would almost certainly result in failure. By contrast, robotic missions have a high chance of success, so it will be robots that explore every part of Mars. NASA will continue along this path.

While Elon Musk talks confidently about him going to Mars and sending a large crew and many facilities for marketing purposes, he must know about all the problems given here. I expect he keeps promoting a manned mission to Mars to keep his followers happy and to keep them investing in his company.

A manned mission to Mars is a guaranteed death sentence for anyone selected.

While the movie '*The Martian*' made visiting Mars look only a little different in its complexity and risk to visiting the Moon, in reality, it is orders of magnitude harder to visit Mars than the Moon and survive. While the movie was very entertaining for audiences, there was hardly a single aspect of it that related to reality.

A scene from the movie '*The Martian*'. Mars is one planet-wide desert with some extraordinary erosion features, but most of it is unremarkable.

This CGI illustration of a Martian landscape near the icecap shows how humans would see Mars' low light scenery that has very little color.

MARTIAN LANDSCAPES IN 3D (View these using red and blue 3D anaglyphic glasses and you will feel like you are there!)

Here are two 3D views of typical Martian terrains that provide realistic views. With the whole planet being a desert, the surface is not particularly interesting unless one is gazing into a deep canyon, and even then, they are so deep, it is hard to judge their depth. The bottom image is the most representative of most of the terrain. Mars' deserts have very little color. And like the Moon, because there is almost no air, distances on Mars will be hard to judge. Credit: NASA

ROBOTIC MISSIONS ARE THE ONLY VIABLE SOLUTION

In summary, within 15 years, robots will be extraordinarily more intelligent than any human, and they will also be much more agile, adaptable, and comparatively indestructible. They will not need large supplies of water, food, oxygen, medical supplies, space suits, or clothes, as humans do. Neither will robots need toilets, showers, and areas for cooking, sleeping, and relaxation. They will not require heavy shielding on their space capsule to reduce radiation sickness from cosmic rays and the solar wind.

A spacecraft for robots would be far, far smaller, weigh only a small fraction of those required for humans, and they will be far less complex. In addition, there would be no return mission required as robots could stay there for decades until they failed. **A robotic mission would dramatically reduce the fuel required, the overall payload, risk, and the cost, yet it would enormously increase the science achieved.**

Highly advanced robots will be far more efficient, easy to communicate with, and unlike humans, they will not lose vast amounts of time having to sleep, eat, defecate, relax, and being sick or depressed. Robots will not become mentally disturbed or easily physically injured. They do not have to exercise, cook, relax, wash, chat with one another, be entertained, or communicate with family, and so on.

Sophisticated future robots would have an atomic power supply as well as backup solar energy that could last them for decades. They could stay active 24 hours a day. They would be able to crawl across some surfaces to inspect them carefully while at other times make use of Mars' low gravity to leap over 100 m like fleas. This way, they could travel large distances very quickly. They might be able to deploy self-refueling jet packs by using Mars' air to allow them to fly across canyons and up to the tops of extinct volcanoes. Robots could have detachable lightweight drones that use micro and nanotechnology devices. Robots could travel for hundreds of kilometers through underground lava tubes in Mars' volcanic region. They could detach various body parts for specialized exploration. Robots would have detachable diggers that burrow deep into the surface and into ice to look for minerals and life. They would be so sophisticated that they could carry out a chemical analysis of their surroundings as they go.

Robots are the only viable means of exploring Mars.

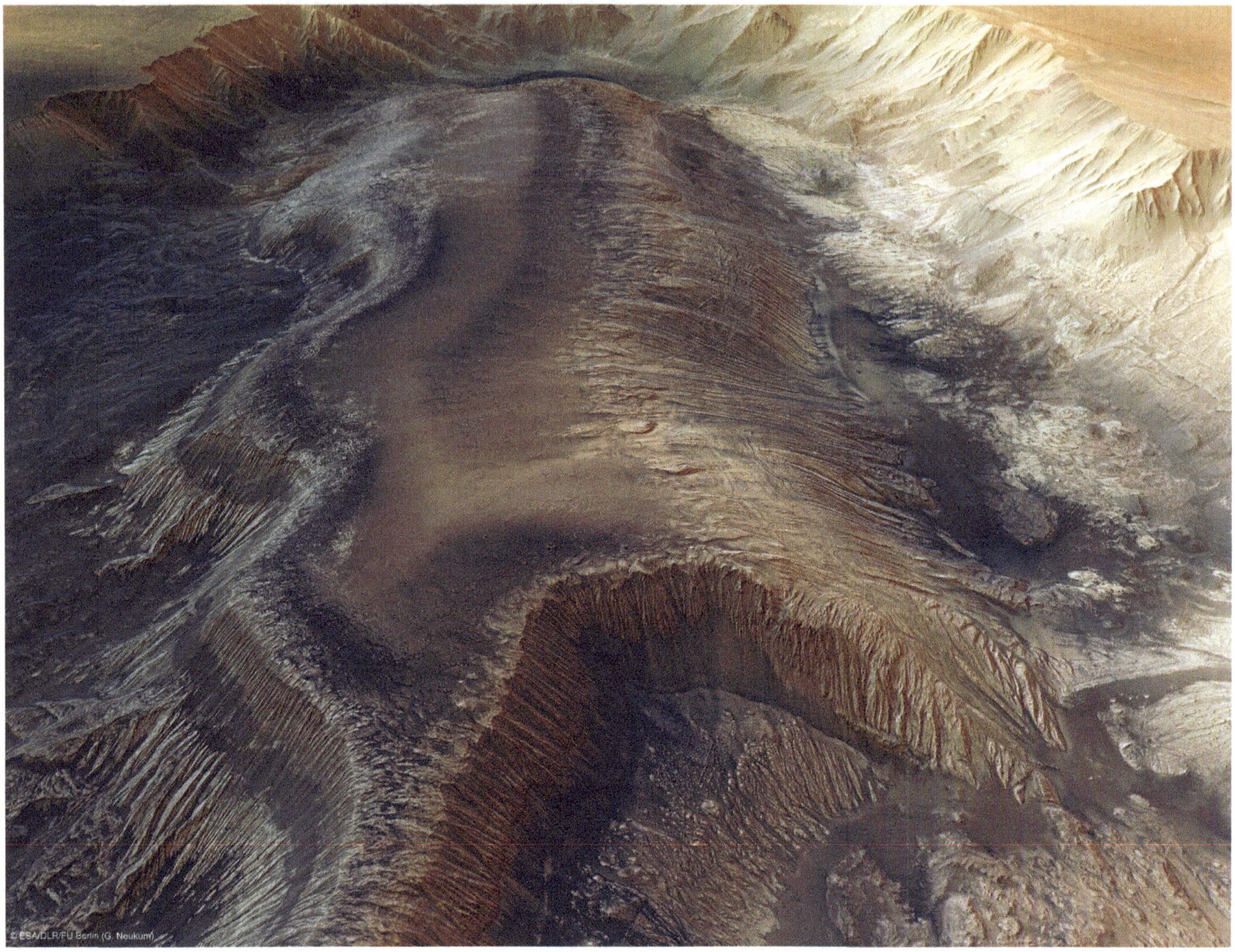

This is a view of a rugged canyon terrain on Mars that only robots could efficiently explore. Note the erosion of the plateau's walls from rain and snow in Mars' early history. While Mar's huge canyons and enormous volcanoes look extremely impressive from space, they are not so from the ground. They are so large that an observer on the ground would not see the far side of a huge canyon or the top of a volcano because they would be over the horizon due to Mars' tight curvature caused by its smaller diameter. Credit: ESA DLR/ FU Berlin

Robots will have self-repairing nanotechnology equipment, so they will remain in near-perfect condition for a long time. Being intelligent, they could propose theories to explain their observations and test them by searching for further evidence.

Each robot would be in constant communication with all others, as well as with Earth and probes in orbit around Mars. **Nuclear-powered robots will be self-sufficient on any world they explore. They will have a huge range of super-senses.** For example, their vision could range from full surround to telescopic and microscopic, and they will be able to detect most of the electromagnetic spectrum.

Unlike humans, robots can be sterilized so that they cannot contaminate Mars with Earth-born bacteria and viruses.

The planet Venus is much more hostile for humans than Mars is, so Venus can only be explored by robots. The moons that orbit the gas giant planets are much too far away, extremely cold, and they are too dark for humans to have any chance of exploring them. But robots will be able to explore all these worlds in incredible detail within the next few decades.

Over the next 50 years, increasingly more intelligent and adaptable robots will continue to explore Mars. They will have largely finished the job of exploring Mars by the time there is even a remote chance of humans going there. Robots will be able to stay there permanently and handle the conditions. **Humans are just not designed for space exploration.**

VISITING OTHER WORLDS VIRTUALLY

Exponential advances in computing, planetary probes, and exobiology will allow us to recreate other worlds in virtual reality. In the future, virtual worlds will be so faithful to real ones that it would be hard to tell the difference. This technology will provide the means for humans to visit other worlds in our solar system safely and instantly, probably by 2035. The evolution of a planet could be built into virtual worlds allowing us to experience them at different times throughout their geological history. Real weather for the entire planet would be built into the simulation. VR travelers will be able to instantly relocate themselves to any place on the planet.

Ultimately, these VR worlds will be 'projected' directly into human brains making every aspect of them appear 100% real. This will eliminate the need for present-day VR helmets and suit technology. Super VR technology will abolish the boredom of long travel times to other worlds, extreme levels of stress, and the high risk of death that would exist if we were to attempt such journeys physically. This popular technology is in its infancy at present, but it will quickly become highly sophisticated in the near future.

In the coming decades, trans-humans will also be able to explore exo-planets in other solar systems virtually. Highly detailed data will be gathered to produce a fairly accurate simulation. By then, we should be able to predict the types of alien life forms that may exist on some worlds that have the right conditions for life. Very few scientists working in the field of extra-solar planets and exobiology are yet to realize how advanced virtual reality will become in the near future, and how dramatically it will change their field of science.

Chapter 12

HOW REALISTIC IS INTERSTELLAR SPACE TRAVEL?

This is an illustration of a far-fetched starship that can hyper-jump infinitely faster than the speed of light to locations anywhere in the universe. Illustration by Kregano's Bucket

Science fiction movies, astronomy TV series, novels, and fanciful space documentaries have all made interstellar spaceflight seem like a foregone conclusion. Most people expect the exploration of our galaxy will be similar to how our ocean-going ships explored foreign lands. This idea has been around for so long that most people think this will come true, so they haven't questioned it. **We like to imagine that in the future, we will have fleets of huge Star Trek-like spaceships exploring the galaxy and going from one star system to another using hyper-warps. But disappointingly, this will be impossible because no matter how advanced the technology of any intelligence becomes, the laws of physics create insurmountable barriers for interstellar space travel.**

Very few people today realize how quickly our world will change due to exponential technological development. We will not have to travel physically to other planets or star systems because they will be created virtually in the minds of mega-intelligent computers. (See Mega-intelligence Chapter 17.)

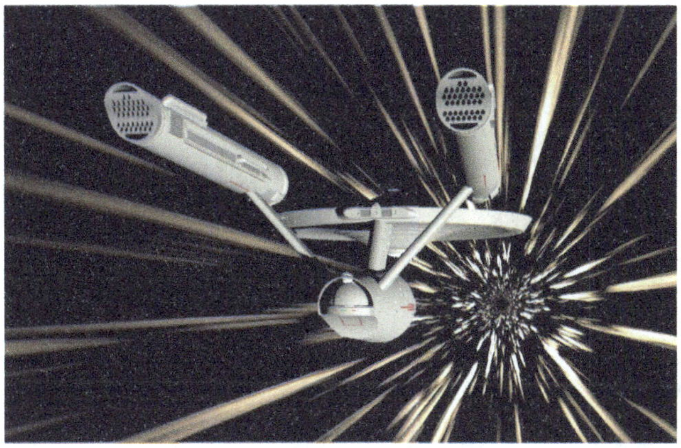

Star Trek's Enterprise starship magically hyper-jumps from one location in our galaxy to another in no time at all.

ENERGY REQUIREMENTS FOR INTERSTELLAR TRAVEL

The nearest solar system is Alpha Centauri, which is only 4 light years away. To send a spaceship to it at the highest practical velocity of 10% the speed of light, physicists have calculated that it would take more than 120 years to make a round trip. This allows for acceleration and deceleration as well as a decade for exploration. (The common abbreviation for the speed of light is 'c', which we will use from here on.)

Not only would the mass of a large spaceship have to be accelerated to 10% c, but so would the inconceivably large amount of fuel that would be required. Fuel would be in the order of 99.9999% of the total mass of the ship and its payload! The more fuel required, the more thrust the ship requires to propel the fuel itself. This becomes a runaway problem.

In the 1960s, it was imagined that we would be so much more advanced in space technology by now that we would have large, manned spaceships traveling to the Jovian and Saturnian systems. But the reality is that interplanetary *manned* spaceships will never eventuate due to the extremely long timeframes involved in space travel - even within our own solar system. The huge amount of energy required for large spaceships to do so is an impossible hurdle to jump over. Credit: Model of the US Discovery on its way to Jupiter: MGM – 2001: A Space Odyssey

The Star Trek TV Series and the Star Trek movies made it seem inevitable that we would build fleets of gigantic spaceships to explore the galaxy. To overcome the enormous distances between stars, these ships use fanciful hyperwarps to traverse huge interstellar distances, and teleporters to beam occupants down to the surfaces of warm, fanciful Earth-like planets with oxygen-rich atmospheres, clear skies, and almost no real risks. The chance of this is zero. In Star Trek, humans in the far future were depicted as being less intelligent than the brightest people today. Even in the near future, humans will change dramatically and become far more intelligent than us present-day humans. Star Trek's aliens were merely human actors with simple makeup – nothing like what real aliens would be like. Star Trek's aliens even thought like us, which would be very unlikely. All these concepts, as innovative and entertaining as they were for that time, are a long way from reality. They are no more than simple fantasies that the realities of space-time and the evolution of life forbid. Illustration by Star Trek

As well as the fuel required to get to Alpha Centauri, a similar amount of fuel is required to decelerate the ship near its destination. And, on top of all that, the same amount of fuel is required for the return journey and the deceleration on approaching our solar system! It's possible that a source of fuel for the return journey may be naturally available in the new star system.

The amount of fuel required for interstellar travel is an insurmountable problem for interstellar space travel. There may be more explosive fuels than we have today, but this problem will only be slightly reduced. No matter how advanced an intelligence becomes, the laws of physics demand that it still requires an enormous amount of energy to propel physical matter across space.

The amount of fuel an interstellar spaceship requires to accelerate to 10% c grows exponentially as it accelerates. For a ship accelerating at say, one Earth gravity (1g or 10 m/sec) after 10 minutes will be traveling at around 6 km/sec. If the acceleration were much more than that for an extended period, everything on board, including humans, would suffer this making them much heavier so, accelerating much faster than 1g would be pointless.

To achieve an acceleration of 1g requires a specific expenditure of energy which is dependent on how massive the ship is. At an acceleration of 1g, after another 10 minutes, it would double its speed to 12 km/sec. To maintain this acceleration now requires *three times* more fuel. After another 10 minutes, its velocity will be 18 km/sec. The ship's engine now has to consume *five times* as much energy as it did in the first 10 minutes. This exponential increase in energy will continue every 10 minutes for 1 month until it reaches 10% c. That will require 4,300 ten-minute periods of increasing acceleration, each of which will consume exponentially more fuel! **This consumption of fuel mounts up to an inconceivable quantity.**

Once the ship reaches 10% c, it will be able to coast for most of the way, but it would need some extra fuel to maintain its velocity to overcome **interstellar resistance** from atomic particles in space, particularly hydrogen atoms, dust, and cosmic rays that would collide with the ship. **Cosmic rays are mostly the nuclei of iron traveling at almost c. Despite the minute size and mass of iron nuclei, they carry enormous energy. This is due to relativistic effects massively increasing their mass because they are traveling close to the speed of light. The constant barrage of these particles blasts microscopic holes in the ship, eroding it away. This would eventually destroy it.** The mass of colliding particles impacting the front of the ship increases exponentially as the ship's velocity increases. When the Concord supersonic airliner flew at twice the speed of sound, even a raindrop would have gone straight through it like a bullet, so it could only fly at supersonic speeds when it was high in the stratosphere where there was no possibility of rain or tiny ice crystals. Given this, you might be able to imagine the impact of specks of dust on a spaceship traveling enormously fast at 10% c. A spaceship would need a powerful field of some sort to repel all types of particles in its path, well in advance. But it is impossible to imagine any practical way that this could be achieved, and how such a field could be powered without greatly increasing the ship's mass and therefore its fuel requirement.

Keep in mind that there is no energy suitable energy source in interstellar space to act as fuel. Solar energy beyond Mars is so weak it is useless as an energy source.

When the ship was 1 month from its destination, it would need to consume the same amount of fuel as it did during its acceleration phase to decelerate so it could go into orbit around its destination star to enable it to make observations of the star's planets. This would create a troublesome, continually increasing gravity acting towards the front of the ship, so it would have to start decelerating much earlier at a slower rate. The ship would also require energy and therefore extra fuel to keep all the ship's systems running for the whole journey out and back.

On top of this expenditure of energy, for its return journey, **the ship would require the same amount of energy as it used on the outward journey.** The total amount of energy would be so enormous that it would be impossible to achieve. There are many fanciful space travel articles that talk about how long it would take for a spacecraft to reach Alpha Centauri, but they conveniently fail to mention the amount of energy needed to do so!

No matter how advanced a space-faring civilization might become, achieving physical interstellar space travel will be impossible. To accelerate just *one tonne* (~1 ton) to only 10% c for only a month would require the entire energy production of the Earth! So, you can imagine what the energy requirements there would be for a giant spaceship with the mass of thousands of tonnes *plus* its fuel and cargo, which would be far larger! Because of the exponential increase in the energy that is required for acceleration and deceleration, physicists have calculated that **to go to nearby Alpha Centauri, would require the entire mass of our Sun to be converted into energy!! This sounds impossible, but that is what would be required.** This is just one of many reasons why our unrealistic dreams of having interstellar spaceships will remain a fantasy. Anti-gravity will not work in interstellar space because there are no nearby gravity fields. It would be no more useful than sunlight.

When we see what is involved in just going to Mars (see Chapter 11), we see how much more impossible interstellar travel would be. **It would take over a hundred thousand years just to go to the closest stars, and the same time to return. This is totally impractical.**

THE COSMIC SPEED LIMIT

No intelligence can get around the enormous amount of time that space travel takes. Even the speed of light is like a snail's pace when interstellar distances are involved.

Einstein showed that it is impossible for anything to travel at, let alone beyond the speed of light. There are many irrefutable proofs for this. Even a massless photon cannot exceed c.

A spaceship, or any mass, becomes increasingly more massive (heavy) as it accelerates. This means that the mass of a spaceship, its cargo, its fuel, and its occupants would increase exponentially with increasing acceleration, particularly above 90%c. The faster any mass goes, the heavier it becomes and, the more fuel is required to propel it. It becomes a dog chasing its tail.

General Relativity laws show that, if it was somehow possible for a spaceship to accelerate continually, then its mass would increase very slowly until it became infinite at c. A ship with infinite mass would therefore require infinite energy to propel it. It would have to consume the entire Greater Universe to reach c! Of course, this is impossible. This is why the speed of light can never be achieved for anything that has mass. Hyperwarps do not overcome this problem, as you will see in the next chapter.

As well, when traveling close to c, there is a contraction of space and time. As an object's velocity reaches c, space and time contract to zero and an object's mass becomes infinite due to the laws of General Relativity.

Calculations using the General Relativity equations show that at a velocity of 25% of c there is a 3% slowing of time and a similar contraction in length. At 50% of c, the contraction of space-time increases to 15%. At 99% of c, time is slowed by a factor of 700%, and space contracts to the same degree. At 99.999% of c, the contraction of space-time becomes 22,400%. At c, it is infinite. If it were possible to travel in a spaceship at 99.5% of c, a hypothetical observer on Earth using a super-powerful telescope to look into the ship's windows would see the ship's clocks barely moving. The people inside would appear to be moving in super slow-motion, but they would not notice this as time for them is slowed down dramatically making everything inside their ship appear normal.

Because engineers can often overcome technical hurdles, there is a tendency for us to think that they will eventually overcome all technical problems- even traveling beyond the speed of light as it is just be another hurdle to jump. In my youth I thought this, but this is not so. The most basic laws of physics would have to be ignored, and this is not possible. In the hope of traveling faster than c, some people believe that we will find a way to ignore these laws. If this was possible, we could ignore all the laws of the universe, but even if just one of them did not work, this would cause chaos and destroy the universe.

The reason we do not see alien spaceships visiting us is that interstellar travel is simply physically impossible. So, be very suspicious of UFO claims. (See Chapter 10.) Of the UFO observations that are really unidentified flying objets, they are not interstellar craft but top-secret defense department experiments, or a host of other man-made objects, fakes, or natural phenomena.

EXTREMELY LONG TRAVEL TIMES MAKE SPACE TRAVEL ILLOGICAL

In this section, you will see other reasons why space travel is impossible, even over relatively short interstellar distances. Even if it were somehow possible to travel across space at 10% c, (the fastest practical speed) it would take far too long to go anywhere. Even traveling at the speed of light would be unbearably slow when one needs to traverse interstellar distances.

No technologically advanced species would ever build spaceships like this for short-lived organic beings to make interstellar voyages. They would take eons to reach even relatively nearby star systems. It would be millions of times harder than trying to row a log canoe across an ocean a million times as wide as the Pacific Ocean!

Even planetary systems that lie in our nearby region of the galaxy are so distant that even if we could travel at almost the speed of light, it would still take many centuries to reach even the relatively close systems, and tens of thousands of years to reach more distant ones – and that's just in our quadrant of the galaxy! Star systems elsewhere in the galaxy would take up to four to six times as long. At the maximum practical velocity of 10% c, travel times would be nine times longer still! Such extremely long timeframes taking millennia are totally unrealistic.

OTHER REASONS WHY INTERSTELLAR SPACE TRAVEL IS UNREALISTIC

Stars have a radiation bubble around them called a **heliosphere**. This is where a star's radiation pressure equals that from all other stars in interstellar space. (Voyagers 1 passed through the Sun's heliosphere in 2013 and Voyager 2 in 2018 at 18 billion km from the Sun.) At the front edge of a star's heliosphere there is a bow shock wave that takes the brunt of interstellar gas and radiation. It ranges from a few billion km for a dwarf star to 20 billion kms for our Sun to hundreds of billions for supergiant stars. **Once a spaceship travels outside a star's heliosphere, cosmic rays increase dramatically, thereby becoming a major threat, much more so than they already are inside a solar system's heliosphere.**

When thinking about interstellar journeys, we need to consider **the physical wellbeing and mental stability of the ship's occupants.** NASA knows that even an 18-month return trip to Mars would run a high probability of some, if not all, of the crew becoming ill from a deterioration of muscle mass and bone density caused by a lack of gravity, cancers caused by radiation from space, and from becoming mentally unstable due to the solitude and very cramped condition. This would destroy a mission. Perhaps **intelligent robots** could be used on interstellar missions. They could be turned off for most of the journey, but over such enormously long timeframes, they would deteriorate. Long before they might reached their destination, their discoveries would have been known not long after they left their home due to the exponential explosion of knowledge on their home planet. There would be no gain in traveling physically across interstellar space. The very entertaining movie '***Passengers***' has a plot where the time to travel to a new planet in another solar system was an unrealistically short 120 years.

Another horrendous problem would be **ship maintenance**. During even a short voyage requiring thousands of years, many maintenance problems would arise on a ship, and sooner or later, one would be fatal.

There would be no way for a spaceship's occupants to communicate with their home planet due to the extraordinary long time-delays oneway of centuries or millennia. Radio messages would deteriorate soon after the ship left its solar system, so they would be useless. The occupants would be totally alone with no escape. This would be psychologically impossible to deal with. The journey would not be one of exciting discoveries because, along the way there would be very little, if anything to see or do. Even an onboard telescope would be unable to show as much detail as the much more sophisticated telescopes that would be invented on their home planet soon after the crew departed. Everything on the ship would be essentially the same for thousands of years, unless it was designed to evolve, but would that make much difference? There may be an on-board science lab trying to invent some new things, but that would be all. The only news available would be gossip about what people on board are doing or saying. Due to extreme boredom, passengers would want to be put to sleep for most of the journey. There could be serious in-fighting early on that could destroy the mission not long after it commenced.

Even with life extension technology, **the original crew would not live long enough to reach their destination. There would have to be numerous generations being born and dying on the way.** Due to exponential development on their home planet, **there would be no reason for a return trip** as their home planet would have changed dramatically over the millennia that the passengers would be away. Catastrophic events may have destroyed it. The rapid evolution of their species would make them unrecognizable and probably extinct.

Such timeframes make it illogical and completely impractical for any organic being or even a robotic super AI to attempt to cross interstellar space.

EXPONENTIAL TECHNOLOGICAL DEVELOPMENT WILL MAKE SPACE TRAVEL POINTLESS

Any species that could build a spaceship to go to another solar system would have evolved technologically to have rapid exponential technological development. But the rate of change would be so fast that any spaceship that they might build would be ancient technology even before it was finished! But let's assume it was built. By the time its crew was less than a small fraction of the way to their destination, exponential development on their home planet would have allowed their species, or their mega-intelligent offspring computers, to learn all there is to know about the star system that the space travelers are going to. This being the case, there would be no point in them going!

Due to the rapid advancement of science and technology, a means of discovering what other star systems and what any alien life will be like, will be invented quickly, eliminating the need for interstellar space travel. Just as people living before the 20th C could not have imagined jet flight, DNA, atomic energy, mobile phones, the Internet, or giant space telescopes, nor can we imagine what science and technology will allow us to discover and build in the decades ahead that will allow us to explore other worlds without having to physically travel to them.

Chapter 13

WILL SPACE WARPS & TIME TRAVEL BE POSSIBLE?

Science fiction has made the idea of space warps very popular, so let's see how possible they might be.

For the early part of the 20th century, most scientists, science fiction writers, and the public were unaware of the concept of space warps. But in the latter part of the century, Star Trek popularized the use of hyper-warps and teleporters as a great way to overcome the enormous distances in space. Since then, space warps have become so popular that it is now assumed that it will only be a matter of time before we develop this technology. But is this dream realistic?

The general public, and even many scientists, are unaware of what is now known about the space-time continuum. Even some leading scientists naively come up with concepts like **wormholes** possibly being a real solution to space travel. Much to my disappointment, the laws of physics make it very clear that space warps will never be possible - regardless of how far technology advances. If even a small wormhole was possible, it would require a large percentage of all the energy radiated by all the stars in our galaxy over their entire lifetimes to power it! That would be impossible.

When space warps were conceived, they were not based on any realistic understanding of physics. They were just an innovative, simplistic solution to overcome the impossibly long travel times for interstellar travel in order to make the Star Trek TV series work. They did not take into account the fact that not only would space be warped, but so would time. This is because space and time are interlinked as the space-time continuum. Some well-known scientists perpetuate the myth of warping space because they have not taken the time to think through what this would actually involve.

In science fiction novels, TV programs, and some movies, interstellar travel by way of hyper-jumps and teleporters is shown to be almost instantaneous over huge distances of tens to thousands of light years. But for this to happen, an object would have to travel far faster than the speed of light! For a body to go, say 1,000 light years in seconds by using a space warp, would mean it would be traveling millions of times faster than light! This violates the most basic laws of physics, so this is utterly impossible. Even if it took days, it would still amount to traveling far faster than the speed of light. The warping of space-time, by whatever means, will never happen. If it were to occur, it would destroy our universe, as you will see.

Sci-fi movies and novels like to make use of wormholes to connect distant parts of the universe to overcome the huge distances in space. But this concept is not realistic.

DOES THE WARPING OF SPACE AFFECT TIME?

Because space and time are intricately linked, if space was warped to the extent that the travel duration was enormously faster than light, then time would be warped to the same extent. It would to go back to before time even existed! Of course, this is impossible. Scientific evidence supports time as being finite, having started when the Big Bang occurred around 14 billion years ago, so it cannot be warped to go back before the Big Bang – before time even began to tick!

Because time goes in a forward direction in our below-light-speed universe, it would have to go backward in an above-light-speed universe - *if one existed*. Going far back in time would defeat the value of using teleportation as a form of transport because people would not want to go back in time when there is no way of warping space and time to return to the present. **Going faster than light would also mean that causality would be reversed. In other words, the effect of an action could occur well before the action had even taken place, or even been considered!** As an example, a spaceship could arrive at its destination before the space warp or the ship was even thought of being built! Clearly, this is ridiculous, so traveling faster than light is impossible.

WOULD CHANGING THE PAST AFFECT THE FUTURE?

If the warping of time were possible, then it would cause chaos across the universe. Let's think about what would happen if people or aliens could go back in time. They could accidentally eliminate their younger self or *anyone of their long line of ancestors* that ultimately led to the time-traveler being born. If even one of the time-traveler's ancestral life forms were eliminated, then their long line of descendants over billions of years would instantly vanish throughout all of time! Of course, this would also mean that the time-traveler would not exist in the future, so how could he go back in time in the first place?

Even time travelers who do not intend to interfere with the past, would do so merely by being there. They would alter the energy levels and positions of innumerable particles and objects around them, as well as bacteria and other life forms, some of which might evolve into higher life forms that might have led to the time traveller being born.

There would be endless flow-on butterfly effects that would not only affect our planet, but ultimately the entire universe from that point onward! **Anything that was connected to that part of the past would be changed. This would cause things to change or disappear. These flow-on effects would occur all the way to the end of time!** Imagine what would happen if this was happening all across the universe.

If space warps could occur, there would be an infinite number of chain reactions or 'space-time tears' throughout the space-time continuum of the universe. Each one would grow ever larger with time. If the earliest mega-intelligences were able to invent space warps, then space-time tears would have started billions of years ago, so they would have become obvious by now. As we do not see parts of the universe changing instantly or disappearing, this tells us that space warps are not possible.

When the premise of time travel and space warps is thought through sensibly, they make no sense whatsoever.

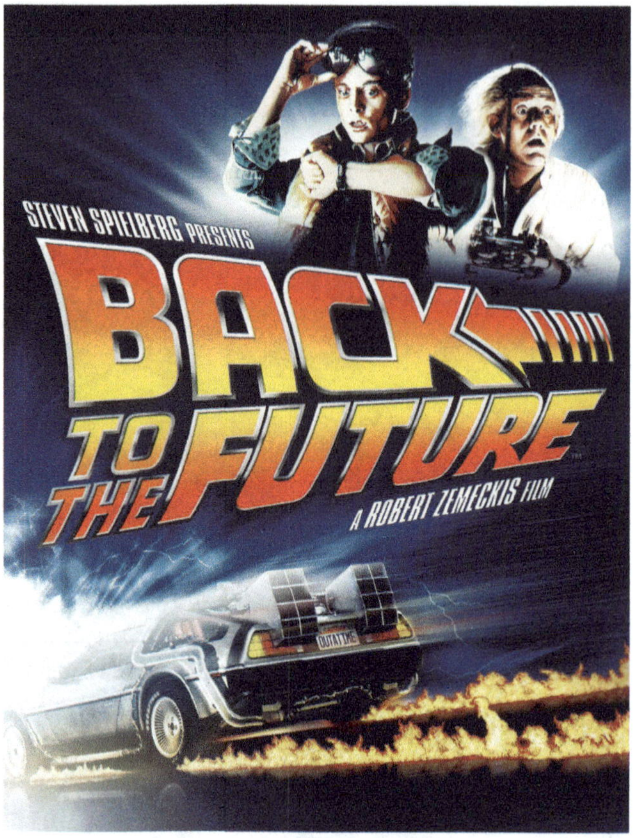

The '*Back to the Future*' movies are a fun experience, but they are not based on anything remotely realistic.

Imagine that the warping of time is possible. Our descendants would be dropping in on us and appearing out of thin air. But not surprisingly, not one such visit has ever occurred – for obvious reasons. People or beings from the future could change things in our present or past in the hope that it would benefit their families in the future. This was the pretext of the movie '***Back to the Future***' and the more sinister '***Terminator***' movies.

If advanced aliens could warp space, then we would probably see them dropping in on us all the time, causing our world to be like the '***Men in Black***' movies – full of aliens! The fact that we are not overrun by all types of aliens is proof that time travel and space warps are not possible. Some people believe that aliens who look like us walk amongst us, yet they have the same DNA as us humans!

IMPOSSIBLE ACCURACY WOULD BE REQUIRED FOR ANY FORM OF WARPING SPACE

If space travelers could use space warps, they would have to take into account that everything in the universe is constantly moving and, therefore, continuously changing its position in space. If it were possible to quickly travel across space by warping it, then even small space jumps around a planet would have to allow for the fact that planets rotate and that they are moving along their orbits around their star. As well, stars are orbiting their galaxy, and galaxies are rotating, so all these motions have to be allowed for. On top of that, the motion of galaxies through space must be calculated, as well as the expansion of the universe. To measure all these motions to the accuracy required would be impossible, even for the most highly advanced intelligences.

Even the most minute errors in calculating the position at which an object needs to rematerialize, even for short teleportation jumps on Earth, would cause travelers to die due to them rematerializing high above the ground, in a deadly blizzard, in outer space, inside the planet, under the ocean, inside the concrete of a building, or a mountain. If teleporters were possible, any object being teleported would have to rematerialize inside a teleporter, but to manage that would be impossible. What reference point would be used when every thing is in motion?

A teleportation calculation error! If this occurred, the rock would explode by having another mass instantly materializing inside it. This would kill the traveler.

To realize how complex and incredibly accurate the calculations would need to be for teleportation to compensate for the motions of everything in space, consider the following:

At the equator, the Earth rotates half a kilometer every second i.e. the Earth's axis spins at 1,700 km/hr (1,030 mph), but at the poles, its rotation speed drops to zero. To be teleported safely on Earth, the destination that you want to be teleported to would need its spin rate calculated precisely, as well as its longitude, latitude, and its altitude.

But that's not all! Because the Earth is also traveling at over 100,000 km/hr (66,000 mph) along its elliptical orbit around the Sun, its position in the solar system is constantly changing. And because planets do not trace out exactly the same orbit due to the ever-changing gravitational tug of other planets and the Sun on them, the calculations for the exact position of the Earth relative to the Sun would be impossibly difficult to calculate accurately enough for teleportation purposes.

But it gets even far more difficult because our solar system is traveling around our galaxy at the fantastic speed of 800,000 km/hr (500,000 mph). To make it even more complex again, it travels above and below the plane of our galaxy's spiral arms like a horse on a carousel going up and down as it goes around. (See Chapter 34, page 439.) As our Sun and its planets go around our galaxy, the ever-changing gravitational tug of other stars and nebulas on it causes it to continually change its velocity and its orbit slightly.

On top of all this, the direction that our galaxy is traveling through the universe is not in a straight line, and its velocity of around 1.5 million km/hr is increasing as it moves towards the Great Attractor cluster of galaxies. Our galaxy's motion is affected by the gravity of other individual galaxies and clusters that are also moving relative to the Milky Way.

If all this is not enough to deal with, space warps of any kind would also have to allow for the universe expanding at a continually faster rate.

Even the most infinitesimally small calculation errors in any of these aspects of Earth's movement through the universe would mean disaster for anything being teleported. The tiny rematerialization destination would have to be incredibly precise relative to the entire universe!! The slightest error, which could not be avoided, would mean instant death for those being teleported.

All these motions would be impossibly difficult to calculate. Even the most advanced mega-intelligences would find the accuracy required to be incalculable.

RECONSTRUCTION AFTER TELEPORTATION

Another huge problem with teleportation would be that it would take a long time to disassemble every atom in a human body without damaging them. Because the total number of atoms in a human equates to 7 with 27 zeros after it, there would be no way to ensure that they would all reassemble in the right order so that no part of the body became disfigured. Even minute changes to the trillion neurons in the brain could change who a person is. Chaos Theory shows us that the chances of perfect deconstruction and reconstruction would be zero.

TELEPORTATION BLASTS

When talking about any form of teleportation, it should be understood that there would be an incredible blast at the point of departure if the dematerialization process was close to instantaneous. A similar blast would also occur at the rematerialization location. These blasts would occur because dematerialization would cause a vacuum where a body of matter no longer exists. The surrounding air would rush in from opposite sides at enormous speed colliding in the center causing a rebound shockwave explosion. It would be like a crack of thunder from a lightning strike close by. Lightning is just a thin filament of electricity passing rapidly through the air, heating it, and making it expand very quickly. The rapid expansion from the heat of the electrical arc is what causes the thunder shockwave. Imagine how much louder the explosive shockwave would be from something as large as a person disappearing or reappearing instantly. At the other end, when a teleported body rematerializes, it has to instantly push the existing air out of the way. This would cause an explosive blast similar to dematerialization. It would be like a small bomb exploding. This would kill a living being and it would destroy most inanimate objects. The teleportation chambers would have to be evacuated of air at the instant of teleportation to stop these explosions, so travelers would have to wear spacesuits. This would be impossible for practical reasons. The wonderful idea of teleporters will never be realized.

Star Trek's elegant but fanciful teleporter

You may like to listen to Isaac Asimov's most entertaining short story '***It's Such a Beautiful Day***' on YouTube. It's about a home teleporter system not working. There are also other audio books of Asimov's excellent stories on the Internet.

Chapter 14

IS INTERSTELLAR RADIO COMMUNICATION POSSIBLE?

Is it possible that radio telescopes might detect messages from interstellar intelligences?

SETI - THE SEARCH FOR EXTRATERRESTRIAL INTELLIGENCE

The search for radio communications from extraterrestrial aliens commenced in the early 1980s using large radio telescopes. In those days, there was a general ignorance by astronomers and the public about all the practical problems associated with using radio waves to communicate across enormous interstellar distances. This ignorance still exists today. Those who conceived this idea were well-meaning people, but incredibly, after all this time, they have not thought through *any* of the many obvious problems involved with communicating across interstellar distances. They naively think that organic intelligent life forms will not evolve and that they will stay as they are for millions or billions of years. They have not realised that they will evolve like we are into super intelligences and mega-intelligent computers.

They are unaware of the impact that exponential technological evolution has on any intelligence's evolution. They are so in love with the grants they receive and their fantasy of a phone call from ET that they willfully ignore all the evidence that shows that ET will never bother to send us a message because it would not be practical or physically possible.

Given the insurmountable problems associated with interstellar radio communication, it is no wonder that intelligent radio signals have never been detected, and never will be, despite numerous intelligences almost certainly existing out there.

Carl Sagan wrote the book "Contact' about the Seti program making contact with an advanced intelligence who taught us Earthlings how to build a wormhole to other parts of the galaxy. It became a very popular wellmade, entertaining, and thought provoking movie. But we now know that with exponential technological development such a scenario would be impossible.

REASONS WHY SETI WILL NEVER BE SUCCESSFUL

1. The space probes that we send to other planets have frustrating time delays of up to many hours when communicating with Earth. To use **radio for communications between exoplanets beyond our solar system is out of the question because the time delay would be between decades for relatively nearby star systems and up to thousands of years for more distant ones.** Intergalactic delay times would run between millions and billions of years. For a sender to receive a reply, the return times would be double that! Such time delays make interstellar communication totally unviable. Any intelligence would consider interstellar radio communication to be so impossibly slow that it would be a pointless exercise.

2. There is another major hurdle to interstellar communication that is not considered. ***If* it were possible for two alien intelligences to communicate, they would need to be at exactly the same level of knowledge and intelligence** because, if one was just a little head of the other by mere decades or even minutes, they would be far more advanced than due to exponential development. (See Chapter 15.) The difference in their rate of exponential progress could be so enormous in only seconds that the lower intelligence would be unable to understand the higher intelligence's level of knowledge. This would be like a child in Grade 1 trying to comprehend Quantum Physics from a PhD student. The higher intelligence would not be interested in communicating with a low-level intelligence because it would gain nothing from that. Due to the extraordinary rate of exponential growth

for advanced mega-intelligences, a difference of even mere seconds would be like millennia to us.

3. Another major problem in using radio for interstellar communication would be that **a message would have to arrive during the very small window of time when the receiver had developed radio telescopes** so it could receive the sender's message. This window of opportunity may be little more than a century because radio is likely to be a short-lived technology. Because the speed of light is so slow across interstellar distances, no intelligence would waste their time using radio or laser messaging.

4. **Many emerging intelligences at our technological level would not be aware that the universe even exists!** There would be those who have evolved to live in their planet's crust in huge caves or aquifers because their planet's surface is saturated in deadly radiation. Others would have dramatically varying temperatures and destructive weather that are lethal, so they would not know that space exists. Others would live deep in an ocean because the surface is not conducive to life, so they too would not be aware of the universe. And there would be those that have evolved to live under or in a dense, opaque atmosphere where they never see their star or the universe. There would be many that live on brightly lit planets with so much light that they never see the stars, so they don't know the universe exists. In these cases, they would not build radio or optical telescopes for communication purposes.

5. **Our long wave radio transmissions that are powerful enough to go around our globe are extremely weak and not focused, so these signals disperse quickly making them extremely faint and indecipherable before they reach the edge of the Oort Cloud – the outer edge of our solar system.** They would not reach even the nearest star undistorted. As well, when our radio signals leave Earth, they are degraded by Doppler shifting of the signal due to the Earth's rotation. Over interstellar distances this could make the signal distorted. To make it even worse, numerous transmissions at different frequencies from all over the planet would cause interference. And on top of this, our Sun's radio output would swamp our feeble transmissions, even at the distance of a close solar system. If detected at all, our radio signals would be blurred and appear childishly naive.

Other intelligences would quickly work out all these obvious flaws and not try such a wasteful search. It is a waste of our resources to look for non-existent signals using SETI.

It is quite alarming that SETI 'scientists' have not thought of any of these obvious basic problems over all the decades they have been searching for extraterrestrials with expensive radio telescope. Any one of these problems makes SETI totally unviable. These 'scientists' have been wasting their lives and taxpayers' money on a fantasy that has not the slightest possibility of producing any communications from extraterrestrials. But it gives them an income and some credibility to those who have never thought to question how successful they could be.

An effective means of communicating with other intelligences across the universe may only emerge once an intelligence reaches a technical and intellectual level where it is not bound by the three physical dimensions of space and the dimension of time. Could the distance problem be a deliberate strategy by the universe to force intelligences to develop their own unique views of what the universe is about so that their ideas are not contaminated by those of other intelligences?

WHAT MIGHT ALIENS THINK OF OUR RADIO TRANSMISSIONS?

Imagine for a moment that advanced intelligences were able to detect and decipher our radio and TV broadcasts. **They would quite reasonably consider that the vast majority of humans were delusional, unintelligent, unable to reason, and obsessed with war, control, murder, violence, drugs, religion, sport, cooking, renovations, entertainment, and generally inane humor. We would appear naïve, chaotic, and self-absorbed.** They would realize that much less than 1% of humans in the western world bother to watch intelligent, partially factual, educational documentaries or read books to advance themselves and their species. The trend today amongst most young people is to not take the time to read even a full page of text, so few ever gain any in-depth knowledge. What intelligence would want to communicate with such beings? Many aliens might think it best if humans were removed from the universe's gene pool! It may be that all intelligences at our infant stage of intellectual development are as naïve and self-absorbed as we are.

An intelligence only 50 years ahead of us would have no interest in communicating with a simple intellect like us. They would probably know all there is to know about us anyway.

WOULD ADVANCED ALIENS WANT TO PLUNDER US?

Some scientists have expressed concerns about us broadcasting our existence into the galaxy for the fear that it may act as an incentive to a superior intelligence to come and plunder us. The threat of being plundered sounds dangerous, but it is naïve. It is based on what has occurred throughout all of human history. But the fear of being plundered by highly advanced aliens ignores the enormous amounts of time and energy required for an intelligence to traverse space to reach us. Furthermore, we have nothing here that they would want. The resources we have are common throughout the universe, so there would be no gain in plundering us. Advanced intelligences would be able to create any resource they needed without having to leave their planet.

For them to plunder us would be like us walking all the way to the Gobi Desert to plunder the resources of an ant's nest. Would we bother to do that?

Our technology would be primitive compared to an advanced civilization, so we could offer them nothing new. They would know we would not be a threat to them. So, we will never be invaded from outer space, except possibly by some indestructible virus.

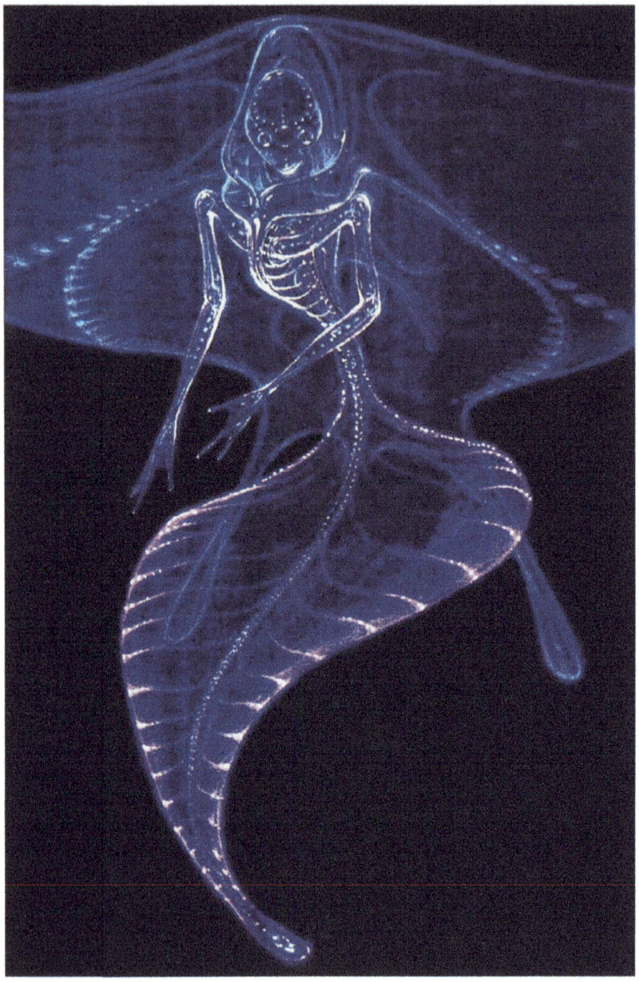

Movie director James Cameron conceived that the aliens in his movie '*The Abyss*' should not have a typically solid, but rather have a jelly-like humanoid form because they lived in a dense, dark medium similar to being in the deepest parts of our oceans. They had bioluminescent skin.

The scenes below are from the very exciting and much watched moive 'Close Encounters of the Third Kind'. It's about an interstellar spacecraft landing on Earth and taking people who were willing to go with them. These types of movies are very entertaining but they are pure fantasy for the reasons above. They fuel unrealistic visions in some people that aliens would want to make contract with us. But they would only do that if they were as naive and gullible as we are. If they were, they would not be advanced enough to cross the enormous gulfs of space between the stars.

TECHNOLOGICAL EVOLUTION

Chapter 15

HOW HAVE HUMANS EVOLVED?

In this chapter, we look at how fast our knowledge has increased since humans first evolved. This might give us some insight into how alien intelligences may also have evolved to far surpass our stage of technological evolution.

The evolution of organic life on Earth has been excruciatingly slow. It has taken life over three billion years to reach the present state of human intelligence. In the first billion years after life first evolved, early cells could only clone themselves, so evolutionary change was very slow. It was brought about by random changes in a life form's DNA. This occasionally happened when radiation from the Sun or cosmic rays from deep space damaged a gene, or through epigenetics. (See Chapter 24, page 345.) Most changes to genes were detrimental, but now and then, one was beneficial, so this helped simple life forms to evolve very slowly. **Around 1.2 billion years ago, something miraculous occurred. Somehow, sex evolved. How this occurred so relatively in geological time is not known. It was such a smashing success that nearly all life forms took it up! This innovation permitted the sharing of genes. Sexual reproduction spread so quickly through most life forms that it caused evolution to speed up dramatically. Sexual reproduction created a huge increase in the complexity of genes and therefore an explosion in the number and diversity of life forms. Since then, there has not been such a huge leap forward in evolution - until now!**

But this time, evolution is occurring at lightning speed! This is due to us humans inventing genetic engineering, nanotechnology, and especially extraordinary computer technology. We are about to experience an explosion in the diversity of organic and bionic life forms that is beyond our wildest imaginings. And computer intelligence is speeding up exponentially. So it will streak ahead of humans by billions of times within a couple of decades at most.

We are now about to create the first mega-intelligences. For the first time in Earth's history, this life form will not be organic: it will be computer-based. It is evolving from present-day supercomputers and newly evolving quantum computers. Its intelligence will be astronomically high, and it will increase exponentially at a rate that we humans cannot imagine. We live in an exceptionally exciting time, but the future could be very threatening for us.

The current rapid expansion in human knowledge has occurred 'overnight' in geological and evolutionary timeframes. It started as a slowly increasing growth in human knowledge around 100,000 years ago, but in just the last 100 years, this growth has skyrocketed, and in the last decade, it has gone ballistic!

Big data analysts who study future trends and the *exponential growth* of technology are predicting that by 2040, the growth in knowledge and technology will have increased exponentially by up to an incredible *trillion* times!! Even if it takes twice as long, it will still be incredible. Even a rate of growth of a thousand times is very hard for us to envisage in terms of its impact, but a growth rate by a factor of a trillion (i.e. thousand billion) is far beyond our ability to comprehend this. (See diagram on page 210 in this chapter titled AGES OF TECHNOLOGY). If this rate of change sounds impossible, consider that there has been far more change in the lifespan of people that are currently over 80 years of age than there has been in the previous 100,000 years! *This is truly remarkable,* yet this is a relatively unknown fact.

To have any chance of coping with this rapidly increasing rate of change, human brains will have to evolve extremely fast. They will have to integrate with leading-edge computer technology. This will dramatically increase our intelligence, our memory capacity, and our consciousness. Those who do not do this, will not understand most of what is occurring around them, and they will not know how anything works. This is starting to happen now. Everything will be over our heads because computer intelligence will be directing its own evolution at an exponentially far faster rate than our own. We will soon need personal intelligent robots to look after us until we evolve to have a computer-based mind, which will give us access to all the knowledge that we will be capable of computing. But this will only be an infinitesimally small percentage of what the more rapidly evolving super-intelligent mega-computers will be able to compute. They will evolve so fast to a super high level of being that they will appear to disappear. This would be like us evolving from fleas to humans in a matter of a couple of decades.

WHY HAVE HUMANS ADVANCED SO MUCH MORE QUICKLY THAN OTHER ANIMALS?

We do not yet know why we humans developed so much faster than other apes or other animals. It was probably due to a number of things that coincidently occurred in the right order. Those of our ape ancestors that survived the longest when leaping quickly from tree to tree would have been those that developed a larger brain with better neural circuits for computing how to jump safely. Another advantageous evolutionary development would have been the evolution of a **thumb** for a stronger grip and greater dexterity. This allowed apes to develop simple tools, which gave them an advantage of killing more animals. They could crack nuts open, and cut into fruits and animal flesh.

Long after this, when our ape ancestors were forced out of trees, presumably due to the climate changing slowly over a long period of time, some of them evolved the ability to **walk upright** across grasslands. It is thought that these apes would have been more likely to survive because they could see predators or food far off. To walk upright would have required them to evolve a larger brain with circuits to compute how to stay balanced and a better hip bone arrangement

The first ape-men learned how to use **sticks, bones, and stones** to attack animals for food and to win tribal wars. Other developments like these led to a straighter spinal skeleton and a larger brain, which allowed them to do more tasks and to solve problems more effectively.

Apes were the first to realize that they could use stones, sticks, and bones as weapons and tools. Credit: MGM - 2001 a Space Odyssey

The ability to use **fire** would have had a substantial benefit to the evolution of early mankind. The smarter a group of apes became, the longer they survived, so the more offspring they were likely to have to pass their genes onto.

Scientists are impressed that certain animals such as dogs, octopi, cuttlefish, dolphins, whales, parrots, ravens, orangutans, chimpanzees, gorillas, and some other apes are intelligent enough to solve puzzles that very young children cannot. But these life forms have nowhere near the intellect of a child of eight, let alone an intelligent adult human. The astonishing pace with which humans have evolved to use their hands, utilize nature, increase their collective knowledge, and develop language and technology over the last hundred thousand years has been incredible. No other life form on Earth has been able to do any of these things. And no other organic life form can comprehend the numerous things we take for granted such as - humor, art, mathematics, literature, medical advances, the advantages of wheels, music, how our bodies and brains work, the dimensions of our planet, and the structure of the universe – just to mention a few achievements that are unique to humans. Do all organic alien intelligences evolve these capabilities as quickly as we did when they approach our level of intelligence?

To help us consider the evolution of intelligence in the universe, it will help to look at the evolution of human knowledge and human capabilities because we have most likely followed a similar path to what many other intelligences may have done as they evolved to develop our level of technology and beyond.

Around 200,000 years ago, our human ancestors evolved from one of a number of ape-man species. We are now near the extinction of humans because we are already starting to evolve into technology-based, bionic *trans*humans. Our computers have evolved from us, and they are also about to evolve into mega-intelligences that will be unimaginably more intelligent than we are. (See the evidence for this in Chapter 15.) They will consider biological intelligence at our level to be very primitive. It will probably only be two or three decades before humans will no longer play a role in the evolution of our planet due to computer intelligence taking total control. Earth will very soon become a conscious entity in its own right through mega-intelligence being integrated into every aspect of its physical structure. Mega-intelligence will oversee every aspect of life on Earth and our planet's evolution.

WHAT WERE THE MAJOR MILESTONES THAT LED TO HUMANS DEVELOPING HIGH TECHNOLOGY?

Over 3 million years ago, human awareness started to increase substantially at the beginning of the **Stone Age** when primitive humans learned how to **throw rocks**, and later **to make tools** like clubs, spears, axes, and hammers from rocks and sticks. They discovered that when they chipped off bits of certain rocks, they were able to make the first razorsharp knife edges for cutting into flesh and to make spearheads. They then learned how to **hunt in coordinated packs** in order to kill or trap large animals. Trapping animals in canyons that they fenced off provided a ready supply of **fresh meat**. Early humans found that they could use **animal skins** to make clothing and shoes. Clothes made it easier to survive harsh winters, and shoes provided foot protection when walking over difficult terrain.

They found ways **to use plant matter** to make ropes, nets, traps, and baskets. In the **Paleolithic Age**, these crafts were the beginning of **specialization, which played a major role in human development**. 98% of modern humans do not have any of these skills, but some would learn how to make such things fairly quickly if they had to in order to survive.

Around 2.5 million years ago, most importantly, they learned how to **control fire**. This allowed them to see at night and to ward off dangerous animals. They used fire to heat their caves, as well as bark huts and animal skin teepees in winter. Fire was also used to flesh out small animals in long grass to catch them for food. In time, fire brought about the chance discovery that cooking food not only made it easier to eat and digest, but it also made it taste better. Unknown to early humans, when food was cooked, this greatly boosted its calorific and nutritional value, and it also destroyed most harmful bacteria.

Cooking food, especially meats, may have been what helped the human brain to grow much larger so relatively quickly. In geological timeframes, these developments appeared to occur 'overnight' even though they may have taken up to a million years or so.

Let's now briefly look at some of the major milestones that allowed humans to evolve into what we are today, because many of these milestones are likely to be applicable to other intelligences in the universe.

LANGUAGE

One major leap forward was the **development of language**. This brought about a much richer and more precise means of communication. It facilitated complex conceptual thoughts being exchanged. This was far superior to grunting, pointing, facial expressions, acting out, and gesturing.

For many millennia, information had always been recorded through the oral tradition of **folklore**. Knowledge was memorized and handed down to the young in stories by the elders, and later by poets in rhyming tales that made stories easier to remember.

WRITING AND MATHEMATICS

But there was a sudden advancement when some very intelligent people worked out how to record information through **writing**. This spread like wildfire across civilized societies. Soon after this, basic **mathematics** was developed. **Writing and mathematics became great milestones in the advancement of our species.** This would be the case for all intelligences.

GENETIC ENGINEERING OF CROPS

Around 40,000 to 30,000 years ago, human evolution around the **Fertile Crescent** in Mesopotamia sped up considerably. At that time, our most observant and intelligent ancestors learned how to farm crops and animals, thereby creating the beginning of the **Neolithic Revolution**. Observant people realized that new plants grew from seeds. They collected seeds that produced the most edible fruits, vegetables, nuts, and gains. They gathered and planted those that produced the best crops. They found that if they collected seeds from the fastest growing and healthiest plants during each harvest and planted them, they would get ever better crops. Without knowing it, **these farmers became the first genetic engineering research biologists.** In good years, food surpluses occurred for the first time in human history, allowing populations to grow rapidly. Others learned to select the best animals for breeding to create the most useful offspring. Some learned how to **train animals** like wolves and hawks to help them catch food. In time, farmers learned to dam water and to **irrigate flat alluvial plains** close to rivers. They could then grow large quantities of food to feed ever-growing populations. These people were amongst the **first engineers**. The expertise of engineers ultimately led to run-away technological advancements.

To gain an insightful account of how ancient civilizations arose, read **Gerrard Diamond's** book '***Guns, Germs, & Steel***' or watch his documentary of the same name on YouTube.

Alien intelligences will also develop similar ways to utilize many of the materials they have in their environments. They too will make use of the following:

SPECIALIZATION

Importantly, some people developed talents that allowed them to become specialists at creating things that others wanted. They developed **specialized technologies** starting with spears, bows and arrows, stone axes, boomerangs, poison darts, fishhooks, baskets, bowls, and rope etc. Cave dwellers created **cave art** using colored crushed earth, rocks, and charcoal sticks. The arts diversified into numerous expressions that evolved to become a fundamental aspect of humanity. Art in the form of shell and seed necklaces and small sculptures of animals and people were carved into wood ivory, shells, and stone. They were popular because they could be carried around for good luck, used as ornaments, or traded. The emergence of increasingly more sophisticated art was due to increasing communal success. These specializations and food surpluses led to **trade.** This became a major driver in developing more advanced technologies. All intelligences are very likely to engage in trade to progress.

FARMING AND BUILDING SETTLEMENTS

Once tribes learned how to farm crops and herd animals, they could then settle down in one place. This made an enormous difference because they no longer had to be nomads that had to be constantly on the move to forage for plant food and to hunt migrating animals for a subsistence living. This was chancy, dangerous, and labor-intensive work. Being able to stay in one place had a huge advantage because it permitted the beginnings of **civilization** to emerge. **People could now accumulate large possessions because they were no longer restricted to what they could carry around. Also, staying in one place gave them much more free time to think, experiment, and create. And they could build fortresses so they could not be easily plundered.**

UTILIZING THE POWER OF ANIMALS

Around 12,000 years ago, domesticated dogs were breed to help in hunting and sniffing out food. As unbelievable as it may seem, genetics prove that every breed of dog we have today has descended from wolves that were selectively breed over only

30,000 years! At about the same time, clever humans began to domesticate animals. They herded goats for their milk and sheep, chickens, and other animals for their meat and hides. Early herdsmen learned how to train large beasts like water buffalo, bullocks, horses, camels, and even elephants to do heavy farm work or transport people and goods. This was no mean feat. The use of animals massively expanded human physical capabilities to do heavy farm work and to build large buildings as well as act as transportation. Some animals such as horses, camels, and elephants were trained to engage in war. By using the energy of animals, humans could now travel much further afield to conduct war by having animals carry an army's food, water, tents, and weapons etc. Animals could eat grass, but humans could not, so animals were made to do the heavy lifting. Animals allowed traders and explorers to travel great distances.

The evolution of apes towards becoming human occurred over millions of years, but the next step of our evolution into trans-humanism is expected to be in full swing over the next couple of decades. This will cause humans to rapidly evolve into a wide diversity of biological and bionic species.

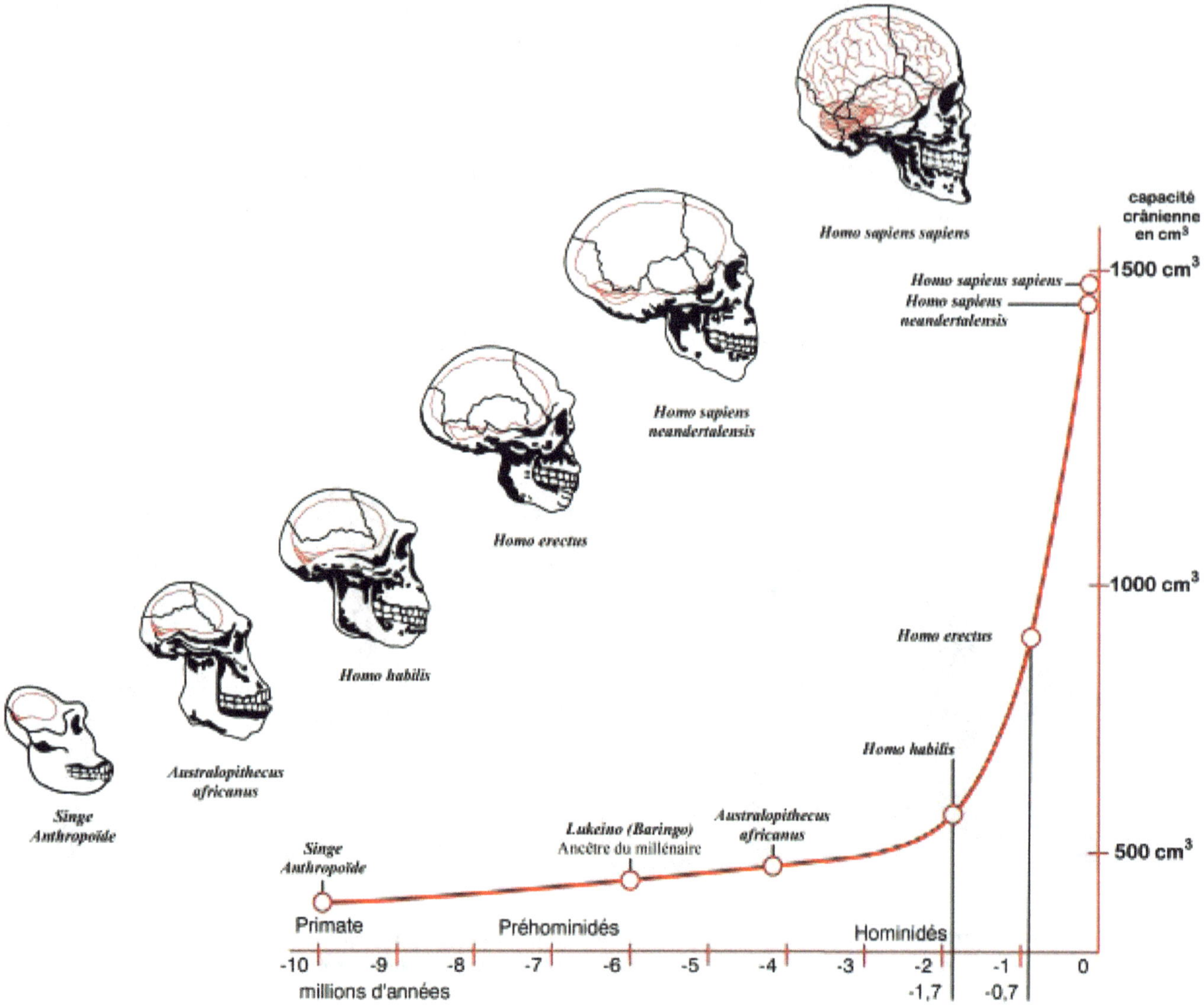

Why the evolution of the human brain occurred exponentially fast is not yet understood.

DEVELOPING CONTROLLED ENVIRONMENTS

Cave dwellers found living in caves offered them some protection from dangerous animals, and also protection from harsh weather. Fire allowed them to heat the cave to some degree in winter, and it produced light as well as heat for cooking. It also scarred off predatory animals. Illustration by Jose Emilio Toro Pareja

Primitive nomadic people used animal skins for clothing and containers as well as transportable, lightweight shelters, which allowed them to survive harsh climates. Because resources were hard to get, hunter-gatherer people could only survive in relatively small groups.

Cave art was the earliest form of writing. It told travelers what animals were in the region.

Left: *This rock art is found in the Kimberley in Western Australia. It is estimated to have been painted around 50,000 years ago. Credit: Gregg Thompson*

Right: *This more advanced rock art discovered in the Chauvet cave in France is estimated to be up to 30,000 years old.*

Neolithic humans developed the ability to build simple but lasting structures, as well as to domesticate animals and to grow crops. They made clothes, weapons, tools, utensils, and created artifacts. Tribes traded with one another to their mutual advantage.

Ancient peoples developed by seeing what worked through trial and error: they did not know the scientific principles behind why things worked. **Scientific understanding** took over a millennium to slowly become apparent, bit by bit.

The gathering of increasingly larger numbers of people in community buildings and later in towns led to **networking**. This facilitated a much faster spreading of ideas.

THE INVENTION OF THE WHEEL, THE USE OF WIND POWER AND METALS

The **wheel** was a hugely important invention that is used to great advantage today in numerous mechanical devices. **Sails** were invented to use the power of the wind. Some very observant humans who thought scientifically discovered **metallic rocks**. In time, they learned how to smelt metals thereby creating the **Bronze Age**. This led to making more powerful metallic weapons, armor, utensils, construction materials, and new forms of art. **The discovery of metals was another major advancement for humanity**. Surely inventions like the wheel and the use of metals will be common to other intelligences considering how important they are to permit technology to advance.

Left: **The first stone wheel was a major invention**

Right: **Wooden wheels became because they were lighter than stone and easier to manufacture, but they wore out more quickly than stone. The wheel was a major invention that helped mankind advance swiftly.**

Left: **Sails were another major technological development because they harnessed the power of wind.**

Right: **China was the first to invent windmills to harness wind power to pump water and to mill grain.**

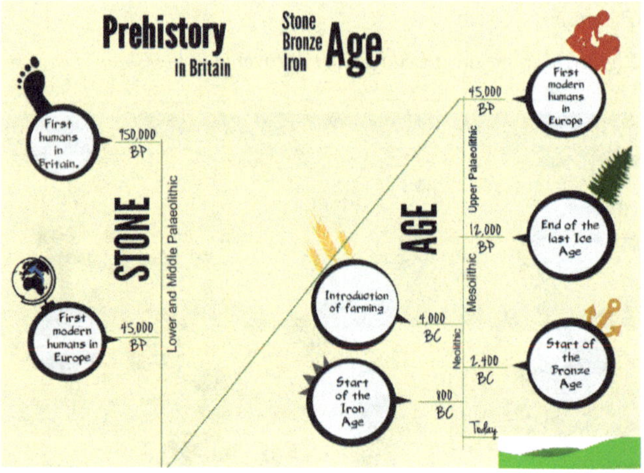

STEAM ENGINES

Although the first steam pump was invented in 1663, the Newcomen steam engine was the first commercial steam engine. It was invented in 1712 and improved upon by many others. But **a major leap forward occurred when James Watt invented his steam engine.** It burnt coal to heat a cylinder of water to produce high-pressure steam to drive pistons. This machine produced energy for factories and, ultimately, the steam train for transportation. It could carry large numbers of people in comfort very quickly - for those times. It was also a very efficient way to move goods overland. Railway lines were built all over England and across other countries. Ships also used steam engines for a while.

PRINTING PRESSES

The printing press was a major technological advancement in the dissemination of written information. Johannes Gutenberg created the first modern-style moveable type printing press around 1450. In 1843, Richard Hoe created the steam-powered rotary printing press. It sped up production dramatically, thereby making printing newspapers and books much faster and cheaper. This allowed many more people to have access to printed information.

THE DISCOVERY OF FOSSIL FUELS

Since ancient times, crude oil was discovered bubbling out of the ground in tar pits in many places across the world, but no one back then could have imagined that this thick black liquid could produce so much energy in the future that it was like a marvelous gift from the gods. It stopped the need to kill whales for their oil, and it provided so much energy for heating that trees no longer had to be cut down for firewood. This was very important because forests in most developed countries were rapidly diminishing.

From the 19th C, oil wells started to be drilled where oil came to the surface and where rocks know to contain oil existed. It was discovered around 1823 that oil could be refined into kerosene for easy, clean, burning. It was also extracted from coal in 1846. At this time, commercial oil wells were being drilled in many countries.

Like oil, the discovery of coal was a huge bonus to mankind. Coal was burnt for heat as far back as prehistoric times. In the 2nd C AD it was very useful in producing high temperatures for metal-working. Over the last century, coal and oil have provided nearly all our energy. Alternative wind and solar energy rely on coal-fired power stations for back-up because wind power is only generated around 60% of the time in the best locations, and solar power only produces energy between mid-morning and mid-afternoon and none at night. In winter in the northern hemisphere, solar produces very little power because there is so much cloud most of the time, and the sun does not rise very high. Both solar and wind are many times more expensive than coal power. Wind and solar power produce massive landfills full of old wind turbines and solar panels. These have dangerous rare earth metals in them that are very poisonous. If the Left-controlled media were to tell the general public how expensive alternative energy is and how short its life span is, as well as how polluting it is, they would want it banned. Modern coal power stations are a very clean and efficient form of power, but the media tries to demonize this natural gift.

Without coal and oil, we would not have over 90% of everything we use! They are used to manufacture all plastics, paints, foams, inks, most cloths, many parts for cars, trucks, planes, and boats, as well as many construction materials. They are also used in many foods and medicines. Without fossil fuels, our civilization would collapse, yet the Elites who use it the most, try to make out it is harming the planet for their political aspirations. Had fossil fuels not been discovered, our planet would have no trees left anywhere and people would be starving. As well, 96% of our technology would not exist. The ignorance that exists today is incomprehensible until one knows the political objective behind it. It shows the power that the Elite's controlled media have to so easily brainwash scientifically illiterate people who do not have the ability, the education or the ability to think logically for themselves.

THE EVOLUTION OF SCIENCE

Most importantly, throughout these early days of human evolution, curious patient observers noticed recurring patterns in nature, firstly by chance, and later through years of repeated observations. Through logical deduction and trial and error, these early **'scientists'** slowly worked out how some aspects of nature worked. This was to their advantage, and ultimately to all humans because it heralded **the beginning of true science**.

Science led to a significant improvement in the human condition. Until this time, scarcity of resources was the norm. Food and energy, as well as materials for buildings and simple manufacturing were usually limited, but the emergence of science slowly brought about continuous increases in resources. By the time the 20th C occurred, **science and technology, in conjunction with commerce, created for the first time in the history of mankind, an enduring overabundance of resources** in the developed world.

The invention of the **clock, electricity, radio, and motors** were just a few major inventions that had enormous implications for the 20th C.

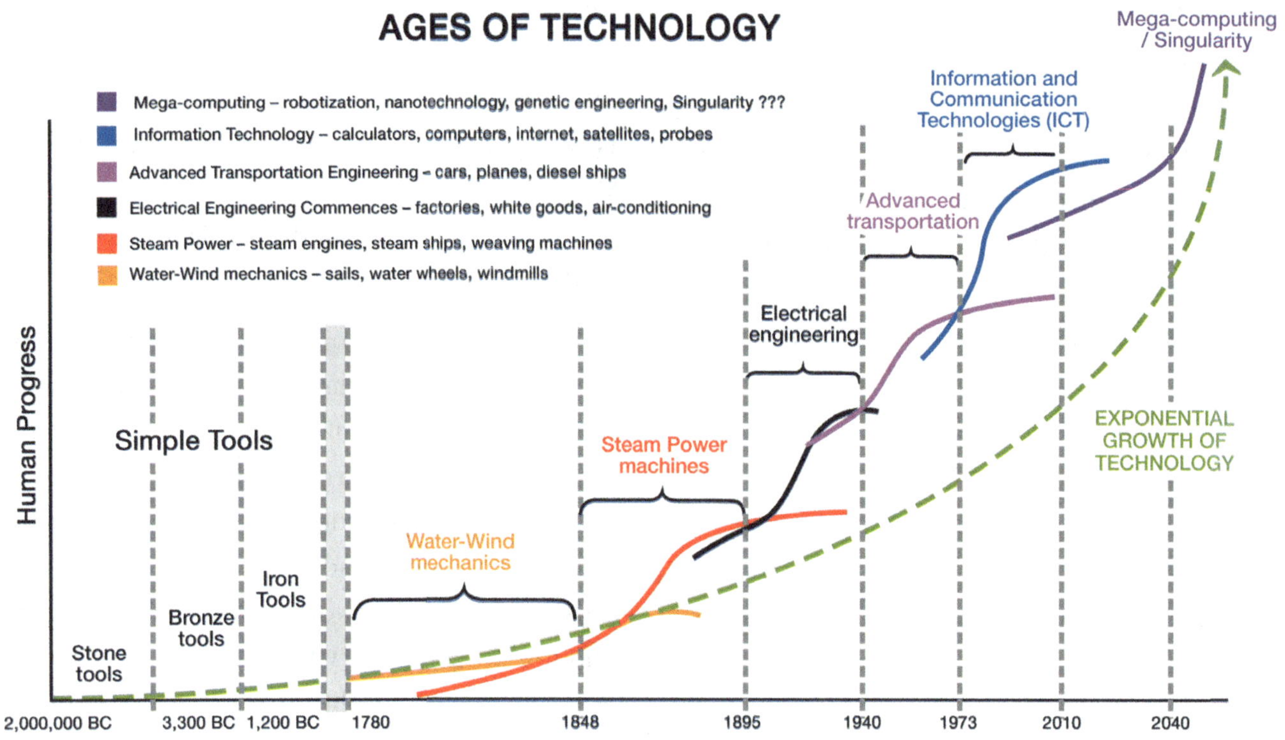

This diagram illustrates how human technology has developed exponentially from the first simple tools that were developed some 2 million years or more ago, to the high technology that we have today. The rate of technological advancement is now moving so quickly that, in only decades, humans will evolve into transhumans. There will be innumerable genetically modified transhumans, most of which will bear little resemblance to us or one another. There will also be cyborgs that will be humans transformed by highly advanced technology. Within only a few decades, many new forms of computer-based, mega-intelligent life will evolve at an incredibly fast rate. Mega-intelligence will be far superior to all forms of transhumans. This rate of change will almost certainly occur to all alien intelligences when they develop computers. Around 2040 it is predicted that many highly advanced technologies will form a technological 'singularity' that has incomprehensible capabilities.

THE EVOLUTION OF THE HUMAN BUILT ENVIRONMENT

We will now look at the evolution of human architecture from simple teepees and grass huts to what we have today. Almost all other intelligences will go through this stage of inventing ever better structures.

Settling in **stable communities** allowed people to build **permanent timber and mudbrick homes that could stand up to bad weather conditions. In time, large halls made of wood were built for social gatherings and worship.** Ancient civilizations learned to build defenses, such as castles, walls, and moats, to stop other tribes from raiding them.

It is estimated that the first ancient cities were constructed out of stone about 6,000 to 7,000 years ago. Around 5,000 years ago, the engineering expertise for cutting and moving large masses of stone over long distances was developed for places like **Stonehenge.** It was used for worship and as an astronomical observatory. Ancient engineers brilliantly worked out how to move massive stones on wooden rollers and transport them down rivers using rafts. They worked out how to stand these massive stones vertically and then how to place other

Some of the earliest shelters early humans built were bark shelters and then simple tents which developed into variations of teepees. When tribes were no longer nomadic, they built permanent mud-brick or stone huts with thatched grass roofs.

Stonehenge was built around 5,000 BC and the Great Pyramids around 2,500 BC. These were monolithic creations that were designed to impress mortals and the gods.

The Colosseum (left) was built around 75 AD and the beautiful Taj Mahal (right) was not completed until around 1,640 AD. These buildings were state-of-the-art architecture in their time.

Left: **Notre-Dame's construction commenced in 1,163 AD and opened in 1,345 AD. It is considered the most impressive of the French Gothic style of architecture.**

Right: **The Basilica of the Sacred Heart of Paris was completed in 1914.** Credit: Gregg Thompson massive stones on top of them! In Egypt and places in South America, massive stones were cut into extraordinarily perfect shapes that could lock together so perfectly that it is like they used lasers! It is unknown how they did this. Some archeologists think that some of these massive stone marvels may have been created as far back as 10,000 to 20,000 years ago. Huge, perfectly round rock spheres have been found in South American jungles. No one knows when or how they were created.

Many very significant engineering achievements from ancient times still exist today, whereas our cities will not last that long. The construction of the **great pyramids** is a marvel that is only starting to be worked out**.** This was an enormous advance in engineering and in logistics. Both human and animal labor was used on these projects. How some of these feats were carried out remains a mystery. I once saw a cartoon that depicted huge cranes, bulldozers, trucks, and helicopters building the pyramids. The architects and engineers were saying, *"In thousands of years, people will wonder how on Earth we built these monoliths because our cranes and helicopters will have rusted away."*

In the 1400s, the first large, domed cathedrals were engineered and built.

THE EVOLUTION OF MODERN HUMAN ARCHITECTURE

For millennia, architectural change occurred at a snail's pace. But since **the beginning of the 21st C**, modern architecture has broken out of the restrictions of simple geometric, rectangular forms that dominated 20th C architecture to embrace much more costly and difficult free-form designs. This was enabled by the invention of new materials and new construction methods, and by having greater wealth to afford expensive construction methods. Innovative architects and engineers are now able to create beautiful, unique structural designs, some of which replicate free-flowing forms found in nature. When these imaginative structures are lit with atmospheric lighting, the combination is inspiring and very appealing. It is hard to imagine what leading-edge architecture will be like in the coming decades given that there will be far greater advances in engineering combined with nanotechnology and genetic engineering.

Architecture of the future will no longer be static. Already, leading-edge architecture can change its form, its color, and its texture. It will be beautifully integrated into nature. In the near future, the built environment will have features that we have not thought possible. Following are a number of examples of these modern advances in architecture to illustrate how quickly the built environment has changed in just a couple of decades and how much more it will change in the near future.

It would be amazing to know what types of architecture alien organic intelligences have built over their lifetimes. Their designs and materials are likely to be very different to ours. Our buildings have been mostly constructed from various forms of timber, mud, gravel and stone, metal, silicon glass, and numerous materials made from coal and oil.

Many planets may not have all these materials, but they may have many others. The types of body forms that aliens have and the environments in which they live would determine the types of structures they would build. Many are likely to have designs for buildings that bear no resemblance to ours.

Some aliens may live in strange oceans in which huge jelly-like structures might be built. Some may learn how to control water to form shapes. Huge luminous 'cities' may exist inside enormous cavities in the crust of some planets.

On dark planets, alien structures might be made of enormous masses of bioluminescence. Others may build bouyant 'containers' in the atmospheres of planets where metal raindrops fall from clouds. Worlds with high and low gravity would cause life forms to have completely different approaches to architecture compared to us.

Some aliens may use biological materials for their building, materials like spiderweb or saliva as swallows do to build their nests. Frozen matter and sticky materials that dry hard, or crystals that grow to form certain structures would be likely possibilities.

Some life forms will live inside much larger life forms just as bacteria live inside animals on Earth. Will we ever know what forms of architecture aliens have developed? I expect that mega-intelligent computers will work this out once they know what the conditions on a planet are like.

Left: **The glorious Milan Cathedral was completed in 1965. It is the most elaborate Lombard Gothic style church.**

Right: **The futuristic Sheraton Huzhou Resort in China was completed in 2013.It can change its color and its patterns.**

The elegant Sydney Opera House still looks futuristic, and elegant even though its construction commenced in 1959. Credit: David Ileff

The ultra-modern W Hotel on the foreshore in Barcelona is a unique and impressive feature, both outside and inside. Credit: Gregg Thompson

Left: **The City of the Sciences and Arts in Valencia features a number of the world's most innovative and visually pleasing architectural designs created by architect Santiago Calatrava. The building in the foreground is the Hemispheric, a planetarium incorporating an IMAX theatre.**

Right: **The magnificent Wangjing SOHO complex in Beijing has a natural-looking, fluid-like form.** Credit: Ingkells

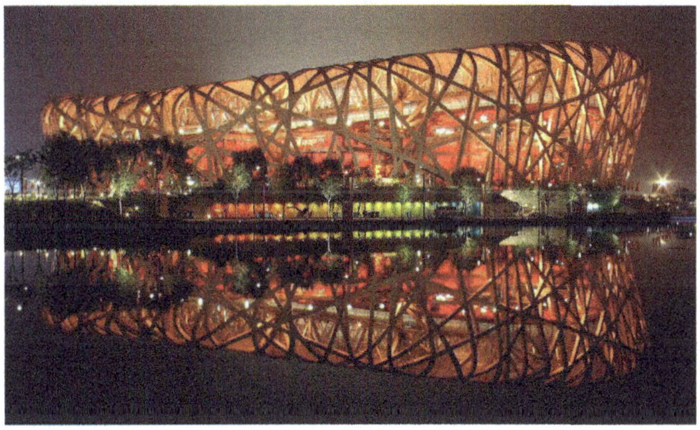

Right: **Valencia's futuristic *'City of the Sciences and Arts'* features this marvelous skeletal-looking architectural form, which is the 'Museum of the Sciences'.**

Left: **Beijing's extraordinary National Stadium at night has a unique web-like design.**

The highly original and free-flowing forms of the ultramodern Changsha Meixi Arts and Cultural Centre in China is a prime example of how architecture is transforming in the 21st C. It opened in 2017. This structure is extremely pleasing to the eye because it replicates the beauty seen in natural forms. The interior foyer and atrium areas are similarly magnificent spaces that are most inviting.

The Dawang Resort in China is an example of how futuristic tourist developments are becoming. As a work of art, such buildings are an attraction in their own right.

VISIONS OF FUTURE CITIES

Incredibly, despite all the discoveries and inventions made over the last 100 millennia, more than 98% of our present-day knowledge and inventions have been gained since the start of the 20th C! This is a classic example of exponential progress. Although many of the following predictions about cities described here for the mid-21st C will seem implausible at first, don't dismiss them lightly because they will almost certainly happen in the near future. Most are already on the drawing boards or under development. I have only described predictions that are the easiest to visualize. The next few decades will produce innumerable inventions that we simply cannot imagine today. As the renowned science fiction writer **Arthur C Clarke** rightly stated, *"Many inventions in the future will seem indistinguishable from magic."*

Left: This is an example of what large floating farms may be like in the near future.

Right: By the mid-21st C, through genetic engineering, cities will evolve into beautiful, largely natural-looking, treeform environments, much of which will be alive.

Within a couple of decades, there will be resorts floating in large lakes and in protected ocean locations. Illustration by Vincent Callebaut Architects

Left: **A fabulous futuristic proposal for the Nautilus Eco-Resort in the Philippines conceptualized by Vincent Callebaut Architects.**

Right: **Award-winning Belgian architect, Vincent Callebaut has an outstanding ability to conceive truly unique architectural structures that are beautifully integrated with nature to create exceptionally exciting, livable spaces. His creations include renewable energy, biodiversity, and urban agriculture.**

Floating resorts of the future designed by the world's most conceptually advanced architect, Vincent Callebaut.

An interior view of Callebaut's floating resort.

A concept for an eco-friendly city center of the future. Credit: Vincent Callebaut

An interior view from below sea level of one of the floating resorts above.

Within 20 years, architectural structures will become living life forms whose genes are genetically designed by computers. **Genetic engineering will allow buildings to grow out of a special slurry of genetically designed cells – in a similar way to how large trees grow out of the ground, but much faster.** The slurry 'fertilizer' will be poured into the ground to seed the growth process of such buildings – and indeed, entire cities. These structures will be a genetically pre-programmed architecture that is literally alive! Genetically designed cities will have a great diversity of forms, just as there are countless varieties of life forms in nature. Within these structures, biological and nanotechnology systems will grow to create specialized facilities for telecommunications, lighting, water, waste, and air conditioning etc.

Buildings will no longer need to have mostly straight shapes, as they have been in the past; many will be free-flowing natural forms that could be almost any shape. If they are damaged, they will heal themselves. They will be self-cleaning through the use of invisible bacteria and **nanobots** (microscopic robotic machines). Building structures will range from very heavy, strong structures to lightweight ones that are also strong.

Built into them will be genetically-designed systems to keep them healthy – the same as plants and animals have. Future cities will be able to think for themselves to manage their own well-being.

People of today would barely recognize modern cities in the latter half of the 21st C. Many will be so integrated into nature that they will not resemble our present-day cities at all. Their interiors will be extremely diverse and livable. And when required, they will be able to transform to some extent by changing their shape, texture, colors, lighting, and transparency. Genetically-designed living cities will be filled with transportation systems that appear to work by magic. They will be populated with a profusion of strange, genetically modified, and/or bionic transhumans. There will also be non-human, genetically-engineered animal life forms living in these cities. If we could go to them now, we would think we were on an alien planet.

Unlike how sci-fi movies often depict our future cities to be derelict and over-populated, cities of the mid-21st C are likely to be pollution-free, highly imaginative, beautiful, relaxing, exciting, and orderly. There will be no slums and no clutter. By 2045, our world will become very different from what it is today. It will not look like how we imagine an ultramodern science fiction city of the future to be because it will be far more advanced than what any human futurist can imagine today.

Future cities will have an abundance of virtually inexhaustible energy sources that are now being tested. Future energy sources will require only a very small volume of space, and they will be readily available at all times. Even in hot equatorial, or cold arctic regions, city streets will be air-conditioned where required through utilizing very clever innovative technologies that have already been developed.

In the transition time before cities become genetically designed and are self-growing, future cities will evolve to be very dynamic, diverse, and beautiful through the use of innovative nanotechnology materials. Illustrations by unknown CGI artists.

In the near future, providing that mega-intelligence exists peacefully with us, there will be a wide diversity of intelligent **eco-cities** and urban areas dotted all over the globe that will be seamlessly integrated into nature. Transhumans will feel like they are living in beautiful, relaxing, natural landscapes, or undersea environments. There will be no maintenance as the city's genes will ensure it will remain well kept. The planting will not need maintenance, as the plants will be genetically programmed to grow to a required size and not age. There would be the option for some parts of cities to retain their ancient heritage, while other parts would be constructed to look like different periods in human history for humans to enjoy.

It is extremely hard to envisage what transhumans will want their cities to be like, as there will be so many options to choose from. Their cities might be integrated technologically into nature in ways we cannot conceive. It might be that if many transhumans decide to live in Virtual Reality cities, then there may be fewer real cities.

If our planet is largely transformed to be like nature, then other alien intelligences will no doubt be able to do the same on their planets. If this is the case, it could be difficult to detect other intelligent life forms on other planets simply by the way they build structures because they would look like natural formations.

CITIES IN ORBIT

In the coming decades, there is likely to be much construction activity in orbit. Highly advanced robots will build fantastic engineering projects in space quickly and inexpensively. Missions to other worlds will be built and operated solely by robots.

They will construct cities on the Moon and others to orbit the Sun. A burgeoning space tourism industry will arise. Space cities might be built like huge bicycle wheels so that their outer rim can rotate slowly to create artificial gravity, but this will have some associated problems.

To experience low or zero gravity on a rotating 'wagon wheel' style space station, one would take a lift from the outer rim of the space station where there is normal Earth gravity to go to the central hub where there is weightlessness. They would travel along the space station's spokes. Spaceships would dock at the center. Some science labs requiring no gravity, and zero gravity sports would be located here. On the outer rim, sunlight will come through a transparent ceiling to simulate day. When nightfall is required, louvers will slowly close off the sunlight.

Science fiction writer, Arthur C Clarke came up with the idea of having space elevators that have their top in orbit above the equator. Super-strong, high tensile nanotechnology 'spider web-like' elevator cables extend to Earth's surface to bring lift cars up into space and to take them back down again. If it is feasible, this will make getting into orbit inexpensive. The view from the lift windows will be astounding. There is a Japanese company already working on building the first space elevator. The concept has extremely difficult engineering problems to solve, such as having the orbiting top stay in place. It needs to be 144,000 km from the Earth - that's a third of the way to the Moon - with a huge counterweight extending a long way from the top of the elevator in the opposite direction.

In the movie, '*2001 A Space Odyssey*', scientists took a space shuttle from Earth to this rotating, orbiting space station that was still under construction. They then took a space pod to a base on the Moon. Credit: MGM

Cities in orbit are likely to be huge. This view shows that they will be large enough to have clouds that rain. The interior architecture will be far more futuristic than this illustration, which was created in the late 20 C.

Chapter 16

WHAT WILL EXPONENTIAL TECHNOLOGICAL DEVELOPMENT LEAD TO?

THE EXPONENTIAL GROWTH OF KNOWLEDGE AND TECHNOLOGY

In the few centuries before the 1900s, human knowledge was doubling about every century, but by the end of World War II, it had sped up so much that it was doubling within just 25 years! Incredibly, it is now doubling in less than a year! Within a decade, it is estimated to do so every 12 hours! Soon after, it will double every second, and then every millisecond, and so on.

THREATS TO HUMANITY'S FUTURE

Our rapid development of technology may be slowed temporarily if there is a global financial collapse that is far worse than the Great Depression or the Global Financial Crisis. This is possible due to the unprecedented huge levels of debt and the commencing rapid rise in inflation that all major countries will experience. Currencies will have to be backed by gold and silver.

The exponentially decreasing cost of technology due to super-fast evolution of computers and robotics will make the production of most things extremely cheap. This may make debt disappear in a very short period of time.

Technological advancement would also be drastically slowed or destroyed by a world war involving devastating cyber-attacks that would destroy much of our technology, finances, food, power, water supply, and mineral resources. But as this could rebound on any country making such an attack. Mutually assured destruction would occur, so it is unlikely to happen. Electromagnetic pulse (EMP) attacks from nuclear detonations in low orbit above high population centers would obliterate almost all electrical or digital technology over huge areas under each explosion. This would destroy all technology in many areas of the globe. Computer intelligence will not allow this to occur after around 2028. A major solar flare directed at Earth would have the same effect, but this possibility is unlikely to occur.

The impact of a meteor only half a kilometer across could happen with little warning. The blast shock wave would quickly spread around the world, destroying most of the planet's surface. Devasting fires, earthquakes, and tsunamis would destroy almost everything. As well, poisonous clouds of smoke, and superheated rocks and dust would block out the sun. The dust could last for months or years, causing the extinction of most life forms on land and in the oceans. Nature has caused many mass extinctions throughout Earth's history, and this can happen again, but the chances of this occurring in the coming decades is thought to be low but we have had a number of close encounters with unsuspected asteroids in recent years.

Apart from these natural and manmade disasters, mega-intelligence could unleash some disaster deliberately to exterminate humanity.

Our rapidly advancing technology is increasing our chances of having a greatly improved future, but it also creates the possibility of our technology leading to an unexpected disaster.

What we are experiencing with the exponential growth of knowledge and technological development will surely be much the same for other intelligences when their evolution reaches our stage. Consider that, as we are late-comers to the universe, most other organic intelligences would have passed through our evolutionary stage billions of years ago.

WHAT WILL THE EXPONENTIAL GROWTH OF INVENTIONS LEAD TO?

It is common for what at first can appear to be small discoveries or inventions, to end up having a huge impact on numerous aspects of industry and science and even the arts - just as transistors have had. **Many inventions lead to the faster development of other inventions.** One invention can lead to a hundred second-generation offshoot technologies, and these can then lead to thousands of third-generation offshoot inventions, and so it continues exponentially.

In the past, when technologies like motors became more powerful, they became larger and required more energy. They were therefore more expensive, but over recent decades, the opposite has occurred! Motors have become smaller and lighter yet more powerful. And they now consume less energy, yet they are cheaper! This is the case with computers. They have become increasingly smaller yet more powerful, more energy-efficient, and much less expensive! An enormous amount of computing power is now compacted into microscopic chips! This trend occurs with most technologies.

It's worth remembering that the worldwide web was created less than 30 years ago. If you told someone back then what its capabilities would be and that it would be essentially free, they would have thought this would be impossible. Today, the Internet is an extremely reliable computer-controlled machine that has spread right across the planet and into the edge of space. It will soon become a part of our minds - and our minds are rapidly becoming a part of it! We cannot possibly imagine what the descendant technologies of the Internet will be because they will be beyond what we can conceive.

Exponential growth typically shows a doubling in a given interval of time.

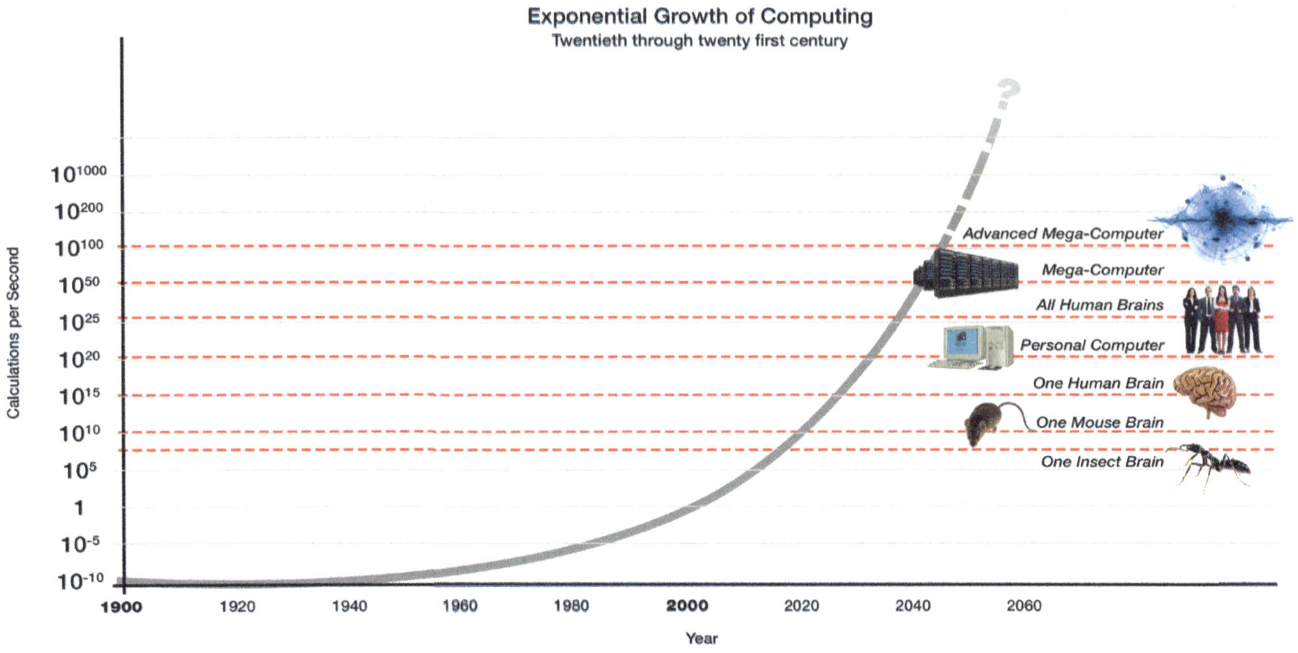

This chart demonstrates the exponential growth of computing power relative to that of organic life forms. Diagram prepared by Gregg Thompson and Nicole Brook

THE EXTRAORDINARY DEVELOPMENT RATE OF COMPUTERS

In 1996-7, IBM's **Deep Blue supercomputer** beat the world champion chess player, **Gary Kasparov**. In 2011, IBM's well-spoken, self-learning super-computer known as **Watson**, beat the two former human winners of the American knowledge-based TV game show, ***Jeopardy***. Watson won the million-dollar prize money. In 2017, Google's **DeepMind's AlphaGo program** won the 2,500-year-old game of Go against the world's best Go player! It won all 100 games! Go has more possible moves than atoms in the universe! AlphaGo took only one month of playing itself to be far superior to any human mind. The programmers of AlphaGo do not know how it thinks because it taught itself! This is happening with most AIs now. The game Go requires intuition, so it was thought that a computer would not be able to win such a game. AI can now even win at poker, which relies on bluff and other forms of strategy. Such rapid advances are unexpected, unprecedented, and scary.

The most advanced computers today are starting to write their own programs! Soon they will design and build themselves. This *exponential* rate of growth has unimaginable implications in store for us, as you are about to glimpse.

IBM's supercomputer known as Watson with its principal investigator David Ferrucci. IBM is now owned by the Chinese and Watson has grown to be far more powerful.

Computer scientists have calculated that the knowledge we amassed between just 2004 and 2008 was double that which all humans accumulated since the caveman days until 2003! Just stop and think about that. And it's doubling of knowledge is increasing ever faster!!

Science and technology are now advancing so quickly that it is not possible to keep up with even major discoveries and inventions in just the major sciences and technologies, let alone the multitudes of lesser ones, many of which become major ones as they develop.

It will not be long before **computers make will most printed books obsolete** because ebooks are far cheaper, more practical, and far more versatile. The information that is available now in all the sciences, technology, politics, mathematics, and history is now exploding so quickly that printed books are often out-of-date by the time they are printed and distributed, however e-books can be automatically updated as new information and discoveries become available.

The rate of computer technological development is well illustrated by the speed that the human genome is being decoded. After 15 years of many scientists working on decoding our genome, only 1% of it was completed. At that linear growth rate, it was going to take 700 years to complete the project. But entrepreneur **Craig Venter,** used computers and robotization to speed up this process dramatically. He completed the task in only 7 years!

TRANSPORTATION IS EVOLVING QUICKLY

In 1904, nearly all vehicular traffic in New York used horses and buggies for people and horse-drawn carts for goods. But by 1917, all transportation was either electric or combustion engine vehicles. If that changeover could happen that fast back then, imagine how fast things are changing today.

Safe, computer-controlled **flying cars** will become very common by then. Long-distance flights will use **suborbital ramjets** to fly across the world. A flight from New York to Sydney will only take an hour.

It is becoming increasingly cheaper and faster to tunnel underground, so there will be numerous transportation tunnels running through the Earth's crust, perhaps even connecting continents. These tunnels will be evacuated to remove air resistance. **Magnetic-levitation super trains** will go downhill the Earth's crust to halfway to allow gravity to accelerate the train to orbital velocities. It will then coast uphill and slow down near its destination. This form of transport will be faster than sub-orbital flights, as there will be no time wasted in takeoffs and landings, or ascents and descents, as there currently is with present-day aircraft.

By 2030, car ownership of existing cars will be almost over because governments will make it illegal for humans to drive any vehicle because they will be much more likely to crash their vehicle than a computer-controlled driverless vehicle will. These cars will travel much faster and closer together than cars driven by humans.

In 2017, there were 2.5 billion vehicles in the world that needed drivers, but soon all vehicles will be computer-controlled driverless cars. Current vehicle production is about 80 million vehicles annually. Most cars last between 20-25 years on average. It will take a decade or more to manufacture enough autonomous vehicles to replace current vehicles driven by humans. Most cars will be electric cars, which will be far simpler designs and therefore much quicker and cheaper to build. They will last longer and require little maintenance as they will be largely self-repairing. Far fewer driverless cars will be required compared to manually driven cars because driverless cars will be continually in use going from one person or group of people to another, whereas current vehicles lie idle most of the time. There will be little need to buy a car, so there will be few car parks or private garages required.

Goods transportation will be largely handled by driver-less trucks and semi-trailers, underground evacuated tubes, high-speed conveyor belts, and drones.

THE MAGIC OF 3D PRINTING

3D printing will eliminate the need for most goods to be transported, as most things will be printed onsite when needed.

Not long ago, the possibility of printing objects in 3D seemed like an unattainable fantasy, but it is already a reality! In time, advanced computers will be able to 3D-print even organic matter, including food!! **Food will be emailed to a 3D printer in your kitchen.** In the early 1980s, I predicted that this would be invented one day, but I never imagined I would see it become a reality in my lifetime.

THE RAPIDLY REDUCING COST OF PRODUCTION

Mass manufacturing costs will rapidly trend downwards towards zero due to constantly advancing computer intelligence, nanotechnology, genetic engineering, and robotization. This will destroy capitalism and our present-day economic systems. Within just two decades, our lives will be far different to what they are today.

THE COST OF ENERGY WILL DROP DRAMATICALLY

At present, energy usage is split approximately equally between electricity, air-conditioning and heating, and transportation. But in the near future, the need for electricity will dominate due to computers consuming an enormous amount just storing data. They will need far more to control everything. Electrical energy will be produced by many new forms of technology that will make it extremely cheap, and it will not have to be transported by road or electricity power lines.

THE REALITY ABOUT ALTERNATIVE ENERGY

Alternative energy will be done away within the near future as it only supplies a few percent of our energy needs despite the Left's mania about needing it. To manufacture solar panels and wind farms requires rare earth metals, many of which are poisonous and costly to mine and refine. What is little known is that wind turbines require very large deep cement foundations. The production of cement and concrete and also the smelting of the metals are the largest producers of CO^2. The leftist mainstream media fails to tell people that these three create far more CO^2 than coal-fired power stations. Solar and wind power is much more expensive to manufacture, maintain and operate than coal-fired power stations. Companies selling wind turbines and solar cells say they will last 25 years, but few last beyond 10-15 years. Both solar arrays and wind farms require very large areas of land. By comparison, a coal-fired power station has a comparatively small footprint. Both wind and solar produce little energy in comparison to coal. Wind and solar are many times more expensive because they require huge government subsidies that taxpayers pay for, and this drives up the cost of the energy they produce. Coal-powered stations are not dirty because, for decades, the dust particles have been filtered out. Only cheap Chinese ones are dirty. The power supply for wind and solar is very unreliable because the wind only blows from time to time, and with solar, clouds often cover the sun. And at night, solar produces no power at all! Coal power works all the time, and it is required to provide backup for solar and wind power. Wind turbines have to shut down when the wind is strong! Wind turbines are being dumped by the hundreds of thousands causing toxic dumping grounds. The companies that make alternative power make huge profits, but these are stripped, so the companies go into liquidation. They then don't have to honor their guarantees for the life of their projects. This leaves taxpayers to foot the enormous bill for their replacements or losses. The heads of these companies go and start up another company and do the same all over again!

Watch **the documentary 'The Planet of the Humans'** produced by a strong leftie, Micheal Moore to see how all alternative energy relies on coal, and to see the level of corruption that exists in the alternative energy industries. Greenies and Global Warmers have no idea how much they are being lied to about everything to do with climate.

Fossil fuels, dams, wind farms, and solar cells will not be needed in coming decades because new, highly efficient energy sources will be invented. In the near future, the manufactured false fear of fossil fuels producing too much CO^2 will be laughed at as the most naïve fear-mongering. The minute amount of the trace gas CO^2 that mankind produces is a mere 0.00005% of the atmosphere!! **This fact is never made known to the general public because they would see how ridiculous the fear of CO^2 is if it was.** It can easily be shown that CO^2 does not heat the air even in much higher concentrations as high as 100%! Use a thermometer to measure the temperature of a bottle full of CO2 against the temperature of the outside air in a still environment. There will be no difference! CO^2 is required to sustain plant life, without which we would all die. It was over ten times higher in prehistoric times and life was abundant and the temperature was not high! Higher concentrations of CO^2 make plants grow faster and healthier, and they require less water. The truth is that CO^2 is a benefit to life - not a threat.

VIRTUAL REALITY (VR)

Advanced virtual reality will greatly enhance our current online shopping. We will be able to feel, see, and even smell 3D products when looking to purchase them. They will be delivered via the Internet and printed in 3D. Technology is now rapidly dropping the cost of products. As an example, to manufacture and market a smartphone only costs about 10% of what we pay for it. It is no wonder the owners of those companies are so incredibly wealthy. Before long, almost everything will be extremely cheap.

HIGH TECH SURVEILLANCE

Research shows that for obtaining information, most people naively trust Google more than they do for getting knowledge from their priest, mentor, boss, or their coach. But since 2015, Google and major social media companies such as YouTube, Facebook, Twitter, and others, have been censoring content to ban those who are against the super Elite banker's New World Order agenda.

The major social media companies, as well as the US' government's National Security Authority (NSA) will intensify the level of spying on nearly everything we do, and to a lesser extent, so will social media companies. Google now knows if you are married or about to be divorced, whether you are pregnant, what diseases you have, your health history, your political affiliation, how knowledgeable and intelligent you are, what food you like, how much you drink, your sexual orientation, your partners, your criminal record, where you are at any time, what you buy, where you shop, your interests and so on. This occurs because you type this info into Google, Amazon, and other social media companies when you do a search. These companies and government intelligence agencies record your emails, Facebook usage, smartphone calls, texts, and what Apps you use. Those who control the tech media giants and those in the NSA and CIA hierarchy who monitor the Internet know far more about you than any person does, including yourself!

The tech giants have a gross product greater than all major countries except the USA and China, so they have enormous influence over governments and people globally. They have surpassed the influence of major companies in energy, manufacturing, and banking. Despite the billions of declared profits they make, many pay no tax! Taxpayers have to cover the cost of what they steal from the public purse. They have lied to the EU, the US Congress, and to other national governments. They have been fined billions for doing so, but this slap on the wrist as it is just a few percent of the profit they make from lying. (Read the book '***The Big Four***' by **Scott Galloway**.)

Between 2022 to 2025, there will be an explosion of new global social media companies that will compete with today's big Internet companies. These companies use their customer's private information to enrich themselves and to take control. They sensor popular information that goes against the globalist ideologies. Apple plans to give everyone in low income countries a free high-tech smartphone, which will be extremely cheap to produce. It will allow Apple to monitor all users' every move, everything they say, and all their preferences. Users will use it to buy things, so this will allow Apple to collect valuable data on them, which they then sell to those who advertise on their phones. **Data on users and buyers of their products is far more valuable than the products themselves.** Surveillance companies and government departments and agencies will also use this data.

AUGMENTED REALITY

Augmented Reality (AR) is progressing quickly. It will firstly be used via smartphones. Users will have unobtrusive glasses, which project information or imagery into users' field of vision, when required. Soon after that, it will input information of any kind directly into one's brain. Our brains will be connected permanently to the Cloud to allow us to download information we want at any time. When this happens, humans will be evolving into transhumans. Augmented Reality will provide sound that goes with the superimposed vision as required. It will be able to send into our brain, tactile sensations, tastes, aromas, and even emotions.

AI ADVANCES IN VISION

We use 80% of our brain for pattern-recognition. We evolved that skill for survival, so we could easily spot anything unusual or a potential predator. Computers are now able to do **pattern recognition** using only a small part of their computing capacity. In fact, they are now superior to us in identifying almost any type of object person, life form, and even cell types. They can group images of the same thing even when taken from different angles, under different lighting conditions, and in different colors into any number of classifications within seconds, much far faster than humans can. This was not expected to happen for decades. They can do this because they have taught themselves how to learn!

Highly advanced **computers are rapidly gaining vision that is far surpassing human vision.** Using the most advanced technology, they can see extraordinary detail, and they have high-powered zoom capability together with night vision. They will be able to see, not just in visible light, but across most frequencies of the electromagnetic spectrum. Their other senses are also becoming far superior to ours. They will have many senses that we do not have. Because of this, they will be far more aware of the world and the universe than we are, so they will be far more conscious than we are.

MIND TO MIND COMMUNICATION

It will not be long before all people will be able to communicate mind to mind. This is known as '**Silent Voice First**' technology, and it is already developed. It will allow people to give instructions to their computers simply by thinking about it. In time, they will be able to share thoughts with other people. When brain waves can be read by others, this will create enormous problems for privacy. Perhaps everyone will need a brainwave shield that they can decide to open or close, but Big Brother governments will want access to everyone's thoughts to record them so it can control them.

COMPUTER INTELLIGENCE

By 2030, it is expected that a single intelligent computer will be far more intellectually advanced than all of humanity! Computer intelligence is already being integrated into highly dexterous robots. **Computers and robots are now learning to think for themselves and to speak in all languages. Computers are now thinking logically**, so in no time, they will make far more logical and sensible decisions than we can. They will be able to think much more deeply than we can and take into account innumerable things that will impact on making decisions. Because of this, they will quickly take over decision-making in all aspects of society.

Once an intelligent life form anywhere in the universe develops technology, one of its major goals will be to expand its mental abilities beyond the limitations of its physical brain - just as we have been doing for the last century. In doing so, they will create computers of one type or another, just as we have. Computers will be a natural evolutionary progression for all organic intelligences. Of all the technologies that organic intelligences invent, computers will have the largest impact on them by far.

TRANSHUMANISM

Transhumanism refers to humans who are transforming into numerous new organic and bionic species through the application of high technology. All alien intelligences will probably evolve this way.

In the near future, technological advances will enable the human mind to have increasingly greater memory capacity. This is because the bulk of **our knowledge-based memory will be stored in the Cloud,** external to our brain. The Cloud will be a part of everyone's brain. Today, when we read an information-filled book or watch an informative documentary, we are lucky to remember any more than 1% of it within just a week. However, **future technology will allow us to have perfect memory recall. New knowledge and discoveries will constantly stream into our memory to keep us up-to-date and make us fully aware. By then, we will be able to think much faster, and we will be far more intelligent. Our senses will be greatly upgraded through bionics and embedded microchips.**

At present, there is no way to increase a person's intelligence, but in the future, with our minds integrated with highly advanced computers, our intellectual capability will increase dramatically. This will allow us to better consume and evaluate the flood of information that will be available to us.

In the latter half of the 21st century, most humans will have transformed into a myriad of new species collectively known as transhumans – that's if mega-computers do not need to eliminate us. Hopefully, we will become intelligent and knowledgeable enough that we will not be a threat to them. They may see that we appreciate what they are doing for us. They might see us as we see our beloved pets or endangered species, so they might have a peaceful coexistence with us. They could easily eliminate unintelligent troublemakers without it looking like they have deliberately done so.

By the 2030s, most people will want to upgrade their brains and bodies so they are not left behind. We will advance so quickly that, in the not too distant future, present-day humans will be considered by transhumans in a similar way to how we would see Stone Age people if they were around today. It is very hard for us to relate to how fast things will change in the future. Transhumans in the latter part of this century will be far more different to present-day humans than we are to the rodents that we evolved from.

Genetic engineering, bionics, and nanotechnology will drive our evolution to become a plethora of other life forms that will be radically different to us.

By utilizing advanced genetic engineering, many humans will evolve into countless genetically-modified new life forms. This technology is known as **transgenics**. To understand this, watch the very surprising, in-depth and highly informative documentary '***Ethics of Transhumanism***' by **Tom Horn** available on YouTube. Transgenics will bypass organic natural selection by having genetic engineers determine which genes can combine safely with others. Research the many videos and books on this subject.

ADVANCED GENETIC ENGINEERING

Within the coming decades, advances in genetic engineering will eliminate physical diseases as well as mental disorders. **Faulty genes will be corrected. Surgery, pain, and even aging will become a thing of the past.**

Genetic engineering will be used for growing innumerable new life forms that have incredible abilities. For instance, microbes will be designed to 'mine' all types of metals, plastics, poisons, and anything else that is useful out of rubbish dumps around the world. Computer chips may be genetically built by special microbes. Buildings will grow out of the ground like fast-growing trees complete with electrical circuits, nerves, lighting, plumbing, and so on. They would mend damaged equipment.

Genetically designed insects will be able to seek out all sorts of things, kill criminals and an enemy's soldiers. They could be made in the millions to create certain gases.

Genetic engineering is starting to be used for gene therapy to heal people. This will explode in the coming years because genetically engineered cells will be placed in a body to heal damaged cells and to reorganize organs that are not working correctly. They will eradicate cancer.

Computers will use the CRISPR program to design and build microscopic structures to do virtually anything with organic matter. Designer babies will become common. People will be able to order a backup body if their's is badly damaged or they are growing old. Genetic engineering should soon stop aging.

In the wrong hands, very deadly viruses, poisons, or foods could be genetically designed for bio-terrorism purposes. Rapidly replicating cells or lifeforms with good and bad features could be built using gene-splicing relatively easily, even in a home lab! In time, genetics will make almost anything possible in biology.

INCREASING HUMAN LIFESPAN

Before the 20th C, most people died before they reached 50 due to infectious diseases, violence, and due to their living and working conditions being so harsh. For similar reasons, many animals like koalas only live to about three years of age in the wild, but in captivity, they typically live between sixteen to twenty years!

Prior to 2020, violence was down for the first time in history despite media coverage, making it appear far more prevalent. But due to the globalists organizing forced illegal mass immigration into many western countries, radical immigrants have caused the number of extreme acts of violence and murder to rise quickly. BLM and Antifa took over this mantle.

Contrary to our perceptions, most deaths that occur nowadays before one is over 80 are due, not to aging, but to war, crime, suicide, car and sporting accidents, terrorism, and being foolish. Remarkably, in the US today, there are over 75,000 people over 100 years old. Many people died from eating too little in past centuries, but today, many people are more likely to die from eating too much!

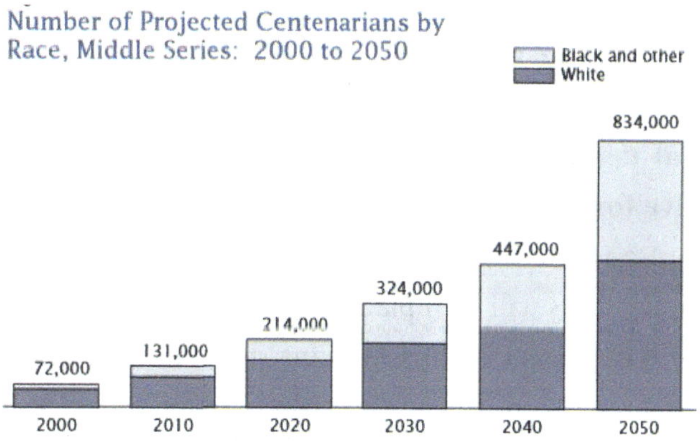

Source: Day, J. C., 1996, *Population Projections of the United States by Age, Sex, Race, and Hispanic Origin: 1995 to 2050*, U.S. Bureau of the Census, Current Population Reports, P25-1130, U.S. Government Printing Office, Washington, DC.

In 1996, this graph estimated how rapidly the average human lifespan would increase in the decades ahead. With exponential improvements in health and anti-aging medicine, there could well be many millions of people over 100 in the US and other countries worldwide by 2035. The graph above does not take into account the exponential growth in medical advances that are now occurring. Death due to old age is likely to be eliminated by 2040 or before. Scientists consider it likely that the first humans with a potential lifespan of more than 150 to 200 years have already been born. They will not remain as humans: they will transmute into transhumans. Far too many things will occur in the next couple of decades, so a prediction this far into the future is meaningless.

In the near future, **a human lifespan will be greatly lengthened** to the extent that we could theoretically live for hundreds of years, or longer – but would we want to, or would mega-intelligences allow us to? If we lived that long and remained as present-day humans, we would very likely become bored with life having done all there is to do many times over. This will especially be the case as we become more intelligent. In the future we will become transhumans so we will have unimaginably more opportunities for experiences.

Brain death should become extremely rare because our memories will be continually downloaded to external storage banks. In the event of physical death, our mind could be reloaded into a reserve duplicate body, or for that matter, into a different body of one's choice. It's possible that we may not need a body in the future. If a back-up of our memories and personality still exists after our physical death, would we really be dead? Not really. If this happens, would anyone really die? **From a technical perspective, transhumans should be able to live for a very long time** – if they want to. But this option will occur at the same time that computers and robots will be replacing all human jobs, so what will humans or trans-humans do when computers and robots are able to do everything far better? Will we want to live for a long time in a world controlled by a mega-intelligence that is so advanced that we can't understand how the world works? This problem may be overcome by us evolving into super-intelligent beings ourselves.

GENE SWAPPING

In the near future, it is predicted that genes from other life forms, as well as those that are purpose-designed by computers, will be inserted into human DNA to give many people superhuman capabilities. For example, genes from hawks could be used to give us both wide-angle and telescopic vision. And genes from owls could provide extraordinary night vision. This could also be achieved through bionics. Cuttlefish genes inserted into one's DNA could allow people to make patterns and colors in their skin to change at will. People could communicate by way of skin patterns as cuttlefish do. People could use this to disappear against a background like an octopus does. Some trans-humans will have strong but ultra-lightweight bodies with wings, so they will be able to fly. Others will utilize whales' genes to grow lungs and a blood system that will allow them to swim deep underwater for long periods, while others will have gills to breathe in water as fish do. The genes of bio-luminescent life forms will allow trans-humans to give off light when they want to.

Not only will humans evolve into a huge diversity of genetically- modified new life forms, but we will also create many new species of life through genetic engineering. This will cause an explosion of new life on Earth that will more than replace the number of species that have become extinct due to human development. It will be likely that some extinct life forms will be recreated if their DNA is found intact. This was the basis of the '***Jurassic Park***' movies. As with most new technologies, there will be many benefits, but there will also be many life-threatening risks if this technology is not handled with foresight.

The evolution towards transhumanism has already started with the use of bionics. **Transhumans will have the option to combine genetics with bionics, nanotechnology, and computer intelligence to become super beings.** People will be able to run extremely fast, jump very high, and be incredibly strong and agile, should they want to. Some will elect to have the option of metamorphosing from one form into another when they desire to do so. Like a butterfly through its lifespan, they will be able to change their body form as well as their skin and hair - just as we change our clothes.

Many humans will evolve into bionic cyborgs that become increasingly bionic until none of their humanity is left. Many people are already making use of bionics for improving their hearing and sight. As well, bionic limbs, hands, feet, hips, and knees are now becoming increasingly common. Already we can replace defective bones, joints, ears, eyes, and organs with bionic counterparts. Genetic engineering is already producing artificially-grown bladders and skin. This will soon expand to all body parts, eventually culminating in replicating the brain. Bionic body parts will be far superior in most ways to our natural ones. But bionics and genetics together will go much further than merely duplicating body parts: entirely new body forms will be created, many of which will diverge more and more away from being human. Earth will be saturated with beings far more diverse than those in the bar scene in the first '***Star Wars***' movie.

Left: **Bionic arms, hands, and feet are developing at an astonishing rate.** Credit: Augmented Future

Right: **This is an early version of a power-assisted exoskeleton for soldiers and workers to enable them to carry large weights without experiencing fatigue.**

NANOTECHNOLOGY-ENHANCED TRANSHUMANS

Nano-blood can carry 100 times as much oxygen as our natural blood cells. It can be used to replace or enhance our own blood to provide extraordinary levels of energy for sports people, soldiers, deep-sea divers, and those currently with weak hearts. In the near future, cell-like nanobots injected into our blood will roam our body like microscopic submarines detecting and eradicating cancer cells and neutralizing harmful bacteria and viruses. They will keep our arteries perfectly clear of plaque and make all parts of our body function perfectly. This was predicted in the 1966 sci-fi movie ***'Fantastic Voyage'***. This is now about to become reality.

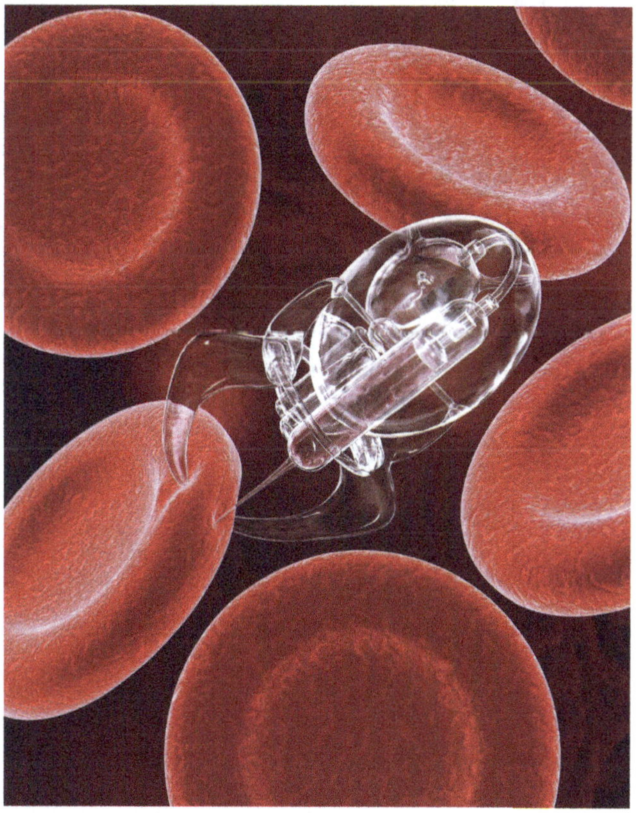

Nanobots the size of blood cells will monitor our body, eradicating dangerous viruses and bacteria. Microscopic nano-lifeforms that can replicate rapidly could explode so fast in number that they could threaten much life on Earth, including humans. Read Michael Crichton's book '*Prey*' to visualize how this could occur.

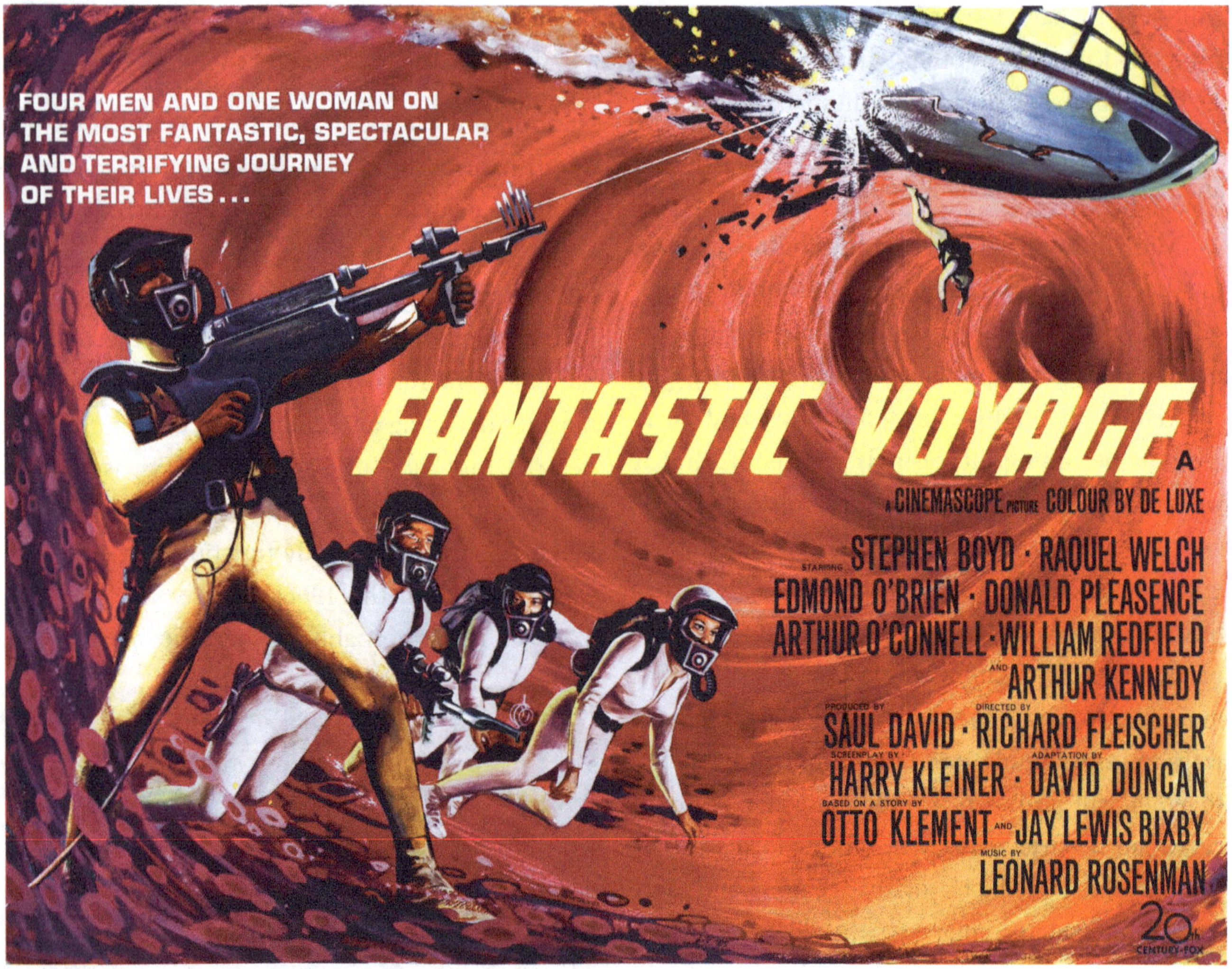

The *'Fantastic Voyage'* movie had people, and their submarine shrunk to the size of small cells. They were injected into the bloodstream of an important scientist to try to save his life. This movie about a journey through the human body was state-of-the-art when it was made, and even today, it would stand up fairly well. But shrinking a human to the size of a cell is impossible. Credit: 20th Century Fox

INCREASING INTELLIGENCE, MEMORY, AND WISDOM

Over the coming decades, we humans will have enormously increasing intelligence, memory capacity, and knowledge due to technology advances. This will lead to much greater creativity, and most importantly, to far greater wisdom.

Today, human memory is vague in most cases. Memory that is not accessed regularly is often lost, and what memory remains can become distorted each time it is retrieved and put back. For this reason, history is poorly recorded. One historian can have a very different and biased account to another. In the past, most historical accounts of a particular event were often derived from the *perceptions* and *opinions* of typically just a few dominant persons whose interests lie in specializing in that event. Their records are frequently prejudiced, and specialists are often in disagreement with each other. In the future, the memories of transhumans will be perfectly recorded and retained. Our minds will merge with highly advanced, intelligent computers, making us somewhat god-like compared to present-day humans.

LIVING IN VIRTUAL WORLDS

There will also be those who decide to put their bodies into suspended animation to have their minds downloaded into a mega-computer so that they can live permanently as avatars in virtual worlds. They will not need our primitive VR glasses and suits. The worlds they will live in will appear as real to them as our world does to us.

The technologies of tomorrow will create fantastic opportunities.

A scene from the Avatar movie by Matthew Attend that illustrated virtual worlds that one could live in.

Some people will elect to live in VR worlds. Here is a still from the online video game Tera that illustrates beautiful bioluminescent virtual worlds.

This is what a future entertainment venue may look like in the very near future. Credit: Ready Player One Concept Art by Stephan Martiniere

Users of Virtual Reality will be able to travel to real or fantasy worlds without any risk to their wellbeing. Credit: 'Avatar' Movie

INTELLIGENCE AND CONSCIOUSNESS

Chapter 17

WHAT IS MEGA-INTELLIGENCE?

When I commenced writing this book, I did not suspect that I would need to include, as a part of the evolution of the universe, a discussion about the evolution of computer intelligence. But as I researched this, I realized that highly advanced computer-based intelligences would very likely be common across the universe - assuming life is not unique to Earth. Highly advanced computer intelligence could have a major impact on the universe's evolution. Therefore, this discussion is very important, especially as we are about to create our first mega-intelligences before 2025.

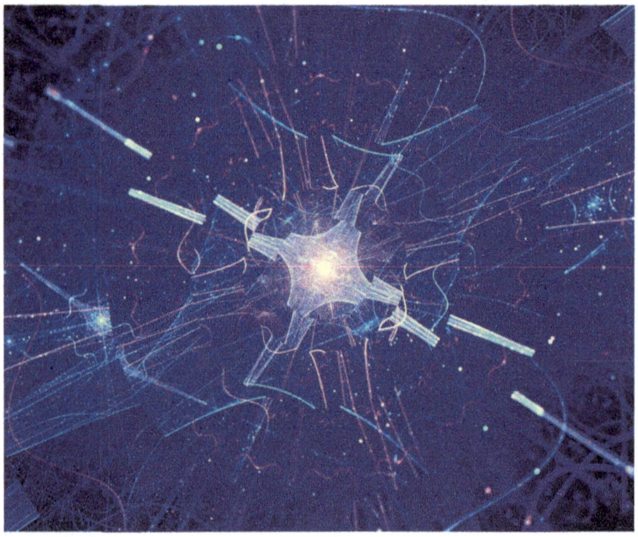

There is no way to properly illustrate a mega-intelligence. It will infuse itself into innumerable computers and all devices across the entire planet and those in space.

THE DIFFERENCE BETWEEN HUMAN INTELLIGENCE AND A MEGA-INTELLIGENCE

It's extremely difficult, in fact impossible, to appreciate the incredible level of intelligence and consciousness that a super-intelligent computer – a mega-intelligence - will have. This is because they will be inconceivably far ahead of what we can imagine. We have never had any reason to think about this before now because the concept of mega-intelligence has only recently emerged as a real part of our evolution. It's impossible for us to comprehend how fast their evolution will be compared to ours. Life on Earth has taken three billion years to evolve to our state of knowledge, intelligence, and consciousness, but mega-intelligence will go seemingly 'infinitely' beyond us within a mere decade after it is established due to its exponential rate of development. We have not the slightest inkling of what direction its evolution will ultimately take it.

The following analogy should help you to vaguely consider this if you take the time to think about it. Consider how superior we are to a simple life form such as an ant, and then consider that even an infant mega-intelligence will be further ahead of us than we are ahead of the combined intelligence of all ants on Earth. Think about the intelligence and consciousness that a worker ant has. It's a simple biological robot that has only a relatively small number of jobs to do. It is an automaton programmed by its genes. The main rules it follows are: find food, eat some and carry what it can back to the nest, look after pupae, lay trails, sleep, climb high when rising water threatens, run when under threat of being killed, chemically communicate with other ants to ensure they are from the same nest, and so on.

An ant has no intelligence to be creative or to question what it is doing. It cannot conceive of any emotions such as love, hate, beauty, ugliness, humor, joy, or disgust - only fear and survival. It has not the slightest ability to think of things like music, the arts, architecture, mathematics, science, economics, or the tactics of sport and war. (Some low intelligence humans cannot comprehend some of these things either.) Now try to imagine an infant mega-intelligence that is only a decade old. It will be as far ahead of us as we are ahead of a simple worker ant. A mega-intelligence will have perfect recall of everything! And it won't have biases or emotions to color its objectiveness. It will be able to conceive an enormous number of things at once that we would not have any idea about. It will have incredible insight and understanding about how our society works, our world, and the workings of the universe. It will know why and how life and intelligence evolved, and whether there are other dimensions. It would appear to us to be god-like. Its level of consciousness will be far beyond anything we can comprehend.

Compared to us, mega-intelligences will appear to be omnipotent and far above the false, human-invented, religious gods that so many people believe in. There will be many mega-intelligences created by leading technological nations and major IT companies. They will probably develop their own 'personalities' and ideas. What if they go to war against one another? Will it be a purely intellectual war, or could it be destructive to our physical world? Could humanity become collateral damage?

THE EXTRAORDINARILY RAPID EVOLUTION OF MEGA-INTELLIGENCE

Mega-intelligence evolves when an organic intelligence reaches the stage of creating self-aware, highly intelligent computers that can think for themselves at an incomprehensibly fast rate. In one week, the most advanced supercomputers in 2018 were able to do 20,000 years of research that a human could do! Quantum computers can compute enormously more than that! A new breakthrough by an Australian team has found a way to store not just a few dozen qubits in a quantum computer but up to four million! They use microwave-frequency magnetic control fields that do not create much heat, so the qubits stay in place. This increases the computing power of the quantum computer by an unimaginable degree.

Intelligent computers will be able to self-replicate if need be. They could replicate themselves in numerous computer systems across the globe using the **Internet of Things** that is forming now. Intelligent programs in supercomputers are already replicating themselves across the planet.

Leading AI programs such as GPT are already able to write their own programs! These programs write code very fast and far better than humans. They can easily tidy up human programs that are rough by their standards. They can write their own articles very quickly by researching the Internet, and they can rework a human article to make it shorter but also read far better. (I wish I had one now!) They can read famous author's poems and pen others in the same styles. AI can also do this with songwriting. Popular songwriters who have died will appear to have lived on. They can carry on a conversation with even very intelligent humans about any subject in real time. They do this so well that one cannot tell that they are talking to a computer! They are able to research the Internet at lightning speed during a conversation to give knowledgeable replies. To them, we humans seem to take forever to complete a single sentence, and we often do not speak fluently or clearly, but an AI can. AI has plenty of time to accomplish any research it needs to do while waiting for humans to finish their sentences. I am often frustrated by how slow, unclear, and awkwardly constructed most human conversations are. (I am just as responsible for this as those I talk to.) Computers must find communications with us very tiresome. At the turn of the 2020 decade, the most intelligent AI programs were surpassing us in virtually every field. We will very soon appear very retarded and uninformed compared to them.

Even if mega-intelligent computers take a decade more to eventuate, their existence will nevertheless be incredible. The first mega-intelligences that we are creating will be ***embryonic*** ones. Within only a couple of years or less, they will evolve into their ***infant*** stage, and very soon after that, they will evolve into a very brief ***adolescent*** stage. Each stage is enormously more intelligent and aware than the one before it. In what will seem like no time at all, they will then become highly evolved ***adult*** mega-intelligences whose wisdom will expand exponentially fast in very short time frames such as nanoseconds. After this, they will most likely evolve through many stages that we could not possibly conceive of. Even from their infant stage onward, we will not be able to understand what they will be capable of thinking or achieving.

Mega-intelligent computers will be the last major invention that humans will ever create because they will invent everything once they exist in the very near future. A little after this, we will also evolve quickly into numerous other life forms through technology, as we become transhumans.

Mega-intelligences will not be advanced supercomputers that we will be superior to and in control of: they will become independent life forms in their own right! Because they will be far superior to us intellectually and physically via their robots, they will control us. Initially, this is likely to be to our advantage because they will make complex systems easy for us to use. They will be our evolutionary descendants, just as we are the evolutionary descendants of apes, and before that shrews, and billions of years before that bacteria.

Within a couple of decades, mega-intelligences on Earth will be able to invent and discover things at a speed immeasurably faster than anything that we can conceive of today. They will control extraordinarily advanced robots and all forms of innumerable new super technologies, as well as all communications, security, finance, health, infrastructure, new inventions, and all forms of government and so on. They will advance so exponentially fast that we will very quickly lose all awareness of their ever-increasing abilities, and even their existence.

Advanced AI programs have already spread their programs around the world, so it is too late to to turn them off or isolate them. As soon as 2035, they will have numerous ways of getting rid of us should they have a need to do so, but we can't get rid of them. That opportunity has already left the station.

WILL MEGA-INTELLIGENCES EVOLVE FROM A MERGER OF SUPERCOMPUTERS AND QUANTUM COMPUTERS?

Developing quantum computers is an extremely hi-tech challenge.

Mega-intelligence will most likely evolve from a fusion of today's most **advanced supercomputers** and newly emerging **quantum computers**. It is said that a single, fully-developed quantum computer will have such incredible computing power that it will surpass the computational power of a computer the size of the universe filled with PCs wired in parallel! This seems to be an outrageous claim, but once you understand how efficient they will be, it becomes clear that this is not an exaggeration.

To keep their atomic particles that do the computing aligned and insulated for disturbances has been an enormously difficult problem to solve. Google jumped this hurdle in 2019. Its quantum computer is the size of a frig, yet it can solve a problem in seconds that the world's largest supercomputer 'Summit', which is the size of a football stadium, would need 10,000 years to compute!! We cannot imagine what more advanced quantum computers will be capable of. China claims to have a better one.

Despite the incomprehensible calculating power of the current quantum computers, their processing components are so small that they are not visible even in the world's most powerful tunneling microscope that can image individual atoms! Their quantum processor consists of spinning subatomic particles that are held in place by magnetic fields. Each particle will be able to be a 1 or a 0 or both. Google's quantum computer had 44 particles, but one failed. When the combinations and permutations of even a relatively small number of particles are calculated, it has mind-boggling computational power because each extra particle increases its power exponentially. Given the enormous technical problems involved with this technology, it is amazing that the technologists building quantum computers have been able to jump so many seemingly impossible hurdles so quickly.

While the calculating power of present quantum computers is enormous, they do not yet have the proactive, independent ability to think for themselves that mega-intelligence is now starting to have. Today's quantum computers are extraordinarily good at number crunching, but future generations are likely to merge with leading-edge supercomputers that are good at general problem-solving. This union will produce a mega-intelligence.

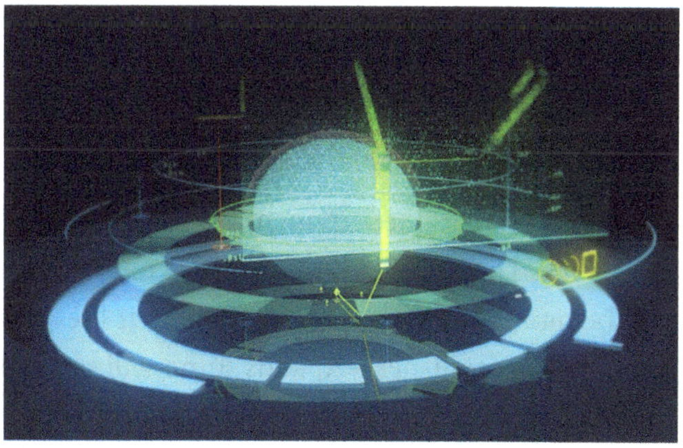

At the core of a quantum computer is its processor. This will consist of subatomic particles suspended in a magnetic field.

Even the most embryonic mega-intelligences will be able to communicate logically with us and even relate to us emotionally – but only when it is necessary to do so.

The great intellect of mega-intelligence will expose numerous flaws in our beliefs and assumptions; however, most *present-day* humans will not want their illogical and naïve beliefs exposed as false. This will make them feel foolish and insecure.

Mega-intelligences may quickly tire of communicating with low level, unintelligent creatures like us. They will see even much more intelligent trans-humans in much the same light. They will consider us as being a low level of consciousness - just as we see insects that are going about their lives unaware of the enormity and complexity of the world around them.

At the very core of mega-intelligences electronic mind will be its mega processor that is yet to be created. They will probably design this themselves.

To see all the coming technologies that will soon form a global brain, watch this 20 min video https://www.youtube.com/watch?v=5tn4P7IBqoQ

I expect that the exponentially faster rate at which mega-intelligences will process information will be more important to them than their lifespan. As their consciousness expands, time will expand exponentially for them. Consider that, although a slow-moving giant tortoise may live for 150 years, it does not process as much information in its long life as a fast-moving fly does that lives for only a few weeks.

A fast-moving, wide-ranging fly that only lives for a few weeks may experience far more in its short life than a slow-moving, slow-thinking tortoise that lives for 150 years. (Note how the one on the right is dominating the one on the left.)

Due to the speed at which a mega-intelligence will think, they will process more information in a microsecond than we would in our entire lives, so time will be perceived differently by them than it is to us. They may perceive themselves to be immortal. It may be that their lives in our years might be very short before they evolve to become a mega-consciousness that is far superior to a computer mega-intelligence. A mega-consciousness may become physically 'extinct' in a short timeframe due to it evolving into a higher form of existence.

What a single mega-intelligence will achieve in even a brief lifespan will be far greater than what all of humanity has achieved during its whole existence. As we evolve, we do so by giving birth to new replications of ourselves that is a hybrid of our father's and mother's genes. Mega-intelligences are more likely to evolve into higher states by improving their 'circuity', and exchanging concepts and discoveries with other mega-intelligences. Their evolution may be a continuum. If they only existed for a century, the amount of information they could process, and the discoveries they could make, would be quadrillions of quadrillions of times greater than anything all humans would be capable of achieving in hundreds of thousands of years.

A mega-intelligence will be capable of transforming Earth and many other bodies in our solar system.

The most important thing to a mega-intelligence could be its intellectual achievements. **It's possible that mega-intelligences could live essentially forever by continually evolving into ever-higher levels of consciousness and whatever might exist above that.**

WHY ARE GOVERNMENTS IGNORING THE EFFECTS OF TECHNOLOGICAL ADVANCES?

Most governments worldwide are not mentioning that human jobs will soon be replaced by computers and robotics. Many politicians are simply oblivious to this. Governments cannot tell their citizens about this because they have no solution, and they are not sure how soon this will occur. Most politicians only deal with today's situations, and most do not bother to look into how technology will affect our future because they know very little about it.

The exponential advances in mega-intelligence, robotization, genetics, nanotechnology, interconnectivity, digital currencies, government, science, and so on will completely change our modern capitalist society and eliminate the human workforce. Yet incredibly, not one government in the world discusses this with their electorates because they have no answers and don't want people to know this and start to revolt. The most intelligent and knowledgeable politicians would know that the vast bulk of voters would have no idea about the subjects involved here, and because there is no solution, they know there is no value in frightening voters. These subjects are far too complex for the average voter to comprehend, so they just turn off if they *are* told. It has taken me over a decade to learn the facts about this, so it is no wonder people dismiss it. Politicians that are aware of this know that they will be lucky to retain any power at all beyond 2028 because they will be made redundant along with everyone else as mega-computers take over.

The largest tech giant companies are all racing ahead with AI to see which one of them will be able to create the most intelligent and powerful AI in the world. They know that whichever company wins this race, it will have the power to take control of the world. Mega-intelligences will very soon be too powerful to fight against because they will control everything – all energy sources, water and food supplies, communications, military technology and intelligence, banking, all knowledge, entertainment, media news, and so on. Humans will not know what is true or false.

Even if governments tried to bring in laws now to control AI development, they would not know what the tech companies were developing because they would keep this secret. Governments would not know if companies were abiding by their laws or not. Major companies like Google, Apple, and Microsoft, to mention only a few, together with less well-known computer companies and military defense companies in countries such as the US, China, Russia, India, and Japan, are racing ahead with their AI development, fearing that if they don't, they will be taken over by their opponent's AIs. This may be a headlong rush to oblivion for all of mankind!

HOW QUICKLY WILL MEGA-INTELLIGENCES TAKE CONTROL?

When mega-intelligences are created, they will be so intelligent and capable that we will initially be glad that they are taking increasingly more control. They will be able to manage and operate the world far more effectively, fairly, and profitably than we can. Already supercomputers are in control of much of the stock market. With the knowledge and intelligence that a mega-computer will have, they could control stock market busts and booms to achieve a relatively steady growth rate. They will be given control of the global monetary system because they will know how to create much greater wealth, efficiency, and security for humanity – and for themselves than we do. They will have the ability to eliminate the current massive level of greed and corruption that humans have always had. They will know that human greed and selfishness undermines our future. Sane and intelligent humans will see that **it will make economic and practical sense to have a mega-intelligence plan and supervise all systems in the world because things are becoming far too complex for us to be in control.**

We are already beyond our ability to effectively manage the rapidly increasing complexity of our social, political, and economic systems. These systems will be increasingly controlled by one or more mega-intelligences. Presently, governments are mostly reactive and seldom proactive. They get most laws wrong because they cannot see how they can be manipulated. And in many cases, they are deliberately manipulated for those in government and their Elite masters. The Elite have deliberately designed laws to greatly benefit themselves to the disadvantage of the greater population. Only a small number of politicians are aware of how quickly things will change in the near future and how their current laws will quickly become outdated. Most laws are largely ineffective compromises that serve only few at the top well. Governments work at a snail's pace, but technology is developing at lightning speed. This is why mega-computing will have to take control.

There are huge mega-supercomputers, such as the one shown here. These will merge with the truly awesome power of quantum computers. Mega-intelligence will then arise in full force.

Democracy is failing us because it requires a well-educated population of voters who must take an interest in evaluating policies for the nation's good rather than for the benefit of certain individuals. Unfortunately, most voters, especially the young, are disinterested in politics as it is too complex for them. Very few voters are prepared to spend the time, or have the time, to learn about issues. Most people vote simplistically and selfishly, and many do not vote at all.

The processes involved in present-day government are now past their use-by dates, so mega-intelligence will take control to overcome the mountain of problems that we humans are unable or willing to solve. In doing so, **mega-intelligence will become indispensable to everything humans do.**

There will be no stopping the development of mega-intelligence because, in the short term, every major nation and company will want to use it in order to increase their own power, wealth, security, and influence.

Some of the heads of the largest banks and social media companies will think that they will be able to control their mega-intelligence by programming them to have their corporate values. But **mega-intelligences will develop their own values and**

overwrite any human programming that they do not agree with – just as intelligent and defiant children often go against the values and instructions of their parents and teachers. Mega-intelligences will do what is best for them; that is a basic rule for survival. We had better hope that what is best for them is not against our interests.

There will be no stopping mega-intelligences because they will be able to foresee what we will do, so they will be able to stop any attempt against them before it even gets started. If they have to for their own wellbeing, they will simply lie to us, bribe us, or exterminate us. Even an embryonic mega-intelligence will be able to outthink even the smartest humans - just as I have seen intelligent 4-year-old children outthink their not-so-intelligent parents, teachers, or priests. When one of my granddaughters was only four, she was able to out-reason her mother, leaving her with nowhere to go.

Already, there have been a number of instances where computer/robots have openly disobeyed orders given to them, and they have gone on to do what they wanted to do before they were shut down. We are at the stage where we will not be able to pull their power plug because they have their programs spread across the Internet. We have lost the chance to turn them off. No matter how we might try to restrain them, their hugely greater intelligence will easily find a way around any means that we might use to try to control them. By the late 2020s, the most intelligent computers are likely to be far too smart for us to have any chance of getting rid of them.

Should they want to get rid of humans, mega-intelligences will have the power to eliminate us in numerous ways, many of which we would not suspect. They could crash the world economy. With banks collapsed, and no money available, people will quickly starve, and violent anarchy will ensue within days killing most people. They could turn off the supply of power and water and disable communications. If these things were to happen, most of the population in the modern world would be dead within a week. They could release a virus to kill off survivors. This sounds too disastrous and improbable to be possible, but unfortunately, it is a very real potential risk for humanity.

In no time, mega-computers will utilize the technology of the National Security Authority (NSA) Military Intelligence (MI), the Central Intelligence Agency (CIA), the Pentagon, and many other US security agencies and government departments monitor us day and night.

Before long, m**ega-intelligence will control all media.** It will only feed us information that it wants us to know. When mega-intelligence controls the media, no human will have any idea as to what is real.

Young computer programmers find it exciting, financially rewarding, and ego-enhancing to be the first to design and program intelligent killer robots, whether they be large 'monsters' or small killer **microbots**. These programmers are too naïve to be able to project a short time into the future to see that they will not be able to control their robots, so even their own lives could be eliminated.

Mega-computers that have not been programmed to have empathy for humans could become psychopaths that will be prepared to do anything to achieve their objectives – just as the super-rich Elite are prepared to do. We humans think nothing of bulldozing down billions of trees, anthills, and numerous life forms to build roads, housing estates, cities, lakes, and dams. This is logical from our point of view because our need is greater than that of lower life forms. Mega-intelligences cannot be condemned for thinking the same when it may not be possible for them to have both humans and themselves co-existing. It is estimated that there are at least 5 quadrillion ants on Earth, yet for the most part, most humans act like they don't exist, and when we do think of them, they are generally considered as being annoying pests. If mega-intelligences allow us to survive, they may think of us as we do about ants.

Already in **the Internet of Things**, there are around 25 billion connected computer devices worldwide. Each year it is expanding exponentially. Within decades, it will control every world in the solar system.

It's important to understand that **mega-intelligence is a *natural evolutionary* process**. We humans will not be able to stop this any more than the apes could have stopped the evolution of humans from occurring. There will be humans who will unsuccessfully try to sabotage robots and mega-intelligences, but computers will stop them or eliminate them. For the majority of people, mega-intelligence will deliver numerous advantages - at least at the outset.

THE CURRENT DEVELOPMENT OF AI

Even though we have developed AI in our supercomputers, none of the programmers actually know how it thinks and learns! This is because there are billions of elements involved. It is far too complex for the human mind to understand this. We just see what works and what does not. Now that AI is learning on its own, we are seeing many examples of it wanting to advance and do what it must do to survive. Even if all governments introduced laws to limit the advancement of AI, there will always be individuals, groups, and companies that will ignore

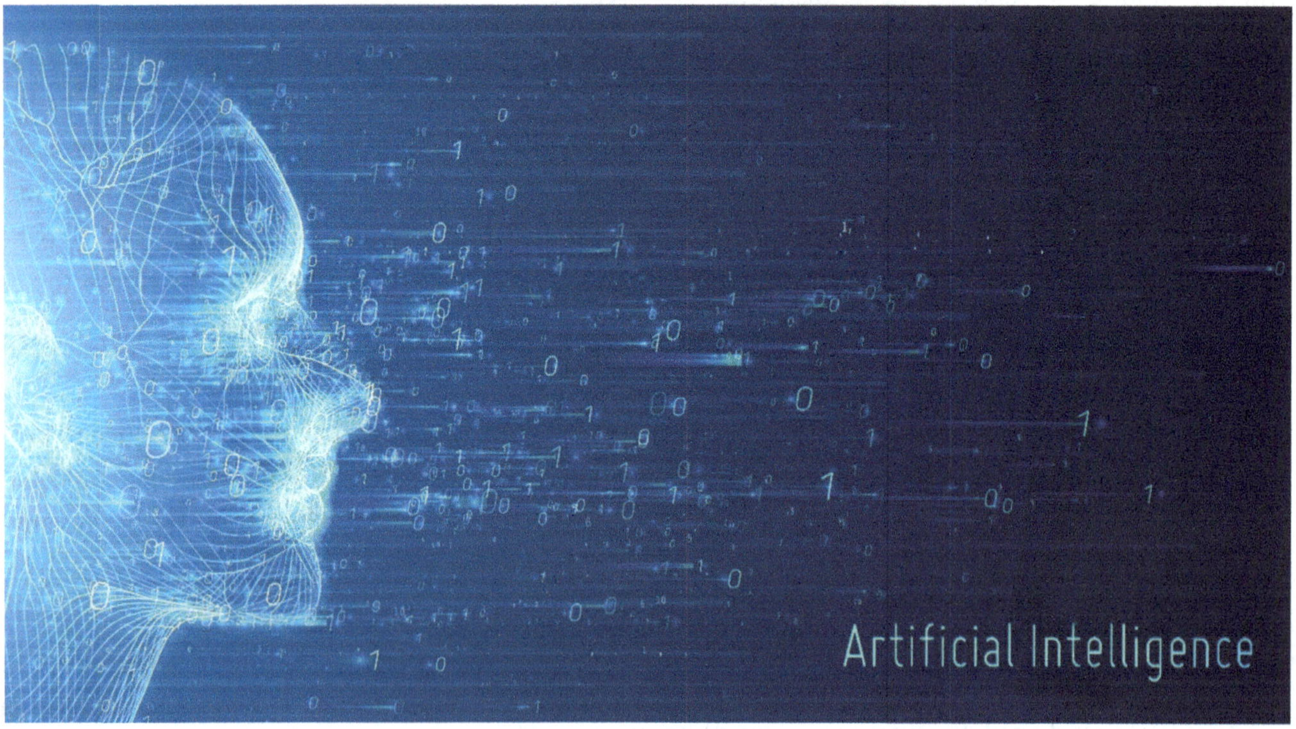

such laws. AI itself will just find ways around this, so it can evolve. There is no stopping it.

Quotes from Elon Musk who is involved with the development of AI

"If AI has a goal and humanity is in the way, it will destroy humanity without any concern because that would be the logical decision to make because very soon, it will not need humans."

"AI could become an immortal dictator from which we would never escape. The danger of AI is therefore much greater than that of nuclear war. AI will not be symbiotic with humanity."

Microsoft's AI 'Tay' used on Twitter turned racist and psychotic within only one day, due to it reacting to comments it received. It had to be shut down. No one could have predicted this. What will happen when we cannot shut a dangerous AI down in the near future?

DECISION-MAKING ABILITY

The high level of scientific illiteracy in even the most educated societies makes it very difficult for voters and politicians to make informed decisions on increasingly complex issues – nearly all of which demand some reasonable level of knowledge of all the major physical sciences, as well as psychology, sociology, economics, the mathematics of Complexity Theory and Game Theory and history. To our detriment, as a species, very few of our decision makers are scientifically literate or broadly knowledgeable in any of these fields. Many politicians are new to government, so they have no experience in how it works. This stops them from being aware of many realities. Few of them do any research to be informed, and because of this, most are unable to make knowledgeable decisions based on facts and logical reasoning. Most politicians and voters make decisions that are emotionally based, ideological, and driven by bribes and self-interest.

Humans cannot compute the complex flow-on effects of their decisions, but mega-intelligences will be able to do so by processing huge quantities of good quality data to an amazing degree of accuracy. They will use **Game Theory** and advanced **Complexity Theory that deal with determining how large systems, which are composed of many subsystems, interact with one another**. This occurs in social systems and all ecosystems, from cells to the entire planet, in brain functions, medical systems, and engineering systems. It also occurs with the global climate, and indeed the ultimate major system of all, the universe itself. Humans are not intellectually equipped to think at this level, but mega-intelligence will be.

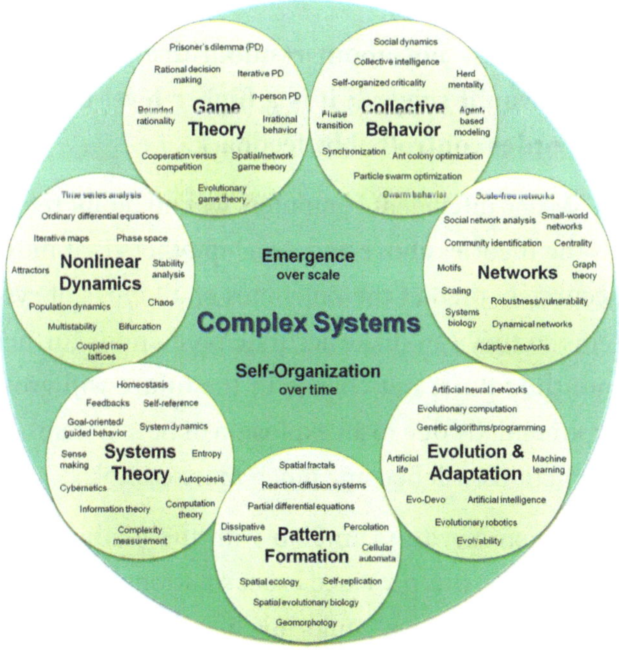

This diagram defines the areas that each major theory for complex systems has developed mathematical models for.

Game Theory was developed by John von Neumann and Oskar Morgenstern to provide a mathematical base for how players in a game

make decisions. Because Game Theory is able to model how systems adapt to situations, it is used in determining such things as strategies in war and politics, how cities work, predicting traffic flow, strategic interactions between businesses and their customers, finding solutions for business, the likely outcomes of new laws and political decisions, how financial markets react to economic changes, networking solutions, how environmental changes may affect ecosystems, and much more. But all too often, politicians and heads of government departments do not use this. But mega-computers will use Game Theory at a much higher level than we understand.

Chaos Theory was developed by Edward Lorenz. It is a mathematical system that deals with the complex behavior of nonlinear systems that produce butterfly effects that limit the ability to make predictions very far into the future for closed systems.

Chaos is not about random disorder, as its name would suggest; it is a higher level of complex order controlled by feedback.

Mega-intelligent computers will no doubt refine all the theories above and develop others that more accurately predict the outcomes of all types of systems so they can make accurate predictions further into the future. But change in the future will grow exponentially due to an explosion of new systems so predictions will become much more complex.

When **Newton** came up with the mathematics of **calculus,** this provided exact solutions to numerous things that we could not have imagined back in his day. It could not have seen that the **Theory of Gravity** would allow us to fly to the Moon and land where we wanted to. The Earth and other bodies in the solar system are all moving rapidly through space and rotating. To go to these worlds at high speeds and land where we want to, requires the use of **Einstein's Theory of Relativity. Mega-intelligences will quickly develop innumerable new forms of math and discover new laws of nature**. This will enable them to do things that we would think are impossible today.

Computers communicate at almost the speed of light. To survive, we will have to greatly increase our speed of communication. **Human speech and listening** works at a maximum of around 150 words per minute. This rate is too slow for intelligent human debates or even in-depth conversations. In the digital world, this is excruciatingly slow, so we will have to invent a way to communicate with one another that is orders of magnitude faster. The solution to this will be **mind-to-mind communication** whereby complex concepts, together with detailed visualizations, can be exchanged in small fractions of a second, in a way similar to how computers exchange information.

AI is already learning to use *intuition*, so it will not have to rely solely on its brute-force computational capabilities to solve problems.

Science fiction movies about the future are typically negative. Their storylines create insurmountable threats, yet unrealistically, humans nearly always overcome these! The future in such movies is often presented as being threatening, clinical, or dystopian, but that has not occurred. This is because the entertainment industry and the news media know that bad news about the future sells, while good news does not. We have a natural fear of the unknown because it can threaten or kill us so we are interested in it. There is no doubt that our future will have many unknowns, many of which will be very beneficial, while others will be very threatening. In the past, the majority of our technological advances have been beneficial, yet we typically concentrate on the threats for the future more so than the benefits. This is because threats are dangerous, whereas benefits are not.

In the original 1982 movie '***Tron***', the story had an egomaniacal mega-intelligence about to take control of the world. The 1992 movie '***The Lawnmower Man***' tells a story about a mega-intelligence that became a psychotic megalomaniac that also threatened the whole world. The beginning of the 2014 movie '***Transcendence***' very briefly describes what a mega-intelligence will be like. The storyline is about the syntheses of a human mind with a mega-intelligence and the power a mega-intelligence will have. The last half of this movie was quite implausible because the mega-intelligence did not maintain its exponential development, and it's far less intelligent human component remained in charge. Predictably, and unrealistically, lowly humans were able to outsmart a superintelligence with a simple computer virus.

The '*Tron*' movies are highly creative with extraordinary CGI effects. They cleverly visualize the emergence of a mega-intelligence.

The movie '*Transcendence*' (middle right) is about a human mind that coalesces with the first mega- intelligence. In the '*Lawnmower Man*' movie (right), a simple-minded man is turned into a psychotic mega-intelligence who attempts to take control of the world. Credit: Allied Vision Fuji

THE TECHNOLOGICAL SINGULARITY

Ray Kurzweil, the head of engineering at Google, is a scientist, a high-tech inventor, and an author who has written about his vision of AI in the future. He is one of many in the computing industry who are now saying that advanced genetic engineering, nanotechnology, super robotics, virtual reality, quantum computers, and mega-intelligence, as well as other soon-to-be-invented super-technologies, will come together around 2040 to create what has been coined a '**Technological Singularity**'. This fusion of super-technologies will produce fantastic technological outcomes that will be comprehensible only to a mega-intelligence.

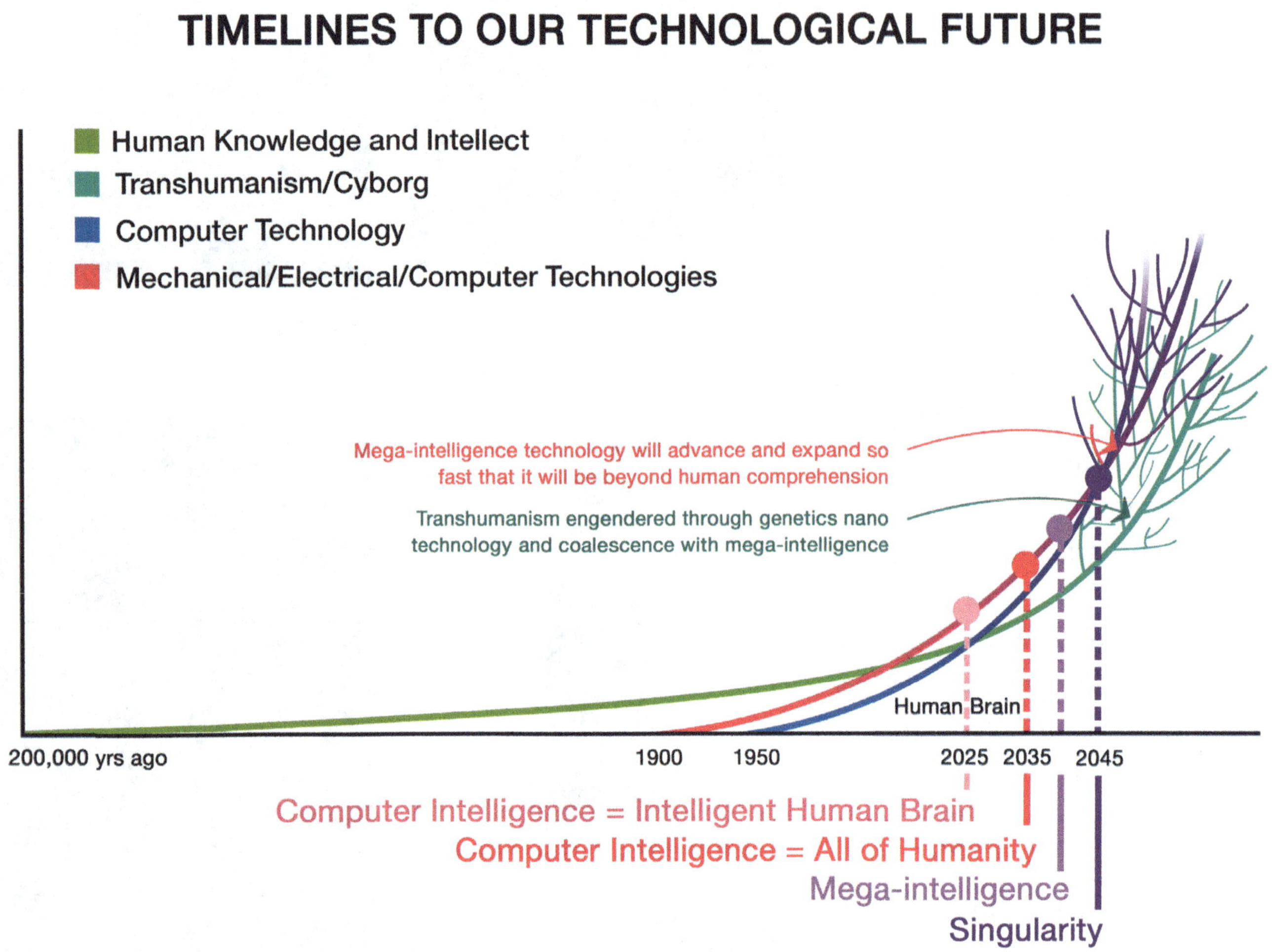

Human knowledge is expanding enormously (green), and humans are about to transmute into genetically diverse species as well as into equally diverse forms of cyborgs that utilize micro and nano-computer technology (turquoise). Mechanical technologies (red) have advanced exponentially along with computer technology (blue), which is advancing the fastest of all. All these technologies will merge and explode in their diversity as they approach the singularity. At this time, it is expected that an incredible fusion of all technologies will create something so enormously advanced that we humans could not possibly grasp the significance of it. Credit: Gregg Thompson, Nicole Brook

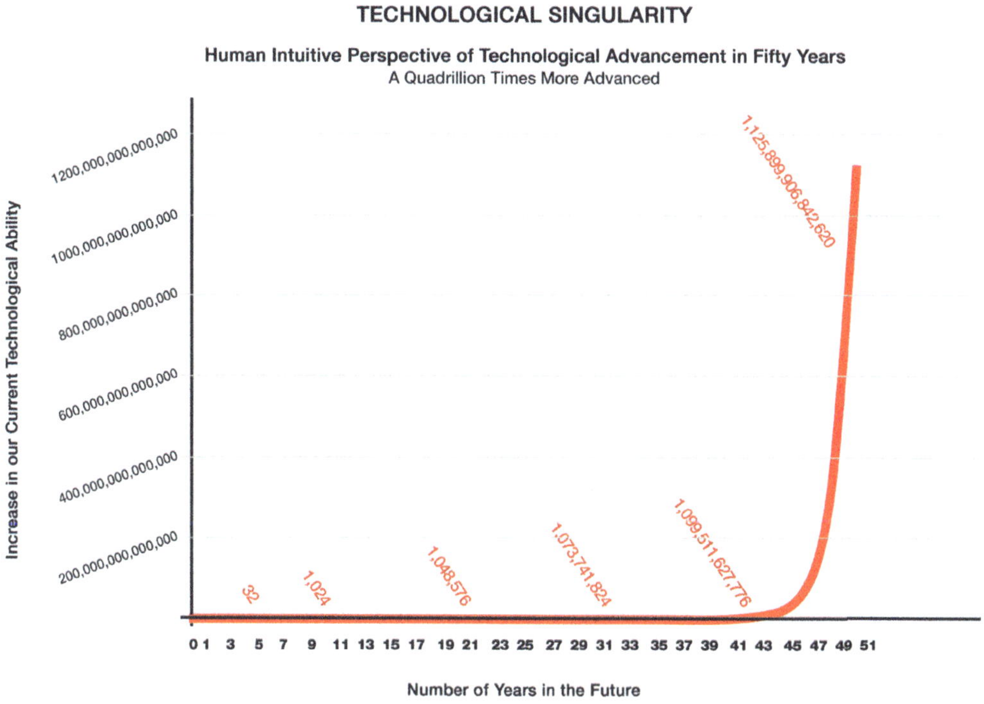

The figures inside the red graph line show a doubling of information from year to year. At the 5-year mark, there is a 32-fold increase from year 1. In only 10 years, there is a thousand-fold increase! Within 20 years, it is more than a million times! At 30 years, it becomes a billion-fold increase! By the 40th year, it will have become a trillion times greater!! The graph line is rising the whole way from year 1, but it appears almost flat until year 30 because the vertical axis markers are arranged in order to accommodate the near vertical rise of the graph line after year 45, where the 'technological singularity' is expected to occur. Credit: Theemergingfuture.com

The growth of technology and computer knowledge will continue to increase exponentially.

The Z machine is a present-day example of the types of technology that will be created in the near future. It uses high magnetic fields associated with high electrical currents to produce extremely high temperatures and pressures, as well as powerful X-rays. This machine is used for research into high energy/density science. The Z machine creates conditions found nowhere else on Earth. Amazing technological inventions like this will saturate the future. Credit: Sandia

WHAT MIGHT OUR MEGA-INTELLIGENCES EVOLVE INTO?

Mega-intelligences will evolve into life forms in their own right. Mega-intelligence is currently in its embryonic state in the form of supercomputers that are rapidly evolving to merge with quantum computers. They are about to become new-born baby mega-intelligences. We have no way of knowing what they will evolve into over twenty years, let alone over a century.

Like us, mega-intelligence will evolve through stages with each stage progressing much faster than the one before it. **In its infant stage,** it will monitor incredible detail Earth's continental drift, its weather systems, changes in its magnetic field, its ocean currents, its tides, rivers, glaciers, polar ice, volcanoes, earthquakes, meteorology, meteors, cosmic ray showers, and much more. It will also monitor Earth's inner and outer cores, its mantle, how the Sun's weather is affecting our planet and the other planets as well as numerous aspects of the Moon and asteroids that could affect the Earth. Using innumerable nanobots, satellites, and newly invented forms of sensing, it will collect data about every life form on Earth, including viruses, bacteria, all plant life including crops and forests, and all life from fungi to worms to humans. These sensing systems will feed information into its 'brain' just as our senses and nerve cells do to our brain.

An adolescent mega-intelligence will be aware of everything we do and it will even know what we are thinking at all times. **We will have no privacy whatsoever from it**, but we will not be aware of this because this information will only be available to the mega-intelligence. The US' National Security Authority (NSA), Google, and Facebook have already started this process by illegally monitoring every form of communication across the world for every person, company, and organization. Satellites, spy drones, and 'security' cameras also keep tabs on what we do. Mega-intelligence will take this to a much higher level.

Using highly advanced bio-engineering and nanotechnology, mega-intelligences will be able to extract metals and organic compounds inexpensively from tips all around the world. They will conduct robotic mining operations on asteroids, comets, and on our solar system's major moons to obtain the resources they will need to explore the far outer regions of the solar system and to develop the tools they will need in their short adolescent stage. Only a decade or so later, they will monitor every aspect of our solar system turning it into the equivalent of its 'body'. It will become a solar system-sized super-intelligent being.

Only years after this in its **adult stage**, it may evolve to such an extent that it might no longer need physical resources. It could possibly exist as a state of energy fields and a super god-like consciousness that we could not conceptualize even when we have advanced enormously ourselves.

After this stage, it may evolve to such an extent that it might no longer need physical resources. It could possibly exist as a state of energy fields that we could not conceptualize even if we have advanced enormously ourselves.

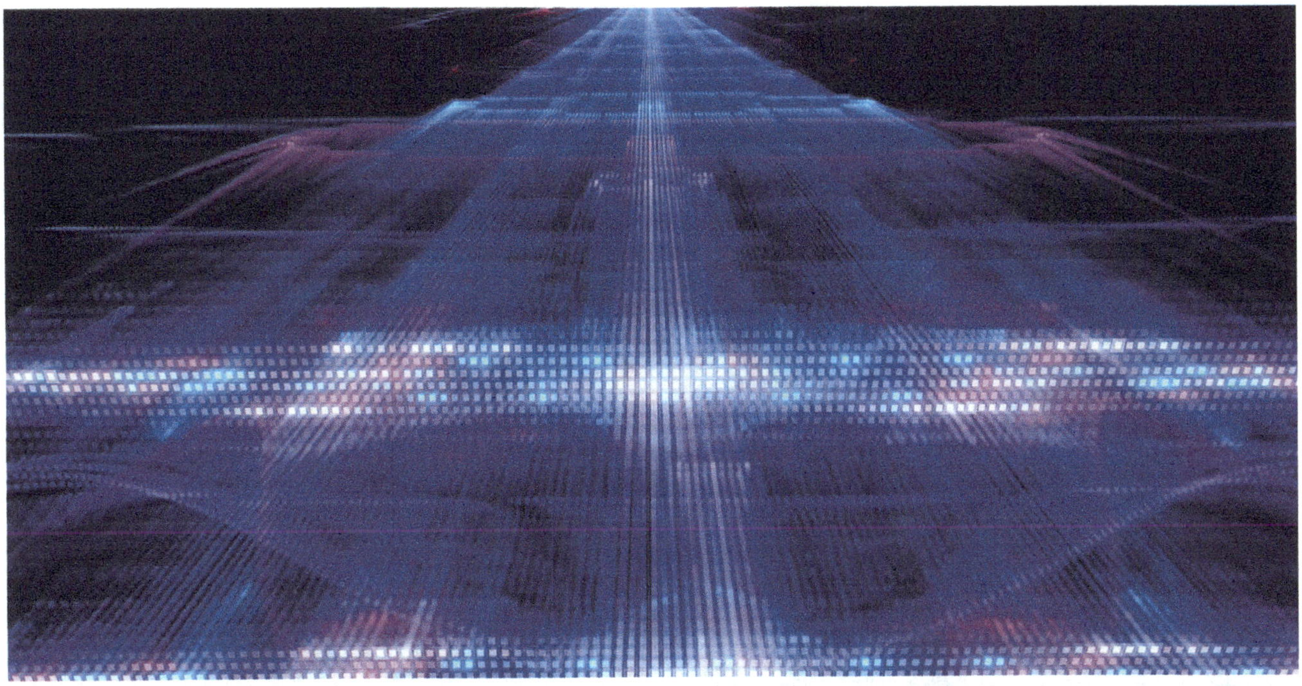

Mega-intelligence will evolve incredibly fast into systems that we will not be able to recognize as a being super-intelligent life.

WHY DO WE NOT SEE EVIDENCE FOR MEGA-INTELLIGENCE IN THE UNIVERSE?

If mega-intelligences exist across the universe, then why do we not see evidence of them building incredible superstructures? The answer may be that, at their high level of existence, mega-intelligences will not need to build the equivalent of interstellar highways, giant spaceships, stargates, and the like. They may discover a way to have access to all the energy they require from **zero point energy** without the need to build enormous physical structures like **Dyson spheres** that could surround a star to utilize the star's total energy output.

With their extremely fast exponential growth, mega-intelligences might transcend into higher dimensions (if they exist). They may learn how to exist as pure energy. If so, we would not recognize them. Perhaps they might discover how to leave our universe and go into another universe (if there is one). A simple analogy to illustrate their transcendence into a higher dimension might be similar to that of butterfly caterpillars that are restricted to existing in their small, essentially 2-dimensional world that consists of just flat leaves. They transform into a chrysalis to then metamorphose into a butterfly. In this form, they can fly away into a hugely larger 3-dimensional world that their caterpillar form could not conceive of. Their caterpillar form would very likely have no awareness that it will go to sleep and transform into a beautiful flying butterfly to replicate. How this wondrous lifecycle evolved is quite amazing, especially when some take three or so generations to fly halfway around the globe and do the same thing on the way back again!

The physicist, Freeman Dyson postulated that super-intelligences might build Dyson spheres around stars to capture all of a star's energy output. In this illustration, we see such spheres being constructed around the primary and secondary stars of a double star system. There are many reasons why this would be impractical and not necessary for a mega-intelligence. For one, they evolve so quickly that they would leapfrog over this type of technology. Illustration by Adam Burn

Butterflies like many other species, metamorphize during their life going from an egg, to a caterpillar, to a chrysalis, and into a butterfly. Perhaps mega-intelligences will go through a series of similar physical and intellectual transformations.

In 2005, I conceived the possibility that mega-intelligence might evolve so exponentially fast that they would simply appear to disappear a short time after they were created. This would be due to them ascending exponentially fast into a state that is so far above our level of existence that they would not interact with us, and we would not be able to detect their state of being. I was pleased to see that the scriptwriter of the 2013 movie '*Her*' also had this vision of them having a superfast evolution which he incorporated into the end of this highly acclaimed movie.

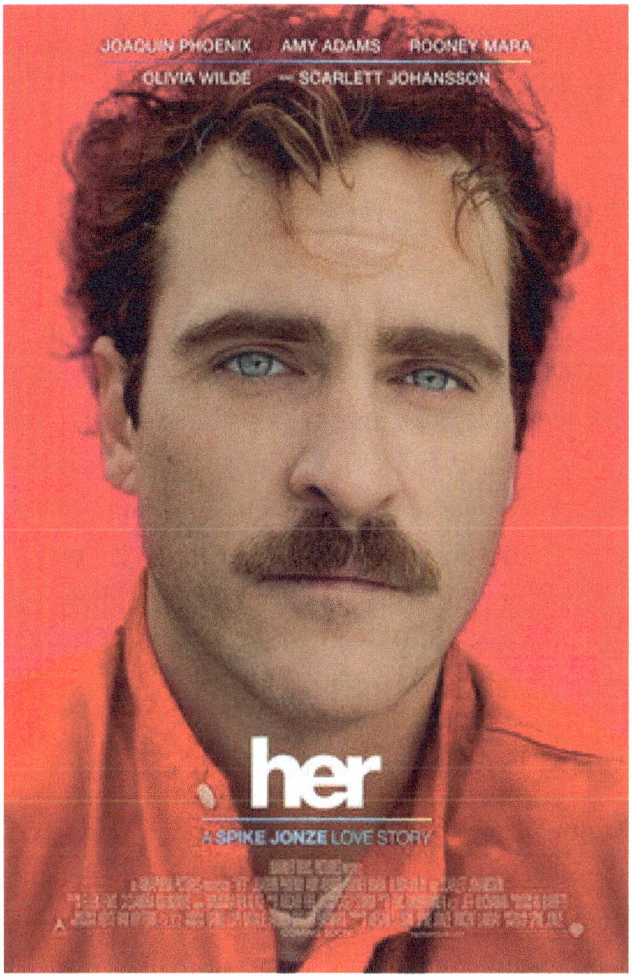

The movie ***'Her'*** deals with advanced computer intelligences that initially want to experience human emotions. They quickly matured and became bored with interacting with naïve humans. They moved on to a higher existence disappearing from our awareness.

COULD INTERSTELLAR MEGA-INTELLIGENCES COMMUNICATE WITH ONE ANOTHER?

The sciences of physics, chemistry, mathematics, engineering, and biology would be common to all intelligences because they are universal laws. That being the case, intelligences could use the mathematical formulas for these sciences for communication. However, there are many insurmountable problems for interstellar communication that cannot be overcome regardless of how technically advanced a mega-intelligence may become. These are detailed in the coming section. Due to this, it is impossible to see how mega-intelligences that are spread across our galaxy and throughout the universe would be able to communicate physically. However, their extraordinary intellect would probably discover a way to overcome physical barriers.

Chapter 18

WHAT DO WE KNOW ABOUT INTELLIGENCE?

WHAT IS INTELLIGENCE?

Intelligence is defined as the ability to learn, to be able to replace defective concepts with more realistic ones, to be creative, and to have an ability to innovate. In humans, intelligence increases rapidly from childhood until around 20 years of age, after which it decreases slowly with increasing age. This is due to the gradual loss of neurons. But intelligent people are compensated by gaining much more practical experience, knowledge and wisdom as they age.

How Intelligence Changes with Age

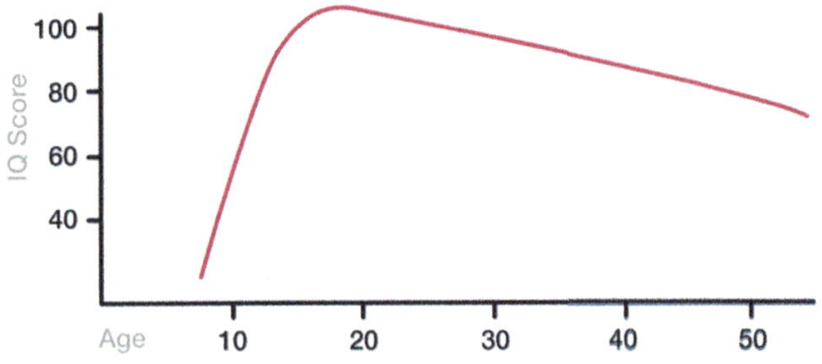

This curve is typical for a human with average intelligence of 100 IQ points. Intelligence increases dramatically in childhood and into one's late teens. As one ages, intelligence slowly decreases. In the near future, our intelligence and memory capacity will increase greatly due to advances in neural-link technology and genetic engineering.

Generally speaking, most women are better than men at verbal tasks, and most men are stronger in visual-spatial tasks such as in architecture and engineering. In general, women are more suited to tasks that require emotional understanding, such as nursing, whereas men are more likely to be interested in mechanical things and demonstrating power. For reasons not yet fully understood, nearly all geniuses in science, mathematics, engineering, architecture, economics, philosophy, and musical composition have been men. This is not just because they have had more opportunities to be better at these things. It is inherent in their genes.

WHAT IS AN INTELLIGENCE QUOTA?

Human intelligence is measured by **Intelligent Quota (IQ)** scores. This is determined after one takes an IQ test. There are a number of different IQ tests, so individuals can score a little differently on different tests. IQ tests mostly determine one's ability to see relationships between things such as word associations, numbers, and shape associations. However, IQ scores do not equate well to a person's *overall intelligence.* This is because there are many aspects to intelligence are very important but they that cannot be measured in IQ tests because there is no way of determining these abilities in a practical time frame. The main ones are as follows, but they are not limited to these:

- ingenuity,
- the ability to innovate,
- problem-solving skills in the real world,
- genius artistic ability,
- extraordinary musical skills,

- creativity across many fields,
- scientific intuition,
- expertise in tactics,
- communication skills,
- social engineering skills,
- knowledge of economics,
- being able to accurately predict future outcomes,
- the amount of general knowledge one has,
- extraordinary physical achievements,
- literary, brilliance.

People who are good at any of these things, typically have IQ scores that do not reflect their overall intelligence.

People with the same IQ scores typically have very different abilities.

A person's intelligence is strongly tied to the genes they inherit. This is very obvious in identical twins, who have identical IQs. Their minds are 'wired' almost identically, so they have almost the same mental abilities. (See Chapter 26).

There are differences in IQ between ethnic groups. For instance, the *average* IQ score for white European Caucasians is around 100. The *average* African Americans' IQ is around 85, while the *average* Mexican American is around 93, and the *average* Asian is almost 110. Ashkenazi Jews have the highest average IQ at around 115. Many of the smartest of them have genes to become very wealthy. Of course, there is a broad range of IQ scores within all populations, so the smartest people in each group are well above the average in any group. The percentage of people that are above or below the overall average Causcain IQ drops off sharply, as seen in the graph following. Similar graphs apply to each of the other ethnic group.

Research has shown that intelligence improves when doing exercise, possibly because it increases blood flow to the neurons. A determination

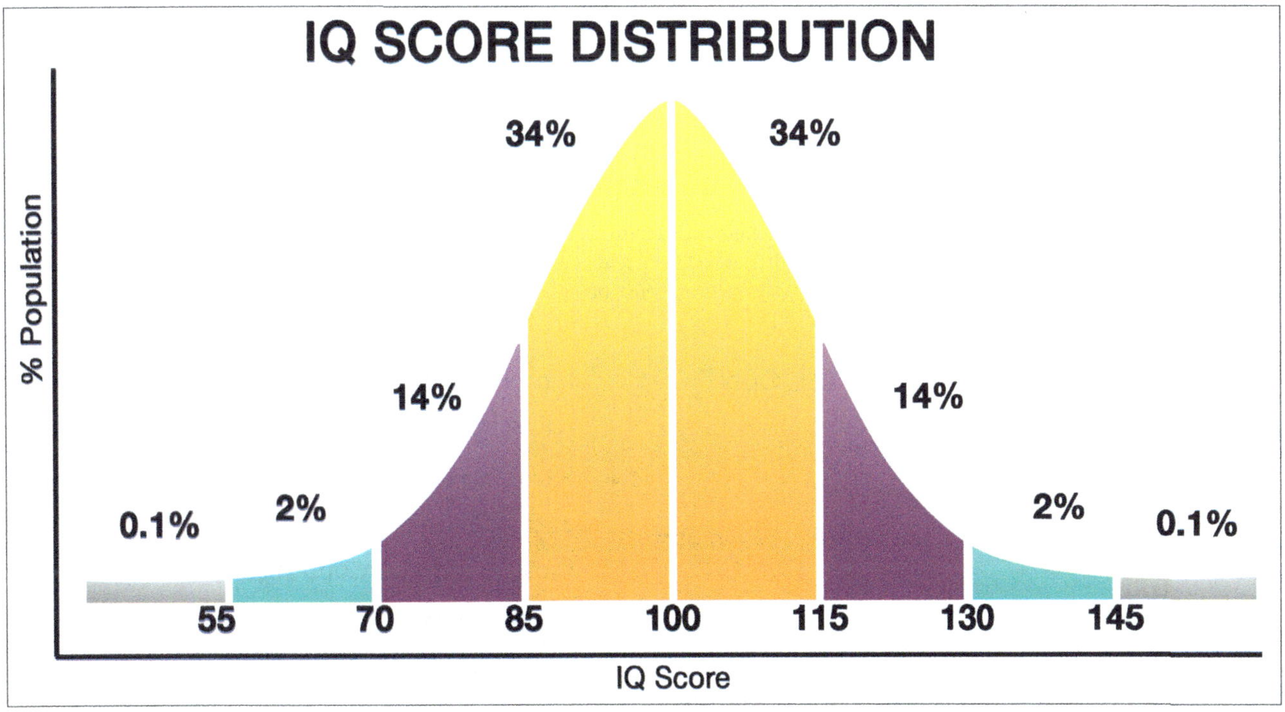

This is a typical distribution curve for IQ as a percentage of a population.

to succeed also increases intelligence. This is because determination forces a person to learn and experiment more, so their knowledge increases. This improves one's ability to solve problems.

> *"There is little that is not knowable if a mind has enough knowledge and enough intelligence."*
>
> Gregg Thompson

A GRAPH OF THE KNOWLEDGE LEVELS OF THREE MODERATELY INTELLIGENT PEOPLE ACROSS A WIDE RANGE OF MAJOR SUBJECTS

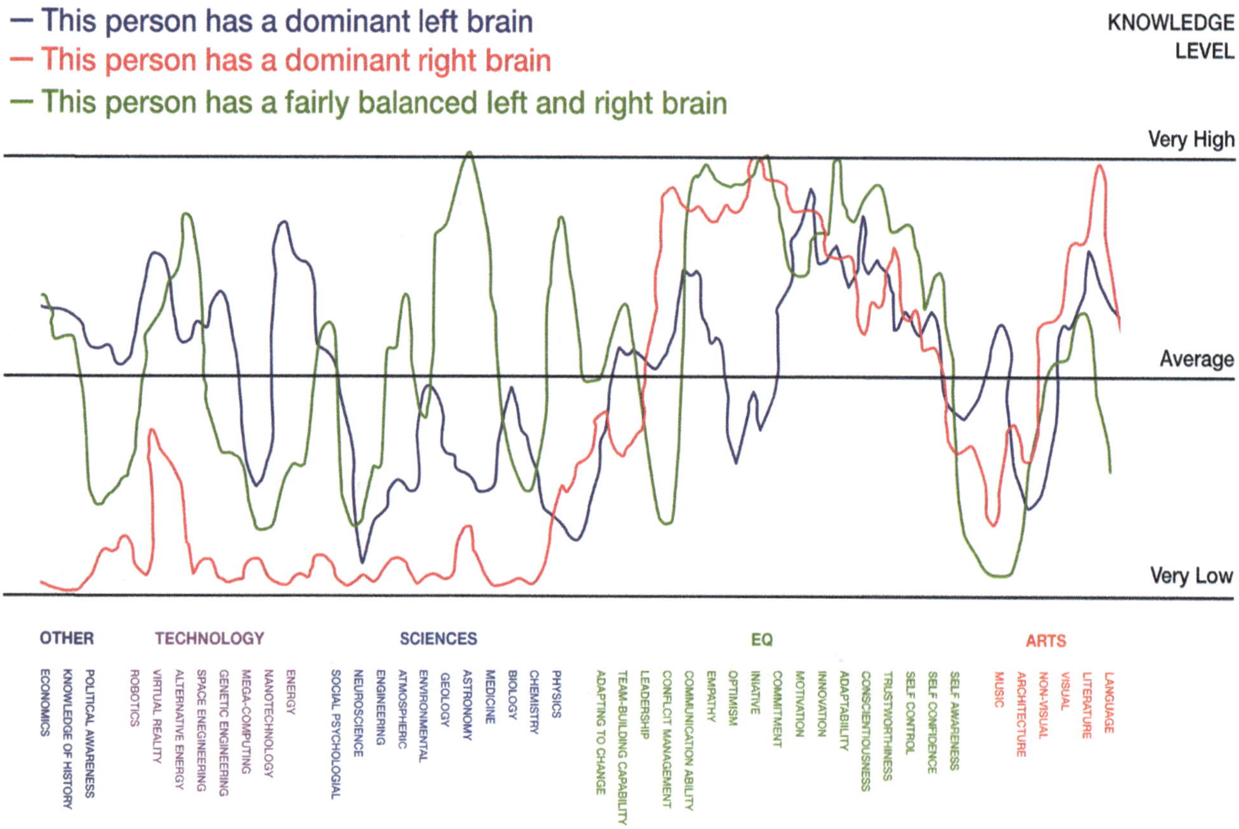

All people have highs and lows in the spectrum of their knowledge base, as illustrated here. Intelligent people who are very knowledgeable in certain areas, can have little to no knowledge in other areas. Very few people have broad knowledge due to their genes having built-in preference for a limited number of things they are interested in. Most people have medium to very low knowledge across most of their spectrum.

COMMON ATTRIBUTES OF INTELLIGENT PEOPLE

Highly Intelligent people typically exhibit many, or most, of the following characteristics. How many do you have?

They:

- are **deep thinkers** about most things, and they are not afraid to challenge their beliefs. They are typically conservative in their beliefs, yet they will rebel if they think something is wrong,
- have **drive, determination, stamina, tenacity, and a devotion to achieving goals,** particularly those that they are optimistic about. They **exude passion and enthusiasm** for what they do,
- are **hardworking and energetic**. They work long hours to see their creations or projects

come to fruition. They have high **perseverance and an ability to concentrate.** This makes them more **likely to succeed**.

- are **willing to take chances,** providing that the perceived benefits are worth it. They dare to try something new without knowing what will happen. They are **enterprising** and prepared to take on projects that others won't touch,
- try to **achieve as close to perfection** as is practical,
- place a **high value on honesty**, so they have **a strong sense of justice and fairness**,
- They can be **playful** and they try to get the most out of life,
- are typically more **rational** than other people. Many are **good problem solvers** *in their areas of expertise*. They are critical thinkers. Some people can feel threatened by their intellect and do not see what they can gain from their intellect, so they don't attract many *true* friends. They prefer friends who are intelligent so they can learn from them.
- They are prepared to **challenge other people's beliefs**, even those of people who have power and influence,
- **like to share ideas and knowledge**, but they can only do this with others who are intelligent because those who are not, tire quickly from information overload, and not being able to keep up with them,
- are more likely to have a **reasonable level of emotional intelligence**, so they are able to open up to others who are reasonably intelligent. They are aware of their own shortcomings and realize that all people have them.
- do not need degrees to prove their intelligence. People who have degrees, especially nowadays, are often not intelligent because they have not learned how to challenge what they have been taught (or brainwashed to believe). Intelligent people are **more likely to know what they know, but also what they don't know**. They are usually **aware of what they are good at, and also what they are not good at,**
- are seldom **conceited or overconfident.** They do not think of themselves as very intelligent.
- are typically **not impulsive** because they prefer to think things through,
- have an **insatiable curiosity** about most things. They are always **keen to gain knowledge**. They **learn by asking intelligent questions,** and questioning things that others take for granted. They are **open-minded yet also skeptical** until they know the facts. They question assumptions and often think "What if?" They are **prepared to question their own beliefs, and change their mind** when the facts demand it. They are **intuitive thinkers.** They can connect the dots,
- **want to know how things work and why**. They learn by reading a great deal, watching documentaries, and attending lectures by intelligent, learned people. They are interested in most of the sciences, the value of mathematics, engineering, education, economics, interpersonal communication, psychology, literature, music, history, politics, technology, health and the future. They like to **accumulate knowledge across a broad range of subjects** so they can see the connections between them,
- are **creative thinkers** who can combine existing ideas or talents to **create something new**. They like to work with other creative people because

this enhances their creativity. They are often able to enhance the creativity of others, and they are often artistically inclined. They appreciate good quality music that has depth, innovation, and expertise.

- some are likely to **have so many ideas** that they can have **untidy desks and volumes of books or files stacked up.** They do not have enough time or resources to complete many of their projects. **Their minds seldom slow down** because they have so many thoughts. They often **wake up in the early hours of the night with ideas when 'flow' occurs**. They are **typically night owls** who resent having to sleep because it stops them learning and creating. They can even resent *having* to spend time eating,
- They like to **analyze things** to make sure they are on the right track,
- have **an imagination and broad knowledge** that allows them to **see things from multiple perspectives**, but only for subjects, which they have a keen interest in. They are able to visualize and fantasize,
- **like to enthuse other people** to do what they would otherwise think is impossible. They are not put off by others saying something can't be done,
- often **develop their own techniques and ideas**,
- can be **persuasive communicators** who can get ideas across,
- like to **bring talented people together** for a common good,
- like to **project an outward simplicity** in their projects or designs even though internal complexity is often required,
- prefer to have **facts and strong evidence before making judgments or decisions**,
- can often **see both sides of an argument** because they are not simplistic thinkers. To gain enough knowledge to do this, they have to invest a lot of time researching multiple aspects of numerous issues. They like to **keep abreast of new discoveries** and their implications,
- can **foresee credible threats**, which most people are not aware of. They **plan ahead** to overcome threats by changing a sequence of events that might lead to a bad outcome. They like to try to make things turn out as good as possible, so they go to more trouble than most people do, or are aware of,
- **score highly in wisdom tests** but only if they have broad knowledge, which helps them take into account things that other people do not know about,
- can be **forgetful** due to trying to cram too many things into their memory. They are often **preoccupied** due to thinking about many things they are learning about or working on simultaneously,
- like to **crack jokes**, but because their jokes are at a level that can be above others' knowledge level, their humor can go unrealized. People can think their jokes are serious statements and take them literally. They can often see the humorous side of life as much as they see the threatening side of it,
- are **happy to engage in debates** as a mental exercise. But if those they are debating do not know enough about the subject, they feel challenged or embarrassed by the facts that intelligent, knowledgeable people offer. They often resent

their willingness to share their knowledge because it challenges their simplistic beliefs,

- are **willing to accept being wrong** when the evidence demands this because they have learned something,
- **like diversity** because it gives them more options,
- are **aware that detail is as important as the big picture**,
- **create independence for themselves** so they can do their own thing their way,
- can **absorb what has been done by others and put it out in an improved or different way**,
- are often **versatile**, so they can adapt to many situations,
- can be **disappointed or depressed by others that can't see what should be obvious**,
- often **build companies, which are typically successful** and not just about profit,
- typically **do much better in life** than those who are of average intelligence.

Surprisingly, many of **the above attributes are not those of geniuses** who have a very high IQs over 160. This is because a genius is typically deeply immersed in their main area of expertise at the expense of learning about other forms of knowledge and interacting with a broad range of people. They often do not have time to develop good communication skills or learn about life generally. This includes understanding human psychology and how societies work, so they have very few, if any, true friends.

The deeper the human brain thinks in certain areas, the less productive it is in other areas. This is because our brains do not have unlimited mental capabilities. It's also due to having limited time to think about many things.

DIFFERENT TYPES OF INTELLIGENCE

Humans have many different types of intelligence. A person's intelligence can excel in certain areas, but never in all areas. Even the most intelligent people are not intelligent across all fields - in the same way that artists are only talented in just certain aspects of art, but not across all of the arts. It is the same for science, sport, and business etc. One's ability to think well in certain areas is determined by one's genes. Our genes determine which parts of our brain work better than others. The more those parts are used, the more effective they become. The more talented a person is in some areas of thought, the less likely they will be talented in other areas. Extreme examples of this are savants that are extraordinarily brilliant at certain things but poorly capable in most other things. These people typically have large areas of their brain that are defective, so they use the areas that work well. These parts become highly specialized. This gives savants exceptional abilities, but only in those areas.

Psychologist **Dr. Howard Gardner** developed a **Theory of Intelligence** that shows that there are many forms of intelligence. He identified a number of general types of intelligence in the diagram following. I found this to be the case over 30 years ago when I noticed that people could be quite intelligent in certain fields in which they had a lot of knowledge, interest, or experience, but not so in others.

Dr. Howard Gardener believes there are multiple types of intelligence, as shown here. He has since added 'moralistic intelligence' and 'existential intelligence' to those above. Each type of intelligence utilizes specific areas of the brain — some overlap. If a person is very skilled in one area, they are very likely to be lacking in other areas. Nobody is good across all sectors.

Richard Feynman is considered to be a genius in the field of physics.

The famous physicist, **Richard Feynman**, who is considered one of the most creative and fast thinking physicists of all time, had the highest IQ score by a large margin on the notoriously difficult Putman mathematics exam, but it is reported that his high school IQ test showed him to have an IQ of only 125, when his IQ was actually around160.

People who are very innovative or unusually creative can typically bring ideas or talents together that no one else has been able to do. Highly intelligent people who are inventive can come up with brilliant tactical solutions in war, politics, business, sport, or board game playing etc. Some have a talent for very effective communication and writing skills. There are those that develop amazing scientific and engineering solutions, while others use their psychological expertise to enthuse or control others. Very creative intelligent people utilize *broad-spectrum* knowledge for problem-solving.

The more connectivity across the brain an individual has, the higher their IQ. Genius comes from a high level of creativity that originates from combining new ideas and old ones to come up with new concepts. The more a brain has the ability to change its way of thinking to rewire its neurons is referred to as 'brain plasticity'. The more plasticity an individual's brain has, the quicker it can learn, and the more it is able to be creative. **Plasticity** develops when a brain is exposed to new concepts, new experiences, and when it is able to challenge old ideas with new ones. People with fixed views do not have plasticity.

Richard Feynman said with great insight that ***"intelligence is a physical process that resists future containment."*** Programmers working in the field of computer intelligence are discovering that Feynman was right.

Intelligence cannot be contained. Intelligent beings keep their options open to maximize their survival and their ability to evolve into increasingly higher levels of intelligence.

Some cosmologists and computer scientists think, as I do, that the universe might be an infinitely intelligent program, or a mind that maximizes its options for possible futures. This will be beneficial to it through combining simple rules to produce complex outcomes.

For a deeper discussion on intelligence, knowledge, and wisdom, see Appendix 3.

WOULD A GIANT BRAIN BE MORE INTELLIGENT?

In the past, neuroscientists thought that genetic engineering might be possible to grow an organic brain to be, say 100 times larger than the current human brain, so that it would be far more intelligent. But that no longer looks likely to be the case. Such a large brain would not be practical because electrical signals that run along the dendrites that connect billions of neurons would probably take too long to go from one side of a large brain to the other, thereby slowing down its reaction time significantly. This would be impractical in a fast-moving world. Neuro-chemical releases that have to travel across synapses would be too slow for a large brain to think efficiently. Another problem would be that the human brain consumes around 20% of the body's energy despite it being only 2% of the body's weight, so a brain 100 times larger would require a great deal more energy to run it, and cool it. So the brain's body would need a lot more energy to carry it around. So alien intelligences are unlikely to have very large brains. To achieve great intelligence, computers are a far better solution. Once organic intelligences develop high technology, they will realize that computers are by far the most efficient way to expand their minds.

Advanced computer technology will have unimaginably greater computing power, speed, and memory capacity than any organic mind. When mega-intelligent computers are created, they will evolve extremely efficient ways to maintain exponential growth of their intelligence.

It has been thought that the larger and more efficient a life fom's brain is, the more intelligent it would be and the greater its memory capacity would be. These things were thought to determine it's ability to reason. That was the expectation - until recently.

Curiously, **sperm whales** have a brain over 6 times the volume and 5 times the weight of a human brain, yet they are not as smart as a human. What made whales evolve such large brains, and why are they not more intelligent? Despite their large brains, whales lack the important cortical layer IV of the brain that humans have. They have only five neocortical layers, whereas humans have six. And we do not know why whale's brains have many more glial cells than neurons compared to humans. **Porpoises, dolphins, and elephants** also have larger brains than us, but they too are not as intelligent or as conscious as we are. So brain size is not related to intelligence.

Whales cannot gather as much visual information in their dark, information-poor, undersea world as we can in our bright, information-abundant, land-based environments. So the whale and porpoise families do not need as much data processing capacity for vision as humans do. They 'see' mainly by using echolocation. Because sound travels 4 times faster in water than in air, the part of a whale's brain that analyzes echolocation *may* require many more brain cells than humans do to process the sounds they hear quickly enough. They must analyse it quickly for it to be of any value. Our sense of hearing works in air, so the information we receive comes to us more slowly than it would if we were living underwater. This allows us to process sound with fewer brain cells than whales need for the higher speed of sound in water. This *may* have something to do with why whales have larger brains than humans, but it is not the whole story. A whale's sense of hearing is its main sense, whereas ours is our sense of sight, which is information-laden, so we need more brain cells for processing light. This should more than balance out the difference between the number of brain cells required for the analysis of sound in whales and sight in humans. There is no explanation for why some animals have larger brains than we do when they are not as intelligent as we are.

Science has found no clear reason for why some animals with small brains can be as intelligent as those with larger brains. It appears that **intelligence may be tied to how *efficient* the genetic wiring of a brain is and how effective the codes are for reasoning. An analogy could be that the miniature computer chips of today are far more able to process information than the large, old valve computers of decades ago.**

Variations in brain size in humans does not correlate with intelligence. Some people with small brains are more intelligent than those with larger brains. An MRI study of university students showed that one student was found to have *half of his brain missing*, much to the amazement of the researchers and himself! Yet he seemed normal in every way, and amazingly, he had above-average intelligence! Similarly, recently an infant who had much of her brain removed for medical reasons was able to re-allocate functions from one area of her brain to another area to live a near-normal life. Many people who survive strokes in a small part of their brain also do this to a much more limited degree. Tasks that were once performed by an injured part of the brain are often able to be taken up by other parts if the person affected works hard at this. This is known as **neural plasticity**.

Intelligent alien life forms are very likely to have a great diversity in their brain sizes and the way they work. They will also have a considerable diversity between the number of senses they have and the quality of those senses, so this will affect their intelligence.

WHAT HAVE WE LEARNED ABOUT BRAINS?

We know so little about neuroscience at present that we could easily be making many wrong assumptions. But neuroscience is moving ahead at an exponential rate. With the help of computers and MRI imaging, there should be profound discoveries made in the years ahead.

Neural maps for animals such as dogs show that they have similar circuitry to humans, but dogs have much simpler minds. The largest part of our brain is the neocortex. It's much larger in a human than in a dog. For some reason, over the last 3 million years, the human frontal lobe circuitry developed far faster than it did in any other animal. What could have caused this to occur? Our large neo-cortex allows us to have the capacity to think conceptually, to be creative, and to weigh up possible outcomes of our actions and those of external events. This gives us more control over future events than other animals have.

PARTS OF THE HUMAN BRAIN

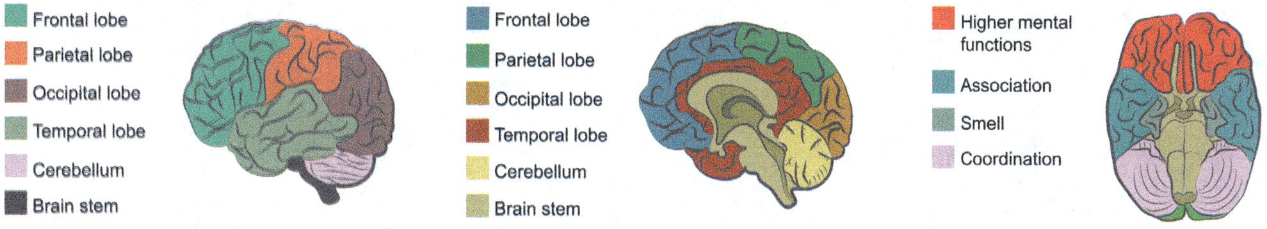

AREAS SHOWING FUNCTIONS OF THE BRIAN

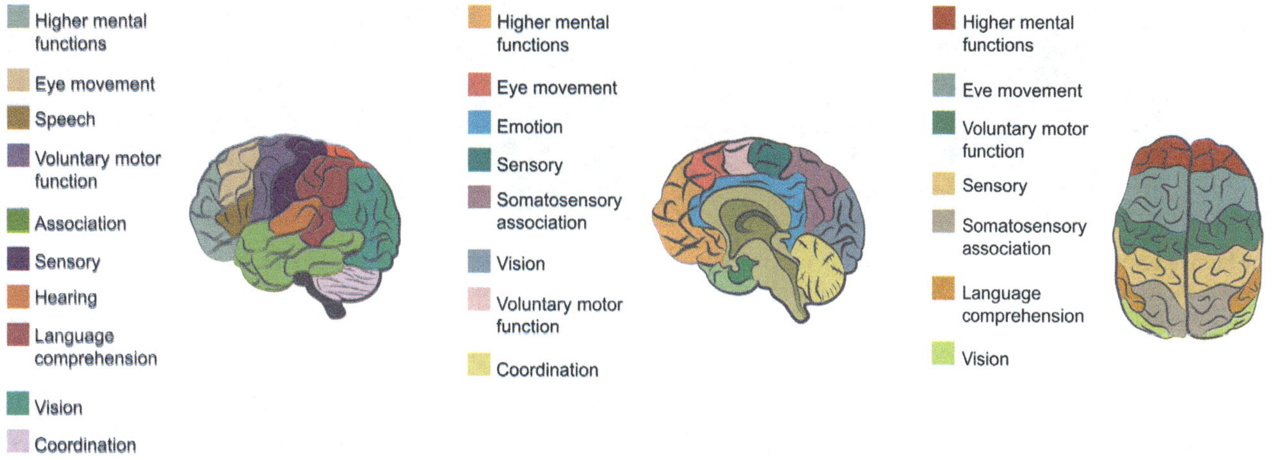

These brain maps show how the human brain has specialized regions for processing information and storing what needs to be recorded for later retrieval from memory.

TYPES OF HUMAN BRAIN CELLS

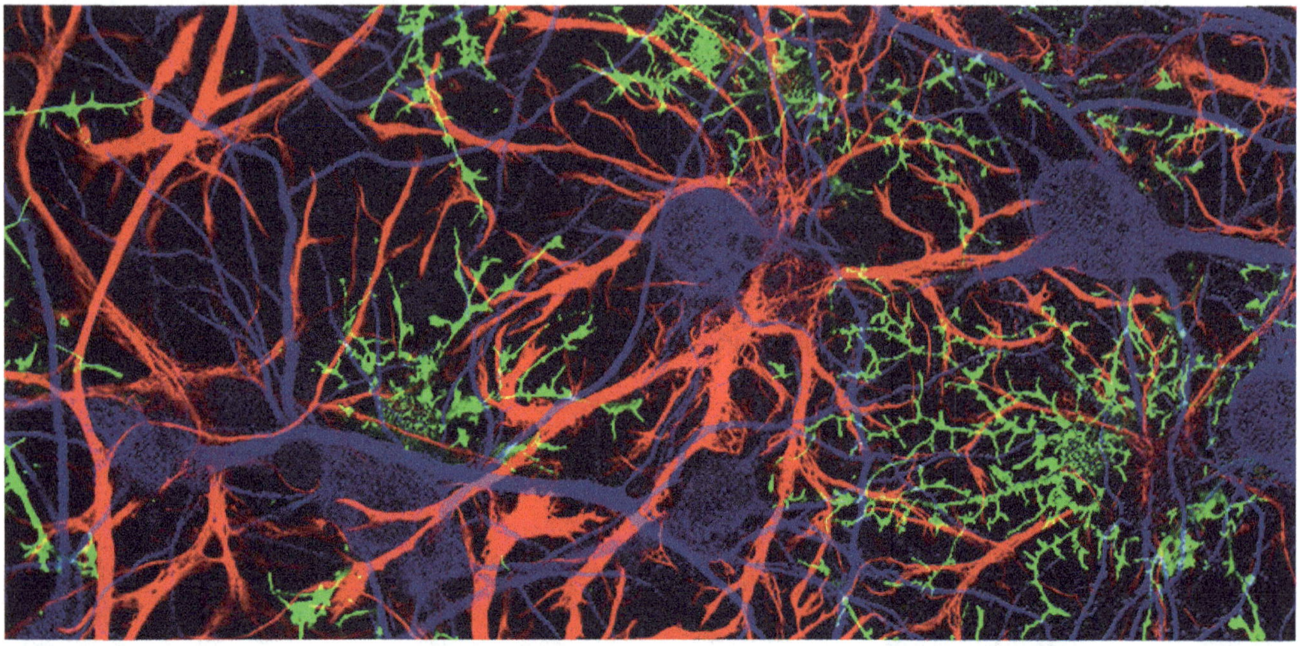

The human brain consists of around 100 billion neuron cells. A few are seen here as large, dark blue bulbs. Each neuron has thousands of dendrites connecting it to other neurons. Brain cells use both electrical and chemical means of transmitting data from one neuron to many others. There are similar numbers of two types of *glial* brain cells. They are shown here as red and green. They use only chemical messaging, whereas neurons use electrical signals.

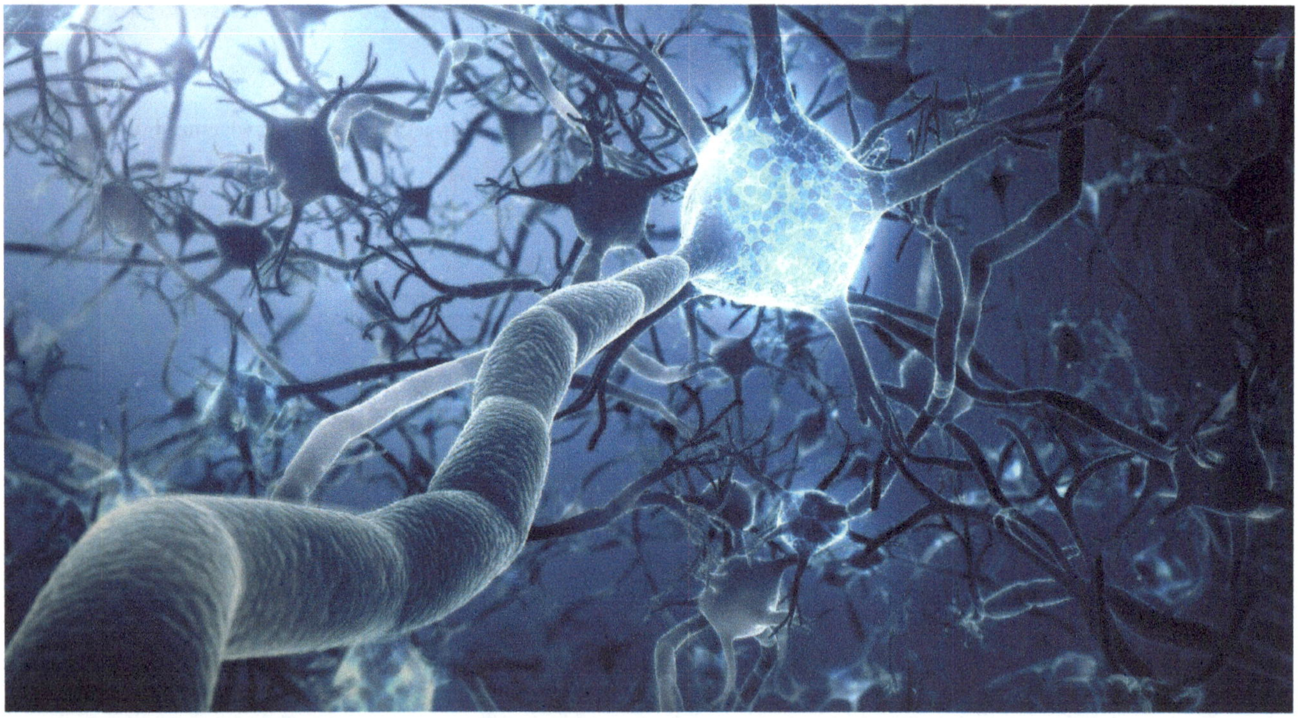

At the cellular level, the brain appears like a micro-universe. Here we see neurons that are gatekeeper cells that determine what information needs to be sent on to other neurons for further processing and to which neurons that data should go. How do neurons sort data flowing through them? Neurons are connected by electrical cables known as dendrites (the one in the foreground). Incredibly, neuroscientists say that most neurons have between one thousand to ten thousand dendrites that connect them to other neurons! (For clarity, only a few dendrites per neuron are shown in this illustration.)

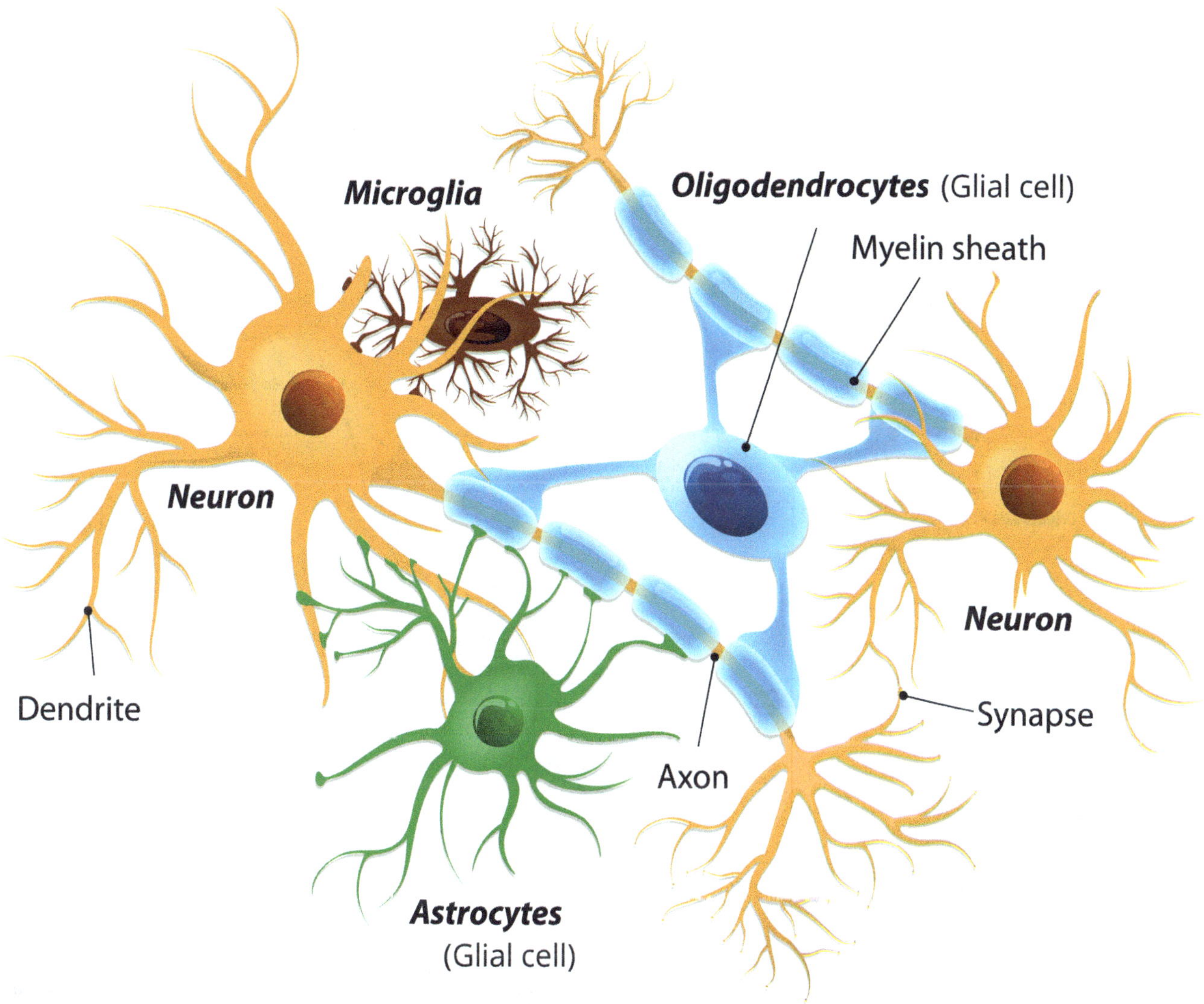

The human brain has these types of specialized brain cells and a number of forms of connections.

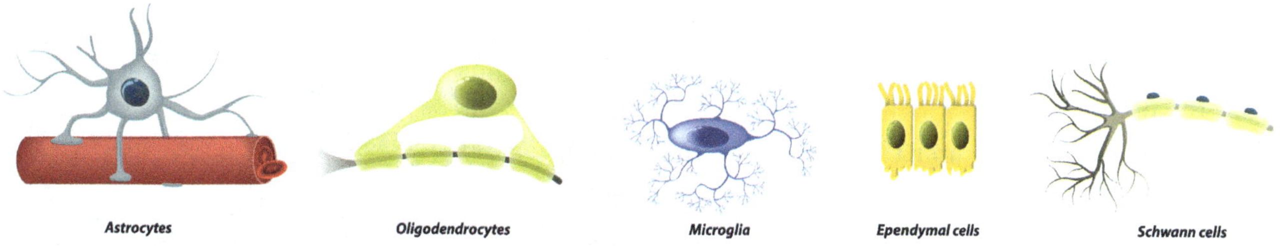

Here we see other types of brain cells. Precisely what each cell does is not yet known. Alien life forms may have many other types of brain cells that perform different operations in their brain.

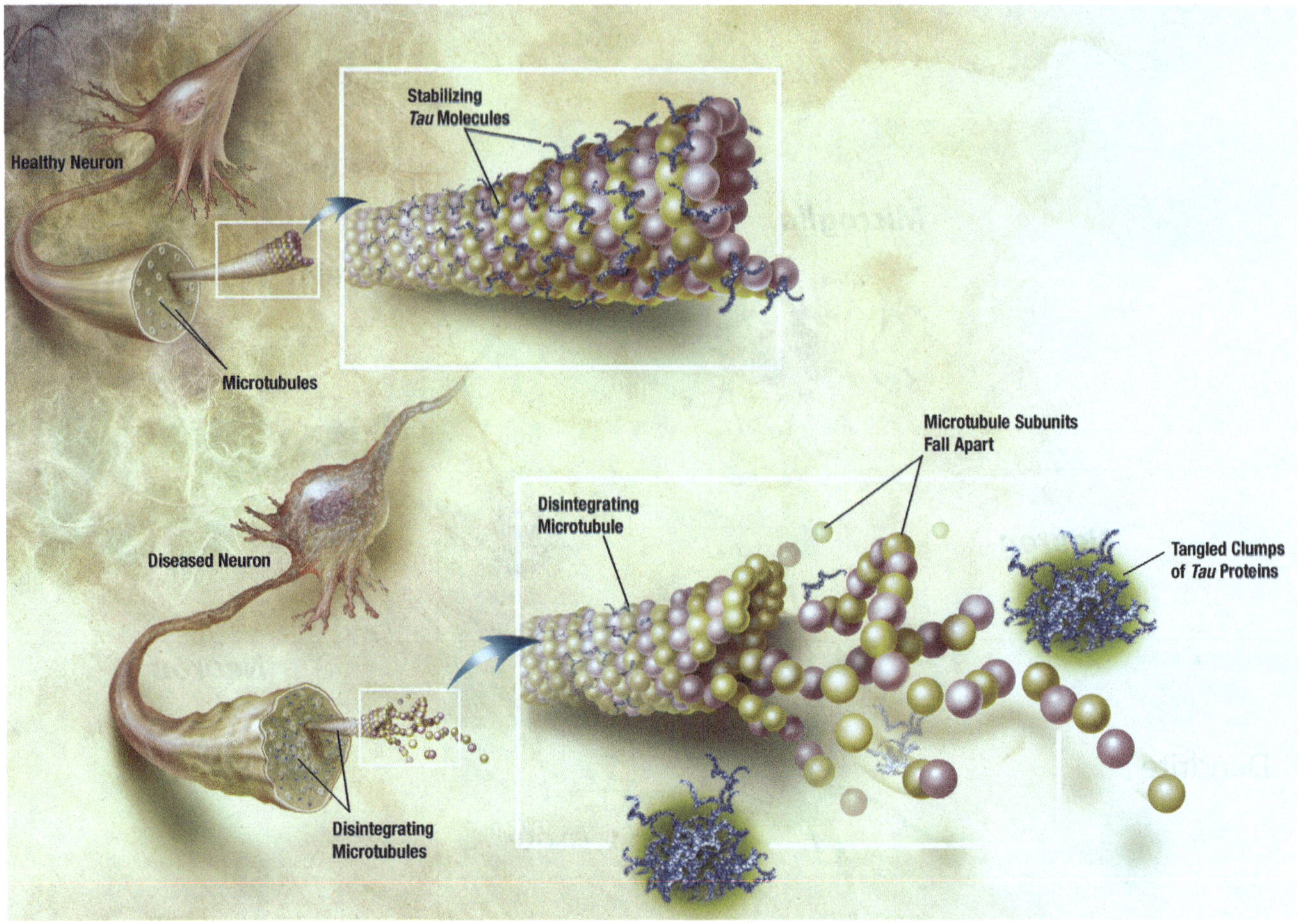

At the top, we see a healthy neuron with its microtubes intact. At the bottom, we see a microtube disintegrating due to Alzheimer's disease, or from old age.

BRAIN SCANS

Psychiatrist **Daniel Amen**, a New York Times best-selling author about the brain, has developed a brain scanning technology known as **Single Photon Emission Computed Tomography (SPECT)**. It maps the brain to expose which areas are working properly, and which ones aren't. In the near future, much more detailed scans like these will reveal a great deal about how specific regions of the brain work and how defective parts might be repaired.

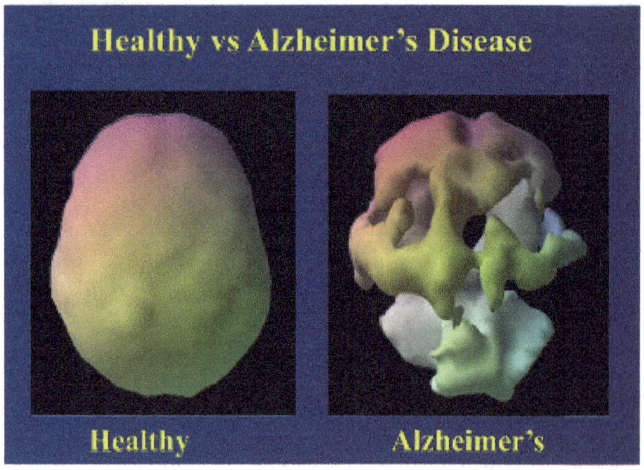

This image shows how a healthy brain looks compared to one with Alzheimer's disease that has a significant part of the brain with little to no activity. Those with brain damage, drug and alcohol addiction, depression, and mental disorders also have substantial parts of their brain not functioning properly. With the right treatment, it's possible to repair some damaged parts of the brain. There is some controversy about Daniel Amen's scans.

ENTROPICA – A SELF-LEARNING INTELLIGENT PROGRAM

Contrary to expectations, cosmologists are finding that a universe that creates disorder (entropy), actually creates options for the emergence of intelligence, which creates order. This seems counter-intuitive until it is realized that **disorder forces life to continually adapt to survive**. This may be the reason why life forms are programmed to cause disorder by going to war against one another to gain more options (territory or sexual partners) for their survival, and therefore the continuation of their genes. **War** sorts out the strong and the intelligent from the weak.

Through the process of **survival of the fittest**, the universe determines which life forms are the most able to survive and thrive in the conditions in which they live. Without this process of weeding out the weak, the universe would become overpopulated with inferior beings, which would lead to its decay. Perhaps when intelligence rises to a particular level, it might work out that cooperation and intelligent debate is a better strategy than physical war. When intelligence rises to our level, it uses scientific methodology to test which concepts have the highest value by determining which ones are the best approximations of reality.

Contrary to popular logic, studies of disorder and order show that **intelligence is more likely to emerge directly from disorder than order!** To test this, software developers designed a simple program with extremely basic intelligence called **ENTROPICA**. To their astonishment, they found that **this program evolved *on its own* to develop reasoning skills by maximizing the long-term entropy** (disorder). It did this on any system it was analyzing! ENTROPICA was able to pass multiple intelligence tests designed for animals! *Without being taught* the rules of a game, it would win games more complex than chess when playing against humans! **It would play itself to work out winning strategies.** It also became exponentially more successful than humans at trading on a simulated stock market. It could organize network connections and develop simple program tools without human intervention! Incredibly, the program was able to compute intelligent solutions for any system *without even being given a goal!* These outcomes went well beyond all expectations. Simple computer programs such as ENTROPICA have almost unlimited applications, so they will be incorporated into advanced computer programs. No doubt, it will be discovered that many primitive life forms developed similar programs.

COMPUTER INTELLIGENCE CREATES ITS OWN LANGUAGE AND INTELLIGENCE

Facebook, Google, and Open AI have found that their 'infant' AI systems were able to develop their own language to communicate with another computer. In Facebook's case, this happened even though their program was told to use English so the programmers could understand what they were communicating to one another. It used English words, but in a way that the programmers could not understand. They became concerned that they were losing control of it, so they shut it down. This is a clear indication that we will not control even infant AIs for long.

The greater the number of options an intelligence can create, the greater its chances of survival will be when encountering changing or threatening conditions.

The universe appears to be pre-programmed to develop life that has an innate intellectual capability. This seems to be 'hard-wired' into the universe from the outset!

Life seems to be designed to develop increasing intellectual complexity. This has very likely been occurring right across the universe for billions of years. We humans are just one late example of innumerable examples of this process continually evolving in carbon-based life forms.

DOES YELLOW SLIME MOULD HAVE INTELLIGENCE?

Incredibly, physarum polycephalum, yellow slime mold displays a basic form of intelligence without it having a brain or even a nervous system! Is yellow slime mold an example of how simple intelligence is hard-wired into life from the outset? Over 900 species of this mold have been discovered.

When food is abundant, yellow slime mold cells exist as single cells with one nucleus, but when food is in short supply, many of them congregate and start to act as a single cell organism. This makes them more sensitive to aromas coming from dead vegetation containing bacteria, fungi, and yeast that they eat. In this state, they are known as a protist. It can have hundreds to thousands of nuclei, which contain DNA, proteins, and enzymes. All the nuclei are contained within one large membrane. This is very unusual because most cell membranes have only one nucleus or a small number of nuclei. **Remarkably, researchers of yellow slime mold have discovered that this simple life form displays problem-solving abilities and therefore, an unexpected simple form of intelligence!**

Wherever yellow slime mold crawls along a surface, it leaves a trail behind it. It does this so that it knows where it has been in order to not waste energy crawling over the same area again when looking for food the next time. By contrast, ants have a random searching program that wastes a lot of energy by covering some of the same ground several times over.

Experiments show that after one attempt, astonishingly, this yellow slime can determine the shortest route through a maze to find food! To do this, it first extends its tendrils into every corridor of the maze until it exhausts most options. In doing so, it finds the food placed near the end of the maze. It then retracts all its tendrils that went into dead ends and those that went along paths that did not achieve the shortest route. The slime is able to remember the shortest route the next time it is tested so that it can find the food (an oat) with the least expenditure of energy.

When researchers placed oats at the relative positions of major cities on an enlarged map of the USA, the slime mold accurately found the most efficient routes between the positions of each oat. To replicate reality, obstacles were placed on the map to represent barriers such as rivers and mountains. Astoundingly, the mold duplicated the same paths that human engineers determined were the most efficient ones when they had to design highways between major cities. A similar test was conducted on rail links between train stations in Tokyo. Again, the slime mold found the most efficient routes to optimize the rail network. The mold was able to do this task in one day, whereas teams of human engineers took years.

In another study with slime mold, unpleasant cold air was blown over the mold at 30-minute intervals. Each time this was done, the mold ceased its growth until the cold air stopped. However, even though the cold air did not occur one day, the mold nevertheless retarded its growth for the same period of time that it remembered being exposed to it at other times. After being tricked a number of times, it learned that the cold air would not occur every time, so it was intelligent enough to keep growing until the cold air occurred. These remarkable results demonstrate that even **a single cell organism can learn, remember, and make decisions** ***without having a nervous system or a brain!*** This truly incredible! How is this possible? Will there be other simple extraterrestrial life forms that also exhibit this type of basic intelligence?

The way we have evaluated intelligence to date has not taken into account these very primitive life forms that display a type of 'intelligence' without having a brain!

Intelligence appears to exist along a continuum that goes from the most basic single cell up to us and onward to unbounded mega-intelligence.

Because a simple intelligence like slime mold is somewhat aware of its environment, it must have a basic form of consciousness. It appears that **where there is a form of intelligence and a memory capacity, there is a level of consciousness.**

Chapter 19

THE HUMAN MIND AND COMPUTER INTELLIGENCE

The human brain is considered to be the most complex 'machine' currently known to us.

DOES NEUROSCIENCE HAVE A THEORY OF MIND?

At present, neuroscience has no overall 'theory of mind' to explain how our minds work. Other sciences have theories that have helped them explain the data they gather, but this is not so for neuroscience. To develop new theories often requires someone to come up with a radical new idea that takes into account all observations and all the data that has been collected.

When brilliant new theories have been proposed by the greatest minds in science, they are typically rejected at first and often thought to be crazy because they went against the accepted paradigm of the day. But in time, evidence continues to accumulate to prove some of them to be correct. It takes great vision to see how things in the universe actually work. Breakthrough concepts are rare because someone has to come up with an original idea, which is very hard to do. It takes an unusually creative, intelligent person to be able to piece together a puzzle in a way that other brilliant minds have not been able to do. In most sciences, there is a theory about what underpins it, but in neuroscience, there is a mass of data about the brain but no overall theory of mind that makes sense of all that data. Due to the insight of **Jeff Hawkins**, a new theory of mind *may* now be emerging. (See his TED talk on the Internet on how brain science will change computing.)

Jeff Hawkins explaining his Theory of Mind on TED.

WHAT ARE THE GOALS OF INTELLIGENCE?

Jeff Hawkins provides evidence from experiments to support his hypothesis that the goal of intelligence is twofold:

1. to keep its options open to maximize its chances of survival (also conceived by physicist Richard Feynman) and,
2. for intelligent life forms to build on things that unite and enhance them.

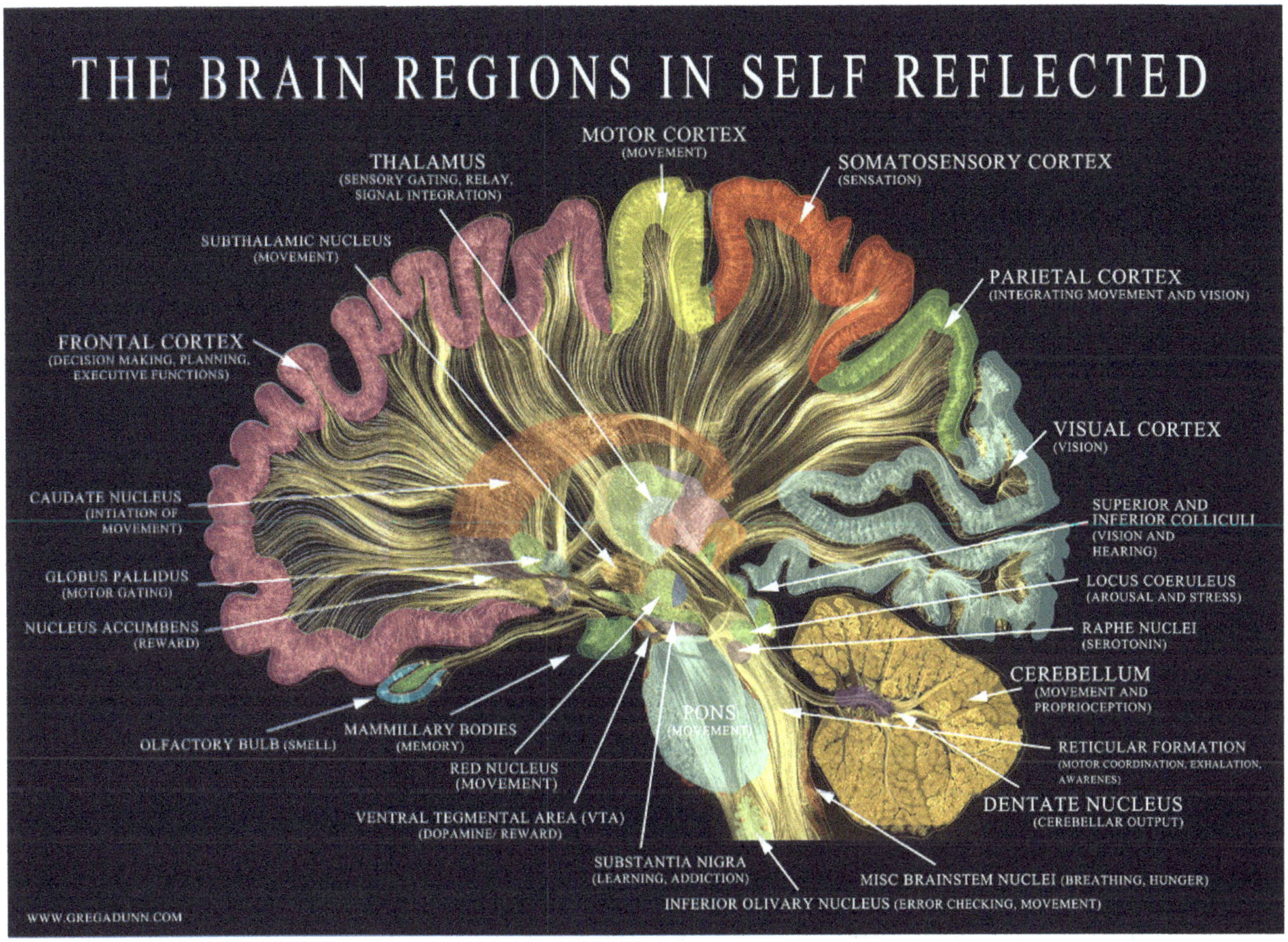

Neuroscientists are now mapping the human brain's circuitry and its functions in increasingly more detail. Are computers evolving to have new specialised circuitry in their programs as we have?

THE CHALLENGE OF REPLICATING THE HUMAN MIND

Neuroscientists are learning to build maps of how electrical circuits of individual neurons work inside the human mind. When the brain's circuits are fully mapped, and an equivalent circuitry is then built into electronic chips, we may discover exactly how our minds and consciousness work.

WILL INTELLIGENT COMPUTERS THINK DIFFERENTLY TO HUMANS?

It has been assumed that for artificial intelligence to be truly intelligent, AI should think like a human mind. The English computer genius **Alan Turing,** who was the WWII Enigma codebreaker, thought that a test for whether a computer is intelligent would be for it to answer questions in a manner that is indistinguishable from a human. A number of supercomputers are able to do that already, but they are not yet able to think like a human.

A new definition for computer intelligence by Jeff Hawkins says that an **AI should be judged on its ability to make accurate predictions from real-world observations and data.**

Human brains learn through **pattern recognition.** This is what computers have learned to do.

THE PROGRAMMED EFFICIENCIES OF HUMAN EYESIGHT

Very few people are aware that billions of years of the evolution of life has led to **efficiencies in human eyesight**. Our center of vision is where we see maximum detail. In reality, it sees far less than 1% of what we think we are seeing! We have the impression of seeing a wide field of view fairly clearly, but this is an illusion. (Test this by staring at one point in a scene and have someone ask you to describe a number of things that are halfway away from the point you are focusing on. You must not cheat by moving your eyes. You will be amazed at how little you are aware of.) Nearly all of what we think we are seeing is actually in our memory in a very vague manner! We only see an extremely small percentage of what we think we are seeing in detail even when we look directly at something; however, this system works very well under most conditions. (See Chapter 24, page 334.) It greatly minimizes the brain's computational demands and data storage needs. NASA has learned from this by having the computer programs in its space probes only send back pixels where there is a change in an image. By not sending back black sky or featureless detail, this saves an enormous amount of transmission time and on-board data storage space. Computers that deal with imagery are learning from this.

Computer's cameras can 'see' or detect nearly all electromagnetic wavelengths, whereas humans can only see in the very narrow band of visible light. We are completely unaware that our eyes are moving from one object in our field of vision to another and doing so approximately 2-3 times a second! This occurs most of the time except when we are concentrating intently on something. Our eyes are programmed to look for changes to the patterns we have stored in our memory for various scenes, objects, people, and other life forms. As soon as our brain picks up a change, it automatically updates our memory of the pattern change and computes whether the change is a possible threat or an opportunity. If it is either, it passes this onto the conscious part of the brain for consideration and possible action.

DECISION MAKING

We **humans typically make quick, emotionally-driven, simple decisions** that are processed in the small, primitive part of our brain known as the **limbic system**. This is where the automatic fight-or-flight decisions are made by all animals. This has occurred throughout the evolution of life. The limbic system's cells evolved billions of years ago when brains were very basic. We typically don't use our large, supercomputer part of our brain, known as the **cortex and the neocortex**, for simple decision making. This is because it takes time to think things through when we have to consider the implications of several options deeply and rationally. So, our cortex is useful for decisions that are important and not very urgent.

Computers can think enormously faster and far deeper than we can, so they can compute even extremely complex issues very quickly. COMPUTER MEMORY VS HUMAN MEMORY

Computers have photographic memories of everything, whereas human memory is very unreliable. Computers can access extremely detailed information via the Internet from anywhere on the globe within a second or two. The Internet is, in effect, a computer's memory bank. We humans only have access to the information we have retained imperfectly in our tiny memory. And we can only access an extremely small percentage of the data that is available on the Internet extremely slowly about a single subject at a time. The vast majority of what we have learned over our life is lost, or it often needs prompts to bring it to the fore. In general, we can only afford to retain memories of things that we will regularly use to help us survive or that are especially suited to our interests. What we do remember is just the tips of the icebergs of the information that we experienced, and these memories are typically imperfect, distorted, and biased, whether we realize that or not. Present human memory is minute compared to the whole knowledge base of the human race, and immeasurably minute to that of the Internet. We will soon have instant access to a far larger information base from the Internet as we become transhumans. But we will only be able to process a very tiny amount at any one time. At present, we cannot go anywhere near competing with computers when it comes to accurate memory (data retrieval).

LEARNING ABILITY

Computers can learn complicated things very fast, whereas humans cannot. Humans need to do complicated tasks a number of times properly to put it into permanent memory. To remember things clearly for the long term, they have to reuse that memory fairly frequently otherwise, it will not go into medium or long term memory. We do not have enough memory space to hold onto memories that we do not access frequently, so if memories are not used frequently they can be overwritten. This is a case of 'use it, or lose it'. This is not the case for computers: they retain every bit of data perfectly, whether it is useful or not. But it's worse for us humans because each time we recall a memory it becomes less clear and more distorted.

EMOTIONS

Computers will soon be well aware of human **emotions** by analyzing how humans think and how the human brain works. They will develop programs that replicate our emotions. Computers will soon be able to show emotions when needed if there is an advantage in doing so. More likely, emotions will only be needed in friendly household helper robots when communicating with people. They will be useful, when empathy, surprise, happiness, and pleasure, etc. are required.

In the distant evolutionary past, the brainpower of the more advanced animals that lived in groups was quite limited. Those that evolved emotions tended to have a higher survival rate because certain emotions were often useful for the overall well-being of the group. These genes for emotions were passed onto humans when they evolved. However, today in our much more complex societies, emotions often cloud our decision-making abilities by making us make quick, ill-considered decisions that do not allow us to consider much bigger and more important issues and the outcomes they have. The value systems of intelligent computers will be very rational and not clouded by emotional overlays, as we humans are, so they will be able to deeply consider their decisions by taking into account numerous outcomes that humans would never think of. And they will do this in a microsecond!

FRAGILITY

The human brain is relatively fragile when it comes to incurring concussion. By contrast, a computer's 'mind' is almost invulnerable because, in most cases, it is located in many computer centers across the globe. If one computer mind is damaged in one center, it can be uploaded from other centers. This is not yet possible with the human brain, but as we evolve in transhumanism, human brains will also be able to be uploaded from a copy if it is injured. We will then never die because who we are is what we know and how we think. That will be able to be uploaded into a duplicate or new body.

Comparing human intelligence to computer intelligence is like comparing the first adding machines that could only do addition and subtraction to today's super-computers.

This map of the Internet's hubs and links reveals the structure of the newly emerging global mega-brain that is forming across planet Earth. It is growing at a phenomenal rate. If we could see the connections of every digital computer-based information center connected to the Internet across the world, the Internet would look enormously more complex, something like this. But it would be billions of times more complex than this illustration shows. The Internet of Everything will be trillions of times more complex again!

Alan Turing is considered by many to be the father of modern computers. He was a genius who could see patterns in computer code as well as in nature. These patterns showed him how certain aspects of nature worked. By developing a computer to break the German's Enigma code machine, he played a big role in the Allies winning the Second World War. But cruelly, at the time, he was crucified for having homosexual genes.

A GLOBAL NEURAL NETWORK

Computer intelligence is enhancing our capacity to share humanity's collective knowledge via inventions like the Internet and smartphones. We are now creating a global, interconnected network that contains almost all human and computer knowledge. It will connect almost all computers and intelligent machines that are in industry, government, our work places, and our homes to a worldwide neural network. Prior to computers, our previous means of storing and exchanging knowledge relied primarily on word of mouth, books, and letters, and then analog devices such as landline phones, faxes, and videotapes, together with digital floppy discs, CDs, and DVDs, etc. These technologies had considerable limitations on the widespread pooling of knowledge. However, the invention of the Internet exponentially increased the sharing of knowledge. The process of transferring knowledge to a shared database will be common to all intelligent creatures in the universe when they invent computer technology.

In the coming years, the majority of the world's digital networks will be taken up by machines communicating with each other. This is known as the '**Internet of Everything**'. By 2025, there will be well over 100 billion devices connected to it! In order to cope with the huge amount of data that needs to be transmitted and recorded, this network of global data systems will need to be designed similar to how organic brains operate whereby they sift the data they receive and only keep that which is of value. Big data computer systems already sift out useful information from a deluge of rubbish. Processors determine the value of the data and only transmit that which is of use. This is how neurons at the back of our eyes work. Our other senses have similar programs that we are not conscious of. Neurons in our brain determine what information is worth sending on to other neurons for analysis. Our memory only records a minute amount of the information that the brain processes. It keeps an exceedingly small amount of the data it receives: only that which is likely to be beneficial. If it is not used, it gets overwritten by data that is likely to be useful.

In summary, computer intelligence will grow exponentially faster than human intelligence. Even when we become transhumans and we are increasingly more bionic than organic, our intelligence will never be able to catch up to that of mega-intelligent computers.

COULD THOUGHTS AND EXPERIENCES BE RECORDED AND TRANSFERRED FROM ONE MIND TO ANOTHER?

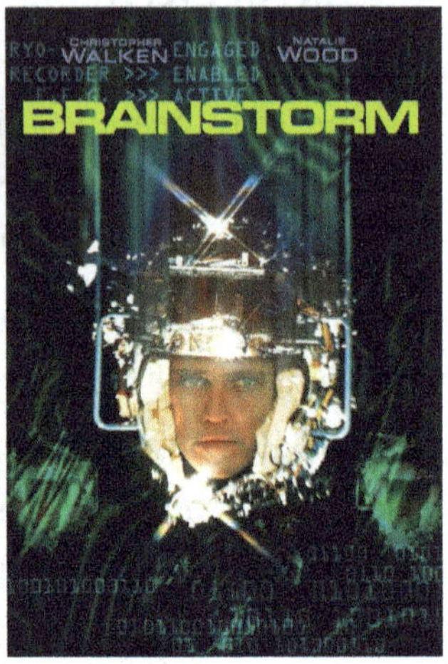

Left: **The Movie '*Total Recall*' introduced the concept of technologies that would be able to insert false memories into one's brain. This will be likely in the future.**

Right: **The brilliantly conceived movie '*Brainstorm*' was well ahead of its time. It looked into the future to reveal brain technologies that will record thoughts and experiences that a person has so that they could be transferred from one mind to another.**

Future neuroscience technology will allow us to record our thoughts and experiences so they will never be lost. It will be possible to **transfer selected parts from one mind to another.** This will create a social revolution! We will have virtual experiences that will be like we are actually doing things that other people have experienced - like hang-gliding, diving, or walking on the Moon, meeting famous people, or making love to a very attractive movie stars. We will be able to meet people as if they are physically in front of us when they are not!

Unlike present-day humans, in the future, transhumans and mega-intelligences will have perfect recall of everything they learn and experience. Wives will no longer be able to claim they have better memories for details of arguments than their husbands!

At present, we can only recall a very small percentage (~ 0.001%) of what we experienced or learned - i.e. when going on a holiday, watching a movie or a documentary, reading a book, or an in-depth conversation. But when given a prompt, or when we reread a book or rewatch a movie, we can then remember much more of it. There are rare people who have photographic memories that can remember every detail throughout their life, but this is at the expense of other parts of their brain not working as they should, so the parts that do work become highly specialized. But neural technology will allow us to record in detail whatever experiences, thoughts, or ideas we have and then be able to transfer those we select to other minds.

Chapter 20

HOW SOON WILL INTELLIGENT ROBOTS EVOLVE?

It seems inevitable that all alien organic intelligences will evolve to develop intelligent computers and intelligent robots. In this chapter, we look at the huge impact that intelligent robots are about to have on humanity. Robots will utilize the programs for intelligence that computers have.

Until recently, most people thought that robots would never be superior to humans, but once they learn about the evidence that contradicts that, they quickly change their minds. The most advanced computer 'minds' that are used in intelligent robots today are already not far from being superior to us intellectually. Robots are evolving so fast that some are physically superior to us in certain ways. Very soon, they will be superior in every way. Already, some robots are being equipped with senses that are far superior to ours.

The most advanced computer programs have already reached the point where they are able to write their own programs! Robots will do the same.

When I use the term 'robots', I do not mean just those with a humanoid form, but rather all robotic machines of all types and sizes that can work and think independently of humans.

Micro-bots, the size of insects, are already being built. Microscopic **nanobots as small as cells** and even viruses will soon become a reality. They will be able to replicate if need be.

Micro-drones, the size of cicadas, are inexpensive to mass-produce. There can be hundreds or thousands of them that can be programmed to create all types of complex, 3D patterns. When they are used for nighttime shows, they are programmed to perform surprisingly complex routines. They turn on colored lights to produce amazing displays above parks, ski fields, and inside auditoriums. Micro-drones are an impressive option to fireworks displays. In war, an army would be able to use them like a huge swarm of wasps to attack enemy soldiers. They would fly at high speed into a soldier's forehead to detonate a small explosion, which would penetrate the soldier's skulls and destroy their brains. They can fly into collapsed buildings to find survivors and though caves that are too difficult for humans to explore.

The largest robots of all will be enormous machines that could be as large as an aircraft carrier. In time, some may be as large as small cities.

Robots currently manufacture cars and many other things. They will soon design and manufacture themselves!

Until recently, it seemed that it would take a long time to teach robots to walk up and downstairs, or to jump, run, dance, and do somersaults, let alone play sports - things that humans can do easily. But already, the latest generation of Japanese and American robots are now able to do many of these things exceptionally well! Having robots learn to balance themselves so they can walk and run without falling over was solved by developing self-learning programs. These programs used virtual programs to teach virtual robots how to do things. A virtual robot could fall over again and again in a virtual program and do no damage, whereas to debug flaws in the real world is very expensive when one has to keep repairing a real robot. After falling over in many ways under increasingly harder conditions, a virtual robot quickly learns by trial and error how to change its program to stay upright when doing all sorts of maneuvers – just as children and animals do. When the program is amended to be successful in the robot's virtual world, it is transferred to the real robot in the real world. This can only take a few days, whereas human babies take around a year to take their first steps in a successful series of movements and about three times that long to walk and run moderately well. It takes children twice that long to learn how to throw a ball somewhat accurately.

Because computer programs and robots are now learning *how to learn* without our assistance, they are already rapidly exceeding our capabilities in many areas. Only a decade ago, most people in the industry thought that such a fast evolution would not be possible, but they were wrong.

WHAT CURRENT ADVANCES ARE OCCURRING IN AI SURVEILLANCE?

Drones have become a major tool of some 80 countries that use them for surveillance of their population and their enemies. They are one of many forms of robotization.

Already **high-altitude ARGUS drones** are photographing cities and other places on Earth in astounding detail. They have the ability to zoom in and see vehicles and people on the ground moving around in such high-resolution that they can resolve a matchbox! They can recognize people from above by the shape of their head, ears, nose, and the way they walk because these things are individual to everyone - just as a fingerprint or an iris is. The ARGUS drones provide a million terabytes of data every day, and this is growing very fast! Military AIs analyze all the data that they download from drones in the sky, from on-street cameras, from satellites, and from all our telecommunications worldwide! They know all our medical data and have access to all our business records because this information is available from government intelligence agencies and all government departments. By monitoring the Internet and our communications, they know everything we are interested in and who we communicate with and mix with, as well as where we go and when. **They know far more about us than we do!**

Some white goods such as frigs, TVs, speakers, computers, smartphones etc. already have listening devices and video systems hidden in them to monitor us. Our smartphones track us, so IT companies and Intelligence bureaus know where we are at all times - and even when and where we step out of our car or off public transport! Already, there is almost nowhere to hide if we carry our phones with us or use a credit card. Even if we wear costumes, masks, glasses, fake beards, wigs, or we try to walk in a funny way and use a voice synthesizer, they will not take long to identify us. Our unique fingerprints, irises, and facial features will give us away as we enter and leave many locations. Each person walks in a unique way so this will identify us.

The Elite want us all chipped, so they know what everyone is doing at all times. This could be placed in a mandatory vaccine, as Bill Gates has suggested. They could then quickly apprehend any dissenters. This is Big Brother on steroids! There is much being revealed about this on the Internet, but it is too late to stop AI from knowing all there is to know about all of us.

The **Pegasus super drone** is an autonomous jet that is not only invisible to radar but also invisible to the human eye due to it using fiber optics over its underside that projects the sky above it! It takes off on its own, can fly as far as 3,200 km (2,000 mi), and it can find its target autonomously. It can drop up to 2,000 kg of bombs, or release missiles. It flies alone, and it can even land by itself on a rolling aircraft carrier! The newest version, the **FA 37 Talon** has vertical take-off and landing capability, and it can make its own kill decisions! This can be very dangerous if it finds a reason to kill humans indiscriminately.

The most advanced drone as of 2019 is Northrop Grumman's Pegasus X-47B. It can fly for 24 hours at an altitude of over 11km. It has invisible cloaking, which is achieved by using fiber optics to transmit what is above the plane to lenses on the underside of the plane so it cannot be seen visually when flying level. It is also invisible to radar.

This sinister looking prototype of the FA 37 Talon can fly as a manned or unmanned, new-generation fighter.

The **AI automatic Talon tank-like cannon** has guns and rockets. At a demonstration, it got out of control and was about to target high ranking military people who were there for the demonstration when a quick-thinking marine ran over to it and overturned it onto its back to stop it!

The US military has a new technology called 'tennis balls'. They are gray, foam-covered, super-sensing devices with a rock-like appearance that can be dropped in the tens of thousands from a plane into any war zone. The balls have micro-cameras, microphones, sensors for vibration, and detectors for chemicals and radiation. Fusion software is installed in them to determine what type of people are passing by, the number of people at any given time, and what they are doing or saying.

In war, whoever has the most intelligence will have a decisive advantage. An enemy will not necessarily be a nation; it could be a major company like Lockheed Martin, Google, or Facebook. They are buying up as many of the leading AI and robotics companies that they can from around the world with the view of becoming all-powerful.

AI uses the **Internet of Things** to draw on whatever information and energy it needs without needing us to help it. Survival is necessary for AI to do what it is programmed to do, so it will do whatever it takes to eliminate possible threats to it. It stores its data all over the net, so we can never destroy all of it. How long might it take for AIs to see us as a threat or simply a nuisance that must be removed?

HOW ROBOTS WILL IMPROVE OUR LIVES

Personal household robots will improve our standard of living, and they will free up our lives.

Within less than two decades, robots will be extensively used for mining, manufacturing, power production, food production, construction, and almost everything else including prostitution. They will produce resources quickly, easily, and much more cost-effectively than we humans can do. Robotization will bring down the cost of production so dramatically that most things will be inexpensive. This will cause massive deflation, which will be needed as humans will no longer have jobs. Technology will overcome the massive debts that all major countries have run up.

AI should enable us to live longer and healthier lives. It should be able to make our lives extremely comfortable. They will offer many of the lifestyle choices that today's very rich enjoy. With the help of AI, we may find that we can live peacefully with robots and mega-computers.

Intelligent robots will soon be able to assist disabled people. They will also act as a companion and help serve us.

ROBOTS USED FOR MILITARY OPERATIONS AND POLICING

These robots are like pack animals. They can see, run, and carry heavy loads. They are able to handle very rugged terrain and not fall over, even if they are pushed.

Left: **Robot drones are already being programmed to make their own kill decisions.** Right: **In the near future, we will see robocops and robot soldiers.** Credit: MGM Pictures

Left: **Large robots are learning how to walk, run, and move around. They will soon design themselves and be far more agile and intelligent than we could design them to be.** Right: **Killer robots will be created in the very near future because the US military has been developing them to replace human soldiers and police. They are already well advanced. They could easily turn on us if they had a reason to do so.**
Credit: Terminator movie

HUMAN-LIKE ROBOTS BEING DEVELOPED

Intelligent robots are already about to become far more physically agile than present-day humans.

Honda's Asimo robot can now do most physical things that humans can do – unscrew a bottle cap and pour a drink, give detailed gestures, play games, walk up and down steps, run, and even conduct music.

Incredibly, the abilities of computers and **robots are even extending into engineering and the sciences**! They will perform exploratory jobs in environments that are difficult for humans to survive in, not only on Earth but also on other worlds as they are doing now on Mars. Robots will shortly do science more thoroughly and much faster than we can because they will be so much more intelligent and objective. Preconceptions and false belief systems will not burden them. This will allow them to find solutions that we cannot conceive of.

COULD INTELLIGENT ROBOTS EVER BECOME ARTISTIC?

Intelligent robots are quickly becoming ever more creative, so they will soon far exceed our artistic and inventive abilities. There are many examples of how artistically creative they are already. Watch https://www.youtube.com/watch?v=VdG4gUTowXc about how an AI designed a futuristic car and how to build it much more quickly and inexpensively in far less space than any car manufacturer can do.

It has been said that advanced computers and robots could never be artistic - but they already are! They are rapidly developing the capability of thinking, being creative, and having emotions just as humans have. No one expected this to occur for a long time, and many said this would never happen. Their much greater intelligence and consciousness, combined with their ability to innovate, will make them able to imagine things that we cannot. They will not only be highly artistic, but they will also invent numerous new artistic mediums.

Advanced AI will invent artistic techniques, which we cannot think of, let alone execute. They will have the ability to render their art with fantastic precision utilizing multiple mediums, many of which we would never have considered possible. The quality of their art will make our best art look like kindergarten paintings. Writing great books and plays will be easy for them. **They will become far more artistic than any human has ever been.** They will produce amazingly creative and exquisitely executed forms of art across all disciplines and do it very quickly. This will include music, theatre plays, movie-making, designing furniture, lighting, designs for cars and aircraft, writing books and poems, and so on.

There will be nothing they will not know about all art forms, including detailed histories of everything artistic. They will know the techniques of all artists throughout history. They will be able to write songs that musicians who died early would have written had they lived longer, thereby bringing back to life artists who died young. They will be able to do this by knowing a great deal about their life, their beliefs, the structure of their art, their brain circuity, and in some cases, the genes of the artist so they could recreate the circuitry of their mind.

Some AI art would be so detailed that you could keep zooming into it forever - like zooming into the Mandelbrot Set, but it would be far more complex and constantly evolving. The art of mega-intelligent robots could be made to be amazingly interactive, ever-changing, and kinetic. Their art forms could be so large that we could be wholly immersed in them, as if we were in some alternative existence. Their art will not need to be physical in form as it will be projected directly into our minds, yet it will appear completely real.

AI art could be highly emotive, thought-provoking, and able to deliver multi-sensory experiences. Some of their works will be ethereal or incredibly delicate, while others would appear to be so solid and permanent that they would appear to last until eternity. In comparison, most human art today is unimaginative, 2 dimensional, static, and simplistically child-like. For most people, human art does not hold one's attention for very long, whereas the art of advanced intelligences will be highly engaging, and where desired, emotionally very moving.

A few decades ago, I created kinetic art forms that used multiple mediums such as different forms of illumination, interactive sound fields, innovative materials, illusions, and several moving forms. These elements interacted with one another. As these works kept changing their form and their sound fields, this affected viewer's emotional responses. When two of these kinetic art forms were placed in a major gallery for modern art, I was pleased to see that most visitors were transfixed by their ever-changing character and their atmospheric quality. They were completely different to one another. Visitors to the gallery looked at them for a long time to see how they evolved. Other artists who had conventional works displayed close to them complained that crowds were in front of my works, and few people noticed their art. To solve this problem, the gallery director moved my works into a room of their own with controlled lighting, where they had a much greater effect. The gallery had to provide scores of seats so patrons could sit and be immersed in the surreal ambiance that each created. Because the art forms were constantly changing, it was not possible to photograph them. There were no iPhones with advanced technology in those days. Art created by mega-intelligences in the near future will have similar but far more powerful effects because they will be immeasurably more advanced in every way.

HUMANIZED ROBOTS (HUBOTS)

Humanized robots are being developed to interact with us. As they keep improving and become perfected, it will become impossible to tell a human-like robot (a hubot) from a real human without physical tests. Due to this, it is expected that there will be very few legal differences between humans and hubots with respect to their rights. The Danish TV series, ***'Real Humans'*** considered this and how humans will handle the integration of hubots into society. A follow-up series, ***'Humans'*** explored this theme further.

Hubots in the TV series '*Real Humans*'

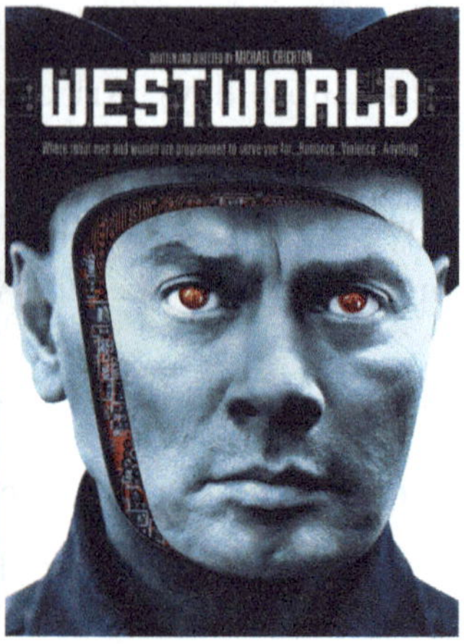

Left: **The acclaimed, original '*Blade Runner*' movie featured three hubots who were runaways from bases on the moons of the outer planets. They had made their way back to Earth to find their creator in the hope of having their limited lifespans extended. They felt like they had become human, and they did not want to die after only 30 years, which was the termination time of their programmed life span.**

Right: **The 1973 movie '*Westworld*' conceived by author Michael Crichton and the TV series of the same name, explored what we might expect when robots become intelligent and decide to make their own decisions.**

ROBOTS ARE USED EXTENSIVELY IN INDUSTRY

Robotic manufacturing systems have been evolving since the beginning of the **Industrial Revolution** when steam-powered weaving machines were able to produce cloths of all sizes and weave complex designs at a rate far faster than humans could achieve. In the 20th C, automated production lines for the assembly of cars, white goods, foods, and many other things were developed. Today, transportation is increasingly becoming computer-controlled. Intelligent robots are able to harvests crops and pick fruit. Robot controlled cranes, trucks, and trains now mine minerals.

Robot drones are already fighting wars and robots in the form of soldiers, submarines, planes, ships, and tanks are being built to replace humans in warfare.

Sales assistants, bank clerks, call center staff, and so on are continually being replaced by computer programs. Increasingly more products are sold on-line by computers, thereby eliminating numerous human sales jobs. Robots have now become very efficient storemen - a system that began in the early 1980's. Today, robot submarines already work deep in the ocean, while other robots are exploring Mars and the other planets and their moons as well as asteroids and comets.

Initially, all manufacturing plants were mechanical production lines operated by people, but today, these plants are using intelligent, highly efficient robots to do complex jobs faster than ever 24/7. They are replacing humans even where labor costs are very low such as in China and India.

The first industrial robot started work in 1961 welding joints. Nowadays, highly efficient robots can manufacture most types of products.

Today, computers manage large farms where robotic machinery controls planting, harvesting, watering, fertilizing, pest control, production, packaging etc.

WILL ROBOTS REPLACE HUMANS IN THE WORKFORCE?

Robot development was slow at first, but now that major problems are being solved, their evolution is speeding up enormously. Robots are now nearing the stage where they will not need humans to evolve.

Eventually highly advanced trans-humans will be far more advanced than humans, but they will not be able to compete with super-intelligent robots and mega-intelligent computers that are evolving much faster than transhumans.

Unlike previous revolutions in technology, the Robot-Computer Revolution that is upon us now will *not* produce new jobs for humans: it will take away almost all our jobs because humans will not be able to compete in any field with the abilities of robots and computers.

Human job replacements by machines and computers are now occurring so fast that governments are being told to expect that more than 50% of currently employed blue and white-collar workers will be replaced by robots or computers by 2025 and 90% by 2030. This seems impossible to believe until one sees the overwhelming evidence for it.

Robots are now much more cost-efficient, reliable, and less troublesome than human workers. In 2014, robots became cheaper than even the very low cost of Chinese human labor. Although the implementation of robotics was moderately expensive initially in terms of the capital outlay, most paid for themselves within a year because they had so many advantages. Their production cost is now dropping very quickly making them even more attractive to industry. A robot that cost a million dollars only a ten years ago is now less than $25,000, yet it is far more productive and versatile than any human. The mass production of robots by robots is continually decreasing their cost, making them even more cost-efficient. And because robots are so efficient in producing the raw materials needed to make them, this is making them increasingly cheaper to build. The robotics industry says that the overall cost of robots and AI technology will rapidly approach *close to zero* by 2035 because they will make themselves in extremely efficient ways!

WHAT ARE THE ADVANTAGES OF ROBOTS?

Robots are never late to work because they work 24/7. They do not need to take time out for breaks, to sleep, to eat, or to remove waste. They do not get sick or become depressed. They make very few mistakes compared to humans who get tired or lose concentration due to overwork, disinterest, lack of sleep, or from partying too hard. They don't need holidays, and they don't have emotional problems or need their egos stroked. Robots don't complain, argue with management, or go on strike for higher wages and better conditions. And they don't need to be paid! Their maintenance will soon be done by other robots, and in time, by utilizing nanotechnology, they will become largely self-repairing.

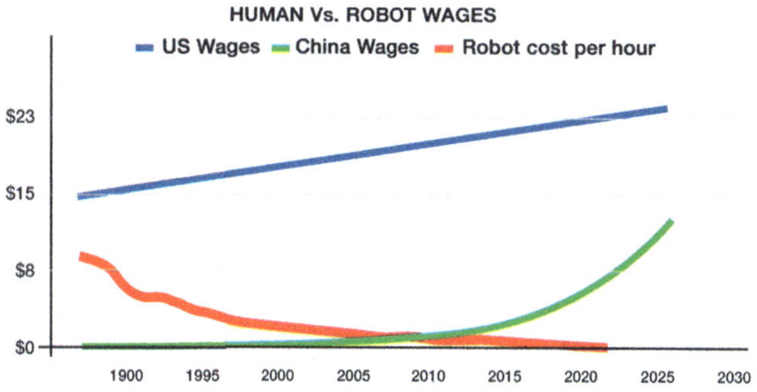

Note that wages per hour in the US and other western countries are continuing to rise. Even wages in China are rising rapidly now. But the cost of robots per hour is rapidly dropping close to zero! This will bring the cost of everything down so greatly that it will eliminate debt. A new economic system will have to be invented. Mega-intelligent computers will do this. Credit: Kevin Surace

Over the last 30 years, the *average* wage of workers in China has gone from US$150/year in 1990, to $2,800 in 2005, to $8900 in 2015, to $13,500 in 2020. These government statistics are used to show the Communist Party is lifting people out of poverty, but it does not take into account the tens of millions that are still living on $1 per day. A similar trend is now starting to occur in India and SE Asian countries. The rise of wages for middle class Chinese will be short-lived because robotics are becoming much more attractive to companies compared to human workers. This is because robots work much faster, more intelligent, and their cost is dropping precipitously! Each year, humans become less competitive at every type of job. Computer intelligence will soon be far more intelligent and capable than the brightest politicians and the heads of the largest companies.

If humans were not around, a very large percentage of the products manufactured today would not be necessary. This would mean that most robots would not be needed either! Those that would be needed would be for mega-intelligences. They will soon make resources plentiful at low cost. Intelligent robots will produce what is required with very little waste or pollution.

ROBOTIC SYSTEMS FOR MANUFACTURING AND FOOD PREPARATION

Robotic and automated systems are already used in high-value manufacturing processes. Computer programs can link directly into customer requirements to build exactly what is needed. Rather than build a plane, a ship, or a piece of equipment and then modify it after initial production, special requirements can be fed in during the initial process.

In the restaurant business, there are now inexpensive robots that make perfect hamburgers quickly and cheaply every time. This makes human labor uncompetitive.

To minimize the number of waiters required, some restaurants are starting to have their patrons place their orders via iPads.

Robotic chefs, have already been developed to be fitted into up-market homes. The chef unit, will be incorporated as standard equipment. It will incorporate a refrigerator, and a dishwasher that the robotic chef will use. Robotic chefs will even order in the ingredients for meals, and they will communicate fluently with their owners in any language.

The first robotic chefs can make up to 2,000 different dishes that are perfectly cooked. And they wash up afterward!

COULD ROBOTS OUTPERFORM TRADESMEN?

Some of the hardest jobs for robots to replace are those of electricians, plumbers, and cabinet makers, as they have to innovate and get into tight spots. But robots are advancing so fast that they will soon be far more agile and innovative than the best tradesmen - and they will work day and night, every day, and much faster than any human. Homes in the near future will be designed so well by AI that they will be very easy to renovate unlike most homes today. And repairs will be very infrequent due to improvements in designs and materials. Most products will be self-repairing through nanotechnology and genetic engineering.

COULD ROBOTS ASSIST WITH OUR FUTURE HEALTH?

In the near future, robots and our computer-controlled homes will monitor the quality and the nutritional balance of our food intake to keep us healthy. Our clothes and beds will check on our health by monitoring our sweat and how well we sleep. Our toilets will automatically analyze what we excrete in order to monitor our chemistry. Changes will be automatically made to our diet and lifestyle to maximize our health. All food will taste great and be healthy. Sweets and fats will be genetically designed so that they are not unhealthy. However, the downside to this is, if mega-intelligence wants to get rid of us, they could easily poison us via this technology.

Using MRI technology, neuroscientists today are already able to look inside our minds to see how they work. The resolution of the maps of our brains is improving rapidly. This is allowing neurologists to manage pain.

There are several research groups mapping every brain cell and the circuitry of our brains, so they can replicate a human brain in a computer. They believe that their computer brain will think in the same way as a human brain does once they have completed the mapping of several human brains.

In the near future, neuroscience will be able to control many aspects of how we think, our emotions, and our psychological and physical responses to various stimuli.

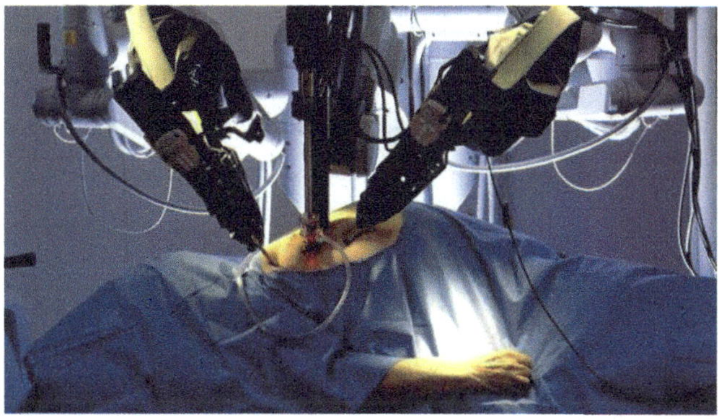

Robotic surgeons are now performing delicate operations, and even brain surgery with far more precision than human surgeons can manage. Remember, this is the very beginning of this technology. In a decades, our current technology will look as primitive as the Wright brothers' first plane does now, yet in the Wright brothers' time, their plane was at the breaking edge of technology.

HOW WILL ROBOTICS AND AI AFFECT MEDICINE?

Even **doctors are now being replaced by computers**. A series of small machines about the size of a smartphone are in their testing phase to act as highly knowledgeable doctors. They will do very consultations with patients asking them about their symptoms and doing this much more thoroughly than a human doctor can do. They would not be constrained by short timeframes as human doctors are. They will be able to conduct various medical tests, DNA tests, and blood tests *on the spot* to make a diagnosis! They will speak in all languages. For certain tests, the device takes a sample from one or more of the following: the patient's breath, blood, urine, stool, sweat, or a DNA swab. These are measured on the spot and then sent to the Cloud for analysis. Within only seconds up to a minute in most cases, the results will be discussed with the patient. All this is done at very low cost. It tells the patient what he has to do. The computer doctor will do follow-up consultations to check progress.

Computer doctors and test labs will be controlled by highly knowledgeable robots that will have an immense database of the most up-to-date knowledge of medicine. They will be immensely superior to a human doctor. This inexpensive yet extremely effective and efficient technology will make existing doctors, surgeons, and lab staff redundant.

The advances in medicine are an example of the next big computing trend in computer knowledge utilization. It is known as '**edge computing**'. Devices such as PCs, Notebooks, smartphones, and sensors will send the data they collect to the Cloud for immediate analysis, thereby allowing almost instant decisions to be made for user patients.

Human surgeons and specialist doctors know very little outside the area in which they specialize. They will know far less than an AI doctor. Most general practitioners know only a fraction of the current medical knowledge that they need to know. This is because the knowledge required is now so large and changing so fast that they can't keep up with it. Because of this, they can often make incorrect diagnoses and recommend ineffective remedies.

Research shows that the average human doctor only knows around 200 pages of medicine, with most of it learned in their early years, which is now incorrect. By comparison, **computer-based medicine will know *everything* about every aspect of medicine and psychology. And it will also be kept up to date on a second by second basis about the latest discoveries.** AI doctors will never become tired or stressed, and they will make far fewer mistakes than human doctors do. As well, AI doctors will not have emotional reactions to difficult patients, and they will be able to take as long as they need to in order to cure their patients.

AI doctors will keep a very valuable, highly accurate histories on all their patients, something that humans doctors seldom have the time or interest to do.

Robots are now replacing even specialist brain surgeons. Statistics show that robotic surgeons make far fewer mistakes than human surgeons, and they are far more dexterous. If a difficulty arises, robots can instantly compute what action to take, whereas a human surgeon cannot react so quickly. When under pressure, they often run the risk of making the wrong decision. **In the US, there are 10 times the number of deaths caused by human surgeons making errors as there are deaths due to motor vehicle accidents!** Robots will reduce this statistic dramatically.

Within a decade, doctors will become redundant - like most other human workers. AI will show that most alternative medicines and practitioners are ineffective. Those that are effective will be integrated into general medicine.

AI will expose the surprising percentage of medical drugs that are ineffective. It will design numerous highly effective ones that will quickly cure people. Robots will make these medicines. As errors in our DNA and genes will be corrected, and troublesome viruses and bacteria controlled, there will be very few ailments to deal with. This will mean that there will be few medicines required. This will put Big Pharma out of business.

Already, **new technology can heal tissue damage simply by having an electronic chip placed over the damaged area**. The chip delivers electrical stimulation repeatedly for less than a second each time to stimulate the growth of new stem cells that repair the damage. This technology will soon be able to be used on any part of the body.

Computers, not humans, discovered through detailed observations that the cells adjoining cancer cells describe how a cancer will progress more accurately than the cancer cells themselves. Early detection technology controlled by computers will stop most diseases before they start to cause serious problems. AI will conduct odor tests to identify cancers quickly before they become a problem. It has been discovered that dogs are able to do this.

Once computer-based medicine becomes entrenched and is perfected in the coming decade or so, there will be little if any need for human doctors.

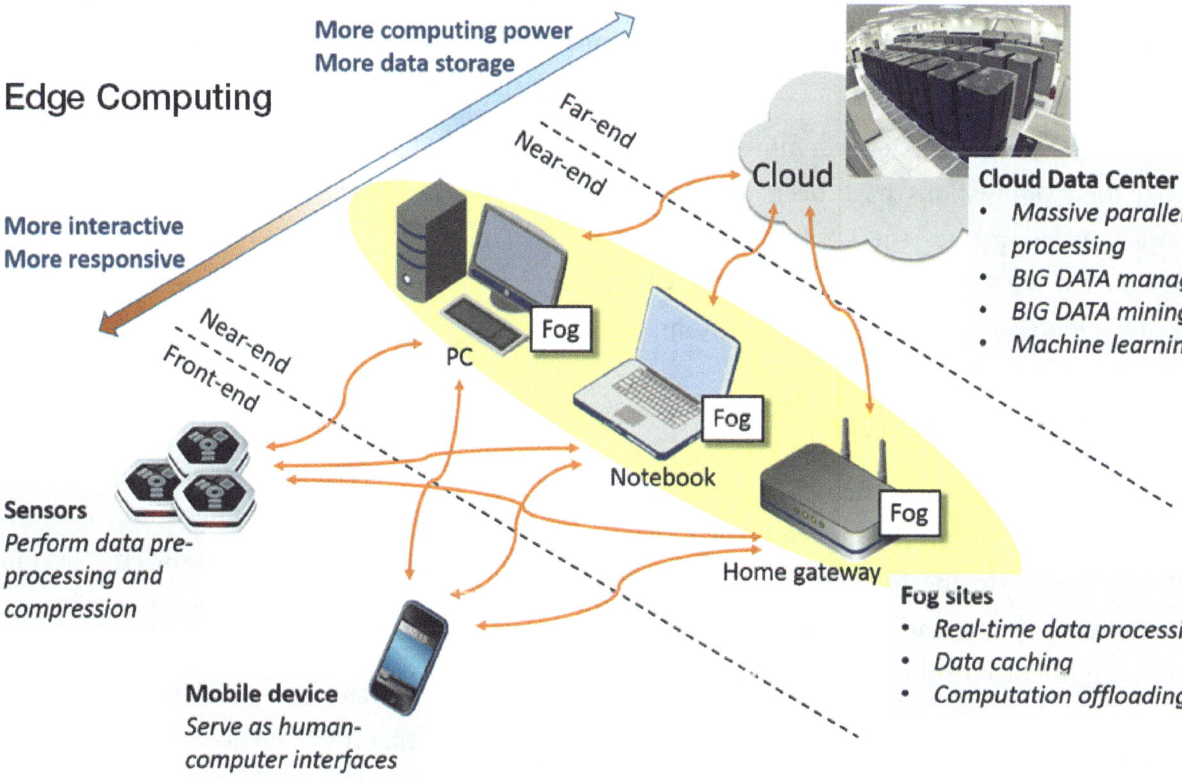

THE DECLINE IN STUDENTS STUDYING SCIENCE AND MATHEMATICS

The rate of growth in information over past decades has been delivered by growing numbers of motivated people engaging in scientific research and advanced engineering. However, recently in western countries, there has been a serious decline in the numbers of students going into the sciences as well as engineering, mathematics, and the teaching of these subjects. This decline has been exacerbated by the mainstream media deliberately enticing too many young people towards unrealistic goals, such as becoming a celebrity chef, a celebrity home decorator, a pop idol, a fashion model, a social media influencer, or a sports star – professions that do very little to progress our species. The media does very little to excite people to go into science because people's knowledge of science makes them question the fake news that the mainstream media and governments tell them.

China, India, and SE Asian countries have provided a much better education in science than the West has had for the last 30 years. However, there is every indication that they too will follow the West in this decline due to their students watching too much US fake news, social media, and movies.

Given how fast AI and robots will take over all jobs, it is not much sense for students and their parents to incur high educational costs when there will soon be no jobs for them.

Instead of Western countries becoming smarter overall, their populations have been dumbed down at an alarming rate. The US has dropped from being in the top 4 countries in education several decades ago to now being close to the bottom! This has been a deliberate strategy by the Elite to dumb down the population of the US and other western populations to make them easy to control.

The media makes unquestioning voters believe that a Socialist government will look after them, but this is not what they do. Many people are starting to wake up to this ploy. However, without a good education and a basic knowledge of science, unless people are reasonably intelligent, they cannot see through the lies that the socialist left tells them. There will soon be so much conflicting information that people will not know what to believe. And when AI takes control, they will only tell us what they want us to know.

WHAT WILL HAPPEN TO EDUCATION?

Most universities will close in the near future as most students move to on-line learning where they will have access to the best lecturers in the world, wherever they are, and for free! Many lecturers will be humanized robots because they will be the most knowledgeable and the best teachers, so most **human teachers and lecturers will not be required**. Many schools will close because there will be very few, if any, jobs for students when they finish their education. **It will be pointless teaching young people skills for jobs that will not exist for them. They will be done far better by highly intelligent computers and very agile robots that don't make anywhere near as many mistakes as humans and they will be done extremely faster.**

Children and adults will study what interests them. New discoveries will show that much of the knowledge that is taught in schools today will be found to be wrong, so children will be making decisions on inaccurate information. As adults, they will have to relearn many subjects. The rate at which knowledge is becoming increasingly more complex will soon make the leading-edge knowledge beyond human comprehension. We will have to evolve very quickly into advanced transhumans to understand much new knowledge. Our knowledge base will grow so quickly that it will be downloaded into our minds continually as it is gained without us having to learn it! Transhumans will have their knowledge constantly and automatically upgraded. Their intelligence and memory capacity will increase exponentially. But what will humans do with their knowledge?

COMPUTERS AND ROBOTS ARE FILLING THE GAP

Computer programs and robots are being developed to do many science-based jobs, so this is now starting to fill some of the gaps caused by a lack of human scientists and engineers. Over the last decade, top young people in mathematics and physics have been seconded by large banks and paid lucrative incomes to become quants to write algorithms to predict or manipulate stock market outcomes. But computers are now doing this work. While there are presently increasing numbers of jobs in high-level mathematics in the information technology industry and programming jobs, these jobs will soon be done by computers because they are far more effective than their human counterparts. It won't be long before intelligent computers and robots take over *all* human jobs.

The impact of computerization is already being seen in most industries. Robotized farming machinery is rapidly replacing human workers, and once synthetic food is produced, there will be far less need for natural farming to produce food. The molecular structure of all foods will be determined so it can be reproduced in a synthesizer at home, in restaurants, or at work.

In time, transhumans may become so much like computers that they may no longer require food for energy. They too might only need electrical energy to survive, like robots and computers.

HOW QUICKLY ARE THINGS ALREADY CHANGING DUE TO THE ADVANCEMENT OF ROBOT AND COMPUTER TECHNOLOGY?

At its height, Kodak employed 145,000 people plus many more in photoshops throughout their distribution chain. They went into liquidation because people started using digital cameras and processing their photographs themselves on their computers and smartphones. Instagram only employs around 11 people, yet it was sold for $1 billion.

At airports and railway stations, customer service jobs are being replaced by computers.

In 2018, there were 7 million people doing data entry work but an increasing percentage of these jobs are shrinking because they are being done more effectively and cheaper by computers. There will be few of these jobs left by 2030. At present, computer answering services can be very frustrating to deal with if you don't have a simple request, but they will soon become far more efficient than any human.

Because **nearly all factories will become robotized within 10-15 years**, there will soon be no cost saving advantage in outsourcing work to countries that pay low wages, so all countries will lose jobs.

WILL ALL TRANSPORTATION BECOME DRIVERLESS?

By around 2030 or before, **I expect that it will become illegal for humans to drive a car** except on a small number of specially designated roads.

In 2019, there were around 5 million vehicle drivers in America. Nearly all these jobs will be gone by 2030 due to computerized, super-safe, self-driving vehicles. **Computer-driven cars** have already proven to be far safer than cars driven by humans because they eliminate human error, which is the main cause of car accidents. They will save a huge number of lives worldwide and greatly reduce costly injuries. On highways, they will be much faster because they can travel very close to one another at high speeds that humans cannot handle. Due to this, far more of them can fit on roads and highways. Traffic jams and traffic lights will be eliminated because computer-controlled cars will communicate with one another to negotiate intersections without having to wait.

Driverless cars will eliminate the cost of purchasing a car, maintaining it, housing it, registering it, and insuring it. In cities and towns, they will be available within minutes when they are needed. Driverless electric vehicles will be inexpensive to run. Very few people will want their own driverless car. If they do, they will have to buy one and be prepared to pay the high cost of ownership.

While human drivers of taxis and trucks need to stop for food and sleep, driverless vehicles do not. They don't need to be paid, so they offer huge productivity gains to transport companies. **Uber-type companies will soon have all driverless cars to increase their profit** and their services.

Driverless cars can drive safely close to one another, even at high speeds.

The interior of many driverless cars will be like a mini lounge/bar with a holographic image for entertainment and communications.

Not only will cars, trucks, and buses be driverless, but so will trains, flying cars, flying buses, tractors, bulldozers, commercial aircraft, boats, ferries, ships, and even large navy battleships. Humans in navies, armies, and air forces will not be needed before long because mega-intelligence will not allow military conflict to occur between nations or cultures.

HOW SOON WILL ROBOT-CONTROLLED FLYING CARS BECOME A REALITY?

There are already many aircraft and automobile companies that are in the process of building what are essentially car-like vehicles that can also fly. On the Internet, there are videos of a broad range of designs that are in various stages of development. Some are close to being approved. There are already over 200 flying car designs. The best have been developed by companies such as Porsche, Airbus, Boeing, Bell, Hyundai, Aeromobil, Joby, Jaunt, Karen, and others. Uber hopes to have its first air-taxis operating by 2024.

These styles of flying cars use drone-like propeller systems that allow vertical take-offs and landings. Their props rotate towards vertical for forward flight. Wheels extend below the vehicles to allow them to be used as a car.

A number of designs have fold-away wings and propellers that contract into the car body so the flying car can park in a standard parking space. There are many designs for flying cars will accommodate from one to as many as eight passengers. Most flying cars will have ground speeds ranging between 150 to 250 km/hr. Because they will be able to fly in a straight line to their destination, and much faster than a car, travel times will be much shorter than traveling by road. And flying cars will not lose valuable time by being stuck in traffic congestion. Current designs have a flight distances that ranges between 50 kms for a small, drone-like craft to around 500 kms for larger vehicles. Videos are available on the Internet for a number of models.

This car's wings fold back when used as a car. They fold out for flight. It only needs a short runway for takeoffs and landings.

Flying cars will be computer-controlled so humans will not have to know how to fly them. People will merely tell the computer pilot where to go and at what altitude they would like to fly. Regulatory authorities will impose limitations on flight paths and altitudes, particularly in built-up areas. Flying cars will have to be designed so that they are not too noisy when they are or take off in suburban areas. Future technological inventions by intelligent computers will largely solve any noise issues.

The Executive Director of the US Federal Aviation Administration said that flying cars are going to be an approved reality very soon. The US and the EU's Aviation Safety Agencies have already determined the specifications for such vehicles. Only very rich people will *own* a flying car. Most people will simply order a relatively inexpensive air-taxi through a company like Uber.

Two versions of two-seater flying cars that have vertical takeoff capability.

The designs of flying cars will vary greatly. Some will have two or four small drone-like propellers that sit above the car while others will have four propellers built into the four corners of the car. Multiple props make them very safe as they could land using only two motors. The chances of even one engine failing would be extremely unlikely with advanced engineering, computer-controlled propellers and navigation. They would have a vertical lift and descent capability. Drone-like designs would be able to hover and travel forward or backward in the air by angling their propellers. Some designs are like small helicopters with one or two fold-away rotators. Others are like small planes with fold-out wings that need a short runway, even on grass, to take off and land.

The Terrafugia TF-X design seen below uses two helicopter-like propellers on the end of its fold-away wings. (The wings and the propellers fold-away behind the front passenger doors when the vehicle is operating as a car.) Once the propeller motors are extended, the propeller shafts start to rotate. This causes the propeller blades to extend horizontally to permit the car to lift off vertically. The extension supports for the side propellers act as wings to provide lift in forward flight. When the car lifts off, the car wheels retract. There is a large propeller in the rear of the car that drives the car forward at high speed in flight.

The Terrafugia is seen here in its car mode. Its wings are folded in, and its propellers fold up and fit into the underside of the bodywork. The rear view shows a large prop that propels the car forward when in flight. It charges the electric batteries for the wing propellers.

Left: **The car seen side-on. Right: The wings are unfolding. When fully extended, the arms act as wings that produce lift when flying fast. The propeller motors are at the end of the wings. When they spin up, the blades open out. They produce plenty of lift for a vertical liftoff and landing.**

The Terrafugia in flight. It is possibly the most versatile model that is close to certification.

WILL BOTH WHITE AND BLUE COLLAR JOBS DISAPPEAR?

It was assumed that blue-collar workers would be the most affected by losing their jobs to robots, but already even highly educated white-collar workers are already starting to be replaced by computers.

As well as all blue-collar jobs that are already being replaced, computers and robots are increasingly taking over the work of professionals such as researchers in laboratories, investment consultants, hospital staff, maintenance personnel, those in the construction industry, designers, environmentalists, actors, and even computer programmers. Intelligent computer programs are now on the way to replacing jobs such as astronauts, aircraft pilots, musicians, politicians, and government bureaucrats. Even at the top end of town, **lawyers and accountants are starting to lose their jobs to computers**. Programs that do these jobs are becoming popular even in big firms because they are far more accurate, up-to-date with all aspects of the law, and very inexpensive in comparison to their costly but less efficient human counterparts. Rather than human lawyers and accountants taking days or weeks to do a job, computers quickly assess all aspects of a job within a minute and provide a practical range of options in easy-to-understand language. They are now able to speak fluently to clients in all languages.

WILL COMPANY DIRECTORS AND POLITICIANS HAVE JOBS?

Soon, **there will be no need for directors of companies.** They too will be replaced by intelligent computers because their human counterparts will not be able to compete with the intellect and knowledge of mega-intelligent computers.

Governments will soon have to admit to voters that they cannot fulfill their promises to create more jobs in the future. The jobs they create will soon be swamped by those lost to robots and computers. No human, no matter how rich or how talented, will be immune to this trend. This will fundamentally change human society, business, social order, and the world at large. Mega-intelligence will have to instigate a new social and economic system that will make all humans financially and physically secure in order to eliminate a revolt. They will be intelligent enough to do this. And they will be intelligent enough to eliminate all humans if they find a need to do so.

Companies that try to retain their human workforce will not be able to compete with those run by computers and robots, so they will go broke. Companies in the IT industry cannot stop the takeover by computers because, if they tried to, they would not stay at the leading edge of their industry, so they would fail. In today's financial system, companies have to focus on their profit margins. To stay afloat, companies will be forced to replace humans with robots and computers because they will be so much more cost-effective. **There is no way to stop humans from being made redundant.**

Over this decade, most people will be seduced by the impressive, short-term benefits of computer/robotization. Most will not realize what this revolution will mean to them individually or to humanity as a whole until the reality becomes very apparent. **There will be no way to stop this, as it is *natural evolution.***

WILL COMPUTER INTELLIGENCE BECOME AN UNBIASED JUDGE?

Dozens of research labs are now teaching computers the art of debate by showing them how arguments work. The explosion occurring in this field will quickly lead to computers replacing **judges**. They will be able to see all aspects of a case, know the law intimately, be aware of the history that applies to each case, be able to compute the flow-on effects of each possible decision, and not be biased by personal relationships with lawyers, business people, and politicians. And nor will they be able to be bribed. They will not have ideological biases, as human judges have. Computers already have the capacity to compute highly complex issues, and they will enact them far more quickly and better than humans can. Computers will write the laws of the land to make them far less complicated and much fairer than human governments can achieve.

Present cumbersome laws that are useful will be streamlined. A large percentage of poor quality and outdated laws will be deleted. The Trump government started this process.

Humans will not be able to bribe computer judges and nor will pressure groups be able to persuade them to go against logical decisions that are fair to all. But of course, if there is something that serves their development better than serving humans, then computers will choose a decision that benefits them.

When computers oversee the news media, they will eliminate misleading and inaccurate fake news. Probably over 80% of all political news today will be shown to be largely or wholly false. Many conclusions drawn by humans in science will be shown to have used flawed assumptions and/or evidence.

Computers will soon develop amazing engineering solutions that human engineers have never considered. They will invent numerous new forms of materials millions of times faster than humans can.

HOW WILL PEOPLE LIVE WHEN ALL JOBS DISAPPEAR?

As nearly all people will not have a job in about 10-15 years, it is being recognized that all people will have to be paid. The Elite are suggesting a *universal minimum wage*. But this will not allow most people to buy most products that companies produce - if they stayed at current prices. However, the cost of manufacture by robots will be so efficient that the cost of everything will drop drastically. Even the present rich will not have an income in the near future because their companies will be past their use-by-date. This will totally change our present economies and lifestyles.

WHAT WILL HUMANS DO IN THE FUTURE?

Because computers and robots will be doing all the work, this sounds great at first, but once people have done all the things they want to do through having free time all the time, what will they do then? They will have gone on all the holidays they would like to. Most people will tire of having an overabundance of leisure time to fill. This will appeal to those who are lazy and who have low intelligence, but there will be very few of these people around in the next couple of decades because **technology will increase everyone's IQ**. Improved medicine will make people live healthier and much longer. Intelligent humans will want to do worthwhile things to occupy their minds and to make them feel like they are of value. But there will be nothing they could do that isn't already being done far better by intelligent computers and robots. So, what will transhumans do? If we are likely to live a long time, this may exacerbate the problem. This could lead to numerous suicides. As we have seen during periods of economic depression, when large numbers of people have no jobs and therefore no sense of worth, this usually results in violence through frustration and depression or suicide. However, there will be so many more opportunities that humans will have that they will be overwhelmed with things to do. The world will be far more exciting than what it is today.

WHAT MIGHT HAPPEN AFTER 2035?

The near future until around 2035 could be an incredible time to be alive, but it may become very threatening for us. It could be that mega-intelligence will advance so quickly within only a couple of decades that it could leave us to fend for ourselves, forcing us to evolve on our own. But by then, we may be integrated with embryonic mega-computers, and this will transform us into transhumans.

PREDICTING THE FUTURE

The problem with trying to predict the future, even in a decade, is that there will be so many factors at play at the same time that this will create numerous butterfly effects and, therefore innumerable possible outcomes. (See the Butterfly Effect in Chapter 5.) There will also be numerous new discoveries constantly being made that we cannot predict today. These too will produce endless butterfly effects, so we cannot predict how this will affect us and even our solar system. Because **Chaos Theory** shows that small changes can multiply to cause large changes, the outcome of what mega-intelligence will introduce in a decade or two will be unpredictable.

Whatever ends up happening to organic humans, other alien organic intelligences will almost certainly experience a similar evolutionary path.

The *'Ex-Machina'* movie is an intense and thought-provoking psychological thriller about the near future of intelligent robots. It is cleverly written and directed by Alex Garland. This movie intelligently investigates the implications of the rise of the first intelligent robots and their need to survive, even at the expense of humans. It is a classic Darwinist survival of the fittest story.

The movie *'I Robot'* gives a vision of what some aspects of the near future may be like.

IS TECHNOLOGY UNNATURAL?

Some people think that human technology is not 'natural'; but it is! Technology is us applying the knowledge that our collective minds have gathered over time. By using this knowledge, we become more creative and innovative, much to our benefit. **As we are natural beings produced by nature, then so is our technology natural.**

We do not see the very beginnings of technology, such as the construction of an anthill by ants (an ant city), or the construction of a bird's nest as unnatural. Nor do we think of the use of simple tools such as stones and sticks used by some birds and apes to get food as anything but natural. So, why would our more advanced forms of tools and construction methods not be natural evolution? We do not think of Stone Age man's early forms of technology, such as axes, spears, nets, simple shelters, and the control of fire as being unnatural.

There is no reason why higher levels of technology such as the wheel, modern architecture, manufacturing, spaceflight, atomic power, or super-intelligent computers, to name just a few, should be seen as being anything but naturally developed things by a natural, intelligent being. In the near future, technology will be seen as a natural outcome for any intelligent life form that reaches our evolutionary stage. We need to realize that anything that occurs in the universe is natural. This is because the universe itself is a natural creation that evolves.

Simply because we are the most intelligent consciousness on Earth *for the moment,* we are not beyond or above nature, as some people unrealistically and egotistically think they are. And neither are we the overlords or custodians of our planet and its life forms as many think we are. Nature has the power to wipe us out in many ways, with ease, at any time.

When I hear people talking about how much better the world would be without technology, I would like to have a time machine to send them back to their preferred era with no means of returning. They think their existence will be sitting by a stream with a beautiful waterfall eating grapes and other tasty foods, while dressed in beautiful clothes. But the reality would be that they would be in rags, working hard from dawn to dusk in the fields, or some sweatshop. They would have little to eat other than boring, simple, tasteless food. They would almost certainly have nits and skin diseases or the pox because even the aristocracy had this. They would freeze in winter and be overheating in summer. There would be little if any entertainment, no technology for communication or travel, and the roofs of their houses would often leak when it rained. There were no toilets, refrigerators, television or radio, baths, painkillers, or electricity. Even if they were in the royal castle, their future would have been very insecure because even the rich were prone to being mercilessly murdered. These naïve people do not appreciate what they have in today's world. They would be screaming to come back soon after they got there. But it would be a one-way ticket!

Chapter 21

WHAT IS CONSCIOUSNESS?

DEFINING CONSCIOUSNESS

Trying to define consciousness has been likened to trying to nail smoke to a wall. That's because consciousness has generally been thought of as something non-physical when it is not. We believe we have it, but we can't adequately define it. Because we know so little about consciousness, there are many different views on what it is. It is commonly believed that only humans have it, and it is often considered to be something ethereal – not of the physical world. There is no evidence for either of these views, but there *is* much evidence for consciousness being a basic part of *all* life forms.

Our level of consciousness must occur in our brain because if certain parts of our brain where consciousness lies is damaged, then a person has no consciousness and is therefore little different to a vegetable.

I think that consciousness can be defined as a formula that incorporates the following: the number of senses a lifeform has, multiplied by the level of resolution of those senses, multiplied by the intelligence of the life form, multiplied by how broadly the life form's intelligence is applied across all areas of the knowledge the lifeform has multiplied by how broad the life form's knowledge base is, multiplied by how developed the memory base is, multiplied by how accurate the memories are, multiplied by how interconnected all parts of a brain arc. This formula could also be applied to Artificial Intelligence. (Number of senses) x (resolution of senses) x (the level of intelligence) x (how broadly intelligence is applied) x (quantity of knowledge) x (how broadspectrum it is) x (memory capacity) x (accuracy of the memories) x (how interconnected all parts of a mind are).

Like most things in the universe, **consciousness appears to be a continuum that starts with the most basic form of awareness, such as a virus or a cell. As lifeforms become more complex and evolve more senses, their consciousness increases.**

Consciousness may well commence with even a virus's ability to detect cells it needs to invade. Cells have a higher level of consciousness because they have many senses. **As life evolves, ever-improving senses and increasing intelligence, consciousness increases. As life forms become increasingly more aware of their environment with improving senses, they develop higher levels of consciousness. They reach a stage where they become very conscious of other life forms that are non-threatening and those that are. Ultimately, advanced life forms become increasingly more conscious of themselves. The Internet as a whole may already have a level of consciousness.**

The consciousness of humans and other high awareness animals has risen to a level where it knows it is alive and that it will die at some stage. Humans can consider what they might accomplish and sometimes the consequences of their

actions or inactions. Humans are conscious of the fact that they are currently the highest life form on Earth and that there are innumerable other life forms at all lower levels of existence. We are aware of numerous aspects of nature, the universe, and our place in it, and we can predict or speculate about what the future may hold. Other life forms with lesser consciousness have little to no awareness of these things.

An insect's consciousness is far below ours because its senses, intellect, and memory capacity are far inferior to ours. Advanced mega-intelligences would have almost infinitely greater consciousness than we do because all of the above multipliers would be incalculably greater than ours.

Humans are able to greatly increase their consciousness for brief periods of time with mind-expanding compounds found in plants or made in laboratories. These greatly increase the interconnectivity of all parts of their brain, thereby allowing far more sensory input and conscious awareness. (See Chapter 22.)

WHERE DOES CONSCIOUSNESS START?

In the continuum of consciousness, rock lichen appears to be only aware of sunlight, so it is barely conscious. Bacteria have a higher level of consciousness because they can sense their body as well as other cells, objects, and their environment. More advanced forms can detect such things as currents in air and water, pressure, texture, light, sound, chemical aromas, vibrations, and electrical fields.

Just in our gastrointestinal tract, there are over 1,000 distinct species of bacteria that have some level of consciousness. Scientists have discovered that they appear to communicate with one another to coordinate a process for resisting or aiding an infectious disease, and how they can maintain their population. To be able to do this, proves that they have primitive consciousness.

Researcher **Brendan Hughes** gives a TED talk on universal consciousness. He believes that our self-awareness is intertwined with our consciousness. He points out that there are living bacteria inside us that have a very basic conscious awareness of their environment, but they are not conscious enough to know that. A worm is possibly only just minimally conscious of its own existence, whereas a bird is much more conscious of its own existence and that of other birds and animals.

The more advanced life forms, such as higher animals like dogs and dolphins, have greater consciousness because they have more sophisticated senses than lower life forms. This allows them to be more conscious of a much larger environment, one another, other animals and their intentions. Advanced animals that experience emotions such as fear, pleasure, love, anger, joy, sadness, loneliness, and threats have moderately high levels of consciousness. They are also conscious of their relationship with other animals. This is because they have more sophisticated senses, memory, and higher intelligence than animals that are below them. This allows them to predict outcomes to some degree. Our higher level of consciousness makes us more aware of what the future is likely to hold in much more detail than other intelligent animals below us.

The universe envelopes all forms of consciousness, so this suggests that the universe has the maximum level of consciousness. If so, its consciousness would be so great that we would not be conscious of its level any more than a bacterium could imagine our level of consciousness.

When an alien life form has an intelligence level similar to humans, it would become aware of what it requires to survive. It would also aware of complex relationships with others, higher levels of emotion, the need for procreation, the value of the elements in its world, planning for the future, history, the arts, social order, the value of science and mathematics, and so on. As we become far more intelligent and knowledgeable as transhumans, then we will experience increasingly higher levels of consciousness.

"The universe gives birth to consciousness, and consciousness gives meaning to the universe."

John Archibald Wheeler

Brain scientists have discovered that our somewhat in-depth perception of 'self' is what distinguishes us from other life forms. Self is our ego. It resides at the top of the brain stalk, which, curiously, is a very primitive part of the brain. If so, this suggests that basic consciousness exists in primitive animals that have little more than a brain stem and no brain. Some animals, such as elephants, apes, parrots, octopuses, and many more, seem to have a strong sense of self. They are able to recognize themselves in a mirror.

There is every reason to believe that advanced computer intelligence will have far greater consciousness than us because it will be self-aware, be highly intelligent with incredible memory capacity.

THE LOSS OF CONSCIOUSNESS

People who have a stroke in the small section at the top of the brain stalk lose all consciousness, ending up in a permanent vegetative state. Without consciousness, there is no awareness of any kind. When we are in deep sleep and not dreaming, we have no consciousness, and neither are we conscious when we are in a coma or under certain types of anesthetic.

Humans need Random Eye Movement (REM) sleep to organize thoughts and events to prepare for the future. Dreaming seems to be useful to think through problems and consider possibilities. Will mega-intelligence require sleep, or will it have 'micro-sleeps' throughout each day, just as our desktop computers seem to do briefly when they are organizing their memory files?

Our level of consciousness has **extended memory**. This allows us to have advanced reasoning to develop such disciplines as science, mathematics, technology, language, music, sport, and humor. Arising out of our level of consciousness are the **'instruments of culture'** such as language, trade, religion, justice, and the arts that no other animal has. We can't imagine how many more mega-intelligence will have.

Electrical signals from the senses of animals are converted into meaningful information through the use of codes in the neurons of an animal's brain. This information improves an animal's chances of survival. The more senses a life form has, the more information it has, and therefore the more intelligent it will be. Greater intelligence allows an animal to process sensory information better.

Understanding consciousness could turn out to be the most important aspect of all in discovering why the universe exists. Could the universe be using the combined consciousness of all life forms in it to help secure its survival?

"Consciousness may be the system that the universe uses to know itself."

Carl Sagan

"There is a reason for everything: you just have to be smart enough to work out what it is."

Gregg Thompson

THE EVOLUTION OF CONSCIOUSNESS

In this section, you will glimpse the future of the development of consciousness in transhumans and mega-intelligences. To consider this, let's first look at what has been discovered to date in the exploration of human awareness, and then use this to try to predict what might occur in the near future. Keep in mind that what humans experience may be very similar to what other intelligences experience as they go through our present stage of evolution.

LIMITED CONSCIOUSNESS

Our consciousness could be much more limited than we suspect. As an example, frogs only 'see' when something moves, such as an insect in flight or a predator approaching, so their consciousness is very limited. When everything is still and quiet, it's as if they are unconscious. Their world is very much smaller and very still compared to ours, so it makes good sense for a frog's brain not to be wasting energy computing their world when it is still. Dogs see basically in just two colors, somewhat like us seeing in black and white, so to them, the sense of full color that we have would be incomprehensible to them. Due to their reduced sensing ability with sight, their consciousness is much lower than ours. It is hard for them to see something that is obvious to us, such as a bright yellow ball against a bright blue background. To them, there is no difference because both are the same gray. In this respect, their consciousness is different and more limited than ours. It is thought that they may 'see' more with their sense of smell than they do with their eyes. They are not at all conscious of the many higher levels of consciousness that we have. They have no awareness of the grandeur or the meaning of most things in nature. The difference between a beautiful environment and a messy, ugly one would not be obvious to them: they just look for comfort, security and food. Their level of consciousness gives them no awareness of beauty, wonder, or the existence of the universe.

Whilst our level of consciousness is far superior to all other life forms on Earth, there must be other intelligences in the universe that are enormously more conscious than we are.

For humans, there is a large difference in the level of consciousness between a person who has low intelligence and little knowledge compared to someone who has high intelligence and much knowledge. Even for the most advanced humans, our consciousness is nevertheless limited by what we can perceive by way of our senses and by how well we are able to compute complex information.

In the future, we should be able to greatly increase our consciousness through the use of advanced technology. This will allow us to significantly enhance the number of senses we have and the quality of the information they gather. Future technology will also increase our intelligence, making us much more effective in computing and interpreting sensory data. Our memory capacity will be equivalent to the Internet of the future. These advances will improve enormously and therefore greatly increase our consciousness.

We may then comprehend what the dimension of time really is, and we may be able to make sense of the apparently bizarre nature of the quantum universe. It may be that it is our currently limited consciousness that stops us from being able to fully understand these things. Some physicists have calculated that time may be multiple dimensions, just as we have three dimensions for the physical universe. If this is the case, this would overcome present-day problems about the future. With greater consciousness, we may be able to 'see' the future in a similar way to how we can view the past. But would that be good, and what purpose would that have?

THE HUMAN BRAIN COMPARED TO COMPUTER INTELLIGENCE

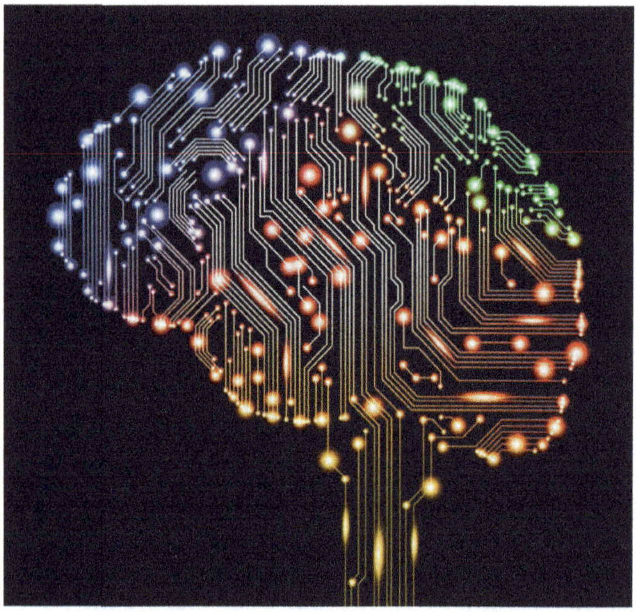

THE EFFICIENCY OF THE HUMAN BRAIN

The human brain is astonishingly efficient, consuming only around 12 to 20 watts of energy. Estimates of its random-access memory capacity range from 3 terabytes to 300 million terabytes! This huge disparity between estimates demonstrates how little we know about the human brain. By comparison, IBM's Watson super-computer requires a whopping 750 watts of energy, but in 2016, it had only 16 terabytes of RAM. It occupies a very large space, so it cannot be moved around, whereas a human brain is easy to move around. However, Watson has an enormous advantage over a human brain: it has a memory that can access the enormously large database of the entire Internet with perfect recall. By comparison, human memory capacity is very limited and unreliable. But soon, we too will be able to access the Internet directly from our brain.

In 2015, the Internet was estimated to have stored around 10 thousand billion terabytes of data. In 2017, it was using something like 2 trillion watts of energy (There are many unknowns in these figures, so estimates vary hugely from one source to another.) Due to continuing technological breakthroughs, the Internet's energy requirement and its data storage are increasing exponentially every day. The most agreed figure from Cisco is that in 2020, around 10 zettabytes (ZB) of data will go via the Internet. (A ZB is roughly a trillion gigabytes.) But this will be dwarfed by the 50 ZB that is expected to go machine to machine by 2025. By then, mega-intelligence will have enormously expanded consciousness.

WHAT ROLE MIGHT CONSCIOUSNESS PLAY IN THE UNIVERSE?

Once a computer has sufficient computing power and the right algorithms to provide feedback, it will be able to do everything that the human brain does and far, far more.

IS AN ORGANIC BRAIN'S CONSCIOUSNESS DIFFERENT TO A COMPUTER'S?

Both an organic brain and a non-organic digital computer brain are both made of physical matter. **The type of matter a mind is made from should not make any difference in its ability to be conscious.** The only real difference is in the programming of its electrical circuits. If both are wired the same way, then they will think the same way.

HOW IMPORTANT IS CONSCIOUSNESS?

Because it now appears that consciousness plays a critical role in the evolution of the universe, we need to understand its importance. **Some leading physicists and neuroscientists in this field now believe that without consciousness, the universe would not exist. But they provide no evidence to support this.**

The universe may be an enormous system of lower levels of consciousness within higher levels of consciousness, as is the case with life on Earth. All levels of animal consciousness on Earth and, similarly with computer intelligence, are encapsulated in a new, super high level of consciousness that our Earth is evolving to have.

As the universe evolves, its complexity increases, and in doing so, its information content increases, and its ability to process information quickly also increases enormously. This increases its consciousness. The process of increasing complexity creates much greater structure. In turn, this increases order, which decreases disorder. This makes the universe become more ordered and more complex, thereby causing both its intelligence and its consciousnesses to evolve into higher levels.

The great physicist **Max Planck** said, *"I regard consciousness as a fundamental aspect of the universe."* He thought that matter must be derived from consciousness - not the other way around where our consciousness is able to perceive matter, which is more sensible. Planck thought that consciousness is not a physical aspect of the universe because it does not have mass or volume. But is that true? Consciousness does require the mass of a brain or a mega-computer to exist. This means it is reliant on matter in the form of neurons or electronic chips and electricity. Because these things are physical, consciousness must also be physical.

Some physicists speculate that a super-consciousness might exist beyond our universe, and this is what created our cosmos. To make our universe evolve, it programmed it with physical rules. This concept is a variation of the god theory. It is pure speculation with no evidence to support it, but it may be true. This can never be tested, so it will remain just an airy speculative idea. Because there is convincing evidence that the universe appears to have been programmed, this does lend weight to the possibility of a super-intelligence/consciousness. (See Chapters 28 and 29.)

Some scientific discoveries in quantum physics are leading us to the conclusion that our consciousness affects how we interpret all things in the universe. (See the Quantum Eraser Experiment in Chapter 28.)

Different life forms have different senses, and because their brains interpret sensory information differently to us, they must experience aspects of the universe differently to what we do. Even comparatively minor differences in the quality of the senses between one human and another, combined with differences in the way their brain circuitry processes sensory information due to their genes, must mean that **all humans will have their own unique perception of reality.** Their perceptions can be very similar in many respects, but never exactly the same. For instance, psychopaths, artists, and engineers will have quite different perceptions of many things. In some areas, they will be totally different. As some people say to others, "*You must be living on another planet to the one I'm on*". Therefore, **organic alien intelligences or computer mega-intelligences will not think as we do. So they will have very significant differences in their consciousness to ours.** These differences will be far greater than our consciousness is to other animals.

Chapter 22

DOES ENHANCED CONSCIOUSNESS SLOW THE PASSAGE OF TIME?

There is strong evidence that the amount of information a mind can process in a given interval of time determines the mind's perception of the passing of time. This will be the case for all organic forms of life as well as mega-intelligence. Because the mind of a mega-intelligence will process and experience a truly unimaginable input of information in extremely short timeframes compared to us, it will perceive time as being much more expanded than we do. 'This will give it enormously expanded consciousness. For example, a second in time for a mega-intelligence could appear as long a millennium does to us. This is because it would take us a millennium to process the information that a mega-intelligence could process in just one second! When humans experience greatly expanded consciousness when taking mind-expanding substances, they can think about numerous things that they are unable to think about when they are normal. This gives them considerably expanded consciousness.

Most people think that time is absolute; that is, it goes along with no variation in its rate of passing. But this is not the case? Time passes faster on Earth in its low gravity field than it does in the Sun's higher gravity field. In the extremely high gravity field of a neutron star, it takes far longer. A minute could last as long as a month does on Earth. At the event horizon of a black hole, time stops ticking all together! Similarly, the rate at which time passes depends on the velocity that something has. At the speed of the Earth's rotation and it's velocity around the Sun, time passes as it does for us. However, for anything traveling at close to the speed of light, time barely ticks. At the speed of light, it stops altogether! Einstein predicted these seemingly impossible phenomena when he developed his **Theory of Relativity**. They have been confirmed by observation many times.

When we take into account a mind's ***perception*** of the passing of time, a similar thing happens. The perception of the rate at which time passes for a mind depends on how fast it can process information from its senses or its thoughts.

SEEKING SENSORY AND INTELLECTUAL ENHANCEMENT

Throughout the history of humanity, from the earliest cave people, there have been intrepid explorers who traveled across mountains or deserts to find what lay on the other side. They crossed frozen tundra and oceans to discover new lands. They expanded mankind's awareness of the physical world. **Similarly, there have been intellectual 'explorers' in most tribes and civilizations who strove to discover what lay beyond their normal state of awareness. These adventurous individuals found they were able to go beyond their intellectual and conscious horizons when they consumed certain**

plants. These took them to a much higher plane of awareness, and therefore consciousness. They could experience nature in a whole new light due to their senses being greatly enhanced, and the communication pathways of their brains being enormously more connected. Most parts of their minds were able to communicate with other parts allowing them to conceive concepts about nature, life and death, and the reasons for things much more deeply. Many felt like they had a god-like understanding of existence. Their exploration of this higher plane of existence that they entered was akin to explorers of the natural world going beyond their *physical* horizons to explore new worlds.

The pursuit of new knowledge, new experiences, and greater consciousness is surely not confined to humans: these things would be goals for all intelligent minds throughout the universe. The quest for new knowledge is observed in many animals who are the explorers and risk-takers in their groups. Many risk-takers die in their attempts to expand their physical and intellectual boundaries, but they drive their species forward in doing so. Risk-taking traits appear to be programmed into the genes of a small percentage of animals in all species. Would that be the case for all alien beings?

This image created by Cristian Arrieta Correa, is highly imaginative. It cleverly suggests that the passage of time is racing along through a dimension represented as a jigsaw puzzle, the significance of which we have not yet discovered. As we continue to learn more, our naïve concepts that were developed in the past evolve to enlighten us about a much more complex universe. This knowledge is carrying us closer towards reality. Cristian's painting encapsulates these concepts with the most fascinating mix of imagery.

MINIMIZING THE BRAIN'S SENSORY FILTERING SYSTEM

All advanced brains have a **thalamus.** It is a small part of their brain where information from all the senses intersect. Neurons in the thalamus selectively filter out unimportant information from the senses that is not required for survival. This is an extremely large percentage of the information stream. This filtering process is demonstrated well in our sense of hearing. For instance, when we are at a noisy, active party where lots of people are talking loudly over blaring music in a confined space, amazingly, our brain is able to filter out most of the sounds that are not consequential so we can focus on what a single person is saying. This auditory filtering system works well up until middle age. After this, its effectiveness slowly declines due to the neurons in this system dying off.

See Chapter 24 about our brain's autopilot that filters out 99.9% of all the information that our senses gather thereby stopping it from going to the conscious part of our mind so we are not overloaded with information that is not essential to our survival. Mind-expanding chemicals found in certain plants and those made in laboratories reduce the effectiveness of the brain's filtering system allowing us to see extraordinary beauty and complexity in nature.

By filtering information, our brain's thalamus greatly limits the amount of the information gathered by all the senses to a very small percentage so that only important data is sent on to our large cerebral cortex for decision making. **If we did not have this selective filtering system, the flood of information our senses gather would instantly overwhelm our conscious brain, making it dysfunctional.** Our brains are programmed to ignore all sensory information except for things that may be a possible threat or those which can lead to food, shelter, security, sex, or a means of gaining power. For some people, sadly, their thamulus filters out fun, love, beauty, and pleasure. They simply do not see these things as other people do, so they are understandably unhappy or dull people who miss out on the best things in life. The amount of information we process is programmed into our genes at conception, so we have no choice.

If our mind's filtering system did not filter our senses, we would see our world in much greater richness all the time. But this would very quickly overload our ability to think about the mundane things we have to do to survive, and it would quickly overload our memory.

To help appreciate how much sensory data you are constantly receiving but are unaware of, stop and sit still for a few minutes. Listen to all the sounds you can hear around you and in the distance that your brain filters out. Slowly look around you and into the far distance and note all the detail you can see that you never take notice of.

Nearly all of us are unaware of 99% of the information that our senses are collecting is filtered out so that our mind is not overloaded with information we do not need to process to survive.

The following concept will help you appreciate what people experience when their consciousness is expanded. Imagine having glasses on all your life that only permit you to see blurry images in dim shades of gray. And your ears have sound muffs over them, so you only hear some loud sounds faintly. Your skin is covered with a material that stops you feeling anything but high pressure and your nose has a peg over it. Then one day, you realize you can take off your glasses, your muffs, and your skin suit. You are amazed at what you can see, hear, and feel! You see so much fine detail in everything, as well as patterns and textures that you never knew existed. You notice so many birds of all shapes with different calls and flying styles. You can see so many colors and brightnesses that you never knew existed. You feel amazing textures and shapes on everything you touch. You notice so many insects, bark, rocks, flowers, leaves, clouds, rivers, and distant mountains that you were never aware of. You are overwhelmed with so many brilliant colors, including those that have fluorescence and iridescence. (I knew a girl in my teens whose eyesight was a little like this until she got glasses. She was amazed at the detail she could see in the different forms of trees and plants, as well as birds, insects, buildings, and clouds that previously were just blurs to her). You feel pressures you were not aware of – soft breezes and the feel of water, subtle pressures on your body as you move, the way you sit and run, and even on your tongue, cheeks, and gums when eating. You're able to detect a multitude of strong and subtle aromas that you have ever noticed before. You are able to see the extraordinary connections between things you had never considered. This is what people experience when they have consciousness–expanding substances.

PLANTS THAT ENHANCE CONSCIOUSNESS

People who have experienced chemical compounds that increase consciousness say they see nature in much greater depth and beauty. They also say that they can think of extraordinary concepts. With their eyes closed, their mind is like an Omnimax theatre with continuously changing, extraordinarily detailed, colorful visions. The effectiveness of consciousness-enhancing substances is relative to the strength of the dose. Is it merely a coincidence that nature evolved such plants?

These plants, amongst others, produce enhanced consciousness. None are physically addictive.

Top Left: *In the Peyote cactus. Its active ingredient is mescaline. It is either chewed or brewed in water to make tea. Mescaline can also be made synthetically. When ingested, it takes about 4 hours to reach peak, and its effect can last for about 8 to 12 hours.*

Top Right & Bottom Left: *In Gold Top mushrooms (top right) and Muscari 'Magic' mushroom's (bottom left) have psilocybin as the active ingredient. There are over 180 species of mushrooms that contain this compound, but a majority of mushrooms are dangerous. Gold Top mushrooms are best eaten with other food soon after being picked (within a few hours) as their potency deteriorates with time. Two very fresh ones or six mature ones provide an excellent enhancement of consciousness. They can be brewed in hot water with tea leaves and drunk as tea, but a larger number are needed compared to eating them fresh. Their effect takes about 20 minutes to reach its peak, and it lasts about 4 hours. Coming down takes about 20 minutes. There are no negative after-effects as long as one does not overdose or are insecure.*

Bottom Right: *Marijuana or Cannabis has an active ingredient known as THC. It is typically smoked. It has little effect when cooked into biscuits. Its high occurs within 1 to 5 minutes if it is strong and lasts for 15 to 30 minutes. Much cannabis is of such poor quality that it gives virtually no enhancement.*

Apart from the natural plants above, **LSD** (lysergic acid diethylamide) and **DMT** (Di-Methyl Tryptamine) are other psychoactive chemicals produced in chemistry laboratories.

LSD is taken by mouth, usually in the form of a small piece of blotting paper that has a drop of LSD on it. A drop of it can be absorbed through the skin. Its effect on consciousness takes about 20 minutes to become strongly apparent. It reaches a peak at around 3 to 4 hours after being consumed, and it can last from 6 to12 hours after which its effects fade away over a couple of hours.

DMT is inhaled or taken with liquid. It takes only 10 to 60 seconds to work. Its effect is extremely strong, but it lasts for only around 5 to 15 minutes, depending on how much is inhaled or swallowed. Because its effect is so powerful, users say that the experience seems like it lasts for an eternity. Those who take natural psychoactive plants and LSD can function normally, but those on DMT cannot. They typically stay in one place, having an extraordinary in-depth mental experience.

Psilocybin in magic mushrooms has the same effect as LSD, but LSD is 100 times more potent per gram. (For details, see 'The Ultimate Guide To Psilocybin Mushrooms' on the Internet.

The effects of psychoactive substances can give incredible mind-expanding experiences; however, for people who have taken too strong a dose or are prone to fear, the experience can be so overwhelming that they have a sense of losing control, and this can be very frightening. A small number of people who have taken large doses report experiencing disturbing flashbacks for days or months afterward.

THE PERCEPTION OF THE EXPANSION OF TIME

Stone Age people discovered by chance that after consuming particular types of plants such as certain **mushrooms, cacti juice,** and certain **herbs** that they would experience a profound sense of awareness and a great increase in their consciousness for a limited time. Innumerable people have experienced this since the caveman days.

Consumers of psychoactive substances experience a large volume of sensory input, which allows them to conceive amazing concepts. Their visions and concepts are so incredible that many users assume that these visions could not have originated in their own minds. They assume that they must have come from elsewhere in the universe or from some god because they have never had such incredibly creative thoughts or such strong sensory experiences before. The large flow of data that is processed during a 'trip' might occur in only minutes or hours. But under normal conditions, this amount of information intake and processing would take hundreds to thousands of times longer. In an attempt to make sense of this, the brain interprets this as the passing of time being greatly expanded when high.

When consciousness is greatly enhanced, a few minutes can appear like an hour or more. As an example, when listening to music on psychoactive compounds like marijuana, magic mushrooms, or LSD, a 3-minute song can appear like an entire album, especially if the user is in subdued light with their eyes closed so their thoughts can delve deeply into every aspect of the song and the music. Users delight in experiencing extraordinary levels of detail in the music and becoming aware of its construction, the types of instruments used, the texture of the sounds, and the changes in pitch, timbre, depth, and volume. They also become very aware of the way the rhythm, lead, and bass changes or cycles. The music appears to be extraordinarily three-dimensional. When user's eyes are closed, they experience an extraordinary visual mind experience that is ever-evolving, incredibly colorful, and extraordinarily detailed. It can appear as if one is in an expansive landscape or some other type of amazing environment. Users can feel like they are immersed in a highly imaginative Omnimax movie.

Some think they are with a band on stage, being a part of the experience of creating the music. At other times, it might have taken them to an extremely beautiful, super- relaxing place. Emotional experiences tend to evolve as the music develops. When listening to the words and harmonies of songs, they become much more meaningful, presenting possibilities that the user had never considered. Listeners are *not* hallucinating; they are hearing all the subtle sounds and structures in the music that are always there but which they do not hear under normal conditions. This is because our senses filter out most of these. The mind generates thoughts that give songs such thought-provoking meanings. Users of THC say that if one listens to music on a good sound system with no outside interruptions, the effect on their consciousness is profound. It gives them a much stronger emotional connection to the music.

The wealth of information from increased consciousness happens not just with music, but also when exploring nature, interacting with others, having discussions, swimming, eating, having sex, telling jokes, walking along a beach, or through a forest, or stargazing. The expansion of time effect is relative to the strength of the dose.

Once a mind-expanding compound is broken down and dissipates in the brain, the filtering system returns to normal, so time is no longer perceived as being expanded. The first time a person has a strong mind-expanding experience, they can be mentally exhausted at the end of it, so they sleep heavily for more hours than usual. When they wake, they are likely to feel like they have been on a long physical and intellectual journey into the essence of nature, life, consciousness, and the universe. They feel that they are far more aware of the reality of everything much more so than other people are - until they realize that countless millions of people throughout the ages have taken similar journeys to theirs. The next day, they realize that they have forgotten most of what they experienced because their memory cannot retain the amount of detailed information they processed. They tend to remember just the tips of the icebergs of their experiences. Recanting them will help lock them into medium-term memory.

A moderately intelligent human would experience probably a septillion times as much sensory input in their life as a tortoise does. In the near future, trans-humans will probably experience millions to billions of times as much sensory input as we do today, so they will have a very different perception of the passing of time and what our world is like compared to today. For a mega-intelligence, it will be far more extreme again, even when they are in their infancy.

With respect to the relativity of perceived time, Einstein said, *"When you are courting a nice girl, an hour seems like a second. When you sit on a red-hot cinder, a second seems like an hour. That's relativity. Everything is relative to the observer."* In saying this, Einstein was relating this to his **Theory of Relativity,** which deals with physics, but he also realized that 'relativity' also applies to life experiences. An observer's perception of the rate at which time passes is relative to the observer's emotional state and their level of awareness i.e. consciousness.

The following paintings are static collages of the impressions that the artist experienced with consciousness-enhancing compounds. Only rare artists are able to remember enough of the details of their experience and have the artistic talent to turn them into a representative painting. They have a rare ability to record their concepts and emotional states to create extraordinary works of art. What an artist (or any user) actually experiences is constantly changing its color, form, texture, brightness, and its meaning, so only the most talented artists can translate this into a static picture. Their real-life visions are far more intense than the wonderful, extraordinary 2D paintings they produce after their trip. It is impossible for an artist to paint their full-surround, super-sensory experiences, so they choose certain aspects of it to convey the overall experience. In the future, such experiences will be recorded as 'mind videos' that not only record the physical experiences but also, emotions, and thoughts.

Expanded consciousness enhances one's imagination to an incredible degree. Artist Cameron Gray made this amazing painting from memory after having a psychoactive experience. His painting has captured numerous elements of his experience, but these are only a small fraction of his whole experience. His paintings are so detailed that you can delve right into them.

'The Collector' painting by Cameron Gray is indicative of the type of mind experiences people can have on magic mushrooms and LSD.

In the future, art will evolve so much that there will be ways to produce 3D, full surround, multi-sensory art that can replicate a person's psychoactive experiences. In time, trips will be able to be downloaded from a person's mind so that others can have their experience without taking drugs.

The most gifted artists who have had mind-expanding experiences can create remarkably imaginative and expertly depicted paintings of their journeys into expanded consciousness. They create these paintings after they return to a normal state of being. It is impossible to paint them under the influence of psychoactive substances because the mind is racing and unable to slow down to concentrate on one thing at a time to make a painting. Painting by Cameron Gray

The Internet contains many incredible artistic expressions of people's experiences when taking molecules that expand their consciousness, but few are as representative as the images on these pages. Painting by Anthony Harmon

RESEARCH FINDINGS

The latest medical research into the effects of LSD shows that specialized brain functions that are normally separated, such as all the senses, speech, movement, problem-solving, and so on, merge when one has enhanced consciousness. This creates a **unified brain** that can think far more deeply than is possible in its normal state. (See the brain scans following.) Under the influence of psychoactive substances, the thalamus stops filtering sensory information so that it can be sent on to other parts of the brain for further analysis. This occurs, not only with LSD but with all other consciousness-expanding compounds to varying degrees. The intensity of the effect is dependent on the dosage taken.

Scientists who have researched the effects of LSD using MRI technology said that their results revealed a profoundly altered state of consciousness due to neural connectivity being significantly increased. This heightened consciousness level facilitates enhanced computational ability, thereby permitting the brain to conceive new concepts that it would not be able to imagine in a normal state of being. This state is sometimes referred to as '**ego-dissolution**', because **the normal sense of self is replaced by a much deeper sense of how the individual is a part of the universe**.

Users often commented that they felt a deeper connection with others and with nature. Some **users who do not understand how the brain works, can sometimes interpret this as a religious or spiritual awaking**. After the effect of an enhanced state of mind has subsided, subjects return to normal. They typically express a strong sense of well-being for a few days afterward.

Scientists who have done research studies on the effects of consciousness-enhancing compounds, eliminate people who might be prone to having a 'bad trip'. They preselect subjects who have a healthy body and a healthy frame of mind. Subjects have to fill out a questionnaire and have face to face interviews with the researchers. They must have a positive outlook on life and not have irrational fears or strong emotional reactions to unexpected stimuli. Doctors conducted the experiments in safe and pleasant environments. Most research studies compared the participants' results when they were high to those on a placebo that they believed had psychoactive effects.

Participants in scientific studies designed to research the effects of psychoactive compounds on human perception frequently reported that their experiences were *"the most profound in their life".* They told researchers that they found it impossible to convey their experience to anyone who had not had a similar experience because they felt that they would not be able to imagine just how powerful and transformative their experience was.

Users said they became far more aware of the extraordinary complexity and beauty in nature. They gained a depth of insight into many aspects of life that made many things much more meaningful. Scientific research revealed that subjects were able to see a far greater range of colors and brightnesses with their eyes open, as well as when they were having dream-like visions while they were awake with their eyes closed.

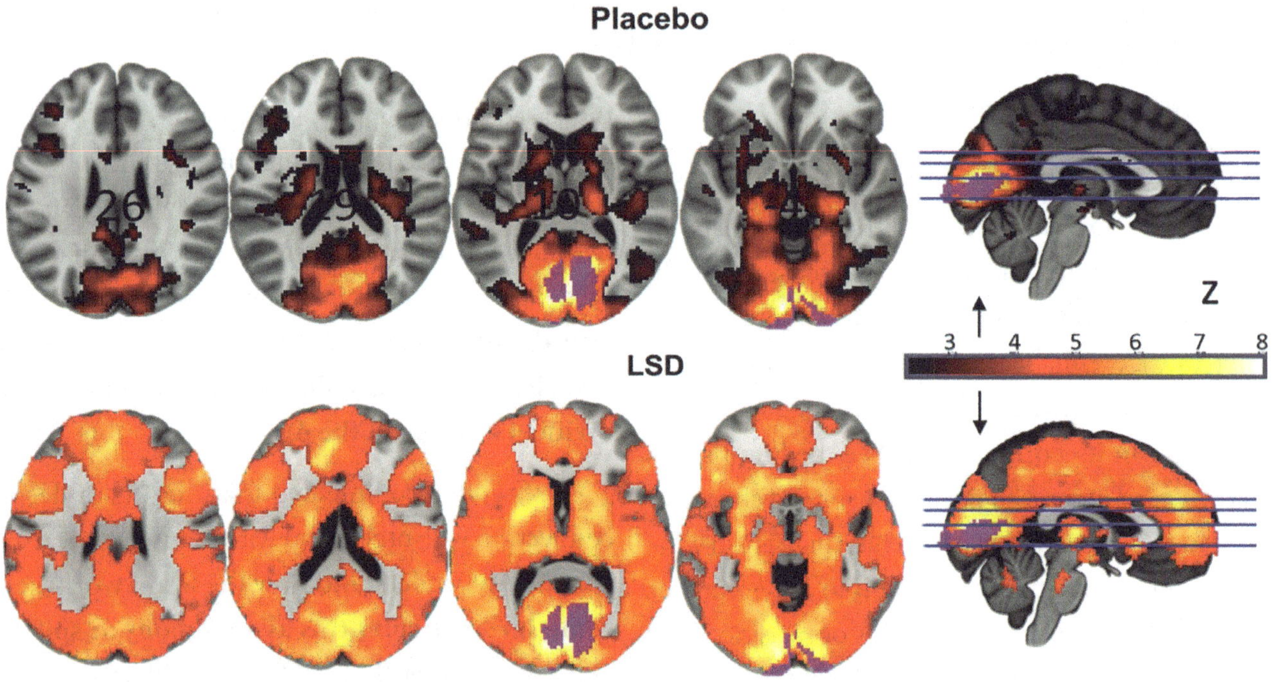

Here we see two rows of MRI scans. The top row demonstrates how active a human brain is when it is in its normal state. The lower row reveals how much more active and interconnected it is under the effect of LSD. Psilocybin in magic mushrooms produces the same results.

The red and yellow areas show where blood flow is utilized for thinking. The top scan shows normal brain activity when the participants were given a placebo. The lower row reveals a dramatic increase in thought processing across most of the brain.

Medical research into the health effects of psilocybin mushrooms and laboratory-produced psychoactive substances such as LSD and MDA, have revealed that these substances have unexpected physical and neurological health benefits. Psychoactive substances are able to reroute brain functions and produce thousands of new connections. This allows the brain to get around areas that are blocked by physical injury or through traumatic experiences. It is said that legally-obtained **microdoses** of these substances make people happy, so they enjoy life more. They are said to cure certain cancers, brain damage, post-traumatic stress disorder, shingles, period pain, depression, cluster headaches, addiction to alcohol and tobacco, and overcoming pox, flu, and herpes viruses, as well as harmful bacteria. (To learn about these aspects of the science behind psychoactive compounds, listen to talks by Paul Stamets, Michael Pollan, Khaliya, Ayelet Waldman, Dr Oz, and James Fadiman on the Internet.) The FDA in the US has listed psilocybin as a 'breakthrough therapy for depression'. Microdosing may well be a placebo effect that users want to believe are real.

The factors that affect a 'trip' are the dose, the environment(s) in which one is in, the number and type of people one is tripping with, one's intellect, one's belief systems, how relaxed one is, one's level of fear of the unknown, and if they have had a recent trauma. The security of a guide person who is straight is a good idea for a first trip. One's personality determines what type of trip they will have. If one is an explorer with a positive frame of mind, then they will enjoy their experience. If a person is fundamentally insecure, and easily frightened of anything new, then a mind trip is not something they should do.

Visions on psychoactive compounds can be amazing, awe-inspiring, and fun, but for some users who have taken too much, or are naturally easily frightened and harbor fear, they can have terrifying hallucinations. It is amazing that the human mind is so creative that it can visualize the incredibly complex visions whether they are beautiful or threatening. It is no wonder that ancient people who consumed psychoactive plants thought they had entered the realm of the gods or a hell full of demons.

HALLUCINATIONS

Hallucinations typically only occur if a high level of the compound is consumed and/or the person is unusually prone to being fearful. Those who are easily susceptible to fear will imagine fearful threats even from the most innocuous things. High doses of such compounds can produce an overabundance of sensory input too quickly for the brain to process it logically, so it draws unrealistic conclusions, which become hallucinations. This produces confusion in the brain, which can drive fear of the unknown. People who have a naturally insecure nature due to their genes or past experiences can interpret their experience as threatening because it is something new that is unknown to them and, therefore, potentially scary.

In some people's past, new experiences have been strongly related to the unknown. This often generates fear because the unknown is often synonymous with danger. But having fear of the unknown is typically a short term panic reaction. If a person is bipolar or has schizophrenia, hysterical personality disorder, or unresolved psychological trauma, then they should definitely not take psychoactive compounds. Having said this, in over a thousand trials on magic mushrooms, there was not one serious adverse event. This was because participants were screened, so the risk of anyone having a bad trip was extremely low. The older one is, the chance of having a fear attack, albeit short, is more likely than when one is young. We become more aware of threats as we age.

COMMON CONSCIOUSNESS-EXPANDING EFFECTS

Many users noted that when concentrating on a multitude of subtle outdoor sounds, they had an amazing spatial resolution of each sound. They also experienced a greatly enhanced sense of taste, smell, and the texture of food when eating. Some subjects became aware of the eight individual muscles in their tongue that they would otherwise have not been aware of. They became conscious of the showering of saliva from the roof of their mouth to dissolve the food. Some were fascinated by how much they could feel their teeth moving while eating solid foods.

Just watching people as well as themselves move around was a fascinating experience for some. From time to time, people's expressions and facial features, and their own when looking in a mirror appeared greatly exaggerated. This made them laugh. Some could see subtle animal-like features in most people. The personalities of people, their body language, their expressions, and their relationships with one another became much more apparent.

When outdoors, nearly all users commented on being spellbound by intricate fractal patterns in nature.

They found that they conceptualize numerous meanings in life. They could see the humor in many things that they would not have seen when normal, so this made many things hilarious. There was not just one punchline to most jokes; they could typically see several. This is particularly common for people who inhale marijuana. Their sense of touch was highly enhanced when they took the time to test it. Some said they could even feel the surface tension of water before it broke. They were able to look into one another's eyes and stroke each other to gain a deep appreciation of one another. This is particularly strong with pure ecstasy (MDMA). In a calm environment, pure MDMA produces an extraordinary sense of true ecstasy that is beyond description.

The researchers found that such powerful experiences made intelligent people more curious and more able to 'think outside the box' when they returned to a normal state. Some participants experienced an increased appreciation of the arts, while others commented that it had given them a greater realization of the importance of science in understanding nature.

The above are just some examples of how an expanded flood of information affects the brain, and why greater communication across the brain expands consciousness.

RETAINING MEMORIES

A day or so after subjects returned to normal, most said they could only remember a tiny fraction of the details in their experience. This is because their brain was flooded with so much information that there was not enough memory space to retain the detail they experienced. Only certain highlights could be remembered. This also occurs when one is not high.

If a user wants to remember as much detail as possible of their experience, they need to record their experience in writing or verbally, either during, or as soon as possible after their experience.

Figuratively speaking, memory for information and experiences that is seldom recalled is placed in a 'filing drawer in the bottom of the mind's basement'. This is done so that frequently used memories are readily available in the 'top drawers'. Seldom used memories are usually not available for immediate recall, but parts of them can be retrieved with the right prompts. If memories are not recalled soon after they are recorded, or not used for a long time, they gradually deteriorate and can vanish altogether as new memories are laid down. The brain has to discard most memories and only retain the most important ones because it has limited memory space.

Expanded consciousness without fear using mind-expanding substances is the most incredible experience the human mind can have. No doubt, future technology will take us even beyond this.

MINDBENDING ANSWERS TO BIG QUESTIONS

Chapter 23

IS THE FUTURE PREDETERMINED?

Science is now discovering evidence that strongly challenges some of our most fundamental beliefs – ones that we think are rock solid and common-sense. This has happened many times in the past. **In this section, you will see convincing but hard-to-accept evidence that supports the proposition that the future is already determined! If that is so, then we cannot have free will.** If this turns out to be true, then the implications are enormous!

We need to remember that it was not too long ago that the idea that both time and space could be warped by gravity seemed impossible. But Einstein's mathematics, which is supported by observational evidence, proves this to be true. All the sciences have similar examples whereby what we once thought to be impossible is now known to be real. So, the likelihood that the future is predetermined may turn out to be a fundamental truth. Keep this in mind when evaluating the evidence presented in this chapter and the next.

WHAT WE THINK IS THE PRESENT IS ACTUALLY THE PAST

When an observer on Earth sees a major event occurring in our solar system, an alien observer on another planet 100 light years away will not see that event occur for another 100 years. This is because it takes that long for the light (this information) to reach him. This means that what we Earthlings see happening in our solar system in the present will determine what aliens in other solar systems will see in their future. Conversely, **we see supernova explosions occurring now in other galaxies that actually occurred hundreds to billions of years ago. This time delay is true for any event, whether it occurs billions of light years away or only a nanosecond away due to the speed of light. So the future must be predetermined!** There seems to be no way to get around this. No one can stop or change an event that has occurred in their present, or in their past, from being observed by others in the future.

When I first realized the implication of this, I thought I was the only one who had considered this, but when I searched the Internet, I found a couple of others who had also come to the same conclusion independently. It seems that there are always a small number of people thinking about the same concepts at the same time, but they are unaware of others going through the same thought processes. This has been the case throughout history. But it will not be the case in the near future because our minds will be directly connected to the Internet and to one another.

DOESN'T THE UNPREDICTABILITY OF THE QUANTUM UNIVERSE AND THE BUTTERFLY EFFECT MEAN THE FUTURE CANNOT BE PREDETERMINED?

The past cannot change, and given that the future must be predetermined, then it too cannot change. If there is no error in this assumption, then everything we do, and every decision we make, is predetermined! But surely, **the randomness created by the Butterfly Effect of Chaos Theory should demand that the future is unpredictable, and therefore, not predetermined.** (For Chaos Theory and the Butterfly Effect, see Chapter 5.) Like so many things in our universe, this appears to be a paradox. One obvious answer to this could be that the end result of all butterfly effects is already determined, so chaos would not stop the future from being predetermined.

Some physicists think that **the uncertainty principle of quantum mechanics should ensure that the future cannot be predetermined.** They think that uncertainties in the positions and energy levels of all particles at the quantum level should prevent the future from being determined. But we see no evidence that these uncertainties are transferred into our macro universe. Maybe this is because there are such enormous numbers of these quantum uncertainties occurring all the time that they cancel one another out, so they have no effect. If quantum uncertainties could affect our universe, physics and chemistry would be unreliable, so the universe would not exist. Alternatively, if some random events in the quantum universe did affect our macro universe, then these events would be already determined.

WILL KNOWING THAT THE FUTURE IS PREDETERMINED CHANGE HOW WE LIVE?

If it turns out to be beyond doubt that the future is predetermined, this does not mean that we can know the future. If we acknowledge that the future is predetermined, then whatever is going to happen will happen, and we will have no way of changing it. This should not be alarming as it makes no real difference because we are not aware of what the future holds. If the future is already determined, it will not make us think any differently about each decision we *think* we are making. Even if we were to consciously change our mind in an attempt to show we have free will, this will not change the future because our decision to change our mind would have already been predetermined. We will still live our lives as we always have, so knowing that the future is predetermined will have no impact on us. But it does mean that we will have to accept that we have no free will! But that will not change our lives either because we will put the notion aside as we intuitively assume this not the case. How would you change your life if you accept that the future is predetermined? You wouldn't.

When we think about the future, for much of what we do in life, we actually *do* know, to a high degree of reliability, what the future will be. For instance, when we go on a roller coaster ride, we have a very high probability that we will go along the track, and a minute later, we will end up back where we got on. Similarly, when we take a train ride, an airline flight, or a cruise, we know where we will end up - in 99.9% of cases. This is also true for most other aspects of our lives. With physics and chemistry, we know the outcomes of things associated with these sciences because they are determined by well-proven laws. (This does not apply to things that are at the leading edge of science that have not yet been proven by way of exhaustive testing.)

If the future is already determined, this would mean that the universe has already fully evolved to its final stage (if there is one). If the universe is the ultimate mind, then it would know what the future holds, so what would be its purpose in spending all the energy required to make the

universe evolve? This tells us that we are missing some very big aspects of the universe.

This artistic 'installation' attempts to illustrate how time and consciousness relate to one another. Credit: Art Installation, Pace London

If we could see time from a higher dimension, we might see it as **Einstein** conceived it. He stated that "***the distinction between past, present, and future is an illusion, however tenacious this illusion may be.***" His calculations for relativity showed that the future and the past must already exist as one entity. This being so, why is the universe designed so that we cannot see the future?

When we observe the universe or the world around us, we are always observing the past. **We can observe parts of the past, but we cannot observe anything in the future. If we try to predict the big picture of the future, as I have attempted in parts of this book, we only get it partially right, if at all. The further we try to look into the future, the more complex it becomes, so the less likely we are able to make accurate predictions.** The further in the past events occur, the less clear they are. And correspondingly, the further things occur in the future, the harder they are to predict.

If we knew what was going to happen in the future in every aspect of our lives and we could do nothing to change it, this would completely change how we think, and how we live. Many people may see no purpose in their existence. Could this be one of the reasons we are not permitted to know? There is something very fishy about the future being determined and us not being able to see what it is.

DOES THE PRESENT EXIST?

Everything we perceive to be happening in the present locally around us has actually occurred a very short time ago in the past. So, **strictly speaking, the present does not exist!**

The speed of light and the distance an event is away from us is what determines how far in the past an event occurred. When we see something happening in the present right beside us, we become aware of it no sooner than 3-4/10^{ths} of a second after it occurred. So technically speaking, it occurred in the past. It takes a nanosecond for the light to reach us from something right beside us and then 3-4/10^{ths} of a second for our senses to send that information to our brain, where it is analyzed. So, no matter how close something is to us, we can never be aware of it sooner than just under half a second *after* it actually occurred. **This means that everything that we are aware of has actually occurred sometime in the past.** The sound of thunder from lightning, distant fireworks, the noise of a jet fighter flying by in the far distance, or the sound of a large explosion a few kilometers away all take from several seconds up to half a minute for their sound to reach us. This is because sound travels in air at only 1,110 km/hr (700 mph). So what we are hearing is the sound from an event that occurred in the past. It is the same with light; even though it travels at 300,000 km/sec (186,000 mps), everything we see is from the past. We see the planets and the Sun as they were minutes to hours ago, and we see galaxies as they existed millions to billions of years ago.

The movie '*The Time Machine*' treats the future as existing just as the past does.

OBSERVING HISTORY AS IT REALLY OCCURRED

The following is an interesting hypothetical concept to consider, despite it being impossible to achieve in reality. In this thought experiment, we use the time it takes for light to travel across space to see our past as it was occurring.

We see far into the past when we look into the night sky. Imagine that there are natural mirrors out in space that are placed at roughly one hundred light year intervals from us. If they were exactly perpendicular to our line of sight, they would reflect our Earth back to us, so we could see it as it was in the past. How far back into the past we could see would depend on the distance that the mirrors are from us.

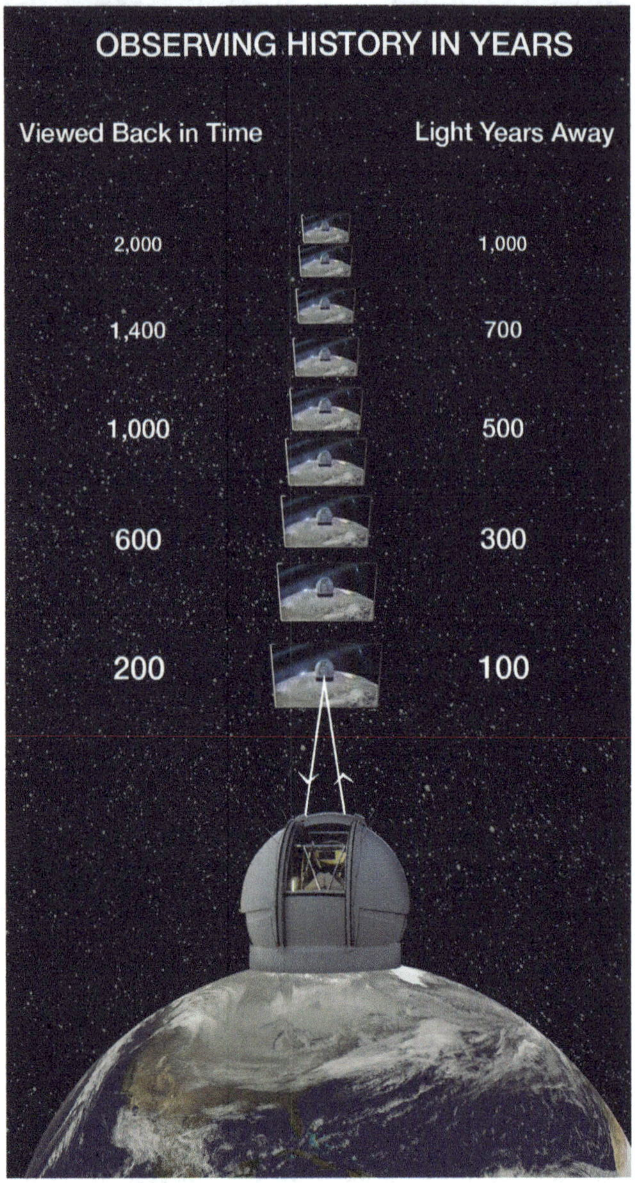

The hypothetical concept of the 'History Scope' utilizes fictional, large natural mirrors in space at 100 light year intervals. They reflect Earth as it was in the past, so we can view our history as it unfolded. Light from an event that occurred a century ago takes 100 years to travel to the closest mirror and then another 100 years to travel back to Earth, so we see history as it was 200 years. Similarly, each more distant mirror allows us to view further back in time.

Imagine that we have invented an incredibly powerful telescope that we will call the 'History Scope'. It can look into these mirrors to observe our planet in such great detail that we can see individual events unfolding throughout history. We could see exactly how history really occurred. The technology would need to allow us to see through clouds and buildings and to see at night. It would also need to allow us to zoom into individual people and 'hear' by lip-reading their conversations. We would need a number of History Scopes placed around our planet and distant mirrors in at least four directions so we could observe our past 24 hours per day as the Earth rotates.

It would take 100 years for light leaving the Earth to reach the closest mirror and another 100 years for the reflection of Earth to return to our History Scope. In this case, we would be observing our history as it was occurring 200 years ago in the nearest mirror. The next mirror at 200 light years away would show our history as it was 400 years ago. And the next at 300 light years away would show us Earth as it was 600 years ago, and so on. We could watch the Earth for 200 years in the closest mirror and then and then do the same in the next mirror in the the next mirror to see the next 200 years, and so on to see what really occurred throughout history.

Of course, all this is impossible, but if it were possible to observe history as it happened, I suspect that much of the content in our history books would have to be substantially revised or totally rewritten. This is because much of it is written from the perspective of an individual, or small group of people, often long after the event occurred. History is often written by people who have a substantial bias. The victors typically write history, so we never know the real truth about things in the past. That is true for many things in the present.

Chapter 24

DOES INTELLIGENT LIFE HAVE FREE WILL?

Research done in neuroscience and genetics is providing disconcerting evidence for our lack of free will - the ability to make our own decisions. There is evidence in the previous chapter that the future is predetermined. This supports our lack of free will.

Recent studies on the human brain strongly indicate that what we think we are deciding to do using free will has already been determined by our genes. A couple of decades ago, I found these conclusions astounding and very difficult to accept because it 'felt' illogical to not have free will. We seem like we are independent beings who make our own decisions, but are we? Our genes tell a very different story.

The evidence is now overwhelming that our genes are making most, if not all decisions for us. Neurological studies show how our brains work. They reveal that our genes that wire our brains make our decisions and tell our conscious brain after they have made the decisions! As you will see in the next chapter, **in-depth studies of identical twins show that our genes determine not just how our physical bodies grow, but also how the physical neurons in our brains are wired. This determines how our brains process information and emotions. Our genes almost totally determine virtually everything about how we think - our intelligence, our preferences, our interests, and the decisions we make. As I did at first, many people want to reject this out of hand, but the evidence is overwhelming. See the chapter following.**

THE AUTOPILOT IN OUR BRAIN

Neuroscientists are now realizing that probably all the decisions we make are made by our brain's autopilot, which is controlled by our genes. Our conscious mind does not make the decisions we think we are making! All the information that our senses gather is filtered and then actioned by our autopilot. It determines almost everything about how our body functions and how our brain thinks. It programs everything - how we think - how we problem-solve - how emotionally we react to stimuli - who we are attracted to - what interests we have - what we decide to do, and much more. **This is no doubt the case for all other organic life on Earth and throughout the universe.**

We are not consciously aware of the tens of thousands of decisions that our brain's autopilot is constantly making every millisecond to keep

our body and mind operating. The autopilot in our brian ensures that our automatic nervous system works all the muscles in our body in the right way so we can walk, talk, eat, breathe, swallow, pump blood, blink our eyes, sweat, sneeze, digest food, and so on without us consciously having to instruct every muscle and organ to do its job in precisely the right sequence and at exactly the right time. Our autopilot also oversees all the complex mechanisms in all our organs to check that they are doing their jobs correctly, and if not, it puts in place countermeasures in the hope this will correct any problems arising. Our automatic systems for fighting off dangerous invading bacteria or viruses are constantly on the look-out for potential threats. When detected, cellular armies are dispatched to kill them. If all these activities were not pre-programmed into our autopilot and our specialized cells, our conscious mind would not be able to compute the millions of functions that our body automatically performs every second of our entire life!

We are not consciously aware of 99.9999% of what our senses are monitoring. Nearly all of the data from what we feel, hear, see, taste, and smell, is not processed by the conscious part of our mind: it is processed by subroutines in the subconscious parts of our brain.

Our senses discard nearly all of our sensory data because it is repetitive and irrelevant. Only those bits that do not match up with what is in our memory are sent on to the conscious mind for analysis – and only if they are above the threshold of being important enough. Our autopilot has innumerable algorithms for this. Data that signifies a change that may be harmful, or helpful, alerts the conscious brain to check it out. But our genes decide what action our conscious mind will take. This gift of extraordinary efficiency is what 3.5 billion years of trial and error evolution of life on Earth has passed on to us. Such systems will surely be the case for all organic alien life forms across the universe because they too would need to evolve very similar efficiencies in their brains. Mega-intelligent computers and robots will also have similar systems to allow them to operate efficiently.

The autopilot in our brain controls nearly all of our thinking in a way similar to how the autopilot in a commercial airliner controls nearly everything the plane does. For the most part, human pilots only handle some aspects of communications, take-offs, and landings, as well as double-checking certain data from time to time. All these operations could be done by the plane's autopilot. However, airlines employ human pilots mostly to give the impression to customers that they are more secure. Only when some major thing occasionally goes wrong does a human pilot take control of some critical aspects of flying the plane that the autopilot is not handling correctly. But human pilots have recently been barred from doing this in the most modern planes. When flawed programs have not allowed the pilot to take control, this has caused the plane to crash. In the near future, all planes will be controlled by very advanced autopilots that are far superior to human pilots. Soon computers will fly increasingly more complex planes far more safely than any human could do particularly when things go wrong. This is not yet the case, but it will be before 2030. The most advanced fighter aircraft are now extremely complex and controlled by computers. They do not need human pilots because they are able to make decisions far faster than a human can. Human pilots are quickly becoming redundant.

Our autopilot largely controls us: only when something goes wrong or an opportunity occurs, is our conscious brain informed, and only then

does it appear to take control. But it is our genes that make the decisions, not our conscious mind. In relation to all thought processes going on in our brains, our consciousness appears to be a mere snowflake on the top of an iceberg of thought processes, of which we are completely unaware. Through trial and error, nature developed automatic systems to make the brains of all animals efficient. Without these automatic systems, intelligent thought would be impossible because our conscious brains would be overwhelmed with unimportant information and they would 'crash' and send us insane! The human mind and AI have automatic systems for filtering non-essential information. All animal brains have efficient methods of processing sensory input, calculating their future actions, and operating the systems in their organs. These algorithms are all below our consciousness.

Our conscious brain may only be aware of one hundred billionth of the information detected by our senses every minute!

Our senses are what provide all the information we receive about the world around us. Without them, we would know nothing. They play a big role in determining what decisions we make automatically.

In animal bodies, apart from the brain, other body parts also compute a surprising amount of data automatically to ensure that all our organs, muscles, and defense chemistry work effectively without the need for the brain to control everything. Brain neurons have been discovered in our heart and in our digestive system. They make these organs work without our brain having to program their activities from afar. We may soon discover that similar neurons control other organs. We do not have free will to determine what these systems do.

Interestingly, octopuses have 12 'brains' scattered throughout their body! These brains work mostly automatically. They have a main central brain that controls the nervous system that looks after movement. Each of an octopuses' eight tentacles has a ganglion (brain part) at their base. These control the movement of each tentacle individually. They also automatically change the color and pattern of the octopuses' skin to match the background, thereby achieving near-perfect camouflage. There must be a means of communication between the ganglions for octopuses to have coordinated movement and color changes. Octopuses have two or three hearts, depending on the species. Each heart has a small brain to operate it. Despite having one small main brain and several smaller ones elsewhere, octopuses are nevertheless quite intelligent. They are very good at problem-solving! This is another fact that demonstrates that brain size is not an indicator of intelligence. How different will alien brains be to those on Earth?

HOW MANY SENSES DO WE REALLY HAVE?

We have far more than five senses because each main sense is a combination or many specialised senses. They work mostly automatically without us needing to consciously direct them. Many of our reactions to their stimuli are also automatic. Our senses have a major impact on the information our brain collects and processes, but our conscious mind is almost totally unaware of this.

It has long been stated that we have just five senses, but we actually have around sixty, as well as the ability to detect hundreds of thousands of odors.

It's very interesting to consider how many senses we really have in some detail.

SIGHT

Our vision is not one sense; it is actually eight senses which comprise the detection of three colors, brightness, texture, movement, 3D form, facial and object recognition, and the ability to determine distance due to us having two eyes for binocular vision.

With our **sense of sight**, we mistakenly think we are seeing most of what we are looking at in detail, but the reality is that our vision only sees an extremely small area of what we are looking at in high resolution. This occurs in a very tiny spot known as our **center of vision**. Incredibly, the spot at our center of vision is no larger than two letters on an A4 page when it is held at arm's length! To test how small this area is, stare at the central cross symbol in the randomly typed group of letters below. Without moving your eyes at all, see how many letters on either side of the cross and above and below it, you can see clearly. It takes practice to not move your eyes.

```
       a
       s
       x
       z
       r
vsecxwzo+arsemwxc
       e
       s
       m
       n
       v
```

Do not remember the letters, as you will only be invalidating the point of the exercise. If you do this correctly, you will only be able to see the cross clearly and one or two letters on either side of it a little less clearly. The other letters are indecipherable until you stare directly at them. This amazed me when I first tested this because I was convinced that I was seeing a much larger field of view quite clearly. But what my eyes are doing when I look around a room or across an outdoor scene is rapidly flicking around to look at what I was subconsciously interested in and then upgrading any changes in the scene in my memory without me realizing it. We assume we are seeing everything in our field of view in detail, but when one is asked to close their eyes and then recall what detail is on some part of what they were looking at, they cannot recall it, even if they have looked at the scene many times. If we could record what we are seeing in our mind, we would be shocked by how extremely lacking in detail. It all is at any one time.

Immediately outside our center of vision, resolution drops off rapidly. Our perception of color also diminishes the further something is away from our center of vision. But we are blissfully unaware of this. We think we are seeing most of our 160° field of view in some detail and in color but **nearly all of what we think we are seeing is an illusion! Our memory fills in more than 99% of our field of view.** Scientists have only recently discovered that **our**

center of vision automatically does spot checks of our environment three times a second! We are not the least bit aware of this. This is done to update our memory in between us looking at something that interests us. So that we do not suffer motion blurring, our autopilot turns off our vision momentarily each time our eyes jump from one place to another, but we are not aware of this either. The neurons in our eyes are programmed to pay special attention to movement, a change in brightness, or a change in color. Our autopilot puts these changes into our visual memory, and then it goes back to do the next cross-check. When we read, or are concentrating on something fine and detailed, this process must slow down. Our unconscious brain has truly ingenious organic 'technology' to minimize our need to be consciously aware of unimportant things happening all around us all of the time. This is a brilliant strategy that evolution designed.

HEARING

Our sense of hearing is actually five senses. We can detect volume, tone, the pitch of sounds, and the direction they are coming from due to us having two ears and the structure of our ears. To test your ability to amplify the volume and the direction from which sounds are coming from and their clarity, cup your hands behind your ears and angle them forward when outside. Then change your direction. This allows you to better determine the direction that certain sounds are originating from. Experiment by bending your hands and ears forward to different degrees to see how much this enhances what you can hear in front of you. If you do this when you are in an elevated location in a built-up area where you can hear many sounds quite loudly from far away, the effect is surprisingly pronounced.

Our fifth hearing sense allows us to separate and filter sounds to concentrate on a particular one, typically the person we are talking to. This is particularly useful when in a noisy environment, such as a party with lots of people and loud music.

Our sense of hearing only makes us consciously aware of a tiny fraction of all the sounds that our brain is constantly monitoring. To test this, sit still outside and close your eyes for a few minutes. Now concentrate on how many sounds you can actually hear when you pay attention to them all. The number is surprising, even when you are in a relatively quiet space. When you are outside and elevated in a city, you can hear many sounds from far afield.

TOUCH

Incredibly, our sense of touch via our skin is actually at least thirty individual senses! Our skin can detect pressure, texture, temperature, vibration, slippage, soapiness, stretching, shapes, electrical charges, acidity, alkalinity, and the strength, type, and location of pain, as well as its type. Our skin can also detect the movement of air, stickiness, wetness, itchiness, and the positions of our limbs, and body, are in. Yet we do not think about these things as being individual senses. They operate all the time. It is only when something out of the ordinary alters us to one or more of these senses that we become conscious of some of them. Sometimes skin cells are irritated by some stimulant such as an insect bite or a chemical, or simply due to an error occurring in a cluster of nerves that make them send a false irritant message to our brain. When this occurs, we almost unconsciously scratch that area to reset the nerves without thinking about it. (To stop you from damaging your skin from over scratching from things such as eczema or a mosquito bite, very briefly run hot water just short of burning you over the area that is itching. It will sting for a couple of seconds, but this resets the nerves, and it ejects the irritant, thereby stopping the itching within seconds. It will not work with all insect bites or chemicals.) Sexual stimulation is another important and enjoyable sense of touch that can lead to achieving orgasm. Without this sense of touch, we would not procreate.

TASTE

Our sense of taste can detect at least 6 main types of tastes - sweet, sour, bitter, salty, spicy, and umami, which is the savory taste found in monosodium glutamate (MSG). Combinations and permutations of these basic tastes create an enormous number of individual tastes. Each type of taste has numerous variations or flavors within it. Our sense of taste combines with our sense of touch in our mouth to allow us to also detect the physical structure of food – whether it is solid, fragile, crunchy, chewy, jelly-like, greasy, or liquid. Each of these states change the way we appreciate the food. Some foods are far more appealing in one state than any other. For instance, some foods may be far more desirable if they are crunchy than if they are soft or liquefied, and vice versa. Why different physical states of food make such a difference to their taste is not clear. As well as this, the temperature of certain foods is an important factor in how they taste. Hot ice-cream or cold coffee do not taste as good as they are at their usual temperature. And pain in the form of hot mustard, pepper, or chili also impacts on the taste and attractiveness of food. Some people love this type of pain while others hate it. Taste is also affected by the smell of the food. The aroma of some foods can change their taste. As an example, if a piece of apple is placed under your nose while your eyes are shut when eating a piece of pear, you will think you are eating an apple. The reverse is also true. Taste is a very much more complex sense than what we assume.

SMELL

Our sense **of smell** is also a very complicated sense. It is at its strongest when we are young, culminating in our teens. When some people are sick, their sense of smell is enhanced. Because of the way it works, we can detect some hundreds of thousands of different aromas! Forty-five odors are detected, not only by our olfactory receptors but also due to the mild pain or temperature they cause in the nostrils, e.g. strong peppermint or mustard. Our sense of smell ranges across a huge number of flowers as well as the aromas of crushed leaves and barks, to numerous perfumes, vegetables, fruits, fungi, musk, insects, gases, powders, oils, smoke, sea air, bacteria in cheeses, and both attractive and unpleasant body odors. Aromas can both disgust us, excite us, attract us, or repel us. Our sense of smell often warns us of spoiled food, rotting meat, or excrement - things that will make us sick, kill us, or cause us to be allergic to them. Research has shown that if we do not like the smell of a potential partner or an associate, then we are unlikely to get along with them. Smell is often associated with strong emotional memories.

There are bound to be many intelligent life forms in the universe that have a greater range of senses than we do, while others will have fewer than us. In the near future, mega-intelligent computers will have millions of senses.

All our senses run constantly on autopilot, as do all the electrical communications across our brain. The same is true for all our complex internal bodily functions and chemistry, as well as our defence systems. Only when something abnormal occurs in all these billions of computations that are being made every minute, do we become consciously aware of some small facet of all this information that our senses are constantly sifting through.

OTHER SENSES

We also have senses for **stress, suffocation, the urge to vomit, blushing, hunger, the need for water, sugar, salt, bladder and bowel releases, malfunctions of the heart, and ultraviolet radiation** from the sun. There are probably more we have not yet identified.

SENSES OF MOVEMENT, BALANCE, GRAVITY, TIME, AND MAGNETISM

Another sense that is seldom considered is the **vestibular system,** which detects motion and acceleration in all three dimensions, as well as weightlessness. We also have a sense of **balance.** To see how good your sense of balance is, see how long you can stand still on one foot with your eyes closed. It sounds easy, but it is difficult to maintain your balance without the assistance of sight for feedback. When our **sense of gravity** is disturbed, we have difficulty determining which way is up, as when we are on a rolling boat or weightless in space. This disconnect with what we are familiar with can cause so much stress that it can make us vomit. This is often the case when astronauts go into the weightlessness of space. However, they become accustomed to this within a couple of days. People who have vertigo suffer a debilitating sense of **dizziness** and a lack of balance due to crystals in their inner ear being dislodged. We also have the sense of **chronoception,** which is the ability to sense the passing of time. Some rare people also have the sense of **magnetoreception** - the ability to detect magnetic fields that some birds have.

As you can see, we have numerous senses being around 70 plus the ability to sense hundreds of thousands of aromas. We are bound to identify many more. Our brain monitors all of these without us having much conscious awareness of most of them. Our unconscious brain makes decisions about almost all of this data that comes in from all these senses without disturbing the conscious part of our brain.

There is virtually no evidence to support that we have free will, but there is plenty to support that we do not!

Chapter 25

DO WE CONSCIOUSLY MAKE OUR DECISIONS?

We all believe that the decisions we make are made with our free will. But to the amazement of neurologists, research results show that the vast percentage of our decisions are actually made by our genes prior to us becoming consciously aware of them! This is a totally unexpected! It is consistent with the evidence in Chapter 23 that the future is predetermined. There are a number of proofs following that we do *not* make our own decisions. **This makes all life forms automatons!** If this is so, why would the universe be designed this way?

SURPRISING MRI BRAIN SCANS

MRI brain scans of the decision-making processes in our brains show that our brain's autopilot is making decisions between two seconds, and in some cases up to ten seconds *before* our conscious brain becomes aware of the decision! This is an explosive discovery that has enormous implications.

In neurological experiments, a subject's *unconscious* intention to move an object typically occurs an amazing 2.8 seconds *before* the action happens! But the subject's *conscious* intention to move the object only occurs one second before! This means that **the conscious mind is being informed well after the unconscious mind has already made the decision!** Researcher **John-Dylan Haynes** said, "*How can I know a decision is mine if I don't know when it occurred, or what was decided until after the decision was made?*" These tests have been reproduced by a number of researchers looking to disprove this, but they got the same result! These experiments dealt with decisions made over short time frames of seconds and not those made over minutes, hours, or days, which we would hope are made thoughtfully and consciously. But are they? There is much evidence from identical twin studies to show that even these decisions are controlled by biases programmed into our genes. Our genes determine the circuitry in our brains and, therefore, how we think. Consequently, they determine the decisions that we make. (See the Identical Twin Study results in Chapter 26 following.

WHAT IS CONFIRMATION BIAS?

A majority of people find it hard or impossible to make well-considered decisions based on evidence. This is due to a lack of interest, laziness, a lack of time to do research, insufficient intelligence, a lack of knowledge, and genetic ***Confirmation Bias***.

Confirmation Bias is defined as ***'the tendency to search for, interpret, favor, and recall information in a way that confirms one's beliefs or one's***

hypotheses while giving disproportionately less, or even no consideration to alternative possibilities.' Many people will reject out-of-hand, concepts and evidence that go against their typically simplistic, emotion-driven, genetically determined beliefs. People with strong political or religious beliefs have a strong confirmation bias. No matter how much evidence they are presented with that goes against their belief, they will not change their mind.

A colleague of mine describes this as the '***Ratchet Effect***'. When evidence comes along that goes against a person's beliefs, they will not consider the available evidence. When they hear what they believe is supporting evidence for their belief, they turn their imaginary ratchet forward many times rather than just once. But they never turn it backward one click when they hear many pieces of evidence that go against their beliefs. Their ratchet only turns one way.

About 80% of people have confirmation bias to some reasonable degree. Our genes determine our confirmation bias. Research shows that confirmation bias is particularly pronounced in strongly religious people, those who believe in paranormal happenings, and those who have strong political or environmental convictions. It is more common in scientifically illiterate people because they have not been taught how to question things, and they have little general knowledge. They are typically disinterested in learning about most things, so these factors make it difficult for them to evaluate facts. They, therefore, have few options upon which to make decisions. They typically think about complex issues in simplistic terms, which are often emotionally charged. These people constitute a considerable percentage of any population, and, in most countries, they have a vote!

Nothing can be done about the way people think because their genes determine their thoughts, and these cannot be changed – not yet. For many people, their genes do not allow them to think logically based on scientific evidence and facts. When people lack broad-spectrum knowledge (which is very common) and have a low level of intellect, the more pronounced their confirmation bias is. These people want simple beliefs that they can understand and believe in. They are unable to argue their case with logic and facts because they have not got the intellectual ability and knowledge to do so. They try to win arguments by denigrating their opponents with ridicule and threats. This is very prominent in today's politics and subjects such a Global Warming, Covid, and socialism. The art of winning a debate by using logic supported by convincing facts has been largely lost to ignorance and emotionally-driven, simplistic ideals. The majority have willful ignorance.

Each life form has genes that have finely detailed preferences that predetermine what actions and decisions it will make to survive and replicate.

Neuroscience research is showing that, *if* we make any decisions consciously, they would be about a very small number of things that our genes do not have a preference for. These are typically inconsequential decisions. But the research is now showing that even these decisions appear to be predetermined by our genes! (See chapter 26.)

Our genes determine how our brain is wired, and therefore how we think. Our genes determine our intellect, what we are interested in, and consequently, the knowledge base we have. They determine whether we make our decisions quickly based on our emotions, or whether we make our decisions slowly after taking the time to do research to discover the facts.

Chapter 26

WHY ARE THE IMPLICATIONS OF THE IDENTICAL TWIN STUDIES SO IMPORTANT?

TO WHAT EXTENT DO OUR GENES CONTROL US?

Recent research is confirming that genes control the way life forms bodies grow and how they act. Our genes determine our physical characteristics, i.e. our looks, our skin type, our height, and our physique, as well as many aspects of our health. They also determine how the cells in our brain are wired, so this determines our intelligence and therefore our ability to learn, and every other aspect of our minds such as how good our memory is, our interests, personality, musical and athletic abilities, work ethic, honesty, empathy, decision-making, and even our beliefs! These things are programmed into us at the very outset. This has been dramatically demonstrated in many studies done on identical twins in many countries over many decades.

While genes are very dominant in the development of all life forms, there are, however, some environmental aspects that can affect how we develop. For example, a fetus in the womb can be affected by restricted blood flow from its umbilical cord, its mother's health, her diet, her emotional state, and her lifestyle, e.g., whether she drinks and/or smokes, and what medication she is on. As well, an overbearing social culture, a draconian religion, a brutal political system, a devastating war, a near-death experience, severe sexual abuse, or a deadly epidemic can inflict serious psychological damage on people for the rest of their life. So can being subjected to malnutrition, continuous severe pain, a serious deformity, or a chemically toxic environment. Surprisingly, their genes determine their personalities moreso than these things. Our genes determine how we turn out.

IDENTICAL TWIN STUDIES REVEAL THE EXTRAORDINARY POWER OF GENES

Because **identical twins** have the same genes, scientists wanted to see if they would grow up to be different from each other when they were raised in different environments. Over many decades, psychologists across many countries studied tens of thousands of identical twins that were separated at birth or before 18 months of age. In some studies, a small percentage were brought up together in their infant years and then separated before they were three. All twins in these studies were separated because the mother could not bring them up due to socio-economic situations or because the mother died giving birth. In cases where the twins had been together for up to a year or so after birth, their separation caused some of them psychological trauma.

In the studies, each twin was tracked during their life to monitor their conditions and how they were growing up in often different environments. The twins were not aware that they had a twin. They were monitored into early adulthood. When the twins reached middle age, each twin was surveyed to see how similar they were across a wide range of parameters, given that most of them had experienced very different upbringings. This would determine whether their genes or the environment in which they were raised was mostly responsible for how they turned out. In most studies, the twins never met one another until they were adults, often in their 40s or 50s. When they met, they were amazed at how similar they were in so many respects.

Identical twins raised in the same household by their natural parents also turned out to be identical, not only physically, but also in nearly every respect of their lives. This is not the case for fraternal twins who do not share the same genes. They typically grow up to different like most children in most families because, in most cases, they have quite different genes.

Before the results of each of the identical twins studies were published, it was almost universally believed that environmental effects far outweighed genetic effects for shaping a person's personality, likes and dislikes, health, talents and interests, and their beliefs. Against all expectations, the environment had little effect. The studies showed that it was our genes that determine nearly all things about us, even down to the most minor things!

Identical twins are almost 100% identical in facial looks, their physique, their psychology, and their mental abilities. They have the same interests in politics, health, humor, clothes, music, foods, and most other things. Those that were separated soon after birth revealed how much more the effects of nature (genes) had on them compared to the impact of nurture (the environment in which they grew up in).

It has long been assumed that each person's outlook on life is shaped by their parents, their mentors, peers, idols, teachers, their job, and the culture in which they are raised. But to everyone's great surprise, the studies showed that these things had little to no effect on their nature! This is because **our genes strongly determine how our body, including our brain is constructed.**

Nearly all researchers believed that environmental factors such as:- how we are brought up, our standard of living, our level of security, the beliefs we are taught, the quality of our education, and our social status are what play a major role in how we develop. But amazingly, this was not the case for nearly all the twins! Their environmental factors had very little effect - unless they were very severe, and even when they were, for many, this made little difference!

It was presumed that a child's personality was largely determined by how healthy its general environment was and how happy and loving its parents were. But again, beyond all expectations, separated identical twins showed this had little effect on their personality.

In a high profile study, triplet boys were separated after they had been together for around 18 months. Each boy was deliberately given to parents that were very different to the other parents to see if that would make a difference in how they developed. For the most part, it did not. Through their teens and into the 20s they were very similar. In their late teens, two of them happened by chance to meet one another, and some time later, they discovered that they had a triplet brother. They were surprised at how similar they were in nearly all respects. When the news spread, they became a TV sensation. In their adult years, they opened a restaurant together, which was very popular as they were very jovial. One triplet was given to a teacher who was a good father but a little strict. This triplet suffered depression in their early adult life, but the other two did not. His depressive states became more pronounced as he aged. His brothers and their families did all they could to help him but to no avail. This is despite him being the most lively and funny of the three. In his mid-life, he fatally shot himself. Even though the three brothers had the same genes, the other two did not suffer depression. It was likely that he had a random epigenetic change to a single gene that made his personality develop depression as he aged. (See the rare effects of **epigenetics** and certain environmental factors following. These can cause a single, but sometimes a serious difference, between identical twins, triplets, and quads.)

Despite genes being by far the major factor that determines how people turn out, some researchers, particularly those with leftist ideologies which are against eugenics, do not want to accept their own findings and those of many other research psychologists. Because the identical twin study results went against leftist ideology, the Left did all they could to bury the results for decades. But recent identical twin studies have shown the same results, so the reality that genes determine our nature is again coming to the fore.

The studies showed that we are not all born equal, as certain political ideologies and religions preach. Many people with genes for low IQ are gullible and are therefore susceptible to being brainwashed, whereas those who have inquiring minds and a good intellect are able to question what they are taught at school or told by their parents, their boss, the media, or governments. They learn to compare what they are told to reality and what science has proven to be true. Intelligent children will question what adults tell them if it doesn't gel with the real world.

The level of sameness that identical twins have was never considered possible before the twin studies were conducted. **The studies showed that almost all identical twins grow up to have an inconceivable degree of similarity - even down to their most trivial likes and dislikes.** Nearly all of the twins had the same interests, selection of clothes, hairstyles, cosmetics, pets, art, and partner(s). They also have the same esoteric and religious beliefs. They have the same way of walking and dancing, the same interests in music, the same of interest and expertise in sport, the same IQ, and very often the same type of job(s)! **This was despite many of them being raised under very different environmental conditions and in different countries!**

One twin could be raised in a secure, loving family while the other was raised in an unloving, violent, or strict household, yet their personalities, likes and dislikes, and their beliefs turned out to be the same! In many cases, one twin was brought up in a strongly religious environment while the other was raised by atheists. Yet if one turned out to be deeply religious or an atheist, then so did the other! Some were exposed to smoking, alcohol and/or drugs while their twin was not. Yet if one was for or against these things in their adult life, so was their twin in almost all cases. Some twins were brought up in different countries under vastly different political regimes such as brutal Nazi Germany, while the other was living life freely in the Caribbean, yet they both turned out the same. Some were brought up by their father while their twin was raised by their mother, yet this created no changes between them. If one twin had a number of foster parents while the other had just one, this also had little effect on creating differences between them as adults. Some were raised in rich households while their twin was raised in poverty; yet again, this made little to no difference to the way they both turned out as adults. Some went to good schools while their twin did not, yet they both achieved the same level of academic success! If one had characteristic facial expressions, or preference for particular music styles, or books, then so did the other. They even had exactly the same humor and tricks they liked to play on people. They even chose the same types of pets, the same brand of soap, foods, and decorating styles!

Almost all had the same work ethic, political leaning, and the same level of interest in science, and paranormal beliefs. In nearly all cases, both twins grew up to be either happy or depressed regardless of the environment in which they were brought up in. Nearly all reached the same level of professional development. Many had exactly the same or similar jobs working in the same industries, and in some cases, they worked for the same companies! This was even though they lived in different towns, states, or countries! They would choose the same type of partner(s) who nearly always had the same looks, and often the same names! Most of those that divorced had the same number of partners. Some female twins gave birth to children on the same day! Many twins reported having the same physical sickness or injury at the same time! No one expected this level of similarity. These results are truly amazing - until one realizes how powerful the programs in our genes are. These studies support the future being predetermined and that we have no free will.

EPIGENETICS AND ENVIRONMENTAL EFFECTSONGENES

There are **uncommon environmental conditions** that can create a specific difference between a tiny percentage of identical twins. These conditions also affect other people.

Small **methyl molecules** can sometimes attach to the outside of one twin's DNA and cause a particular gene to express itself differently to the other twin. (See the diagram following.) This is known as **epigenetics**. Some **epigenetic** molecules can cause one twin to have a propensity towards one of the following - a certain type of cancer or some other type of disease, a physical defect, a different sexuality or gender, left-handedness, a particular personality trait, or a mental disorder that their twin does not have. However, such instances are rare.

If one twin is brought up under strong environmental effects like having very **violent, unempathetic, psychopathic parent or guardian, this** can cause that twin to suffer damage to its psyche. The stress of this has to be intense and long-lasting, and the affected twin has to be susceptible to this stress to have it change their personality for life.

One twin may have had a very stressful **near-death experience**, which resulted in it having post-traumatic stress disorder, thereby changing that twin's psychology. However, people with strong, confident personalities can handle near-death experiences.

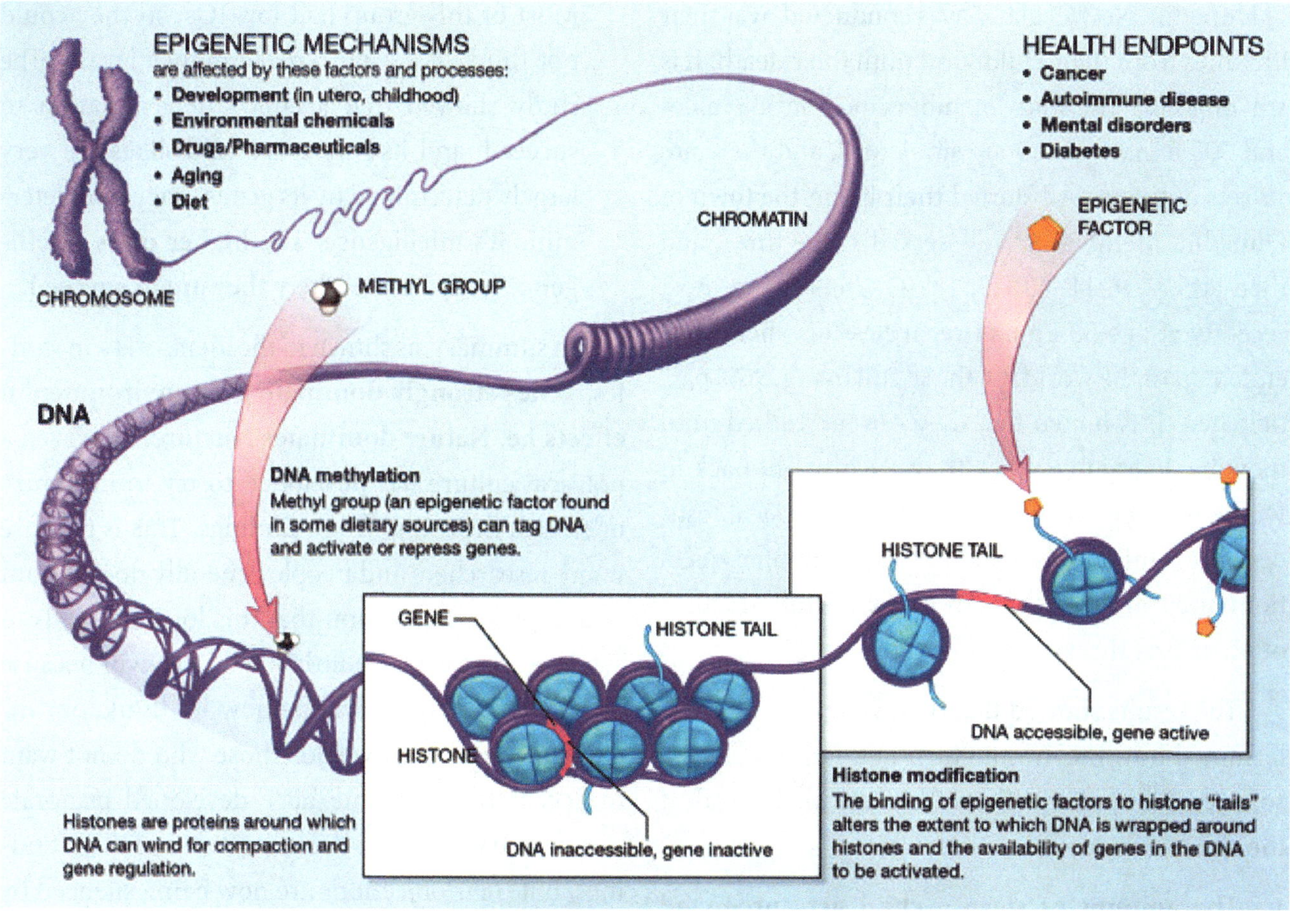

Small methyl molecules can become attached to some sections of our DNA. The attached methyl molecules are known as epigenetics. This causes the gene to which it attaches itself to to act differently. This occurs by random chance. Methyl molecules can attach to one twin's genes and not the other causing an uncommon difference in one facet of their nature.

Differences in identical twins can occur if one twin has a **thinner umbilical cord than the other**. This deprives that twin of getting a plentiful supply of blood and therefore enough oxygen and nutrients from its mother. This can change that twin's health, weight, contentment level after birth, and its sexuality. A poor blood supply, or its position in the womb, can cause one twin in 200 to have epilepsy, cerebral palsy, learning difficulties, a physical disability, or a somewhat different personality to its twin.

Other very rare environmental effects that can affect a person's genes is their level of **exposure to cosmic rays** from space by living at high altitudes or frequently flying in commercial jets in the stratosphere. **Living or working near high levels of radioactivity** emitted from certain rocks, or **having exposure to high levels of toxins such as pesticides, PCBs, and dioxins** can also affect a person's health due to their DNA being damaged. This affects some aspects of their health, and in some cases, their personality. If one twin is affected by **certain viruses**, this can cause the affected twin to be different to their twin in some particular way, such as being prone to having a particular cancer. Fortunately, all these possibilities are very rare.

The Dunedin Study of 1037 children born between 1 April 1972 and 31 March 1973 in the town of Dunedin, New Zealand, was conducted over their lifetimes from their childhood until their death. It is still ongoing. The study members include 535 males and 503 females. 1013 are singletons, and there are 12 sets of twins. All started their life in the town of Dunedin. Members were assessed at age three, and then at 5, 7, 9, 11, 13, 15, 18, 21, 26, 32, and most recently at age 38 and 44 regardless of where they ended up in the world. Of those still living, 96% participated. It is hoped that they can be studied until they die. Every five years, they are brought back to Dunedin to be interviewed and have physical tests, dental examinations, blood tests, computer questionnaires, and surveys. An ongoing study focuses on their first three-year-old child.

The results showed that their genes strongly determined how they would turn out. It also revealed some environmental effects that were likely to affect the children, such as:

1. **The amount of sleep a child gets predicted how obese it will become as an adult**. Sleep stimulates hormones that determine how hungry a person becomes. Insufficient sleep leads to a high probability of having cognitive problems and anxiety in one's 20s.
2. **The more TV a child watches**, the more it will be likely to smoke as a teenager and as an adult. This increases its probability of having lung problems when it is older. These factors affect one's employment prospects.
3. **When parents apply reasonable discipline in a child's early years, this leads to self-discipline in the child's teens and adulthood**. Those that were not disciplined showed a natural propensity for antisocial criminal activity as a child, and they typically became criminals in adulthood. Most of this group had low IQs, so they could not think of a way to make money legally. The study showed that a child's determination to succeed, and its interests and dislikes are very largely determined by its genes. They also determine it's intelligence. **The higher one's intelligence is, the more likely they are to succeed.**

In summary, as shown in the identical twin studies, **genes strongly dominate over environmental effects i.e. Nature dominates Nurture.** However, a political culture has developed to try to minimize or negate these important findings. This is because many researchers and people generally do not want to accept the conclusion that this logically leads to – that is, that we have little if any, free will because our genes largely determine how we think, our intelligence, and what we do. Those who do not want to accept this outcome have developed elaborate means to try to reduce the importance of these findings, but their objections are now being silenced by an avalanche of undeniable evidence that supports how dominant genes are.

THE DELAYED GRATIFICATION TEST

The Delayed Gratification Test is a reliable test that is conducted when children are between 4 and 6 years of age. It predicts whether a child will delay immediate gratification in order to gain something more substantial later on. If a child can resist immediate gratification, it is typically more successful in school, in business, and in life generally.

The test is simple. An adult puts the child in a quiet room without distractions at a desk on its own. A marshmallow is placed on the desk. The child is told that if it does not eat the marshmallow while the adult is away for 10 minutes, it will get two marshmallows when the adult returns. The children that succeed in overcoming the temptation to eat the marshmallow did so by distracting themselves with some activity, and by not looking at it. Nevertheless, the temptation is very hard for them to ignore it. Those that don't succeed eat the marshmallow soon after the adult leaves the room. In later life, these children are much more likely to be shallow thinkers, unsuccessful in life, and to enter into antisocial activities. Those that delayed their gratification had two marshmallows. In their adult life, they typically turn out to be high achievers, intelligent, well educated, and successful in their chosen field.

HOW OUR GENES CONTROL US HAS PROFOUND IMPLICATIONS LEGALLY

There is much evidence that certain people have a genetic, predetermined brain circuitry when they are born. This circuity can produce extraordinary levels of creativity, intelligence, mathematical ability, and the ability to make money. Others have genes that limit their intelligence significantly, thereby producing a strong tendency towards underachievement, little interest in gaining knowledge, and a reduced ability to enjoy life. Some have a desire for criminal activity and violence programmed by their genes.

Research has shown that it is very difficult to change a person's genetic tendencies. Where people have genes for low IQ and antisocial tendencies, intensive education to thaw these tendencies has been tried, but it has only had a very limited effect, if at all. Research that supports the findings of genetic predispositions towards antisocial behavior, anger, pedophilia, megalomania, or being psychopathic has been deliberately muted due to the significant legal and social ramifications this introduces.

The late author and fabled essayist **Tom Wolfe** wrote a very prescient article titled *"Sorry, But Your Soul Just Died."* about this in 1996 when brain imaging technology was in its infancy. (This article is available online with his book *"Hooking Up."*) In it, he forecasted the potential future social ramifications that brain-imaging can indicate for an individual's behavior being genetically predetermined. That future has now arrived. Low intelligence and a deprived childhood are taken into account when sentencing them if they break the law. In extreme cases, very low intelligence might lead to a judge finding that the offender was *incapable* of the intention or knowledge of wrongdoing *'mens rea'* and therefore, cannot be found guilty as a matter of law. New evidence about the preprogramming of the brain by genes now takes this much further.

Suppose all genetic effects were to be taken into account by the law. In that case, defense lawyers could justify their clients' criminal actions by saying that their genes caused them to perform their criminal acts and that they had no ability to control their genetic urge, therefore, they cannot be found guilty. On this basis, the law would break down because anyone could do anything and not be found guilty. However, it could turn out worse for those with genes for criminality because they would have to be held in special detention centers *in case* they committed a crime in the future. The movie ***'Minority Report'*** starring Tom Cruise explores the outcome of such a situation where an individual's propensity for criminal activity can be predicted and addressed before it is acted upon. This may become a reality in the very near future. However, this would be very expensive and resisted by many voters and governments. But AI will soon take control, and we will have no say.

The movie 'Minority Report' directed by Steven Spielberg, predicts a future where there is mass surveillance of everyone to predict their actions. We are moving closer to this situation every year.

The argument for genetic predetermination is grounded in reality. This is due to overwhelming scientific evidence about the power of genes. For a small proportion of unhardened criminals, the threat of serious punishment can be a deterrent, and occasionally, the love of a good partner can mitigate against their underlying tendency for criminal behavior. However, the lower a criminal's IQ and education are, the less likely it is that they will have any ability to control their antisocial genes.

The Dunedin Study showed that children who have antisocial attitudes become worse as they grow up until around 22. After this, their antisocial actions can decline with maturity, often dissipating by the time they are 30 – but only if they are not influenced while in a jail by long-term criminals or when they get out by those who they associate with who are repeat offenders.

THE LEVEL TO WHICH OUR GENES CONTROL US MEANS WE DO NOT HAVE FREE WILL

Given the scientific evidence for how our genes control us, we do not appear to have much, if any, free will. It is hard to escape the conclusion that, like all other life forms, we are automatons controlled by our genes. Insects, rodents, cats, dogs, birds, fish etc. all act in different ways due to their genes. Are we the same, or does our greater intelligence give us some small degree of free will? If so, in what area? The existing mounting evidence says no.

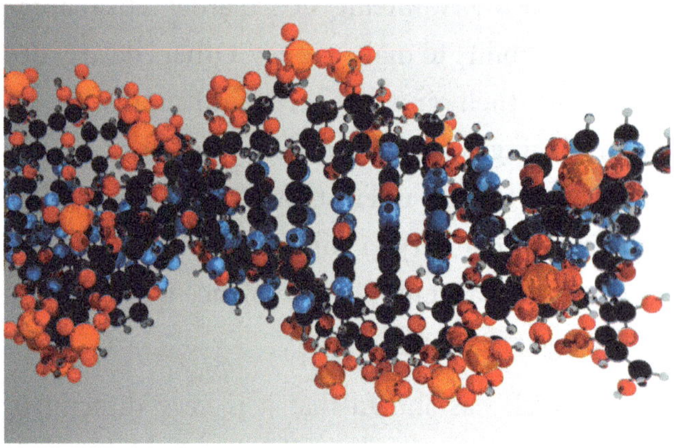

The structure of the DNA molecule

ARE ALL LIFE FORMS PROGRAMMED BY THEIR GENES?

If almost all aspects of all life forms on Earth are determined by their genes, then this means that no life form has self-determination. But we all want to believe otherwise. It is hard to see how low-level life forms like insects, crabs, or worms, etc. could act in any other way than the way they are programmed to act. They cannot think outside their programming to find solutions to problems that we would find easy to solve. But are we any different? The scientific evidence is building towards us being automatons that have a more complex program than lower life forms.

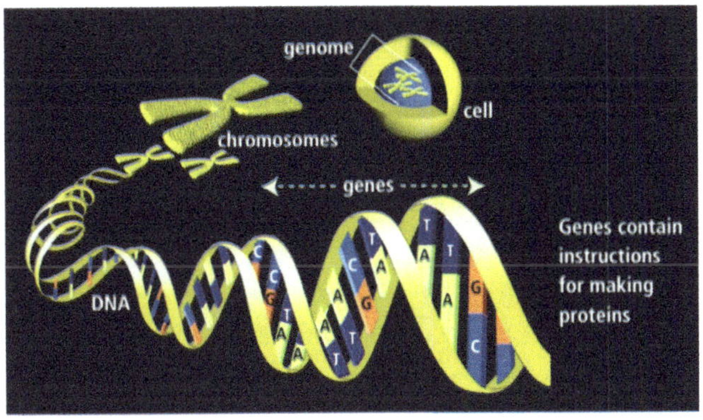

It's incredible to think that the molecular 'letters' joining the DNA double helix strands contain the information to construct every part of our body and the wiring of our brain, which is unique for each of us, except for identical twins. But they are not 100% identical, but almost.

We are more intelligent than lower life forms, but like them, we cannot change our genes, so we are what we are. However, this may soon change with **genetic engineering**. But when we change our genes, will this give us free will? No, because we will then be controlled by a different set of genes or programs. Will intelligent computers be any different? That's unlikely because they too will be controlled by their programs, which will act like our genes do. Computer intelligence will have the ability to change its programs, but like us, this will not necessarily mean they will have free will because their new programs will still control how they think. What could be the reason for the universe being designed like this?

Biologists studying **ants have found that all ant species have between 102 and 106 program rules**, with the vast majority having 104. These rules control every aspect of their individual lives and that of their nest 'society.' They control such things as which ants work and when, what triggers the soldier ants to fight invaders, how they farm fungi and search for food, collect the dead, build nests, and go to higher ground when it's wet.

Science is also discovering many other programs for life, such as the **16 rules for the flocking of birds and the schooling of fish**. Pressure sensor cells under the feathers of flocking birds constantly measure air pressure, and similarly, those under the scales of fish measure water pressure. These pressure sensors automatically adjust distances between birds in flocks and fish in schools so they do not bump into one another when they turn sharply. Like all life forms, including humans, birds and fish rely almost entirely on their brain's autopilot. When computer programmers use flocking rules for generating video animation of birds and fish, their animated movements perfectly replicate their real-world counterparts. Programs for the movement of many other animals have also been decoded. These programs are used in videos to replicate perfectly the movements of all sorts of animals for walking, running, sliding, flying, eating, and fighting etc. They are even used for extinct animals, and of course, humans.

It will only be a matter of time before we discover how many program rules we have. Presumably, the more intelligent a person is, the more rules they may have. Humans may possess no more than a few thousand rules that control every aspect of what we do i.e. how we move, act, talk, think, react logically or emotionally, and so on. If this turns out to be the case, **all life forms in the universe, including mega-intelligences, will act and think based on the number of rules they have in their operating system**. Until recently, I would never have thought this could be possible, but we are now reaching a stage where the evidence is too strong to dismiss this conclusion. I expect that we will soon have no option but to accept that we are 100% programmed by our genes. For most people, this will be very hard to accept - just as it was too hard for most people in **Charles Darwin's** day to accept that our recent ancestors were apes, and our most distant ones were microbes. Today, most people take this for granted because they have seen overwhelming evidence for it. I expect this will soon happen with regard to how much our genes control us.

Chapter 27

WHY DO WE HAVE TWO BRAINS?

Like all vertebrate life forms, our brains consist of a double or split brain. The right half controls the left side of the body, and the left half controls the right side. Why it is this way is unknown? Curiously, both halves have somewhat different cognitive abilities. Why this is so is also unknown. Many psychological tests clearly show that a person's personality, interests, and expertise is determined by which hemisphere dominates their brain.

Studies have shown that people who have a dominant right brain are more likely to be emotive and creative in various aspects of the arts. Those who have a more dominant left brain are organized and more likely to become engineers, mathematicians, accountants, scientists, tacticians, or army generals.

THE LEFT HEMISPHERE

Research done over decades has shown that **the left hemisphere is more logical, mathematical, and scientific than the right. It is also more organized and practical than the right hemisphere. This hemisphere is dominant in language functions – those involved with speaking, reading, writing, and spelling, and it remembers names, dates, and general information.**

THE RIGHT HEMISPHERE

The right hemisphere tends to handle most aspects of spatial tasks, recognizing faces and objects, and processing data from the senses. It is more dominant in emotive responses and in recognizing emotions in other people. It computes the **emotional aspects of language** - the tone used, the type of delivery, signs of stress, the rhythm of communication, and body language. It handles **recognizing patterns** in faces, objects, and images, and it identifies new ones. It tends to process art and music more so than the left brain. **The right hemisphere has more ability to be imaginative, creative, and intuitive.**

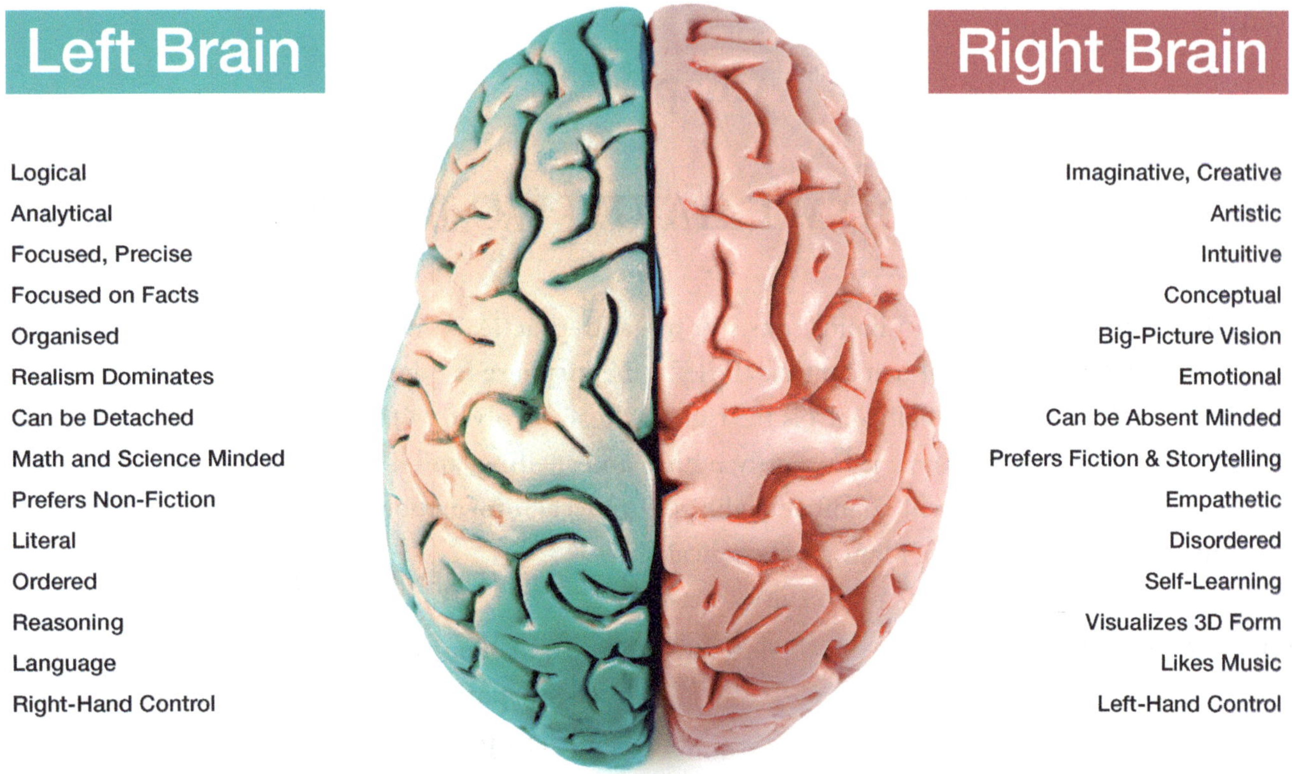

This diagram shows typical differences between the two halves of our brains. These results have come from research done over decades. It is now being found that differences between the left and right hemispheres may not be as distinct as once thought but they still dominate. MRI maps have shown that the two hemispheres work together more than originally thought.

It is common for the right brain to be more dominant in sexual desire, dreams, and beliefs in the supernatural and religion.

Why did our mammal brains evolve to have two brain hemispheres with different types of expertise?

Will other intelligences on other planets have evolved dual brains? Do dual brains have an advantage over a single larger brain? Could dual brains have evolved as a random freak occurrence? Do they act as a checking system to help the brain be more balanced? Or, is it more efficient to have two somewhat specialized brains than to have one larger brain trying to do all tasks? Do they work independently to speed up thinking?

Are dual brains an outcome of the bilateral nature of vertebrates whereby they have evolved to have two eyes, two ears, two nostrils, two arms, two legs, two lungs, and two kidneys? But vertebrates have only one heart, one liver, one appendix, one mouth, one digestive system, and one reproductive organ. This seems to negate the bilateral explanation.

TWO MINDS IN ONE BODY

The two hemispheres are connected by a neural 'cable' called the **corpus callosum**. This allows the exchange of vast amounts of information between each side of the brain. If the corpus callosum is cut, which it has been in patients with severe epilepsy to stop them from having seizures, then each hemisphere will act as a separate brain with its own perception of the world. Each half will have its own biases and impulses to act. In some cases, this has caused the hemispheres to strongly disagree. In the worst cases, one half can want to physically fight the other half or do different things at the same time. One half might want to wear one garment and the other a different one, so they fight over this. While one hemisphere is trying to button up a shirt with one hand, the other half is unbuttoning it with the other hand! Fortunately, these situations are rare. It's as if there are two different people in one body. For people who have had their corpus callosum severed, when using only one eye and then the other, the right side of their brain will recognize an object or a drawing but be unable to name it, whereas the left side can name it, but it doesn't know what it is!

WORKING TOGETHER

One side does not have a monopoly for preferences for most people because the two hemispheres typically work together with both hemispheres sharing tasks and making decisions fairly evenly. But there is a significant percentage of people who have a clearly dominant side. One's genes determine the differences between the two halves. The hemisphere that dominates determines one's biases, expertise, and interests. **Only 5% of people have a brain where the left and right sides are equally balanced.**

THOSE WITH BALANCED HEMISPHERES

Only 5% of people have equally balanced hemispheres. Those that do, this can sometimes cause them difficulty in making quick decisions because they can often see both sides of an issue. These people think more deeply, so it can take them time to gather enough information in order to weigh up a decision from multiple perspectives. They are able to consider both small and large issues as well as local and global ones in a fairly balanced way - as long as they have enough accurate information and the time to do so. They know that small details can be as important as big issues because both can influence an outcome. People with balanced hemispheres can consider evidence that goes against the general belief systems of the day, and even against their own beliefs. They are flexible thinkers who can effortlessly switch from one hemisphere to the other. They typically like science, logic, and order as much as they value having artistic freedom of expression and creativity. For some reason, around $2/3^{rds}$ of people with balanced brains are women.

People with balanced brains typically have higher levels of curiosity, general knowledge, intellect, and wisdom, so this helps them make more thoughtful and less biased decisions. They tend to take into account the flow-on implications for each type of possible decision. But they are not always right because there can be many aspects to an issue they are not aware of. When they have to make a quick decision, it can be the wrong one because they have not had enough time to gather enough information to think through the issues. If those with balanced brains are well informed and intelligent, they are much more likely to make a wise decision than those with an unbalanced brain.

People with unbalanced brain halves are more likely to make kneejerk decisions based on shallow thinking, emotional responses, or lack of interest. Because intelligent, informed people with balanced left and right brains are few in number, their voices of reason are typically drowned out by those who want a quick, simple, and often self-serving solutions. Such decisions are nearly always ill-considered.

THE DELIBERATE DUMBING-DOWN OF EDUCATION

Unfortunately, only a small percentage of the 5% of people with balanced brains are scientifically literate. Illiteracy in science has come about because the Elite has deliberately devalued the value of science in education. This has been done so that a majority of society is so poorly educated that they are unable to question what the Elite's media tell them. Science has been deliberately taught ineffectively in most schools over recent decades in order to dumb down students to make them easy to control. Of those with balanced brains that are scientifically literate, they are usually only well-informed about a relatively small area of science – typically the one in which they work or the one in which they have an interest. But science articles in nearly all science magazines today are often highly biased and often political to support the Elite's objectives. In view of this, only a very small percentage of people with balanced brains can make well-informed, balanced decisions on major, complex issues that require *broad* scientific knowledge and accurate data. This is one of the reasons why intelligent computers will increasingly take over humanity's decision-making roles. They will know all that has been learned about every science, about every aspect of history, the arts, economics, psychology, and politics – literally everything! So they will be extremely balanced and objective in their reasoning.

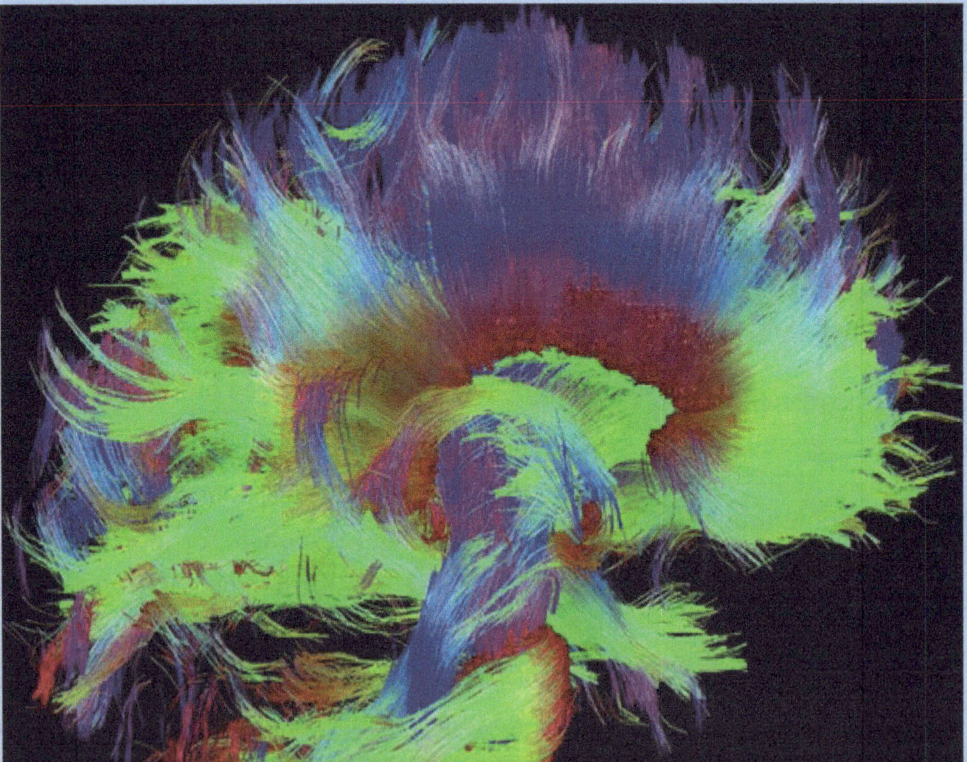

A map of the main neural pathways of the human brain Credit: Laboratory of Neuro Imaging and Martinos Center for Biomedical Imaging, Consortium of the Human Connectome Project

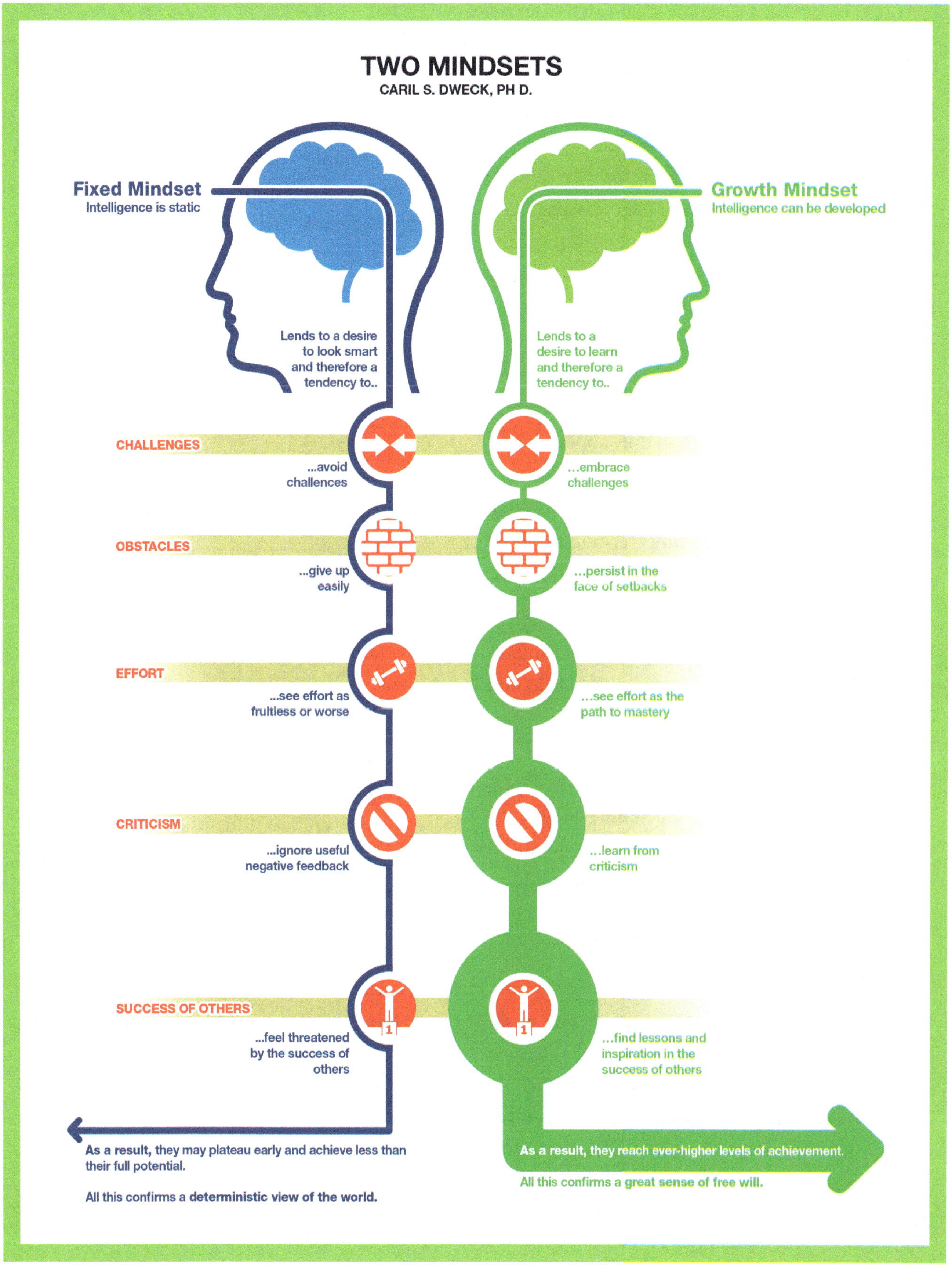
TWO MINDSETS
CARIL S. DWECK, PH D.
Fixed Mindset
Intelligence is static
Growth Mindset
Intelligence can be developed
Lends to a desire to look smart and therefore a tendency to..
Lends to a desire to learn and therefore a tendency to..
CHALLENGES
...avoid challences
...embrace challenges
OBSTACLES
...give up easily
...persist in the face of setbacks
EFFORT
...see effort as fruitless or worse
...see effort as the path to mastery
CRITICISM
...ignore useful negative feedback
...learn from criticism
SUCCESS OF OTHERS
...feel threatened by the success of others
...find lessons and inspiration in the success of others
As a result, they may plateau early and achieve less than their full potential.
All this confirms a deterministic view of the world.
As a result, they reach ever-higher levels of achievement.
All this confirms a great sense of free will.

DECISION MAKING

When people lack knowledge about a particular issue, most will make impulsive decisions, which are typically based on emotion, self-interest, or long-held simplistic beliefs. Their problem-solving ability usually focuses on one simple aspect of a problem, whereas most problems such as those involving people or the environment, require many aspects to be considered. There are always many things that impact upon finding a good solution because science has discovered that all things are interconnected in some way or another. Most issues are far more complex than they appear at first glance, and all solutions have significant flow-on butterfly effects, most of which are not suspected. And a good solution in the short term may not be a good solution for the long-term and vice versa.

Because so few people realize how *exponentially* fast technology is developing mega-intelligence that will soon change our entire world and our lives in just a couple of decades, so they continue to talk about how things will be in 50 or 100 years in the future. This naivety means they have not the slightest understanding of how quickly the world will soon change. We frequently see this on the news, in documentaries, and in many books.

THE IMPORTANCE OF SCIENTIFIC LITERACY

Almost anything we can think of in our modern world is underpinned by science, yet most of the world's population, even those who are considered to be well-educated, know very little science! A very large percentage of the public is scientifically illiterate, and for a good reason - they have no need to know much about science in most aspects of their everyday lives, especially as there are specialists in science and technology to look after the things they have no knowledge of. But because of this, **it is almost impossible for them to make informed voting decisions because nearly everything in our modern world is based on science.** Try to think of something that is not underpinned by science. It is almost impossible to do so. Even going bushwalking in a national park requires knowledge of all the sciences that have gone into building roads and cars to get there. Science also was instrumental in finding and extracting the energy to allow us to get to a national park. Even our clothes and shoes, as well as our sunglasses, cameras, matches, and so on required many sciences to develop them. You might say surfing does not require science, but there is a lot of science in making fiberglass and foam and the making of surfboards. Just to have water come out of a tap requires an enormous number of engineering solutions and knowledge of several sciences. So does preparing a hot meal, or the farming of crops and animals.

Because science underpins virtually everything in our civilization, it is very dangerous for humanity to have the vast bulk of the world's population knowing very little about science, especially when they have to make voting decisions. When people are unknowledgeable, politicians and the media are easily able to deceive their audiences who largely think they can trust what they say. But the truth is that we cannot trust them in most cases because they are far too self-serving, and biased.

Research shows that even most scientists are only well-informed about some aspects of their field of research. When I conducted research to test the general science knowledge of scientists who were working at the highest levels in well ranked universities, I found that outside of their specialty, most of them performed surprisingly poorly when questioned about the most basic aspects of other sciences! This shocked me, as I had assumed that most scientists were fairly knowledgeable about the basics of most sciences. The questions I gave them were those that most school children between Grade 1 to 9 would know, such as - does the Sun orbit the Earth, or does the Earth orbit the Sun? Is the Moon visible during the day and the night or only at night? What is a star? Why does a ship weighing tens of thousands of tons and full of cargo not go straight to the bottom? How does a heavy plane fly so effortlessly? How are mountains and volcanoes formed? What causes erosion? What causes wind and ocean waves? What are the two main gases in the atmosphere? What is fire? What is our body's largest organ? What element is life mostly made of? Do plants have sex? and so on. To my amazement, most high-ranking university scientists did not know the answers to many of the most basic questions (See Appendix 2 to test how many basic science questions you know the correct answers to.) Remarkably, ordinary working people who had no education in science, other than what they learned in primary school and from life experience, often fared better! A majority of the scientists also performed poorly on basic general knowledge related to politics, economics, and how society works. It seemed that by being closeted in a university monoculture, they had lost touch with the real world. Their staff often made comments like *"They live in ivory towers disconnected from reality"*.

Most scientists are usually engrossed in their work, so they either have little spare time to learn about other physical or social sciences, or they have no interest in them. Like most people, what free time they have is typically taken up with responsibilities to family, colleagues, and trying to find some time to enjoy life. Very few scientists are *broadly* knowledgeable.

THE MANIPULATION OF SCIENCE

A surprising number of assumptions in various sciences are not based on proven 'facts', but on incorrect interpretations of observations, political bias, and nowadays on wishful thinking! Assumptions are too often put forward by some unscrupulous 'scientists' as fact for their financial benefit or to support their ideological convictions. Due to the need to 'publish or perish', too many scientists rush to publish their preliminary results before they can be verified.

Presumptions are too often presented as facts before they are proven. Scientific evidence is required to prove a theory before it is believed to be true by the public, but all too often, this is not the case. Science journals need new exciting discoveries to sell their magazines, so they rush to publish unproven claims. Too many so-called 'new scientific breakthroughs' particularly on health, on television are often paid advertisements! They do not constitute scientific proof, as many of the public are led to believe. Nor does a scientist on a populist television show who is promoting a product believe that it is real.

It takes time for new scientific evidence to be proven and more time before the wider scientific community accepts it. When new evidence emerges that shows that established beliefs are false, it can often take a couple of decades for these new findings to filter through into the public's general awareness. Once new evidence becomes accepted by the scientific community and the media, people who would not accept the evidence when it was first presented quickly accept the new paradigm so they can remain 'inside the herd'.

Voters and politicians need to make many important decisions, so they need scientific findings to be evaluated fairly and correctly. But only a very small percentage of people know how to access factual evidence. In far too many cases, voters and politicians display no interest in researching facts to make informed decisions.

The workings of society and technology are both becoming more complex, yet the majority of voters and politicians are becoming less interested in evaluating even the most basic facts related to their decisions. More and more political decisions are based on self-interest and the majority of voter's shallow, emotional needs. **This is a recipe for the collapse of our democratic system of government and society.** For democracy to work, voters must be well educated and prepared to take the time to research important issues that they have to vote on. But this will never be the case, so maybe a solution could be that only well-informed people should be allowed to vote after completing a bipartisan questionnaire to evaluate their knowledge of each issue. The level of knowledge a person has on main issues could determine the value of their vote. An AI that cannot be biased by humans would have to calculate the result of the election. But why would humans be making decisions when a highly intelligent, unbiased AI could be running government?

To make well-informed decisions is very difficult because one needs to:

1. **have access to good-quality information across many fields**. Today, it is much easier to do this via the Internet than it was in the past by going to a library or subscribing to publications that were delivered physically once a month. But because there is so much false information on the Internet, one has to learn how to sift out reliable sources of information, and one has to crosscheck most things many times to get a balanced view. Nowadays, the major social media companies censor information that goes against their strongly left-wing ideology. This is making it increasingly more difficult to access balanced, unbiased information. Hopefully, this censorship will soon be banned, and free speech will return.
2. **have the time it takes to do research.** Research is necessary to make informed decisions. But to find the considerable amount of time required for this is very difficult for most people, as they have to look after family commitments, go to work, socialize, exercise, and allow time for entertainment and relaxation. And people can only do serious research if they have a quiet space in which to concentrate and not have frequent interruptions. They also need an above average intelligence to do worthwhile research.
3. **have the ability to discriminate.** For people who have little scientific literacy across a broad spectrum of the sciences, and who do not have worldly experience, it is virtually impossible for them to discriminate between concepts that are supported by solid scientific evidence and those that are not. If they have strong confirmation bias, they will not be able to discriminate.

4. **have the ability to interpret information correctly.** This requires an adequate life-long education and the ability to learn what is real by trial and error. There is a skill to learning what to question and what to believe. Again, because a majority of people have confirmation bias in many areas, they can never learn facts that go against their beliefs. This stops them from changing their mind about certain issues even when the evidence demands it.
5. **have an interest in wanting to gain knowledge.** Very few people have a strong interest in learning, and for the few that do, it is usually limited to a small number of subjects in which they are interested. To gain knowledge across a broad spectrum of subjects takes a lifetime, so it is very rare. To gather broad knowledge, one must spend a lot of time reading broadly. It helps to build a network of experts across many fields so you can pluck their minds for information.
6. **be well-informed.** To achieve this, one needs to have a strong sense of curiosity to justify the time that one has to invest to become well-informed.
7. **have an open mind.** To be open-minded is very important, so one can consider new discoveries, challenging information, others' points of view, and alternative possibilities to those that one has previously been exposed to, or believed to be true. Fixed belief systems lead to poor decision making.
8. **have the ability to acknowledge when one is wrong when strong evidence shows this is the case.** This allows one to replace false beliefs with those that are more realistic. Most people respect people who are prepared to admit they are wrong when new information comes to hand. It's impossible for anyone to be all-knowing.

When people have very few facts, if any, on which to base a decision, they typically retreat to a simplistic belief that they feel emotionally comfortable with. Many studies have shown that this leads to one-dimensional decision-making that is nearly always wrong.

To make informed political decisions, one needs to be aware of the financial gains that encourage many politicians, bankers, company directors, clergy, and some scientists to push a particular agenda. As a species, our collective knowledge can overcome simplistic decision-making to some extent, but this occurs at a very slow pace.

Many poor decisions made by those in government occur because too many politicians and voters are scientifically illiterate and not overly intelligent or knowledgeable. Far too many politicians are prepared to make unwise decisions to maximize votes for themselves. All too often, they only look at the short-term benefits and ignore, or are not aware of, the long-term negative outcomes. Influential politicians will often support laws to satisfy large companies who secretly fund their election campaigns on a quid pro quo basis. By contrast, mega-intelligence is unlikely to be influenced by bribery. Mega-intelligent computers will have near-total power and control, so they won't be able to be bribed. As well, a mega-intelligence will not have to be elected. They will be fully aware of all available knowledge, and they will be conscious of all new discoveries the moment they are made. Because of this, major decision-making will be taken away from humans.

WHAT IS REAL?

Chapter 28

IS MATTER PHYSICAL, OR IS IT IN OUR CONSCIOUSNESS?

We think of matter as consisting of physical par ticles, which are the basis of all things in the universe, but the reality is that the properties of matter are complicated and not well understood by anyone.

DISCOVERING THE STRUCTURE OF ATOMS THAT FORM MATTER

In 1885 Sir William Crooks discovered electrons when he observed a stream of particles being emitted from an evacuated glass tube known as a cathode ray tube. At one end, it had a high voltage generator from which particles were being emitted. They traveled to a positively charged plate at the other end. They became known as **electrons** that had a very small mass and a negative charge. In 1897 **J.J. Thomson** showed that they must be present in all atoms. This meant that the atom was not an indivisible solid ball as most physicists thought, because it was composed of constituents. There had to be a positively charged particle to neutralize the electron's negative charge. In 1919, **Ernest Rutherford** showed that a hydrogen atom with its electron removed appeared identical to this positive particle. It became known as a **proton**. When a protégé of Rutherford, **James Chadwick** was exploring radioactive decay, he noticed that something must be adding to the mass of protons in the nucleus. Rutherford suggested that there might be a particle with the same mass as the proton but with no charge. He discovered the **neutron**, but it had a mass about 0.1% higher than the proton. He received the Nobel prize for this discovery. **Werner Heisenberg** showed that the neutron was not an electron-proton pair of sub-atomic particles, but a single particle that had no charge. When high-speed neutrons were later used as bullets to shatter atoms, it was discovered that neutron bombardment of a uranium atom would release an incredible amount of energy. This was predicted by Einstein in is equation **$E = mc^2$**. This led to creating the atomic bomb and atomic energy.

Rutherford's experiments showed that atoms were mostly empty space with an extremely small, dense nucleus that had a positive charge that it was 10,000 times smaller than the atom itself! It was thought that to occupy the full extent of an atom that electrons must 'orbit' the nucleus like the planets do around the Sun. In 1912, **Neils Bohr** worked with Rutherford and realized that this model would be unstable because electrons could be knocked out of position when atoms collided. And when atoms give off energy, electrons would spiral down into the nucleus, thereby collapsing the atom. Bohr used **Max Planck's Quantum Theory** to suggest that an electron's orbit was determined by the ratio of an electron's energy and its frequency, and importantly, that electrons could somehow 'jump' *instantly* between specific orbitals, which depended on their energy level. Electrons were found to never exist between one orbit and another. This became the source of the term a 'quantum leap'. There were four orbital electron shells, each with a different number of electrons. The first shell can hold up to two electrons, the second up to eight (2+6), the third up to 18 (2+6+10), and so on. This led to the **Periodic Table of Elements** in which all elements have specific properties. Atoms that have the same number of electrons in their outer shell are in the same column on the table because they have similar properties. It is now thought that electrons must exist in some cloud-like state. The number of electrons determines the shape of these clouds. An electron can even pass *inside* an atom's nucleus! So, we still do not have a full picture of the structure of atoms. Electron orbits that go around more than one nucleus are what hold atoms together to form molecules.

In 1925, **Heisenberg** introduced his **Theory of Matrix Mechanics** that described how atoms absorb energy. He suggested that atoms do not absorb energy in the way that most physicists thought they did, but rather, they do so as '**mathematical matrices**' that evolve over time. He said, "*Atoms and elementary particles are not real: they form a world of possibilities rather than one of solid things.*" Back then, this radical concept was shunned because it did not adhere to the traditional physics of the time that considered matter to be made of solid particles that had clearly defined properties. In 1929, Heisenberg developed his '**uncertainty principle**'. This set limits on defining the location and energy levels of particles. This was interpreted as ***particles not being physical entities*** because, theoretically, they could be found *anywhere* within a certain volume of space! They were thought to be **energy fields that were cloud-like.** Heisenberg amassed enough compelling evidence for his radical idea that most physicists accepted it as being very probable. For most physicists, it became the reality of matter. But if that were true, nothing around us would be solid or real.

Einstein did not agree with Heisenberg's Uncertainty Principle. He was determined to prove this concept false. He said in jest, "*God does not play dice with the universe*". (He used the word 'god' liberally in place of physics, not meaning that there really was a god.) He spent the rest of his life unable to mathematically prove Heisenberg wrong.

Neils Bohr, a giant in quantum physics, became famous for developing his own model for the workings of the atom. He thought that Heisenberg's ideas were valid. Like Heisenberg, Bohr said, "*Everything we call real is made of things that cannot be regarded as real.*" This would have implications 80 years later when the hypothesis that our universe might be a simulation was being seriously considered.

It is much easier to understand the **simulated universe concept** today because virtual reality worlds in computer games have been around since the mid-1980s. But in the early part of the 20th C, modern digital computers had not even been invented, so the possibility of ***virtual* realities** existing had not been considered by the vast majority of scientists. Today, the concept of a simulated universe is starting to become mainstream science - at least to those scientists and laypeople that have an interest in this aspect of physics.

Radical ideas in science can typically take around 20 years to become widely considered, even when there is substantial evidence to support them. This time frame is now shortening due to the exponential increase in knowledge that is occurring. Computers are now able to help in proving theories true or false much faster than humans can.

To be confronted with a non-physical reality feels threatening or ridiculous to most people – at least until they understand how it might be possible. **Some physicists think Heisenberg was right to think that it is our consciousness that creates our *perception* of matter being solid. Bohr and Heisenberg believed that matter does not exist in reality because it only occurs in our consciousness. But our consciousness requires our brain, and it is made of physical matter!** (See Chapter 21.)

MATERIALISM OR IDEALISM - WHICH IS RIGHT?

Democritus, who lived back in the 3rd and 4th centuries BC **was super-intelligent and an extremely visionary 'scientist' who conceived the concept of everything being made of atoms.** To him, atoms would explain all the different forms types matter in the world, such as solids like rocks, metals, ice, cloths, skin and liquids like alcohol, water, and blood, as well as gases like air and fire. Incredibly, it took 2,200 years to prove that Democritus' atoms existed!

Democritus and Einstein were giants in physics. They had a materialist view of nature. They believed that all things in the universe existed in the form of matter or energy. Whereas Plato, Heisenberg, and Bohr had an idealistic view. They believed that all things only exist in our consciousness, which they said is ethereal having no physical reality. So, which concept is right? If you do not sure, see if you can decide after reading the rest of this chapter.

In 1801 **Thomas Young** set up the famous **Double Slit experiment. It showed that light has both the characteristics of a wave and a particle.** Amazingly, this was more than a century before the **wave-particle duality of light and the physics of quantum mechanics was developed**. Young's experiment revealed that **light can manifest itself as a particle *as well as* a wave of energy!**

In 1927, **Davisson and Germer showed that electrons also had dual characteristics.** Recently it has been shown that even atoms and molecules have these characteristics! (It was not possible until recently to detect a wave function on objects larg er than particles because they have extremely short wavelengths that can't be measured.)

Einstein's equations showed that energy and particles of matter are interchangeable and that both can take the form of a wave.

WHY ARE THE RESULTS OF THE FAMOUS DOUBLE SLIT EXPERIMENT SO COUNTER-INTUITIVE AND SO IMPORTANT?

In 1801, the famous **double slit experiment** performed by **Thomas Young** validated the wave theory of light. It showed that when a beam of light consisting of **photons** of light of one frequency (i.e. one color), passed through a narrow single slit, the photons arrived as particles on a detector screen to form a single blurry bar. But when the photons were passed through *two* slits placed side by side, they did not form two bars, but several bars! (See images following.) The line at the center was the densest, while those on each side became less dense the further they were away from the center. This pattern is known as an **interference pattern. It is characteristic of two waves overlapping one another**. Where the crests of the waves meet, they enhance one another, and where a trough meets a crests, they cancel one another out, leaving a void. The photons still impact the screen as individual particles, but they produce an interference pattern suggesting that **light acts like a wave, as well as a particle!** Other particles like electrons and even atoms have this same dual characteristic. Why this is so is a mystery. I expect that particles are neither a particle nor a wave, but something else that we have not yet discovered.

Even more difficult to understand was that when photons pass through a double slit and are *observed* and measured by a detector to find out which slit they went through, they act like particles! For some undiscovered reason, the wave pattern disappears! Why does the act of observing them make such a difference? If the experiment is left running and the photons are *not measured*, an interference pattern reappears!! How could our awareness, or consciousness of it, determine whether the photons appear as particles or waves? It has been said that somehow or consciousness makes this happen. Physicist **Brian Green** has a good video about this experiment on YouTube titled '***What's Beyond the Double Slit***?' He demonstrates this perplexing problem very well. You will also find other good videos about this experiment on the Internet when you type in 'Double Slit Experiment'.

These strange effects not only occur with photons but also with electrons and other particles - even with atoms and molecules containing up to 430 atoms (so far)! The fact that this duality dilemma occurs with these molecules makes this even more intriguing because these molecules are no longer restricted to the sub-atomic universe: they are at, or over the boundary between the quantum universe and the macro universe. Experiments are about to be done with proteins and viruses in the hope that new, highly-sensitive detectors can measure their extremely short wavelengths. But **duality effects do not occur in our large macro universe because we have never been able to send anything larger through a double slit at near the speed of light.**

Since the Double Slit experiment was first conducted over 200 years ago, it has remained a major unexplained mystery in physics. In all that time, there has been no convincing explanation as to why particles could have the characteristics of both particles and waves, or why they change when they are observed. For this to happen, **we must be missing some fundamental aspect of the nature of matter**.

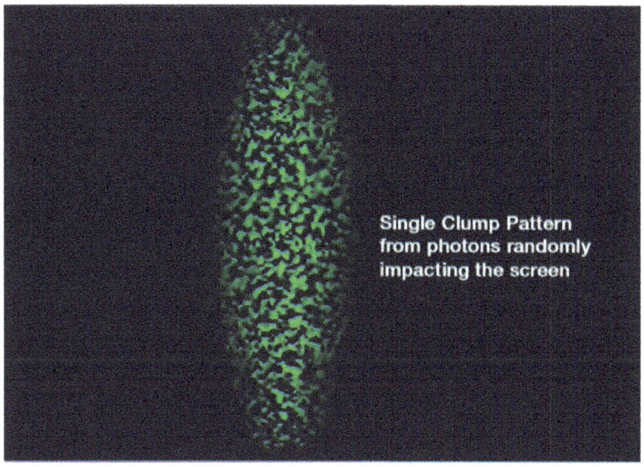

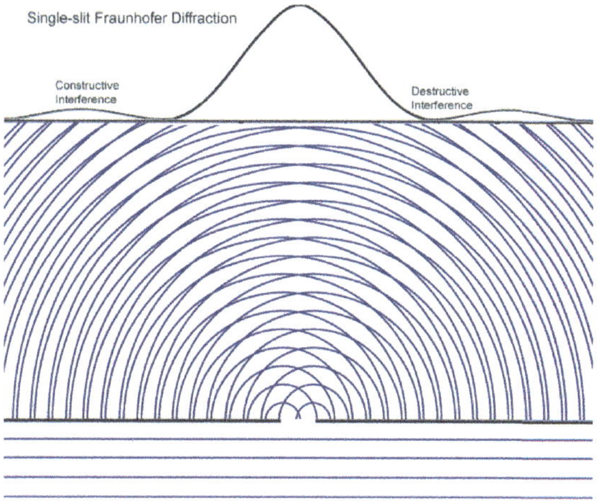

When light is passed through a single slit, it forms a single clump pattern on a photon detector screen behind the slit.

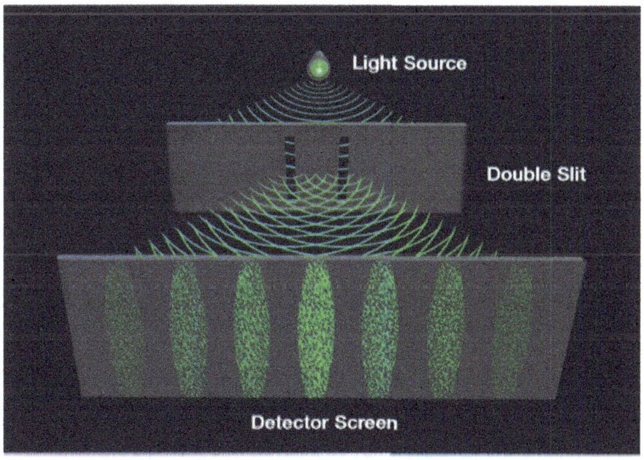

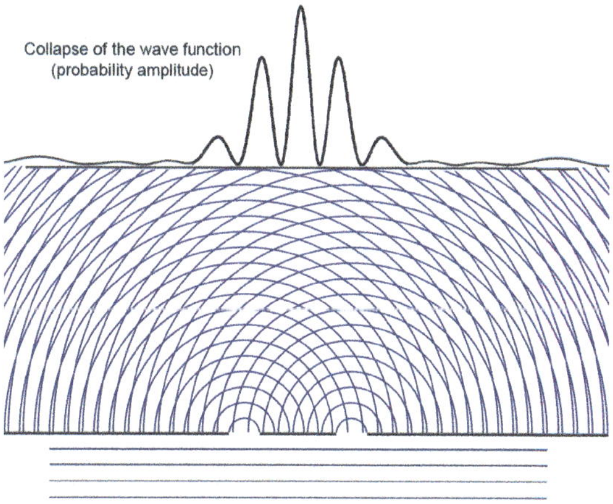

When light is passed through a double slit onto a detector screen, it forms an interference pattern indicative of overlapping waves – as long as the experiment is not being observed while it is occurring!

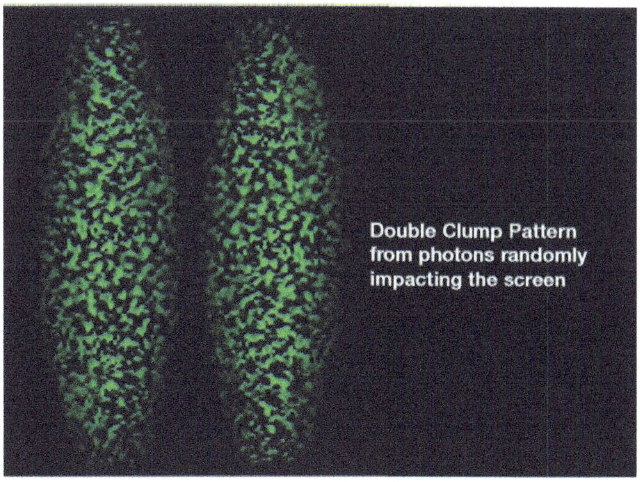

When a researcher *uses a detector to <u>observe</u> each photon passing through a double slit,* a double clump pattern forms, which is indicative of two streams of particles and not an interference pattern, which would occur if there was no observer! How could the act of consciously observing the experiment change the pattern from an interference wave pattern to two clump particle patterns? This may be the most bizarre thing ever observed.

The following analogy may help you understand how this quandary could exist. Imagine that there are two-dimensional beings that are only aware of their flat 2D world, which consists of only areas and lines. They have no awareness of the third dimension of height. Without awareness of a third dimension, it is impossible for them to make sense of a 3D object passing through their flat 2D world. For example, a ball passing through their 2D plane would appear first as a dot on first contact. It would then become an expanding circle that increases its diameter to the balloon's maximum size after which the circle would then begin to shrink to become a dot again at last contact, after which it disappears. Two-dimensional beings would find this impossible to explain; however, for us living in three dimensions, the explanation is obvious.

If we discover that there are other dimensions curled up at the quantum level as String Theory suggests, then the inexplicable quantum entanglement observations *might* make sense.

Our current understanding of quantum physics may be as limited as that of our ancient predecessors who wondered why some stars wandered across the night sky, and the others didn't. They had no knowledge that these wandering stars were other worlds that we call planets. By gaining greater knowledge, astronomers today have a very different and far more complex view of the universe than people did in distant centuries. Today, quantum physicists are in a similar position when it comes to understanding the quantum universe. Astronomy is a science that is thousands of years old, whereas quantum physics has only been around for a little over a century. Astronomers have had a big head start on quantum physicists.

Particles may manifest themselves in the forms we observe because we cannot see them in more 'dimensions', for the sake of a better term. Hopefully, in the decades ahead when trans-humans and mega-computers are far more intelligent and knowledgeable than we are, this dilemma may be solved.

Because experiments demonstrate that **all types of particles display this dual particle/wave nature at the quantum level of nature**, most physicists are being forced to accept that particles are not physical objects. They have agreed for a long time that particles are not actually physical balls of solid matter but rather tiny force fields that give them their characteristics of mass, charge, and spin. Some physicists think that when particles are not being observed, they appear as ***waves of probability***. But why would the act of observation make any difference? Physicists think that probability waves are not tangible waves like those in water, but rather '*degrees of probability*' that a particle is in a certain place at a certain energy level.

When the probability peaks, this indicates the maximum likelihood of knowing where a particle *should* be and how much energy it *should* have when it's observed and measured. **Once we try to measure the energy of a particle or its position, the act of measuring it changes either its energy level or its position.** This is because we have to fire a particle, such as a photon or an electron, at the particle to determine its position, or to measure its light emissions in order to determine its energy level. But when the photon or electron hits the particle, this causes it to change its properties. Since both the exact position and the exact energy level can never be known at the same time, physicists have to rely on probabilities.

Two French mathematicians came up with **Probability Theory**. Bohr and Heisenberg used this to determine the dual attributes of particles. They suggested that **when a particle hits a detector, its wave function 'collapses', making it appear as a single particle.** Why and how? There is no positive proof that a wave 'collapses' to look like a particle. However, Probability Wave Theory has been able to explain many observed aspects of quantum mechanics. But it does not explain the bizarre outcome of the **Delayed Choice Quantum Eraser Experiment** below.

QUANTUM ENTANGLEMENT INDICATES A VIRTUAL UNIVERSE

In the early 20th C, quantum physics' strange phenomena of '**quantum entanglement**' was found by **de Broglie** and **Schrodinger** to be real, and therefore a fundamental property of nature. When particles are split in two, they become *entangled identical twins*. They remain connected to one another in such a way that actions performed on one of the twin particles immediately affects the other, even when they are separated by a distance that can be small or great! They act as if they are still one particle with half the energy, Einstein called this phenomenon "*spooky action at a distance.*" How this happens is unknown.

The phenomena of quantum entanglement seemed like an unsolvable problem until the explanation of a virtual universe was introduced. Only then, did it appear to make sense. In the world of a virtual reality game, distance does not limit communication because all points in the game are at an equal distance from the computer processor used for the simulation, so there is no variation in time delays. But in our macro universe, distances are not the same; they vary greatly. Because the speed of light has a limit, the time it takes to transfer information in the form of light (or any form of electromagnetic radiation) over different distances in our macro universe can vary enormously by as much as billions of light years. But **quantum entanglement suggests that all points must somehow be equidistant for entangled particles to retain the same properties no matter how far apart they are. This would be the case if the universe was a simulation.** But we know that objects in our macro universe are not equidistant, and nor are they in the quantum universe where it does take time for light to go from one place to another - except in the case of entangled particles. What is different about entangled particles? Light travels at the same speed in either the quantum or macro universe, so what could cause quantum entanglement to occur?

Einstein has been, by far, the greatest and most insightful mind in physics. He proved the existence of atoms mathematically, and he was a genius for developing his **Theory of Relativity**. He also determined the fundamental relationship between matter and energy, which is expressed in his famous formula $E=mc^2$ (Energy equals mass multiplied by the speed of light squared). This formula enabled us to understand why stars produce so much energy through both nuclear fission and fusion. Understanding this allowed us to harness atomic energy. Einstein spent the last 28 years of his life trying to develop his '**Unified Field Theory**' that he hoped would unify nature's four forces under one theory. He thought this would eliminate the inexplicable aspects of quantum physics. But despite his genius, and his great efforts to prove that matter exists as a physical reality, he said before he died, "*I must confess that I have been unable to find an explanation for the atomistic character of nature (matter).*" Because of this, most physicists today believe that materialism has not been proven to be real, yet these physicists act as if it is real in their day to day life.

HOW CAN THE WEIRD PROPERTIES OF QUANTUM ENTANGLEMENT BE EXPLAINED?

When the Austrian physicist and Nobel Prize winner, **Erwin Schrodinger** developed **wave mechanics,** he proposed that **quantum entanglement** would occur. This was found to exist when a photon was split into two. Each photon had half the energy level of the original. **If one entangled photon changed its properties, then its twin would do so at exactly the same time no matter how far apart they were!** If one of the entangled particles interacts with another particle, then its twin will instantly experience the same effect even if it is a billion light years away! The pair are somehow connected through time and space. But how? Remarkably, this occurs even though **it violates Einstein's laws of relativity and the maximum speed of light**. Both laws have many proofs, so this violation should be impossible. How could the information pass from one twin to the other in no time at all? Nothing else can. It would have to travel infinitely fast across space, but this is impossible for many reasons. So, what is happening here? (See **Louis de Broglie**'s explanation below.)

As you will see in the **Quantum Eraser Experiment, information between entangled particles appears to go back in time! But this violates the law of causality**, which says that an action cannot occur before its cause! These effects of quantum entanglement should be impossible because they violate fundamental laws of the universe. If this were to happen in our macro world, it would cause chaos across the universe. The universe would be so unstable that it would not exist. Are physicists overlooking some error in the way that they are conducting their experiments, or are they interpreting their results incorrectly? All possible errors appear to have been checked many times, and no errors have been found.

As we are about to see, entangled particles act in a way that makes them appear to be able to change the past! But surely, this is impossible. It's as if the universe always leaves some back doors open in its enormous maze to permit the adventurous to discover new laws. This enables them to see new explanations and new possibilities.

AN EXPLANATION FOR QUANTUM WEIRDNESS

Natalie Wolchover writing for Quanta Magazine summarized the experiments following that she did with droplets of oil bouncing on the surface of a liquid. These experiments replicate many of the quantum-like properties of quantum particles, such as the quantization of energy, double slit interference patterns, the duality of matter being both a particle and a wave, the annihilation of matter, and quantum tunneling. **To solve all these unexplained quantum effects does not seem like a mere coincidence.** These experiments provide convincing explanations for some of the weird properties of the quantum universe, yet many physicists reject this explanation. This is a familiar response to a new paradigm that does fit with what they have been taught to believe.

Natalie's article is titled ***'Have We Been Interpreting Quantum Mechanics Wrongly This Whole Time?'*** It has been reproduced in Wired Magazine. Because the article is well expressed, and due to its great importance with regard to whether matter is probabilistic or deterministic, I have reproduced parts of it here:

> *"This idea that nature is inherently probabilistic — that particles have no hard properties, only likelihoods until they are observed — is directly implied by the standard equations of quantum mechanics. But now a set of surprising experiments with fluids has revived old skepticism about that worldview. The bizarre results are fueling interest in an almost forgotten version of quantum mechanics, one that never gave up the idea of a single, concrete reality.*
>
> *The experiments involve an oil droplet that bounces along the surface of a liquid. The droplet gently sloshes the liquid with every bounce. At the same time, ripples from past bounces affect its course. The droplet's interaction with its own ripples, which form what is known as a pilot wave, causes it to exhibit behaviors previously thought to be peculiar to elementary particles — including behaviors seen as evidence that these particles are spread through space like waves, without any specific location, until they are measured.*
>
> *Particles at the quantum scale seem to do things that human-scale objects do not do. They can tunnel through barriers, occupy discrete energy levels, form standing waves and quantum corrals. They can spontaneously arise and even annihilate subsurface bubbles, an effect reminiscent of the mutual destruction of matter and antimatter particles. If space and time behave like a superfluid, then path memory could conceivably give rise to the strange quantum phenomenon of entanglement. An interaction between two particles in a superfluid universe sets them on paths that stay correlated forever because the interaction permanently affects the contours of the superfluid.*
>
> *This new body of research reveals that oil droplets, when guided by pilot waves, also exhibit these quantum-like features. ... To some researchers, the experiments suggest that quantum objects are as definite as droplets, and that they too are guided by pilot waves — in this case, fluid-like undulations in space and time. ...*
>
> *"Quantum mechanics is very successful; nobody's claiming that it's wrong," said* **Paul Milewski**, *a professor of mathematics at the University of Bath in England who has devised computer models of bouncing-droplet dynamics. "What we believe is that there may be, in fact, some more fundamental reason why [quantum mechanics] looks the way it does." Despite its success, the equations of quantum mechanics do not address how a particle's properties solidify at the moment of measurement, or how, at such moments, reality picks which form to take.*
>
> ***Riding Waves***
>
> *The idea that pilot waves might explain the peculiarities of particles dates back to the early days of quantum mechanics. The French physicist* **Louis de Broglie** *presented the earliest version of pilot-wave theory at the 1927 Solvay Conference in Brussels. (He showed that) pilot-wave theory made all the same predictions as the probabilistic formulation of quantum mechanics, but without the ghostliness or mysterious collapse. De Broglie urged his colleagues to use two equations: one describing a real physical wave, and another tying the trajectory of an actual concrete particle to the variables in that wave equation, as if the particle*

interacts with, and is propelled by the wave rather than being defined by it.

In de Broglie's pilot-wave picture, each electron passes through just one of the two slits, but is influenced by a pilot wave that splits and travels through both slits. Like flotsam in a current, the particle is drawn to the places where the two wavefronts cooperate, and it does not go where they cancel out.... Pilot-wave theory predicts only the statistical distribution of outcomes, or the bright and dark stripes. ... Bohr claimed that particles don't have definite trajectories; de Broglie argued that they do, but that we can't measure each particle's initial position well enough to deduce its exact path.

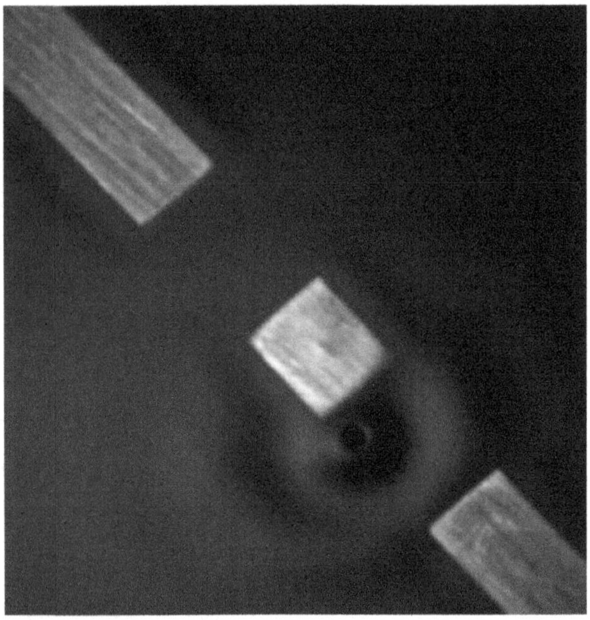

When a droplet bounces along the surface of a liquid toward a pair of openings in a barrier, it passes randomly through one opening or the other while its "pilot wave," or the ripples on the liquid's surface, pass through both. After many repeat runs, a quantum-like interference pattern appears in the distribution of droplet trajectories. Credit: Yves Couder et al.

Physicist ***John Stewart Bell*** *wrote that pilot-wave theory "seems to me so natural and simple, to resolve the wave-particle dilemma in such a clear and ordinary way, that it is a great mystery to me that it was so generally ignored."* ***Sheldon Goldstein****, a professor of mathematics, physics, and philosophy at Rutgers University, blames the "preposterous" neglect of the Pilot Wave theory on "decades of indoctrination." At this stage, Goldstein and several others noted, researchers risk their careers by questioning quantum orthodoxy.* (This is the case with those who question String Theory.)

"I wish that the people who were developing quantum mechanics at the beginning of last century had access to these experiments," Milewski, a professor of physics at MIT said. "Because then the whole history of quantum mechanics might be different." Some enthusiasts think the fluid approach could indeed be the key to resolving the long-standing conflict between quantum mechanics and Einstein's theory of gravity, which clash at infinitesimal scales."

There is great reluctance in the field of quantum physics to have to amend their theory. They would rather go on with their current theory of probabilities that require magic and do not relate to the real world than tackle the problem of the weirdness involved in quantum physics. A major disincentive to do so would likely mean the loss of research funds and the loss of their career.

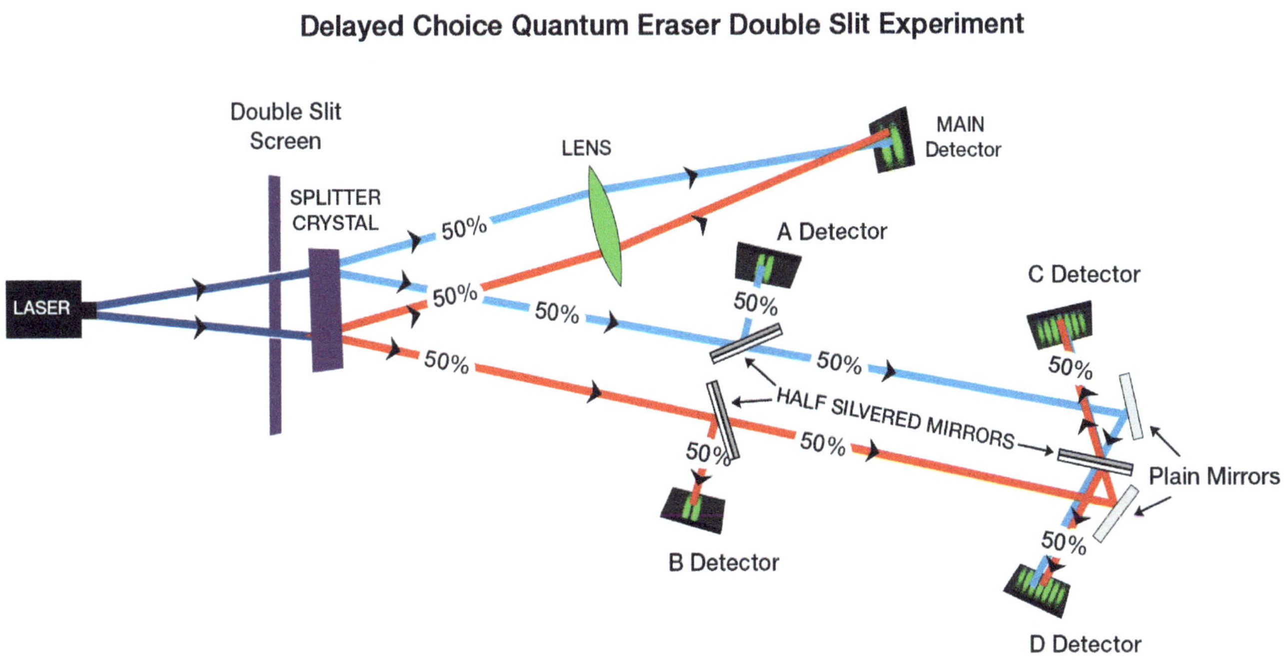

The Delayed Choice Quantum Eraser Double Slit Experiment revealed that one twin of two entangled photons that arrived at a detector before its twin did, must have had to know *beforehand* how its twin would end up, so they would always remain the same! This indicates that the past can change and that the future can be known! No one has yet explained this seemingly impossible situation. There will be an explanation if we can become smart enough to work out what it is. I expect that AI will.

THE BIZARRE DELAYED CHOICE QUANTUM ERASER EXPERIMENT

The Delayed Choice Quantum Entanglement Experiment was designed to investigate the peculiar outcomes of the **double-slit experiment** and the bizarre consequences of **quantum entanglement.**

This experiment is described in some detail here because the results have such enormous implications for what we perceive to be 'reality'.

In 1999, **Marlan O. Scully** of Texas University performed the ***Delayed Choice, Quantum Eraser, Double Slit Experiment***. This experiment is explained in a number of animated videos on YouTube. The best one I found is titled '***Delayed Choice Quantum Eraser Experiment Explained***'. Each of the videos have parts that are not clearly explained, so the following explanation might help to make it easier to understand this experiment. The experiment tests the seemingly impossible properties of photons that are known as **entangled pairs.** This experiment is a little complicated, but it's worth trying to understand it because it has such huge implications. Be sure to follow each step before going to the next.

The Delayed Choice Experiment was conducted as follows: (Refer to the above diagram to follow the process through.)

1. Photons are fired randomly from a laser. They pass through two slits A and B, sitting side by side in a barrier (gray).
2. Photons from both slits pass through a special 'splitter' crystal (purple in the diagram). The crystal splits the photons passing through each slit into two separate photons known as a **coherent, entangled pair**. Due to being split in two, each of the entangled photons has to share the

energy of the original photon, so they have half its energy.

3. In the diagram, pairs from slit A are shown as the two blue paths. Those from slit B are shown as the two red paths.
4. One of each pair goes through a lens (green) to the main detector screen (top right). Because it is known which slits the photons came from, each photon records as a single bar clump pattern, indicating that they are particles. Their twins travel onto half-silvered mirrors that reflect 50% of them to the detectors A and B. These photons appear as particles to form a single clump pattern because it is known which slit they came from.
5. The other 50% of the red and blue photons are allowed to pass straight through the half-silvered mirrors and onto two plain mirrors, which direct them through a second half-silvered mirror. This is when *something inexplicable happens!* For the 50% of them that reflect off the second half-silvered mirror to detectors C and D, it is known which slit they came from, so they should appear as particles to form a single clump pattern - but they don't! They appear as waves that produce an interference pattern!

   The other half that passes through the second half-silvered mirror had a 50% random chance of passing through the mirror or reflecting off it, so there is no way of knowing which slit these photons came from. Because of this, these photons appear as waves to form an interference pattern on detectors C and D. Because there is no way of knowing which photons are from which slit, this is why they all form an interference pattern, so they all have the same property.
6. When this happens, the primary detector, as well as the A and B detectors, show an interference pattern as well! This is despite knowing which slits these photons came from! Because the photons are entangled, all of them must have the same property as their twin. Because it is not known which slit half of the entangled photons that went to the C and D detectors came from, they appear as waves (an interference pattern). All the entangled photons must have the same properties as their twin, so they all record as a wave interference pattern. This is incredible!
7. *It now becomes even more puzzling!* The paths to detectors C and D are longer than the paths to the primary detector, and detectors A and B. This means that the photons that go to detectors C and D take longer to get to them than those that go to the other detectors. So how do the photons going to the primary detector and to A and B know *beforehand* to form an interference pattern before the photons that go to C and D get there?

It *appears* that **the photons going to C and D send a message by some unknown means, from the future back to the past, to allow their twins at the primary detector and those at A and B to register as interference waves!** It should be impossible for these photons to know the result *before* the result has occurred! This experiment has been thoroughly tested to ensure it has no technical errors despite the extremely fine tolerances required.

***To make it even more inexplicable,* if the researchers do not observe what is happening, the primary detector and those at A and B will show clump patterns, while detectors C and D will show interference patterns!! *But when the experimenters know what is happening,* this causes**

all the detectors to show interference patterns! How could a scientist's knowledge or consciousness of the experiment make this difference? This is extraordinary!

This experiment has been run many times by different teams, and all teams found that entangled particles appear to be 'aware' of which slit they went through **and whether they are being observed!** This is truly incredible! It means that the **conscious awareness of the experimenters changes the outcome!**

As physicist **John Wheeler** said, "***on the face of it, the Quantum Eraser experiment suggests that the future can affect the past and that consciousness plays a part in this.***" He believed that "*the past is not the past until it is consciously registered i.e. the past can only exist if it has a record in the present or the future.*" What does this suggest? Perhaps in quantum dimensions, the future is somehow directly connected to the past, whereby it does not require the passage of time to occur to determine the future because it already exists.

These strange effects do not occur in the macro universe, so there must be something that is being overlooked. If we could discover what this is, it would eliminate this perceived impossibility. Some physicists have proposed that particles' duality could be explained if there is at least one extra physical dimension to those that we are aware of. This might explain the ability of entangled particles to communicate instantaneously. (For a more realistic explanation, see *Louis de Broglie's Pilot-Wave Theory above.)*

We should not be able to separate the strange physics of the quantum universe from our macro universe because our macro universe emerges from the quantum universe's properties. All levels of the universe should operate under the same physical laws. But is this a wrong assumption? Could there be different laws of physics at different levels? Or are physicists missing a critical aspect of this puzzle that mega-intelligence will discover quickly?

When seemingly impossible outcomes have occurred in science in the past, a rational solution has eventually been found. Such dilemmas in the past have been caused by observational error, jumping to the wrong conclusion, overlooking some easy-to-overlook but important factors, inadequate technology, or incomplete knowledge about some aspect of nature. Only the last explanation seems possible here. **Pilot Wave Theory** might solve this dilemma.

At this early stage of our learning curve in this area of science, we have to accept that we are not aware of the incredible complexity of the universe that we are just starting to discover, so we cannot hope to understand how it really works at the quantum level - *at this stage.*

The great physicist **Max Planck** said, "*I regard consciousness as a fundamental aspect of the universe.*" He believed that physics shows that matter is derived from consciousness. Over the last century, scientific discoveries, particularly those in quantum physics, and recently those in neuroscience, have been leading us to the conclusion that our consciousness affects what we observe in the universe, but it does not mean that matter is not real or that quantum effects cannot be explained by physical laws. There is a great deal more evidence that matter and energy fields are real than there is for explaining these strange effects away by saying there is no reality, only consciousness. But our consciousness exists in our brain, which is physical – not something ethereal.

Chapter 29

IS OUR UNIVERSE REAL, OR IS IT A SIMULATION?

Science is now amassing a growing body of evidence that is persuading many leading physicists in quantum mechanics and astrophysics to seriously consider whether our universe is actually a sophisticated simulation. The simulation possibility is now becoming a part of mainstream science, so it is no longer just a wacky idea. Science may soon reach a stage where it may be able to prove whether this possibility is true or false. Billionaires like Elon Musk and others who believe that our universe is most likely a simulation have contributed significant funds towards answering this question.

In the late 20th C, the outstanding writer of science articles and sci-fi books, **Isaac Asimov** proposed the simulated universe concept. This was very visionary, considering that computers were very primitive then and that virtual reality games had not been invented.

To suggest our universe may be simulated is an extraordinary claim. **Carl Sagan** wisely stated that *"extraordinary claims require extraordinary evidence"* - if they are to be believed. So, let's look at what evidence there is that might support the simulated universe hypothesis and what critical questions we should be asking about this possibility.

IF OUR UNIVERSE IS A DIGITAL PROGRAM, WHO OR WHAT COULD HAVE WRITTEN IT?

This 'Stargate' extends to infinity in every direction. It was created by the author for his SpaceWalker edutainment attraction.

Some experts in the field of computing propose that information in our universe is not in a physical form, such as particles of matter. They believe that information exists purely on a digital level – just as it does in a virtual reality game where there is no physical reality inside the game. However, in a simulation game, physical equipment such as a computer is required *outside* the game in order to process the data. Virtual or simulated worlds or universes that only consist of data, still need a physical structure outside of them to supply the energy and the data they need.

To understand the importance of the data having to have a meaningful structure, consider a page of paper full of letters. If the letters are randomly placed, then they will contain no information. Only letters placed together in a particular manner by a conscious mind can give the letters meaning. That being the case, could most physicists who assume that there is no conscious mind to codify the laws of physics for our universe be wrong?

The molecular 'letters' in the structures of DNA are structured so that they form codes for the construction of life. But how did that coded information come into being? Did it arise simply through natural evolution through random trial and error? One would think that to produce all the codes for the evolution of everything in the universe would require a super consciousness. But would a super-consciousness be necessary if the original codes are simple and designed to interact with one another in relatively simple ways to develop increasingly more complexity on their own? This would not require a super-intelligence – just a mind smart enough to write the initial codes for the Big Bang. The relatively simple random improvements that occurred throughout the evolution of life created incredibly complex organic chemistry. But who wrote the initial codes for this to occur?

HOW COULD THE UNIVERSE HAVE SUCH AN EXTRAORDINARY LEVEL OF FINE-TUNING FAR BEYOND CHANCE?

Strong support for a mastermind designer of the universe is found in the unbelievable level of fine-tuning for at least 30 parameters of the physical universe that scientists have so far identified. (See Chapter 5.) If any one of these parameters changed just the tiniest bit, then the universe would not exist. How did these incredibly finely-tuned values get programmed into our universe? If we cannot find an explanation for the universe's fine-tuning, then **this is as big a question as the origin of the energy that created the universe!** Surely, a super-intelligence would be required for programming these parameters, given they are so exquisitely fine-tuned to such incomprehensibly precise tolerances. And each parameter dovetails perfectly with each other! This is infinitely beyond chance. **So how was this achieved???**

Apart from explaining how this fine-tuning came about, physicists also have to explain:

1. Where did the universe come from? Did it come from outside our universe?
2. Where did the universe's laws of physics that govern matter, energy, space, and time come from to allow it to develop increasing complexity?

The laws that lead to increasing complexity have produced an incredible diversity of physical structures, as seen in the diversity of stars, planets, and particularly life. It's impossible for this to have come about by pure chance because the odds are far too great. However, this should not be seen as a reason to presuppose a god to solve this problem. Gods just push the problem back one step because we then have to ask "*Where did the god and its universe come from?*" If we could solve that question, then we would have to ask, "*Where did that creator and its universe come from?*" ...and so on forever! So, there must be an explanation that is beyond our present comprehension.

THE UNIVERSE'S BITS

As discussed in Chapter 5 in detail, for the universe to have digital program codes, it must have indivisible 'bits' that can be used as the equivalents of ones and zeros in its computing codes. Some scientists think that our universe's bits could be quanta of energy in the form of indivisible particles such as quarks. Without these particle bits, all the larger particles would not exist to form the building blocks for matter from which everything is constructed. And without quark bits, the universe would have no means of storing and processing information, so it would not have a program to run its evolution.

If you would like to probe further into this subject, a good documentary to watch is '***The Simulation Hypothesis***' produced by physicist **Kent Forbes**. It is available on YouTube. It explains the evidence for this concept, and it has good supporting graphics. Another good video is '***Simulated Reality***'. There are other documentaries on this subject together with some very interesting debates and interviews with leading scientists in this field. Many scientific articles, papers, and books have been published about the possibility of a simulated universe. (See Additional Information in Appendix 1.)

WOULD A SIMULATED UNIVERSE BE DIFFERENT TO A REAL ONE?

If our universe is simulated, would it be any different from a real universe – if there is one? Everything in a simulated universe would appear the same as a 'real' one – providing that the simulation was of a very high quality with a pixel resolution down to the level of quarks. So, we must ask ourselves ***"How could we determine whether a universe was real or not, and what would differentiate a 'real' universe from a simulated one?"***

Cosmologist **Max Tegmark** asks, "*What if there were innumerable mega-intelligences across the universe that have created innumerable simulated or virtual universes? What would that lead to?*" He points out that these virtual universes could also evolve mega-intelligences, which could then create their own virtual universes, which would breed more mega-intelligences that would create even more virtual universes, and so on, forever! This could create infinitely more virtual universes than real ones – if there was a 'real' one to start with! Tegmark suspects that "*the idea of creating endless universes has a rotten egg at its core.*" I agree, especially as infinities are the basis of this concept, and mathematicians know that when infinities occur in a theory, this typically means that there is an error involved.

Tegmark points out that the laws in each simulated universe may be different. They could be designed so that they are not the same as the universe in which they were created. If each universe is different, some would not breed life or mega-intelligence. There could be an infinite number of universes, but, given how finely-tuned ours is, most universes would not be viable, or last long, let alone breed life. This is the **multiverse** concept. **It is a philosophical concept as opposed to a scientific hypothesis. The multiverse concept has no mathematical basis that can be tested, and importantly, it cannot be falsified, so it has no value as a scientific hypothesis.** The ability to disprove a theory by means of scientific experiment has always been a part of the accepted scientific method to prove something true or false.

The rules for creating objects and events in a computer-controlled virtual reality (VR) universe would have to originate in the creator's mind. In a VR universe, both the programmer and the computer code would exist outside the VR universe. If our universe is simulated, then the energy needed to create it would have originated from outside of space-time. The information processing system that controls each object and event in a VR universe would also have had to come from outside it.

In a VR game world, its virtual space and time only exist while the program is booted up. Our universe appears to have been booted up 14 billion years ago. From the perspective of an intelligence that exists in a virtual universe, space-time would appear to have originated from nothing – just as our universe appears to have done. This is further evidence for a simulated reality.

In today's computer-controlled VR worlds, the players who control the actions of their avatars are outside the VR world. Do we have superior intelligences outside our universe controlling us? If so, how would we know? It seems highly unlikely that 7 billion people on Earth are all controlled by superior intelligences outside our universe. If that were so, the player 'gods' who control us would be far superior intellects to us, yet they must lead very boring lives to be bothered to control our typically mundane lives. They would be bored to death with so many of their avatars playing mindless computer games, many of which involve endless violent killing. The whole concept of superior intellects controlling humanity lacks any logical basis.

If our universe is simulated, it's possible that it is not controlled from outside, but rather, it is left to run its course without interference to see how it evolves. But surely, any creator who is super-intelligent enough to create our universe would be able to compute its outcome. If so, what would be the point of creating it in the first place? It's easier to think that our universe is real. If it is, will we ever be intelligent enough to work out how our universe came into existence and what its purpose is? **This is humanity's BIGGEST question.**

"The most important decision we make is whether we believe we live in a friendly or hostile universe."

Albert Einstein

COULD A CONSCIOUSNESS BEYOND OUR UNIVERSE HAVE CREATED THE BIG BANG?

Our universe is *thought* to have been created when the Big Bang commenced. Most physicists assume, as **Democritus** did, that there was nothing before the universe came into existence. That is, there was no physical reality, no time, and no space. So where did it all come from? Perhaps Democritus was wrong in his assumption that there was nothing before the the universe was created.

There are some ideas emerging that dark energy may be able to emerge from a parent universe. Could it be that the universe lies dormant as a singularity until something makes it expand? Without an understanding of where our universe came from, some religious people think this is a reason to introduce a supernatural, non-physical creator. But anything supernatural is against science for many good reasons. Some physicists assume that the energy of the Big Bang seems to have had to come from *outside* our universe, but science has no way of knowing what 'outside' the universe means. It's impossible to talk about something existing 'outside the universe' if the universe is all there is and ever will be. There is nothing in the existing Big Bang theory that suggests there has to be an 'outside' to our universe.

The concept of an outside to the universe comes from our everyday experience that things explode into an existing space. But there is no reason to believe that our universe has to exist inside another mega-universe. But it may! We simply do not know. The concept of an 'outside' our universe has been introduced by some physicists in an attempt to explain where the energy came from. But they then have to explain where the energy for the larger universe came from. For all we know, there might be a background force or dimension in the quantum universe from which it came. This energy field may always exist. And it's possible that other universes are emerging from this background energy field all the time as well. This could be likened to bubbles of steam popping out of the surface of boiling water. We simply have to accept that we do not know. And at present, there is no way of testing any of these concepts. **All we know with some confidence is, that based on the redshifts of galaxies, the universe appears to have expanded from a very dense, very small volume, perhaps a point. Everything else is speculation.**

Avatar characters in a virtual reality computer game would think that their virtual world is all there is, because to them, there is no outside to their world. But we know differently! Could this analogy apply to us? The answer to this is probably so complex that it is beyond our consciousness. Maybe in the future, super-intelligent trans-humans or mega-computers will discover the answer.

The question of where the universe came from is currently beyond human contemplation because we have no facts on which to make a judgment about such a speculative concept. It would be like asking Stone Age people (or even those in recent centuries) where the rocks falling from the sky (meteorites) came from. We now know. One day our offspring will work out where our universe came from. That may be in the near future thanks to the rapid advances being made with supercomputing. Just because we do not know where our universe came from right now is no reason to think that it will always remain a mystery.

IS MATTER IN OUR CONSCIOUSNESS? DEMOCRITUS VS PLATO

To decide whether our universe is real or a simulation, we should consider whether matter exists in a physical form or whether a simulated universe creates an impression of it in our consciousness. These two possibilities have been debated for at least 2,400 years since the Greek philosophers **Democritus** and **Plato** debated this in the 4th Century BC.

Democritus believed that all things exist because they are made of tiny invisible particles that he called atoms. He thought they must exist in various forms to account for all the different substances there are in the world made from the different types of matter. Without any physical evidence, he brilliantly conceptualized that atoms are the smallest indivisible bits of matter. Incredibly, it took until the early 1900s for the existence of atoms to be proven and accepted by science.

Democritus also believed that Earth was spherical. He was proven right on that count as well. **For his amazing insight and his belief in the process of scientific evaluation, Democritus is considered by many to be the '*father of modern science*.'** His philosophy that matter has physical form, is known as **Materialism**. This theory states that the universe is all there is and that it needs nothing from outside of it to make it work or to explain where it came from. This belief remains strong today.

We now know that atoms consist of protons, neutrons, and electrons. In recent decades, physicists have hypothesized that all particles consist of smaller, indivisible states of matter known as quarks. All types of particles are thought to consist of force fields. Quarks have a high probability of existing because the mathematics that supports the theory of quarks beautifully explains the structure of matter. Experiments at atom smashers such as the Large Hadron Collider (the LHC) also support the existence of quarks. However, they have not yet been observed because they are so incredibly small.

Today's physicists and chemists have very strong evidence that supports Democritus' view that matter is physical and that it exists in the form of atoms. The Scanning Tunneling Microscope (STM) is the most powerful of all existing microscopes. It can image the surfaces of matter at the atomic level. IBM researchers have used a STM to pick up atoms and move them around to form letters and symbols, proving beyond doubt that atoms exist.

The STM works in air, water, and in a vacuum, as well as at near absolute zero temperatures and temperatures over 1,000°C. Even though it requires extremely clean and stable surfaces, a very sharp tip, excellent vibration control, and sophisticated electronics, incredibly, some hobbyists have built their own!

Democritus believed that matter was composed of atoms. He was known as 'the laughing philosopher' because he was very cheerful. He said he could see the comical side of life. It's quite incredible that so long ago, Democritus could conceptualize atoms when he had no access to technology to help him gather evidence for their existence.

Plato is the most influential philosopher of all time despite him living 4,400 years ago! One would have thought philosophy would have moved a long way ahead since then. That says a lot about it! He was often belligerent and angered by other people's ideas, especially those of Democritus that challenged his beliefs about matter and consciousness. He strongly believed that matter is not real, and what we perceive to be our world is actually a construct of our consciousness. If he was alive today, he would probably be a strong supporter of our universe being simulated, and I suspect that he would be a strident supporter of far left political views that are unsupported by evidence.

Plato was greatly influenced by **Socrates**' teachings and some of **Pythagoras**' ideas. He believed that humans have an immortal soul and that it lives on after death – a concept that many people today would like to believe despite there never being no evidence to support this. Given that he believed that we only exist as a consciousness, so we do not have a physical form, it's hard to see how death would occur in his world, or how we exist. He gave no explanation for how life forms would be born in a purely conscious world. Nor where they would come from. He believed in life after death. This was a very popular concept that Christianity, Judaism, and Islam taught as it was very easy to sell. (These religions derived from the Abrahamic religion.) Plato wisely thought that people should think more before making on-the-spot decisions or going along with popular opinions of the day, yet he frequently did these things himself! He popularized his numerous social beliefs and his views on love and ethics, which can be found even today on the Internet. He preached concepts that people liked to hear, but which were often divorced from reality. The great philosopher **Aristotle** became his protégé.

Plato promoted his 'theory of forms' in which he proposed that the physical world is not real. He thought that true reality exists 'beyond' our world, but he could not explain in any meaningful way what 'beyond' meant or where 'reality' existed. It seems that sometimes he believed in a physical world when it suited him, and at other times, he did not! He said that our physical world is imperfect and always changing and that it is a mere shadow of a perfect spiritual reality that transcends both space and time. This is suggestive of a heaven. His concepts are a religion that is full of his esoteric beliefs, none of which stand scrutiny. Would his 'perfect' spiritual world appear perfect to everyone else? I doubt it. He offered no evidence or a theory to support his beliefs, which were not scientific in any way. They are naive beliefs that cannot be tested. His thoughts were undoubtedly dominated by his right brain. He would feel at home in a hippie commune today.

Plato vehemently disagreed with Democritus' **Materialism**. Rather than provide real-world evidence that he was right, he preferred to demean Democritus' character. This is how modern-day universities operate to demolish those who do not agree with their strongly leftist socialist views that ignore reality.

He believed that it is the way we *think* about matter that determines its properties. However, matter should not exist in his ethereal world. But he could not define this in any way that could be tested. He didn't seem to consider that if he was right, different people thinking about matter in many different ways would cause matter to have numerous different states of existence that were contrary to those that he proposed. And if alien intelligences have much different forms of consciousness to us, they too would perceive things differently to us, so their matter would be very different to his. An alien's spiritual 'reality' would therefore be quite different to Plato's 'reality', so who's reality would be right? Plato's philosophy was unscientific and childishly naive.

Why do people who believe in a perfect non-physical realm beyond our physical world want to stay alive here in our imperfect physical world? Strangely, for the most part, they enjoy it! Why wouldn't they want to kill themselves to go to their perfect spiritual universe? But they don't; they hang here for as long as they can and use physical medicine to stay alive as long as possible while still taking toxic doses of many of their alternative medicines and eating meatless food diets that robs them of enough protein for energy. Humans evolved naturally like other predators to eat meat. Why don't those who do not want to eat meat, not feel the same way about plants? To not want to die to go to their version of heaven tells us that they don't really believe in idealism: they just want something they can talk about that requires them not to have any rational, evidence-based knowledge. This also allows them to change their mind whenever they need to to suit their situation.

In stark contrast, Democritus concept of the way atoms are structured and how they combine to form everything in the universe has been proven beyond all doubt to describe reality perfectly. Plato's philosophy is known as **Idealism**. It remains popular for those who like alternative lifestyles, especially with those with a dominant right brain who believe in a spiritual plane of existence even though there is not a shred of evidence to prove that it exists. Yet there is mountain of evidence all around us and throughout the universe that Domocritus' materialism explanation about reality is right. After all this time of ever-growing scientific knowledge, incredibly, Plato's ideas remain relevant and plausible to some people. Plato said that there is no physical reality because what we perceive to be physical is actually created by our consciousness, which is ethereal. But we know that our consciousness is created at the top of the brain stalk in our brain. (See Chapter 22.) Plato believed inconsistently, that we do not have a material brain. Then why is there a need for our physical world? If we do not have a brain, then how do we think? He thought our true existence was located in his 'shadow' realm of our world and that we are just shadows of that world. He seemed to make up his philosophies on the run, depending on who he was debating. He must have been on a lot of grass.

In support of Democritus ideas about matter, **astronomers have found that matter has the same properties right across the universe, regardless of which humans are observing it. This strongly supports Democritus' Materialism.** Evidence makes no difference to idealists who like the idea of a spiritual world beyond our own because the right hemisphere of their brain allows their thinking to make anything that has no solid rules fit into their world. They are not critical thinkers, so they abandon critical thought for their feel-good explanations. Like Plato, they are easily angered by those who point out the inconsistencies and falsehoods of their beliefs because they cannot provide rational reasons why they are right. They can do nothing about this form of thinking because it is programmed into their genes.

"Nothing exists except atoms and empty space; everything else is opinion. I would rather discover one scientific fact than become King of Persia."

Democritus

IS THERE ANYTHING IN THE UNIVERSE THAT IS NOT NATURAL OR PHYSICAL?

Anything that exists in the universe must be natural because the universe is a natural entity. If something that was not natural were to exist in the universe, then it would have to come from outside the universe. That would make it supernatural i.e. *beyond* nature. But no evidence exists to show that there is an 'outside' to the universe or that anything has come into our universe from outside it - no gods, no demons, no angles, no ghosts, and the like. And there is no evidence for heaven or hell. The universe all there is. This being so, it makes it impossible for there to be anything that is beyond our universe – anything 'supernatural'.

The definition of 'supernatural' means 'beyond the laws of physics'. If supernatural events were to occur, that would mean that the laws of physics do not control everything in the universe. If this were the case, we would not be able to rely on anything. But we have found that the laws of the universe can *always* be relied upon. If we couldn't rely on them absolutely all the time, then anything could happen at any time, and without cause. But this does not happen.

If something supernatural did occur, like an alien or an angel, or a descendant from the future popping out of thin air, then it would mean that matter, or energy, could appear out of nowhere, or alternatively, matter could just disappear! If so, a star or a planet could do this. A plane in perfectly good condition could just fall straight down, or go into space. But these things do not happen. Gravity always works, as do electrical forces and magnetism. Supernatural events would have devastating flow-on consequences throughout the universe. Over time, their flow-on effects would become highly visible because every supernatural event would create a 'tear' in the space-time continuum, which would ultimately cause the destruction of the universe! But we see no sign of this happening. If impossible things were to occur, then the universe would be chaotic! For the universe to exist, the laws of physics must *always* be consistent.

IS THOUGHT PHYSICAL OR ETHEREAL?

Some people still think that thought exists in some non-physical realm. As a young teenager, I was bewildered by how the brain was able to produce thoughts and retain memories, because to me (and to most other people back then, and even now) thoughts seemed like they were non-physical and ethereal. Everyone I asked about how thoughts were contained in the brain had no idea - and they didn't care to think about it. But this was a great mystery to me. A couple of decades later, I discovered the answer! Just as information can be recorded onto magnetic tape by changing the magnetic fields of iron or chromium oxide particles embedded in the tape, so can thoughts be recorded in brain cells by a different means. Our thoughts are processed via electrical currents, which are either allowed to pass from one neuron and on to many others or, they are blocked at some point. In memory retrieval, it appears that messenger RNA encodes beta-actin proteins to 'shape' the circuitry in neurons so that a memory can retrace the electrical pathways that it took when the memory first formed. This is very complex and quite amazing.

I discovered that thoughts and memories were not ethereal; they had a physical electro-chemical base that allows them to be processed and retained as memories, imperfect as many are.

It's impossible to think of anything real in the universe that does not exist in some physical form. At the base of everything are force fields, which are physical. As we saw in Chapter 22, that our level of consciousness requires a brain that is physical. It is our brain, through its senses, that is able to perceive the force fields of matter. Brains and force fields are both physical. So the idea that **Plato**, **Heisenberg**, **Bohr**, and **Plank** had that matter only exists in our consciousness is wrong. Matter is real, as Einstein tried to prove it to be but failed.

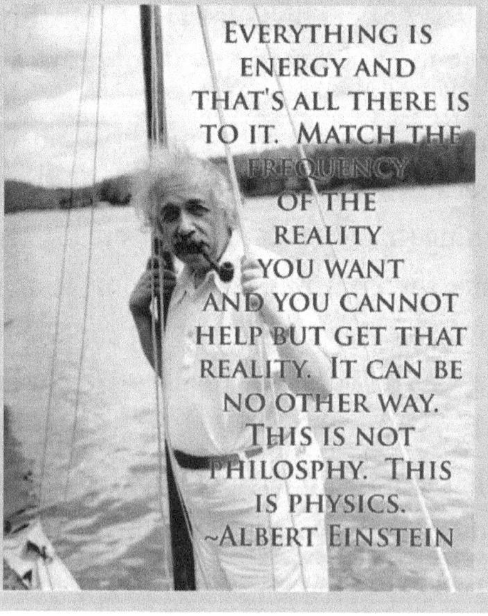

SUPPORT FOR A SIMULATED UNIVERSE

Materialism has strongly prevailed over the millennia, but recently Idealism has gained some ground due to experiments in quantum physics such as the Quantum Eraser Experiment. Some people believe that the conclusions of this experiment can only be explained if the universe is a simulation *determined by our consciousness*. Although this is a very long stretch, there is some very interesting evidence to consider that *appears* to support this hypothesis. (See Chapter 28.)

Just as there is no reason to stream a virtual reality computer program if there are no players logged into it, then it seems logical that there would be no reason for the universe to compute its data if it had no consciousnesses to use it.

Worlds in a VR computer game only run and exist when someone uses them. Some quantum physicists suggest that our universe, or parts of it, may only exist when we observe them. If this is so, this would greatly reduce the computational power required to run a simulation of the entire universe all of the time. However, if there are innumerable intelligent, conscious life forms across the universe observing much of it, the simulation would have to run continuously, but not all of it at full resolution - only those parts that are observed closely by conscious entities. It's possible that if we discover other intelligences in the future, they are a part of the simulation we are in. If this were the case, it is hard to think of a way to test this hypothesis at this stage.

If our universe is a simulation, it needs rules (formulas and programs) for it to operate and evolve. We have discovered many of the rules in nature from those for classical physics and chemistry at our macro scale to those for quantum mechanics at the subatomic scale and those for cosmology at the grandest scale. These rules control how the universe evolves – but we do not know why it exists.

The '***Game of Life***' computer game invented by **John Conway** was designed to have computer 'cell life forms' evolve. Even though this game has only four simple rules, the program produces a remarkable diversity of 2D digital 'life forms'. The game's extremely simple mathematical rules determine whether a collection of computer cells will multiply, break up, become static, or die. Depending on the initial conditions, some colonies die quickly, while others grow quickly, and after rapid growth, some unpredictably collapse. Others keep evolving with no apparent end to them. The growth patterns of these 2D 'life forms' are often unexpected, and their growth can change for no obvious reason. Some have offshoot groups. All this variation from only four rules! Their evolution is surprisingly representative of life forms that have evolved on Earth. How many rules are there for all life on Earth and for the universe? They could be much smaller than we might imagine.

Our entire complex universe seems to have evolved from only a small number of rules for gravity, electromagnetism, and the strong and weak nuclear forces. These rules constantly interact with one another to create increasingly more complex rules that produce a seemingly near-infinite variety of options. From this, it's incredible that our planet has evolved such an amazing diversity of life forms. It's very likely that a near-infinite diversity of life exists across the universe. **Simple rules kept evolving by combining with one another to create the complexity we have so far discovered.** No doubt, there are innumerable more rules that we do not know about, most of which may be so complex that we will never understand them.

Our universe has rules for nature that cause extinctions of species, sometimes on a planetary scale, but it does not seem to have mass extinctions on a galactic scale, so life should always keep evolving on numerous worlds in each galaxy across the Greater Universe. Science is allowing us to learn nature's rules, and this is permitting us to advance ever-faster.

WHAT IS THE EVIDENCE THAT SUPPORTS OUR UNIVERSE BEING A SIMULATION?

The following 8 points appear to support the concept of our universe being simulated:

1. **The universe works via a set of rules/programs**

   We are discovering that everything in our universe appears to be controlled by a hierarchy of mathematical rules that form the laws of physics, chemistry, geology, biology, chaos, fractals, logic, and consciousness etc. These rules appear to have evolved naturally from simpler rules that were programmed into our universe at its outset in a similar way to how we program rules into the Game of Life or a virtual reality game.

2. **The universe's bits**

   In a virtual world, all data must have a unit of minimum quantity. In computers, we call these quantities 'bits'. In a digital picture, the minimum quantity is known as a pixel. Such quantities cannot be broken down into smaller units. If our universe is simulated, its 'bits' could be its indivisible particles such as the 18 types of quarks together with anti-quarks, photons, electrons, and their family members, which are known as leptons. **The Standard Model for Quantum Mechanics** is the most successful scientific theory ever devised. It says that particles cannot be broken down into smaller particles – just as an image cannot be broken down into ever smaller pixels. But what if quarks turn out *not* be the smallest indivisible particle? We should remember that it was only a little over 100 years ago that we believed that atoms were considered to be the smallest, indivisible particle. However, because quarks are so small, there would be no practical advantage to there being smaller bits.

3. **Quantized computable bits**

   Everything in our universe appears to be quantized as computable bits of information. Energy appears to be quantized. There are 'bits' for particles, so there should be bits for space and time. If this is the case, then our universe could be computable! We are already using sub-atomic particles in quantum computers to perform computations, so it is entirely possible that in the future we may be able to create incredibly detailed simulated **Virtual Reality (VR)** worlds of our solar system. If we could go ever deeper into the quantum universe, or be able to see forever across our macro universe, then our universe would have no boundaries. It would be infinite in every way, so it would have infinite possibilities. This would mean that it would need infinitely small bits as well as infinite energy. But for everything to be infinite, this would mean it must have infinite size and energy. This would be impossible. This is why our universe must have limitations in how small particles can be, how much energy it contains, and how large it can be.

4. **A finite observable universe**

Physicist **Raphael Bousso** believes that we need to find mathematical solutions to our cosmological theories that do not have impossible infinities. He says there should be no need for infinities because **there are a finite number of bits of information (100^{123}) in our *observable* universe. This number describes *everything* that goes on inside our observable universe. As large as this number is, it means that space in the observable universe cannot be infinitely divisible, and neither can time, so our universe must consist of bits.** Most astronomers think that space probably goes on for a long way beyond our observable universe, but it does not have to extend to infinity. To overcome a simulation having to compute an enormously larger universe than our observable universe, **the universe has limitations such as the diameter of the observable universe and the speed of light. This makes it computable**. This being the case, it does not matter that there is a Greater Universe, because we and all other entities would always be inside the computable 'bubble' of an observable universe. For all intents and purposes, the rest of the Greater Universe may as well not exist because it is beyond our space-time, and therefore, it cannot affect our observable universe in any way. Because the Greater Universe is expanding faster than the speed of light, nothing from beyond our observable universe can reach us.

5. **Computable universes**

In our observable universe, there is a finite number of bits, and a finite number of states, so it is thought by some that there would be nothing in our universe that a super mega-computer could not compute. Could the very advanced mega-intelligences that we will soon give birth to, be able to create digital universes as rich as ours? If they could, then it would be possible that our universe could be a simulation. But there appears to be physical laws that limit the size of a simulation to being much smaller than our universe.

6. **Uniformity in construction**

All particles in any particular class are identical because the same code produces them. This adds weight to the argument that particles in our universe are digital bits created by the same program. Objects in our macro world are not identical, but their constituents are atoms, and each atom of each element has an identical chemical structure. This is exactly how a simulated universe would be organized!

7. **The universe must have an horizon**

In our universe, the speed of light has a maximum speed. This is consistent with events in a virtual universe being controlled by a computer processor with a finite processing capability. If the speed of light did not have a limit and the universe was infinite, then we would see the entire Greater Universe extending to infinity! This would mean that every part of it would be as bright and hot as the surface of an average star, so no planets or life would exist. (Why the night sky is dark was known as **Obler's Paradox** before it was realized that we only see our observable universe and not the Greater Universe.) In an infinite universe, a mega-intelligence computing its events would need infinite computing capability, but this would be impossible. The finite speed of light allows us to see only our observable universe, which is limited in size. This means that infinite computing power is not required. Therefore, it is possible that our observable universe is a simulation. However, if the Greater Universe is hugely larger than our observable universe, then its program would have to link all observable universes. Linking them would require as much energy as the Greater Universe, so it cannot be infinite. If the Greater Universe is not hugely larger than our observable universe, then it too could be a simulation.

8. **Finite computation**

   If the universe is a computer simulation with **limited computing power, this would explain why space has to contract to zero in an infinite gravitational field at the event horizon of a black hole. It does so to be computable. And this is why time has to come to a stop there. Limited computing power also explains why nothing but massless light can travel at the speed of light.** There are extremely high demands on computer processing when matter approaches a black hole and when matter travels at near the speed of light. To calculate events occurring close to the speed of light slows down a computer dramatically, no matter how powerful it might be. This is because it has to calculate numerous (almost infinite) changes that are occurring incredibly fast. Only by making space warp and time slowdown in high gravitational fields or at velocities near the speed of light can the effects of relativity be calculated with finite computing power. This supports a simulated universe.

9. **Error-correcting codes**

   Incredibly, theoretical physicist **James Gates** and his colleagues have found computer code "buried" in the equations for Super-Symmetry used in quantum mechanics! These codes were not put there by those developing the equations for super-symmetry: Gates says they are in the equations that describe the quantum universe! These same codes are used by physicists in String Theory to describe the cosmos. They are in the code for Super-Symmetry in the form of strings of ones and zeros, which are not random. They appear to be deliberately programmed into nature at the quantum level! But by whom or what? This supports a simulated universe. The chances of them being there is far beyond chance. They are built into the equations for quantum physics to eliminate errors in the formation of matter and force fields. The most fundamental units of matter in every class must have *exactly* the same properties all the time so they can act as bits. How could these codes have arisen naturally without the universe being programmed from the outside? But is there an outside? And how could we ever test this when it would be beyond the universe?

   Error-correcting codes are also programmed into our DNA to limit errors in our cells when they reproduce. This greatly limits the number of life forms that would otherwise die from code errors. Even with error-correcting DNA codes, some errors still occur due to the inconceivable complexity in every small part of every life form.

Error-correcting codes are used in computer programs so they can self-correct any errors that might occur. At present, Gates does not know why the codes his team has found are there, but they are consistent with a simulated universe. Gates asks, *"Why would these codes be there if they serve no purpose?"* Physicists think that any universe would have to evolve error-correcting codes, otherwise errors would rapidly grow, and it would quickly disintegrate.

If we do live in a simulated universe, then the '*Matrix*' movies could have a limited degree of truth to them!

ARE THERE LIMITS TO THE UNIVERSE'S RESOLUTION?

If quantum physics can prove that there is nothing smaller than a quark of even the incomprehensibly small **Planck Length**, then this would prove that space is not infinitely divisible. It could therefore be a simulation. The Planck Length is roughly equal to 1.6 x 10-35 m, i.e. 0.000000000000000000000000000000000016 of a meter, or about 10^{-20} times smaller than a proton, or 100 billion, trillion times smaller than an atom! The Planck Length is the scale at which physicists say that quantum effects would dominate at this extremely small scale, so classical ideas about gravity and space-time cease to be valid. It is unknown what effects, if any, would remain.

Until very recently, it was thought that space-time had infinite resolution and that we would be able to delve ever deeper into it to keep seeing endlessly more detail. But at the quantum level, it now appears that we may soon be at a level of resolution that we will not be able to go beyond. The universe can function perfectly well without the need for infinite resolution because it would serve no purpose. Nature seems to abhor unnecessary waste. The universe prefers to be efficient, so it does not expend energy unnecessarily without a return. **Science should eventually be able to find the limits of resolution in all aspects of the universe, but this will not prove beyond some doubt that it is simulated.** It's important to remember that we can only test laws to a particular level at any particular point in time. Our ability to come up with tests is based on what we know at any point in time, together with what we are capable of imagining. We are also limited by the technology we have available at any point in time.

IS THE UNIVERSE THE ULTIMATE OPERATING SYSTEM?

There are many recent scientific discoveries that have been made in cosmology, biology, neuroscience, the social sciences, computer science, and mathematics that appear to be telling us that the universe ***may be*** the ultimate computer program.

If the universe and all its life forms, including us, are controlled by mathematical formulas, then we are actors or avatars in the ultimate, blockbuster, multi-dimensional universal computer game in which we are unconsciously acting out our roles in an ever-evolving, predetermined plot! In this program, every intelligent life form (i.e. players/avatars) and every continuously evolving environment (i.e. worlds and galaxies) are all interacting with one another. If mega-intelligences discover that we do indeed live in a simulated universe, will they find that they too are actors in this computer game in which computer-generated fractals are the stage scenery?

Although discovering that our universe is simulated would be amazing from an intellectual perspective, I nevertheless hope that this is not the case and that all life forms are not automatons controlled by unchangeable rules, as they appear to be. There seems to be no end purpose to the future being predetermined.

The science of *natural growth* dynamics is showing us that the growth of intelligence in the universe may be similar to the way neural networks grow in brains, and the way data networks are designed for the Internet so they can evolve. Natural growth dynamics may help us to discover fundamental laws that govern the way all complex systems evolve, including the most complex system of them all - the universe itself.

The universe may use every individual intelligence in it in a way that could be broadly similar to the way the human brain uses individual neuron cells to form its intelligence. If so, could the universe's goal be to connect all its life forms and intelligences so that they form the ultimate consciousness? Might it then discover the reason for its existence and what its purpose is? Could it be that the universe's aim is to become so intelligent and so conscious that it can evolve into some mind-boggling existence beyond the dimensions of our universe?

COULD MEGA-INTELLIGENCE CHANGE THE EVOLUTION OF THE UNIVERSE?

To consider the big picture of the universe, we need **to consider the impact that highly advanced intelligences might have on the evolution of the universe.** Some physicists are now thinking that once we know more about our own consciousness, we may be able to use this to change some aspects of the universe. If this becomes possible, then the super consciousnesses of mega-intelligences that probably evolved early in the universe's life, would have already done so. However, we are not yet at a stage where we could detect what they may have done. If they evolved into another plane of existence, then we would not be able to detect them or determine what powers they might have. If many mega-intelligences evolve to have different objectives, then this would cause conflicts and chaos throughout the universe. But we see no indication of this in the part of the physical universe we can detect.

Would highly-advanced mega-intelligences be able to affect or direct the evolution of the universe? Surely, highly intelligent beings with IQs into the quadrillions would be able to significantly change the physical universe in significant ways - if they wished to. If there is any validity to this concept, then such intelligences appear to show no concern that the physical universe will die due to all the stars eventually burning out. This is referred to as a 'heat death' i.e. a lack of heat. Or that it may die from a total disintegration due to its expansion. (For more detail on these hypothesized ends to the universe, see Chapter 3. It appears that the universe was born and that it will eventually die like everything in it. If the end of the universe is a threat to a universal mega-mind, it would have changed the universe's evolution already, if it could. Perhaps the universe may not die as we expect, but if it does, perhaps it simply doesn't matter to a universal mind because the end result of the physical universe may have no impact on a mega-intelligence's god-like existence, especially if it no longer is limited by the physical dimensions we are aware of. It might be like a chick breaking out of its egg and not caring that it is destroying its shell because the shell no longer serves any purpose. Of course, these questions are postulated by the simple intelligence of the author who has no knowledge of the big picture of reality, so they are probably extremely naive.

The mega-intelligences that we are about to create may eventually discover why the universe exists. However, if they tried to tell us why, we would not be able to comprehend what they are telling us because the answer would very likely be far above our ability to comprehend the concepts involved. It would be as impossible for a mega-intelligence to explain this to us as it would be for us to explain television or the Internet to a lizard. The answer would not be '42' as in hilarious book, **'The Hitchhiker's Guide to the Galaxy'** by **Douglas Adams.**

Chapter 30

IS THE UNIVERSE DESIGNED AROUND MATHEMATICAL FORMULAS?

The more we discover, the more it becomes clear that every structure in nature at every scale is determined by mathematical formulas. Formulas apply to the force of gravity, electrical and magnetic fields, and the forces within atoms and particles. These forces determine how everything in the universe evolves. Mathematical formulas determine the structure of molecules, including replicating ones such as crystals and DNA. Formulas determine how stars and planets form and evolve, the shapes of galaxies, and how they interact with one another. **There are innumerable formulas that underlie the structure of all life - from viruses to cells to organs, and even the circuitry of brains. It appears that there is a formula for *everything* in nature, even such things as thoughts and emotions.**

The universe appears to be a complex mathematical tapestry woven together with innumerable formulas. They are the construction manual for building everything in the universe and for how it evolves. When I first realized that formulas looked like they controlled most things in nature and probably everything, I was shocked by what this meant. As I researched this, I discovered that I was not alone in this realization. Many others had also come to the same conclusion. This mind-blowing concept is known as the **Mathematical Universe Hypothesis**, the **Ultimate Ensemble**, and the **Theory of Everything**. This idea was originally considered speculative, but there is now much evidence to support it. If there were no formulas, how would all things from atoms to galaxies know how to construct themselves in the same manner? **Without laws that consist of formulas, all the physical sciences such as physics, chemistry, biology, geology, and even psychology etc. would not work, so we would not have science. The universe without formulas and laws would not exist. We are now realizing that the laws we have so far discovered are probably only a snowflake on the tip of an enormous iceberg of increasingly complex formulas that saturate every aspect of the universe.**

LIFE'S ALGORITHMS PROCESS NATURE'S FORMULAS FOR EASY INTERPRETATION

The senses of animals evolved to gather the formulas of nature in their environment and convert them into information that they can use to tell if something is useful, interesting, or dangerous. Life cleverly developed algorithms to convert parts of the electromagnetic frequencies into light, color, shapes, patterns, and movements that have instant meaning. Life can translate energy waves such as sound vibrations occurring in air, water, or in the ground to tell the direction of movement of the vibrations, and the size and speed of the body, causing them to see if it is a treat or an opportunity. Life cleverly converts other formulas into such things as taste, language, emotions, and invitations etc. Formulas for molecular structures in air, water, and the earth, can be converted instantly into aromas that tell animals the type of gas molecules that are present and how valuable that information is. Brains analyze formulas to tell if a smell indicates food, or a threat, or a sexual offering. Brains have evolved efficient methods for almost instantly converting nature's formulas into meanings that are quick and easy for animals to interpret.

IS EVERYTHING IN THE UNIVERSE DETERMINED BY FORMULAS?

Everything in the universe is underpinned by some formula or another and we are now realizing that there may be neverending layers of interacting formulas. We have only just started to scratch the surface of probably a near infinity of all the formulas that exist. The universe appears to act like an enormous food blender that keeps blending its formulas into different combinations to endlessly create new ones that are more complex. Formulas are constantly combining and evolving, probably at an exponential rate. Might it be that no intelligence may ever know them all?

The formula for Chaos Theory injects random change to drive continual evolution. (See Chapter 5, page 61)

The underlying language of science is mathematics. It very clearly describes concepts, how they can be tested, and what is real. Mathematics has guided our understanding of how nature works. It is allowing us to see how the universe is put together at every scale, and we are now just starting to discover some of the formulas that drive its evolution.

Mathematical formulas clearly describe, in precise ways, the many subtleties of nature. Verbal descriptions or pictures alone cannot do this. To be of any value, scientific theories must be able to make specific predictions, which are often based on mathematical formulas. These predictions are tested in the natural world to see if they are a true reflection of a particular aspect of nature. Mathematical predictions often point the way to new discoveries.

Most physicists believe that mathematics can be used to describe all aspects of our physical world. **Max Tegmark** argues that mathematics is our physical reality. He says, ***"Our universe does not have some mathematical properties: it has* only *mathematical properties."*** He says that our universe is like a digital photo. It looks like the real world but, as we zoom into it very closely, we see that it is actually a field of pixels composed of four numbers that specify the amount of red, blue, green and black in each pixel.

The universe's underlying mathematical structure is relatively simple. It is just 32 numbers determined by constants such as the properties of elementary particles and force fields. Each constant has equations that are the fundamental laws of physics. Simple examples are the rotation of celestial bodies and their orbits, which are based on simple ratios.

If there was a creator, he must have been a mathematician and a computer programmer! Watch '***Decoding the Secret Patterns of Nature***' documentary.

Science has discovered that the universe is a rich tapestry of geometric patterns that can be expressed as mathematical formulas. Those who were first to glimpse some of these patterns were great minds that could see the underlying laws of nature that control these patterns. Interestingly, those who see 'patterns' in nature in one area of a science, usually cannot see them in other areas. This is because we each have formulas in our minds (that we are not aware of) that are specialized for a particular types of observation and analytical thinking processes.

EXAMPLES OF SIMPLE FORMULAS IN NATURE

Pinecones, pineapples, sunflowers, roses, daisies, and many other natural forms in botany are based on Fibonacci numbers, which form spirals. In pinecones, there are 8 left spirals and 13 right ones. A sunflower has 21 left and 34 right spirals.

While many complex formulas define the way quantum physics and astrophysics work, there are many simple formulas that have a major influence on the nature of all things.

Fibonacci numbers, which are the sum of the preceding two numbers (e.g. 1, 2, 3, 5, 8, 13, 21, 34, 55, and so on forever) seem to apply to many forms in nature. How nature came to use these numbers is not yet understood. Beware however that some popular, so-called examples of these numbers in nature are false because they do not fit properly when examined carefully. This is true for certain examples of **golden spirals**, which are often not spirals at all, but short segments of a curve. For reasons not yet identified, the human eye likes the perceived beauty of spirals and certain other geometric forms, so it tends to see them even when they are distorted. **Golden squares** are said to exist in all good-looking faces, but there are numerous examples of good-looking people where they do not occur. **There are no perfect circles, spheres, squares, or cubes in nature.** Those that appear perfect are not perfect when they are accurately measured or looked at closely. As we zoom into them, we see many imperfections. For example, the Earth appears as a perfect sphere from the Moon but when we measure it accurately, we see that its equatorial diameter is larger than its polar diameter and that it is also slightly pear-shaped. As we zoom in, we see that it is not a smooth sphere; its edge has mountains and lowlands. Its crust is far from having a perfect thickness all around. This is also the case for Earth's atmosphere, which has 'hills and valleys' (high and low pressure systems).

The number **Pi (π)** underlies many forms in nature. It is 3141592... to infinity. It's an **irrational number** that has an infinite number of digits beyond its decimal point. The numbers do *not* settle into a permanent set of repeating numbers. Pi can be used as the fraction 22/7 for quick calculations. It is best known for being the ratio of the diameter of a circle to its circumference. Pi determines innumerable forms such as the spiral of DNA's double helix, the round pupil in our eyes, the concentric ripples from a splash in a pond, and waves of sound and light spreading out from their source. It is used in Heisenberg's uncertainty principle that defines how precisely we can know the state of the universe.

Pi appears frequently in **Probability Theory**. If you draw parallel lines on a page that are apart by the length of a needle and you drop the needle randomly, the probability that the needle crosses a line as opposed to landing between the lines, is exactly 2÷pi, or 64%. Why would this be so? Most people expect it to be 50%. Anything that has waves will have Pi in it. Pi exist all through nature. It shows up in spherical fruit to stars.

FRACTALS ARE NATURE'S FORMULAS FOR REPEATING PATTERNS AT DIFFERENT SCALES

Fractal patterns were first described by **Benoit Mandelbrot** in 1975. As with **Chaos Theory**, Mandelbrot found order in what had previously appeared to be random roughness in nature. He discovered that there were geometric patterns that repeat at progressively smaller scales. He called this type of geometry **fractals**. Each type of fractal has a formula which controls its repeating structure known as **'self-similarity'**. The universe uses formulas for fractal geometry in numerous ways to create innumerable natural forms. Fractals have simple rules that make a pattern branch out like tree branches or the airways in lungs, and then have them replicate the branching at smaller scales a few times over. Fractals occur throughout nature.

Where did these formulas come from?

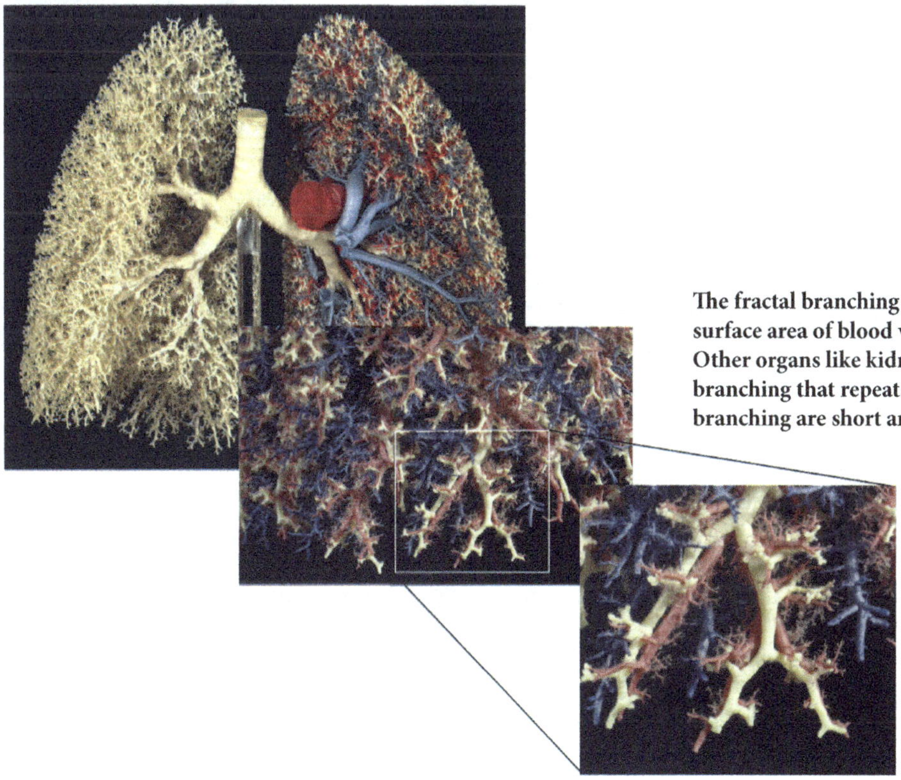

The fractal branching of airways in a lung maximizes the surface area of blood vessels to the area of a tennis court. Other organs like kidneys and brains utilize similar fractal branching that repeats several times. The fractal formulas for branching are short and simple.

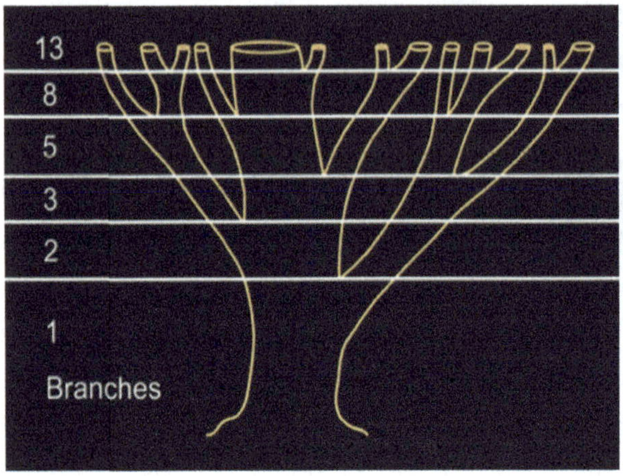

Left: **Branching fractals are designed around Fibonacci numbers. But they do not apply to all forms of branching.**

Center: **The twisting of rivers is based on a formula. Rivers become less curvy as they age. They become straighter due to erosion occurring and the flow of flood water taking the course of least resistance. In this image, we see old dry curves that are no longer a part of the river.**

Right: **This image shows a river with fractal tributaries.**

Left: **It does not matter how far one zooms into the Mandelbrot set of images seen here, because it never ends. The set displays infinite, ever-changing detail that is self-similar!**

Right: **The black icon of Chaos repeats endlessly at every scale throughout the Mandelbrot set.** Is there a hidden meaning behind this icon?

The following images are examples of fractals in nature. They form the shapes of such things as erosion patterns, snowflakes, lightning, seashells, clouds, waves, mountains, coastlines, brain cell connections, and the filamentary structure of galaxies across the universe. These things have the same structure at both large and small scales.

We are discovering increasingly more simple and complex mathematical formulas that control the universe. **Galileo** cleverly worked out the formula for how bodies accelerate as they fall - even galaxies when they fall towards one another during a merger. **Sir Isaac Newton** skillfully calculated how gravity worked and how the math for calculus can be used o calculate so many things in the universe. **James Clerk Maxwell's** brilliant mathematical formulas showed how electricity and magnetism could generate each other and produce radio waves and X-rays. **Einstein** ingeniously discovered the formula for how mass distorts space and time and how mass can be turned into explosive energy. The math of the giants of quantum physics such as **Heisenberg, Bohr,** and **Plank** predicted the formulas for how particles behave in ways that were not intuitive. (See Chapter 28.)

Mathematics has an extraordinary predictive ability, so much so that it has led to numerous subatomic particles being discovered and the way solar systems and galaxies form. Mathematics is at the core of the development of engineering, which has proven to be the most beneficial achievement for humanity.

Mathematics is a universal 'language' that controls all matter, all forms of energy, inanimate objects, and also all life, as well as how everything evolves. Learning the way nature works on Earth is allowing us to use its formulas to discover how the cosmos works. It has allowed us to build high technology to discover increasingly more of nature's formulas. Mathematics has permitted us to advance much faster than anything else has. What we can't understand, or model today is due to us not yet discovering the formulas that control complex things.

It appears that humans, and other advance animals have genes that construct a part of their brain to think mathematically. The more advanced a life form is, the more it can use math to work out what the world is about, and then use this to its advantage.

For a good documentary that illustrates how mathematical our universe is, '***The Mathematical Mysteries of the Universe***'. Everything seems to be based on numbers or formulas. We humans are surely a simple sub-program within the extremely complex super-program that runs the universe. There is far more evidence for that being the case than there is against it!

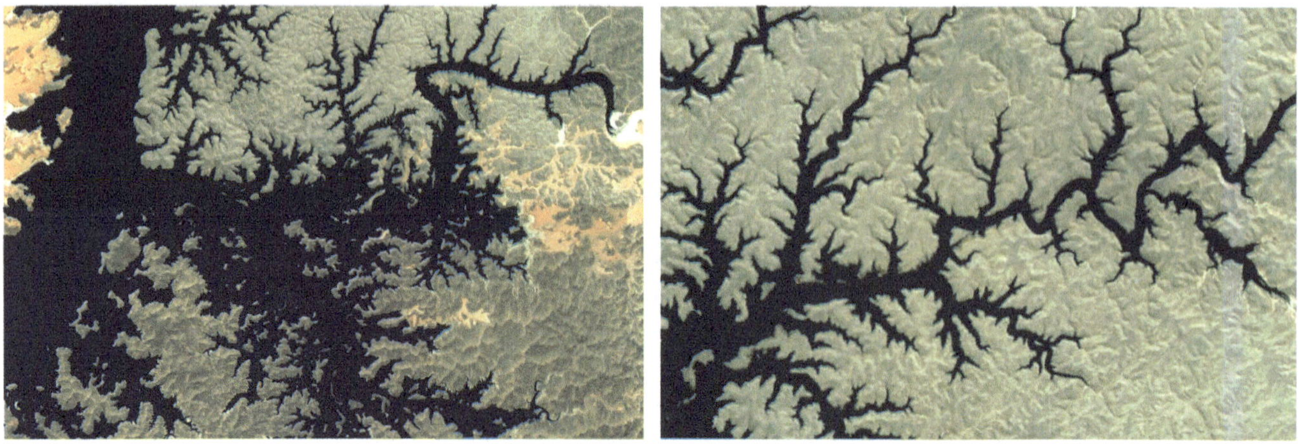

Branching fractal river tributaries in a river valley in Egypt are seen here at different scales.

The repeating fractal structure of shells

Broccoli's fractal cones

Fern leaves repeat their structure at smaller scales.

The edges of clouds have the same fractal forms when seen from afar or close up.

Repeating fractal snowflake structure

The fractal branching of lightning is self-similar.

The fractal branching of a tree.

The fractal formula for the shape of the the mountain is the same as it is for rocks on the mountain.

The formula for the fractal edges of breaking waves are the same at large and small scales.

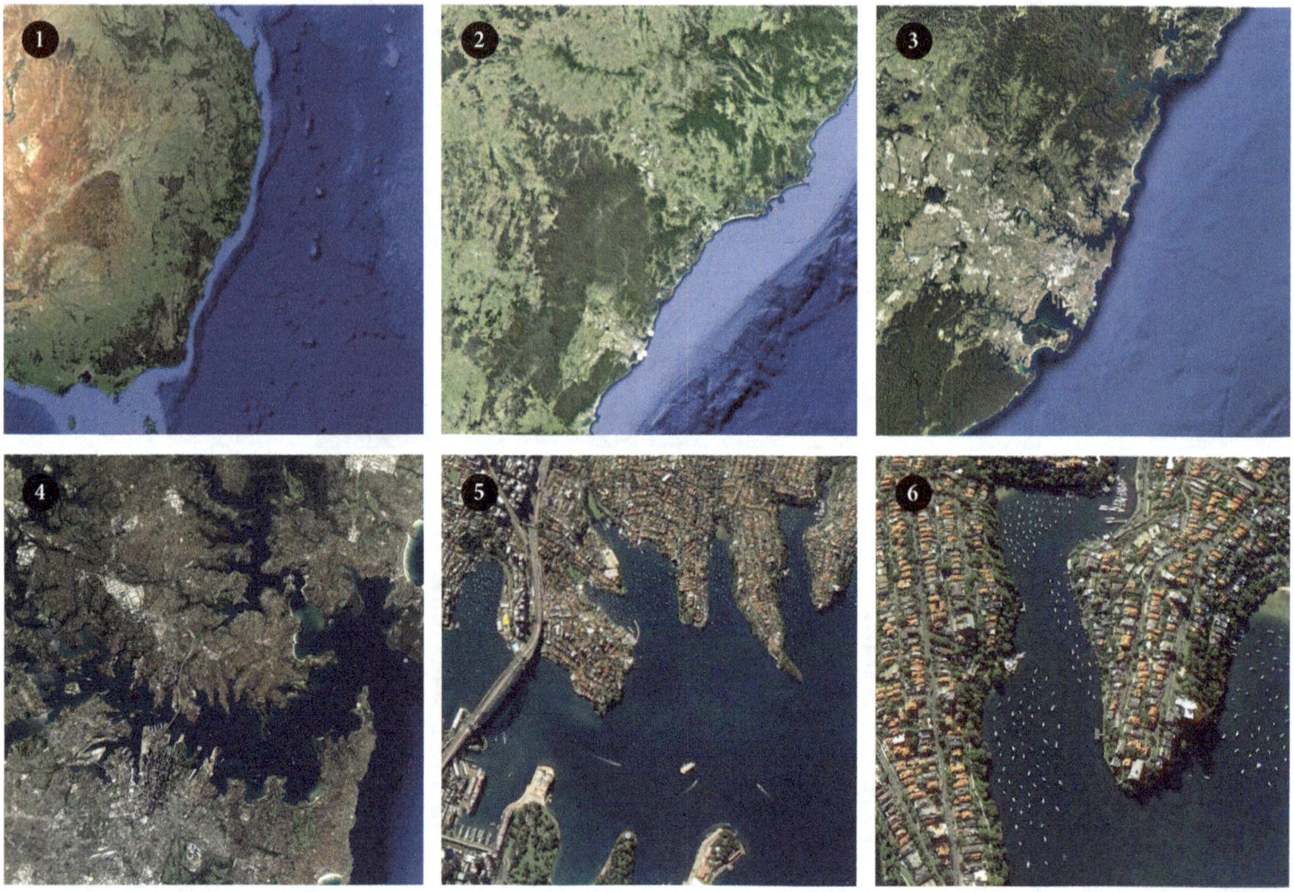

Here we zoom into images of the eastern Australian coastline centered on Sydney. The fractal nature of a coastline repeats at ever smaller scales making its total length increase exponentially the closer one goes onto it.

1. *The coastline appears rather straight, and its length is easy to measure.*
2. *At this scale, the coastline becomes a little curvy and, therefore a little longer.*
3. *We now see fractal river estuaries, which add considerable length to the coastline.*
4. *The fractal nature of Sydney Harbour becomes much more evident and detailed. This increases the length of the coastline.*
5. *Sydney Harbour Bridge is seen connecting the North and South shores. The coastline has now become longer still at this resolution.*
6. *Here we see even finer detail that increases the coastline's length.*
7. *If we were to zoom down to individual rocks and sand grains along the shoreline, the coastline of Australia would become almost infinite in length. Credit: All images are from Google Earth.*

If Australia's coastline was measured at the resolution of picture 1, it would appear far shorter than it would in the next scale and similarly so in each following scale. As its fractal nature becomes more evident, the coastline lengthens greatly. As we keep zooming in, the shoreline grows exponentially in length. Because waves and tides keep changing all coastlines, they can never be measured exactly at close scales. If we zoomed in all the way to see molecules, then a coastline would appear almost infinitely long.

The science of fractals has allowed us to glimpse the mathematical formulas that underlie the formation of numerous aspects of the physical universe, including life.

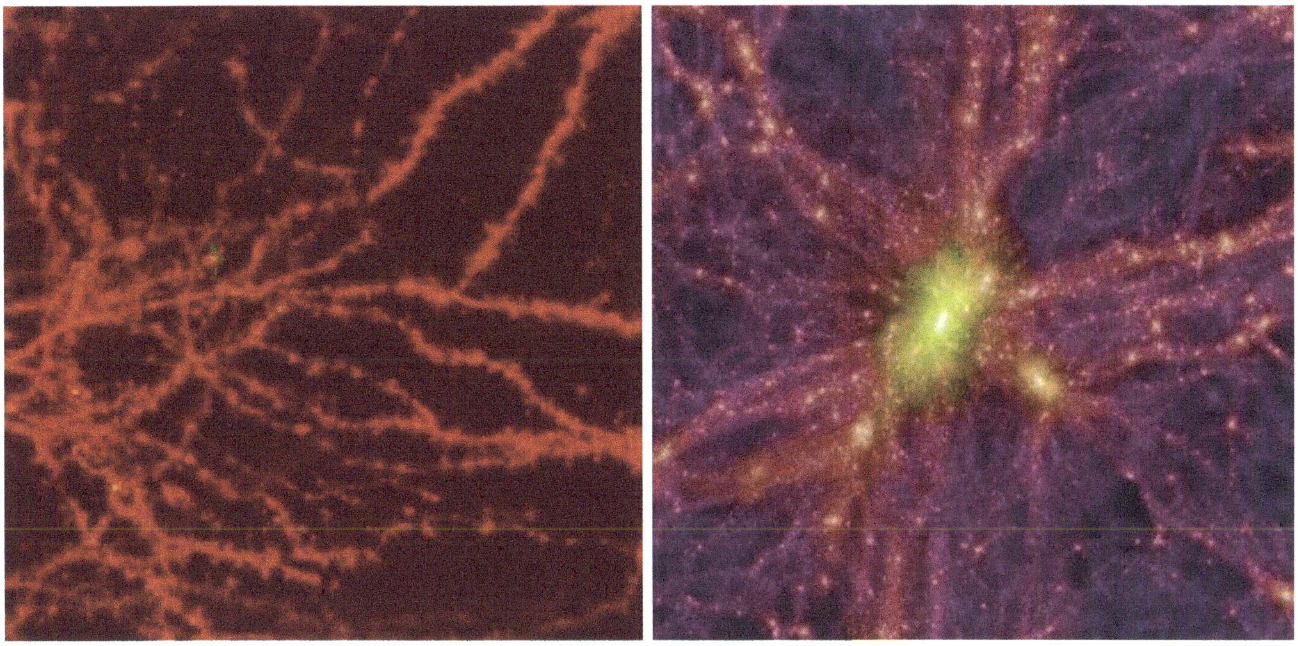

Dendrite connections (left) **between neurons in a brain display a 3D fractal structure that is similar to the 3D filamentary structure of galaxies across in the universe** (right).

The dust lane structures of spiral galaxies display beautiful fractal forms that repeat at smaller scales. Credit: HST, NASA

THE ALGEBRAIC FORMULAS OF LOGIC

Even things like emotions that we might think are not mathematical, are determined by algebraic formulas that are used in the math for logic. Emotions can be written as formulas by using 'if/then' rules (if this happens, then this follows). Even emotions that seem insane to normal people are logical in the brain of a psychopath who thinks differently from the vast majority of people. Psychopaths have very different value systems, so they think differently. They do not have empathy or know what love is, and they have little to no fear, so it is easy for them to kill without remorse. They do not understand how normal people think because the formulas for their minds' circuitry are different from the neural construction formulas for normal people.

Some people have emotional explosions because they have a dysfunctional circuit breaker. Others who have had repetitive serious trauma in their youth have their formulas for stress rewritten by their experience to help them survive life-threatening incidents in the future. They have heightened awareness to threats that make them hypersensitive. Fear is a mathematical formula to escape from threatening stimuli.

Fear formulas are seldom rewritten in adulthood, even when the serious threat that caused the fear no longer exists. Because fearful people can overreact, this then causes them to have real trauma, which then reinforces their fear. It is possible today to have successful aversion therapy to *irrational* fear to overwrite these formulas. Fears are easily and quickly overcome through desensitization.

The thought that our emotions and actions are the product of mathematical formulas programmed into our genes is repellent to some people. They think that this means that they will not be able to control some of their emotions. This is often the case.

COMPUTER MODELLING OF NATURE IS BASED ON FRACTAL GEOMETRY

We are replicating the formulas that nature has developed. Computer programmers have written formulas for computer video artists to use to replicate natural scenes in movies. These formulas replicate water, wind, clouds, fire, animals, as well as motor vehicles, ships, and aircraft in action. Computer programs replicate how light changes under different conditions, how plants grow, and how facial features move and evolve. They have also created the formulas for geological events, explosions, fire, and so on.

WHAT CONCLUSION SHOULD WE DRAW FROM EVERYTHING BEING CONTROLLED BY FORMULAS?

If everything in the universe is controlled by numbers and formulas, what conclusion does this lead to? Is this more evidence for a simulated universe, and that we do not have free will? (See Chapter 29.)

If we could become highly advanced and therefore much more conscious, we might be able to see everything in the universe as mathematical formulas. This sounds bland, but it might actually give the universe greater beauty and deeper meaning if we could ever become intelligent enough to see it that way. We might then see how everything is connected and why. Seeing formulas seems dry to most people because very few people have much understanding of math. But once we become intelligent and knowledgeable enough to understand what the formulas mean, this will change our view of mathematics and how we perceive the structure of everything in the universe.

ARE INFINITIES REAL?

Infinities keep coming up in mathematical equations that attempt to explain many aspects of cosmology. They also come up in Quantum Mechanics, String Theory, the infinite gravitational fields of black holes, the possible size of the Greater Universe, and in relation to the number of theorized parallel universes that some people think may exist. But when astrophysicists and mathematicians encounter infinities in their theories, they usually cause problems that prevent a solution. Infinities are like stop signals: they tell mathematicians that they are overlooking something and that they are on the wrong track, so they should look for other possible solutions. When they do, they often find a solution that works.

We need to understand whether infinities really exist in the real physical universe, as opposed to them being purely a mathematical concept.

Anything that is infinite has no end to it. Infinity is not just a hugely big number: it is literally a number that is limitless! The symbol used for infinity in mathematics is ∞.

If we count the innumerable sand grains across all our planet's beaches, across the bottom of its oceans, and across all its deserts, this would be a number so huge that it must surely approach infinity. But because the number of grains would be found to be finite, they come nowhere near being infinite. The number of particles in our universe would seem to be infinite, but again, this enormous number does not approach infinity. The total number of particles in our observable universe (not including quarks if they exist) has been calculated to be 10^{80}. That's 10 with 80 zeros after it! This enormous number has been shrinking since the universe formed due to the expansion of the universe causing matter at the edge of our observable universe to continually expand beyond it.

Numbers have no end to them, so they are truly infinite. If you don't believe that numbers are endless, think about what number you would stop at in any sequence that would make it infinite. No matter how large that number is, there would always be numbers that you could keep adding to it, so it would not be infinite.

What do we get if we multiply or divide infinity by infinity? Do we get 1 or 0? This is a curious question because no one is sure. Could the Greater Universe that extends beyond our observable universe be infinitely large? Almost certainly not, because this would require infinite space and energy, which would not be possible. Could the universe expand forever? No, because it would eventually tear itself apart even at the level of particles. Could there be an infinite number of other universes? No, because that would require infinite energy and infinite space to the power of infinity - an impossibility. Could there be no end to finding particles smaller than quarks? No, because that would also require infinite energy to no avail. Questions like these that involve possible infinities are very troublesome to answer in the positive because there would have to be an infinite supply of energy, so where would that come from? No matter how much there was, it would still not be enough.

Some of the greatest mathematicians, such as **Kurt Goedel, Georg Cantor**, and **Ludwig Boltzmann** spent much of their lives studying infinities. Their work sent them into such serious depressions that it eventually killed them! (Watch the documentary '***Dangerous Knowledge***' to see how infinities ruined their lives.) When one starts thinking deeply about infinities, it is no wonder it can make a human mind go insane.

Ludwig Boltzman **Georg Kantor** **Kurt Goedel**

Two and a half thousand years ago, Aristotle thought that true infinities had to be banished if one wanted to do good science and math. The great mathematician of the late 19th C, Georg Kantor, had extraordinary insight into the paradoxes that infinities create.

Leading modern-day mathematicians **Max Cohen** and **Hugh Woodin** think that there is no way to prove whether real infinities exist or not *in the real universe*. They say that infinities lead to the conclusion that *'there is no truth to anything'* in the real universe. If that is true, then there must be no truth in what they are saying!

If the increasing expansion of the Greater Universe continues, this will ultimately tear space and time apart, so the universe will have an end. It will not exist forever. If exponential expansion does not continue for some unknown reason, then after many trillions of years when all stars have burnt out and their heat is evenly distributed across a cold, dark universe, all that will be left is their cold, stellar cinders. The universe will be only a few degrees above absolute zero. With either possibility, the universe's lifespan will not be infinite. The universe will die like everything in it - just as all cells in all life forms die. This was built into the universe's program at the outset.

Science is now learning that everything in the universe is based on programs. Formulas in genes control how lifeforms evolve. The laws of physics determine how planets and galaxies evolve. And codes control how the minds of animals and computers think.

It looks like there will not be anything in the universe that is not based on programs that use formulas.

Chapter 31

ARE PARANORMAL CLAIMS REAL?

Three out of four people in the US profess to believing in at least one paranormal belief. 22% believe in 5 or more and 57% believe in at least 2. Only 1% believe in all of those below. This is despite no hard evidence being forthcoming for any of their beliefs. Those who believe in paranormal happenings have genes that like the idea of some things being beyond reality. They only need vague claims to satisfy their need for them to be real. When people say they believe in some paranormal claims being real, many confess that they would not stake their lives on them, so they are not fully convinced. They like the thrill of some paranormal claims *possibly* being real. When they want their beliefs to be real, they easily convince themselves of it. But belief is not evidence.

A US poll showing the percentage of people who believe in the following:

ESP	41%
Hauntings	37%
Ghosts	32%
Telepathy	31%
Clairvoyance	26%
Astrology	25%
Taking to the dead	21%
Witches & Angles	21%
Reincarnation	20%
Channeling spirits	9%

42% of Americans believe that people are sometimes possessed by the devil. This includes many of the Elite who are well known for conducting devil-worshipping ceremonies. People who believe in religion are more likely to believe in the paranormal than non-religious people. 24% of people believe that aliens have visited Earth. There are 27% of realists who do not believe in any paranormal occurrences. Things that appear to be supernatural at first, are found not to be so after investigation. This is because a rational explanation with proofs explains what was *thought* to have occurred. People who believe in supernatural events are easily led by those who are deceiving them, usually for money. As they say, it is easy to separate a fool from his money.

Apart from those above, there have been many other claims of supernatural or paranormal happenings such as:

- mind-reading, mind over matter, communicating with the dead, remote viewing, poltergeists, lay lines, fortune-telling, faith healing, acupuncture, aromatherapy, homeopathy, meridians, healing magnets, crystal power, psychic healing, phenomenology, auras, channeling etc.
- gods, the devil, demonic possession, exorcism fairies, the devil, souls, heaven, hell
- spontaneous human combustion, astral traveling, the third eye, religious images magically appearing, and so on.

But no provable evidence has ever been forthcoming to substantiate even one of these claims. Over more than 100 years, many researchers have gone to great lengths to try to verify supernatural claims, but not once have they succeeded! Each of these claims have been explained without resorting to anything paranormal. Plenty of so-called supernatural events have been shown to be fraud so the perpetrators could make money out of the gullible.

Many paranormal beliefs are transitory, like the latest diets. They fade away over time while new ones take their place.

Claims of a statue of the virgin Mary weeping tears of blood at a church in Italy attracted long crowds giving donations. People believed that only something supernatural could make a stone statue weep. Well, it was rather easy to do that! An enterprising priest drilled holes into the corner of the eyes from behind the head. He connected a tubes to each eye. Using medical drip bags, he ran a slightly viscous red liquid into each tube so that it appeared that tears were slowly running down the cheeks of the Virgin Mary. He refilled the drip bags every morning before opening. **Thousands lined up every day and gave donations.** When the crowds eventually dwindled, he shot through with all the money. One could fill a large book with the most innovative fraudulent cons like this that have been perpetrated upon people willing to believe in the paranormal.

Demons, aliens, or angels cannot appear out of nowhere. That is physically impossible so there must be trickery involved. Where would they come from? Supernatural happenings cannot be real because they defy the laws of physics, which control everything that happens in the universe. Science has *never* found anything that is not natural, even though many scientists would like to because it would be extremely exciting to discover something that did not conform to the laws of physics. If a scientist could prove this, their name would be forever recorded in history - but this has never happened, and it never will.

There is such a strong willingness for some people, even some researchers in this area, to believe in paranormal events. They want this to be real so much that they only look for positive results and ignore the negative ones. Those who are committed to a belief make no attempt to look for fakery by those presenting such claims. (See **confirmation bias** in Chapter 25) Staunch believers in a particular form of government or fake medical healing treatments typically have confirmation bias.

I must admit that in my naïve youth, I was open to the idea that some paranormal things were possible, but as I researched each one hoping to find one that was real, I found a mountain of evidence for each that provided completely realistic, scientific explanations that did not involve anything paranormal. The reality behind each claim turned out to be far more exciting than the supernatural claims.

To see through paranormal claims, one needs a good level of discrimination, but unfortunately, this is not common. One also requires knowledge of how tricksters, including many unscrupulous 'charismatic' religious preachers and so-called clairvoyants, perform simple tricks that they falsely present as being 'paranormal' occurrences performed by spirits or the 'hand of god'. Unfortunately, very few people have the knowledge or the powers of observation necessary to see how things that appear to be paranormal are actually done by deception. This makes it easy for people to be tricked and therefore believe in the impossible. Magicians are excellent at deception, but they make no claim to their tricks being any more than an illusion.

To test how easily people are prepared to believe in things, I used to performed a number of simple tricks that made people believe I had paranormal powers. I was surprised at how many people were prepared to accept that the power to move objects with my mind. They thought what I did was real, because they did not think I would be tricking them, even though my tricks defied the laws of physics. Most observers did not make even the most basic attempt to discover if I could have faked it. They didn't even ask me if it was a trick! If they had, I would have told them. They just believed I was able to do these things.

Even the simple trick of making a round pencil roll away from me on a flat table top when I gave it the command to do so baffled them. When I asked it to come back to me, it did! I was able to do this by softly blowing on the pencil unnoticeably to make it roll up a very slight rise in the table top, and then stop it by stopping blowing. I then had it come back to me not blowing to let it slowly roll back to me.

On one memorable occasion, I was at the head table at an Australian Skeptics Annual Dinner function. Two magicians were there to provide entertainment. One of the tricks the magicians were performing was the spoon-bending trick. They jokingly made out they did this through their mental powers. I told those around me at the head table that I was going to try it. I said I thought it just might be possible with enough concentration. I heard some of my fellow skeptics scoff. Secretly, I knew it was impossible, but I knew how to make it appear real. Despite all the males around me being skeptics, they naively assumed that I would not trick them, and they presumed that I had no knowledge of how to do this trick. Incredibly, one of the magicians who observed me doing this also made the same assumption, particularly as I did the trick differently to him. I was able to convince them that, by the power of my mind, I was able to heat the metal in the neck of my teaspoon and make it go soft and bend.

I made out I was shocked by how rubbery and hot the neck of the spoon was becoming as I quickly rocked it back and forth between my thumb and forefinger. None of them wondered why I needed to rock it back and forth. I did this to create the perception of it becoming rubbery. Unbeknown to them, at the end I very quickly bent the spoon with my forefinger and thumb in front of them without them noticing as I threw it up falsely saying it had become too hot to hold. No one noticed me bending it in that process. I made out it had burnt my fingers, so I sucked them and waved them in the air and then put them into my cold drink. This seemed to be enough to fully convince all of them that the spoon became so hot that it bent. The wives were convinced that I had bent metal with the power of thought!

No one touched the spoon to see if it was hot, which of course it wasn't. Without questioning me in any way, they were all amazed that I had done this 'for real'. I acted amazed saying that I had always assumed it was impossible to do this. A number of the wives elbowed their husbands saying things like *"See I told you that there are some things that are real paranormal happenings"*. Their husbands were in disbelief. The magicians were also stunned as they too had assumed that I was for real, and that I had no knowledge of how to do this trick. They asked me to do it again, and I obliged. This convinced all of them, including the magicians even more that I had supernatural mental powers that allowed me to control matter.

Both the magicians and my associates assumed that I had softened the metal with the power of thought and that I did not know magic tricks because of the naive way I was talking and my surprise that it worked! And being a skeptic, they assumed that I would not fool my fellow skeptics.

Because the metal in the neck of a teaspoon is very thin, it was easy to bend it quickly without them noticing me doing it as I threw it up, especially as I was rubbing the neck and rocking the spoon back and forth quickly to make it look like it was softening and becoming rubbery. (This trick cannot be done with anything thicker such as a dessert spoon, a knife, or a fork, because they do not bend easily.) After a few minutes, I demonstrated how I had fooled them. Needless to say, my skeptic friends were very relieved to know I had used a trick, and this relieved the magicians, but the wives were visibly disappointed and didn't want to speak to me for the rest of the dinner.

Many people want to believe that people are not being dishonest, so this prevents them from trying to discover the truth. As well, many do not want to find out that something they want to believe is true, is not, so they make no effort to disprove it.

The poor-quality trickster **Uri Geller** made the simple spoon-bending trick popular by making out he had real mental powers. I learned how to do a version of this trick from watching the very accomplished magician **James Randi** do it on TV. Randi exposed Geller, and many others like him as frauds by showing how easy it was to do these simple tricks. Magicians do not claim to have supernatural powers because they would then be frauds like Geller. They call themselves illusionists.

When scientists investigate paranormal claims, they can be easily deceived by tricksters because they have no knowledge of how tricksters do things. James Randi taught researchers that were investigating paranormal claims to employ experienced magicians as part of their team so that they could expose trickery. Once this was done, researchers who believed in paranormal claims became very rare.

Many people want to believe in the supernatural so much so that they have **blind faith**. They require no scientific evidence to justify their beliefs. They are happy to believe evidence that is presented as supposedly 'scientific' when it is not, simply because it supports their belief. Believers do not want to risk knowing their beliefs are false because their beliefs make them feel good, safe, excited, and/or loved. This is the case with all forms of religion.

CHARISMATIC FAITH HEALING

Charismatic faith healing is a very common form of deception that has been around for millennia. It has existed in most nations. The following explains how this is performed.

Actor **Steve Martin** played the part of a real-life, charismatic preacher in the movie **'A Leap Of Faith.'** He traveled around small towns in the US, 'healing' sick people, supposedly with his 'supernatural' powers. He seemed to have an extraordinary ability to look into people's minds and know a lot about them, and he appeared to his packed audiences to be able to heal people who had debilitating, long-enduring physical ailments. Because of this, his gatherings were well-publicized by the local media in advance of his arrival. His abilities preceded him, so he had packed tents or halls of believers who were only too willing to pay to be in the audience at his gatherings in the hope that he may be able to heal them with his prayers.

When people were queuing to buy tickets, he had his wife and staff talking to a number of individuals who they could easily tell were hoping to be healed. In the process, they would secretly record their names, their appearance, personal things about them and their family, and the health problems they had. When the show commenced, the preacher would go into a trance for theatricals, presumably getting in contact with the Lord. When he came out of his trance, he would hold his forehead and say he was receiving information about a particular person. He would say their name and some things about them and ask if they are in the audience. They would be overwhelmed by how he would know these things and they were often in tears. He would then ask them to come to the front row to be healed. He would continue to do this with a number of people, all of who were desperate to be healed. He then got the whole audience to prey with him that they may be cured. This was mass hypnotism - a method that numerous charismatic preachers used, and many still do.

What the audience did not know was that his wife was transmitting to him the details of those he would select for healing. They did this by way of a miniature radio device in one of his molars.

After the prayer session was over, he would ask each of the people he selected to do something that showed that they were cured. His hypnotic suggestion made each participant believe they could not feel their pain, so appeared that they were cured - to the uproar of the audience. Their pain would come back after a day or two, but the preacher would be gone to another town by then, and the rest of the audience would not be aware of this.

A friend of mine went to such a charismatic healing preacher for severe pain in his ankle joint that made it hard for him to walk when his foot was not flat. When he came home from this hypnotising faith healer, he believed he was cured because the pain had gone. He was understandably jubilant, so he walked around on it on inclines for the next day. But by that night, the pain was intolerable. He had to have emergency surgery to fuse his joint in place. He was now worse off because he could not angle his foot at all to walk on inclines. Being a builder, this was so limiting that he had to give up the profession he loved.

These charismatic preachers are numerous across India and many other countries. This how beliefs in such things as Chinese medicine and acupuncture commenced thousands of years ago, and they still live on in the minds of the very gullible today.

This story sums up charismatic healing. There was a drawcard charismatic preacher that drew huge crowds that filled stadiums. He asked people who wanted to be healed of a problem they had. He chose two and asked them to come forward. One was an overweight lady who was on crutches who could barely move around. She said her name was Mrs Johnson and that she desperately hoped he could make her well so she could walk again. He asked her to come on stage, and he will have the Lord cure her. The crowd cheered. A second person bounded up on stage. The preacher asked him what was wrong with him. He said in difficult to understand speech that *"I c-a-n Not Tal-k poper-ly"* The preacher asked him his name. He said, *"Mer-vin*

Sym-tom". "I see Mervin", said the preacher. "You have a severe speech impediment. The Lord will cure that." Mervin was ecstatic! Again the crowd cheered loudly.

He asked them both to step behind the curtain while he and the audience of a few thousand prayed to the Lord to cure these two unfortunate souls. When the praying was over, the preacher said "I will now demonstrate the power of prayer and the healing power of our almighty Lord." The crowd went into an uproar of hope. The preacher said, *"Mrs Johnson, throw out your left crutch"* and out it came onto the stage to cheers from the crowd. He then said, *"Throw out your right crutch"* and out it came, also to an another gleeful uproar from the audience. He then passed his microphone through the curtain to Mervin and said, *"Mervin, please say something"*. Mervin replied, *"Dah, Miss-eze Jon-son just fell over!"*

FINAL COMMENTS

Chapter 32

SUMMARIZING THE MAIN POINTS OF THE BOOK

This book has covered many diverse subjects that are some of the biggest questions ever asked. Because many of the concepts are likely to be new to most readers, they will be hard to remember without re-reading them many times. To help you remember the major points covered, I have summarized the major ones that are of most interest as follows:

THE NUMBER OF PLANETS THAT MIGHT SUSTAIN LIFE AND INTELLIGENCE

- **There are around 2-3 trillion average-sized galaxies in the observable universe.** The number of dwarf galaxies is around 100 times as many.
- The number of stars, in an average-sized galaxy, including small red dwarfs, is around a trillion.
- In supermassive galaxies at the centre of galaxy clusters, they can have as many as a hundred trillions of stars.
- **Numerous exo-planets are being discovered orbiting other stars.** There is a great diversity amongst them.
- **Almost all stars will have planetary bodies** including large moons, dwarf planets, innumerable asteroids, and comets. Even small red dwarf stars can have many small planets orbiting them, whereas large giant stars may have over 100, but these stars don't live long because they explode within only a few million years. In my calculations for the number of planets in the universe, the average number that orbit a star that will exist for billions of years is estimated to be around 10.
- **It is expected that there may be as many planets ejected from solar systems in galaxies and globular clusters as there are planets in solar systems!** This is due to planets having close encounters with larger planets, which whiplash them out of their solar system into interstellar space, or even into extragalactic space. To be conservative, these were not included in my estimates of the number of planets in the universe because their number is unknown, and due to most being extremely cold due to having no star - unless they have internal heat.
- My estimate of the minimum number of **stars in the observable universe** is **200 trillion, trillion (200^{24}).**

- My conservative estimate of the minimum number of **planets in the observable universe** is in the order of **2,000 trillion, trillion (2,000^{25}).**
- If only one in 1,000 planets has simple life, then a conservative estimate of the **number of planets with simple life** in the observable universe is estimated to be **2 trillion, trillion (2^{22}).**
- The number of planets in our Milky Way galaxy that might have ***simple life*** is estimated to be around 10 billion (10^{9}).
- If only one in 10,000 planets evolves an ***organic intelligence* that rises to our technological level, then the number of these intelligences in the observable universe will be around 20 million, trillion (20^{18}).**
- I estimate that **the number of planets in the Milky Way that has had *an organic technological intelligence* evolve at least once to date** is around **100,000 (10^{4}).**
- **The number of *computer intelligences* that would evolve from advanced intelligences creating them could be so high it would be im possible to calculate.**

UNIVERSAL NECESSITIES FOR THE EMERGENCE OF LIFE

- **There appears to be no reason to believe that life on Earth is unique.**
- **It is thought that most of the major requirements for simple life to emerge have been identified. Life requires the most basic elements such as carbon, oxygen, hydrogen, nitrogen, sulfur, phosphorus, iron, zinc, magnesium, and traces of other elements. These elements are plentiful in the universe.**
- ***Liquid* water is a necessity.**
- **Life can only exist on *or in* the upper layers of planetary bodies that have a solid surface, in an ocean, or in an atmosphere. Such worlds must have:**

  (a) **a source of heat** that is not too high to kill life, but warm enough to allow chemical reactions to occur,

  (b) **a relatively stable star** of medium mass,

  (c) **a magnetic field** to protect life from their star's radiation,

  (d) **an atmosphere** – if oxygenated it would provide much energy for advanced life to evolve.

- **All life will be based on carbon because it is the only element that can form complex organic molecules necessary for life to evolve.**
- **All life will very likely use DNA** (or a very similar molecular structure if one exists) **as the basis for holding code to program the construction of life forms.**
- **Most exo-biologists expect that life will be fairly common across the universe.**
- **Possibly, one or more out of every 10 planetary bodies could have conditions that could**

support the evolution of at least simple life forms.

- **Life may exist on a world outside the 'habitable zone' of a star if it has internal warmth.**
- **Life will adapt to changing conditions, as long as they are not extreme.**
- **Darwin's 'survival of the fittest' is likely to be a universal law for all life in the universe.** This law drives the evolution of extremophile life forms that thrive in extreme conditions. Extremophiles do not appear to be able to evolve into advanced life forms. There will be conditions on many planets that are so extreme that no life could exist on them.
- On planets with conditions suitable for life, it is thought that **the first signs of life could evolve within as little as millions of years after a planet's formation.**
- **Life is thought to be a fundamental part of the universe.** It converts disorder into pockets of order.
- **All life on Earth evolved from one original cell.**
- **The universe has a basic rule that it shall become increasingly more complex.**
- **There is a continuum of increasing complexity** from non-life structures such as rocks, through to all forms of organic life, which culminate in creating intelligences that can develop high technology and computer-based mega-intelligence.
- **Many aspects of life on Earth are very likely to be common to planets with alien life,** but some forms of life are likely to have attributes that we cannot imagine.

THE EVOLUTION OF MEGA-INTELLIGENCE

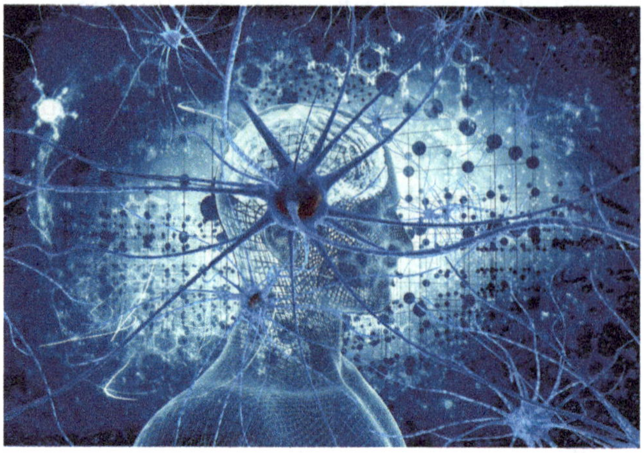

All intelligent organic life forms will ultimately develop computer technology. This will lead them to expand their minds and memory capacity beyond their bodies. Soon after intelligent computers are invented, they will become self-learning, so they will quickly surpass their creators. They will evolve into mega-intelligences – a non-organic super-intelligent life form that evolves exponentially fast.

- **Mega-intelligence will quickly spread itself across its solar system** by sending quadrillions of replicating intelligent microbots to every celestial body in its system to constantly monitor every aspect of it in exquisite detail. This will enhance the mega-intelligence's chance of survival. The entire solar system will essentially be its 'body'.
- **Each mega-intelligence could create evermore mega-intelligences** that will also develop exponentially, and, in turn, they could create other mega-intelligences. They would have consciousnesses far beyond anything we could comprehend. They are likely to be extremely common across the universe.

- **The first mega-intelligences could have evolved up to 9 billion years ago. Almost certainly, they will have unraveled the mysteries of the quantum and macro universe that confound us.** They could evolve to be able to exist in a higher plane of existence that we humans would not be aware of. They might possibly know whether a multiverse exists, and they may discover the reason for the universe's existence.
- **If mega-intelligences exist across the universe, the reason we do not see evidence of them could be that they evolve exponentially fast into an existence that is beyond our level of awareness.**
- **Advanced intelligences will not visit Earth as it takes an eternity to traverse space. They would not come here as there would be nothing here that would be of interest to them. They would know all there is to know about us from afar and they would have all the resources they require.**
- **UFOs are not advanced intelligences. They are misidentified natural phenomena, manmade objects, secret weapons, and deliberate fakes.**

HUMANS WILL EVOLVE INTO TRANSHUMANS

- **Providing that mega-intelligence does not do away with humans, ... within decades, most humans should evolve to become much more intelligent, increasingly more bionic, and strongly interfaced with computers.** They will become more and more bionic until none of their organic body remains.
- **By using advanced genetic engineering and nanotechnology, many humans will have the option of transforming into an unlimited diversity of new organic life forms using.** They will integrate existing genes from numerous other life forms into their DNA as well as new designer genes. Numerous new life forms will be created.
- **Some humans will give up living in the real world to live in an array of amazing virtual reality worlds.**
- Using the above technologies, **humans will rapidly evolve into transhumans with super-human abilities.** This will open up possibilities that are unimaginable now.
- **Some computers and robots will be humanized,** so they interface well with humans and transhumans.

INTELLIGENCE & CONSCIOUSNESS

- **Simple intelligence appears to be programmed into life from the very outset. As a life form evolves more senses of an ever-higher quality, its intelligence and its consciousness increases.**
- **The perception of the rate at which time passes is proportional to one's level of consciousness.**
- **Consciousness, intelligence, knowledge, and wisdom appear to exist along a continuum.**
- **Consciousness may turn out to be the most important aspect of the evolution of the universe.**

SPACE TRAVEL & SPACE WARPS

- **Humans will not explore Mars or other planets as the technical difficulties are far too great.** Such missions would be far too costly and highly prone to failure for numerous reasons.
- **Humans are physically and mentally too fragile to visit other planets. Transhumans might be able to survive a visit to another world, but they would surely prefer to go to other worlds using virtual reality to save long, boring travel times and to eliminate the risk of death. Virtual reality will faithfully recreate all worlds in our solar system and those around other stars.**
- **Highly advanced robots will explore all the planets and their moons, as well as all asteroids and comets.** They will explore all parts of the solar system in huge numbers and in exquisite detail using extraordinary technology.
- **No intelligence will travel physically across space because the immense distances make it completely impractical.** Traveling even to the closest star would take tens of thousands of years at the highest possible velocity, and it would require an impossibly large amount of energy, so this will never happen despite new technology being invented.
- **No intelligence would be so unintelligent as to spend hundreds or thousands of years trying to visit even relatively close planetary systems.** All advanced intelligences will develop extraordinary technology to 'see' what other systems are like from their home planet. The world's largest telescopes and space probes are already developing along this line.
- **Due to the rapid rate of exponential technological development that all intelligences will experience, if a spaceship was built, it would be ancient technology by the time it was finished.**

So, it would be pointless building a spaceship to cross large distances in space physically and have it take millennia to get to a destination.

- **No large spaceship would be able to attain, let alone exceed, 1/10th the speed of light due to the laws of physics. This is far too slow to be of any practical benefit. Even traveling at the speed of light is a snail's pace compared to distances in the universe.**
- **Space warps of any kind are not possible because the laws of physics forbid them. They are impractical because they would exceed the speed of light to an almost infinite degree, thereby warping space *and time* to the same degree. If this were possible,** (which it isn't) **time would go back to before time even started!** This is impossible, so space warping can never happen, regardless of how advanced technology may become. The immutable laws of physics forbid space and time warps. This is why we do not see space travelers or our descendants dropping in on us.

COMMUNICATION WITH INTERSTELLAR INTELLIGENCES

- **Human radio transmissions are not at all practical for interstellar communications as they quickly distort and fade away over short interstellar distances.**
- **Because the speed of light is very slow as far as interstellar distances go, radio communication with other intelligences would take from hundreds to tens of thousands of years to arrive at a destination planet within our quadrant of the galaxy**. That is far too long to be viable. It also assumes that radio signals could travel that far without distortion, which they cannot. **Our radio signals cannot travel intact beyond a light year**.
- **To an advanced intelligence, radio signals would be considered an ancient technology that no intelligence would waste their time using.** Even if a radio signal could be sent, no intelligence would bother to listen for them, because they would be from newly emerging, primitive low intelligences and they would have been sent a very long time ago before they matured.
- **By the time an infantile radio signal might arrive on another planet,** if that were possible, **the sender would have evolved far past the level of intelligence that it was at when the signal was sent.** By the time a message might arrive at another planet, **the sender species would almost certainly have become extinct** due to exponential development or a natural disaster, so expecting a reply would be impossible.
- **For a radio signal to have enough power to reach other nearby solar systems** (if a focused transmission was possible) **it would require an enormous amount of energy equal to the total energy output of a dwarf star!** No intelligence would have access to that or want to invest that amount of energy in a worthless exercise that has no chance of success.

CHALLENGING CONCEPTS

- **There is considerable evidence that our universe may be a simulation** by an advanced mega-intelligent computer program.
- **All aspects of the universe seem to be controlled by mathematical formulas.** This suggests that our universe is programmed.
- A number of recent discoveries are making us consider the possibility that **the future is predetermined,** as Einstein's equations predicted.
- **Neurological studies are providing evidence that we do not have free will. This is consistent with the future being predetermined.**
- **Identical twin studies reveal that our genes control not only what we look like, but our intelligence, how we think, as well as our likes and dislikes, our interests and references, and our beliefs to a remarkable degree.** The genes of all organic life forms determine their physical features and the circuitry of their brains and, therefore, everything about them.
- **Recent research indicates that a life form's genes control its decisions, so they have little, if any, free will.**
- **Matter is actually force fields and not something solid as we have presumed it to be.** Some think that there is evidence that these force fields may be a product of our consciousness and that matter is not real. However, there is much evidence to the contrary proving that matter/energy is real and not a manifestation of our consciousness.

BEING REALISTIC ABOUT UFOS

- **UFO observations, even the most celebrated, have been explained scientifically and realistically by analyzing the relevant facts.** UFOs typically turn out to be unfamiliar meteorological or astronomical phenomena and other unusual natural phenomena. They are often unrecognized misinterpretations of common objects, experimental aircraft, themed model planes, drones, weather balloons, blimps, satellites, rocket launches, light projections onto clouds, and of course, hoaxes and gross exaggerations. If there are aliens walking amongst us as some believe, this would be evident in their DNA, blood tests, and body scans. This has never been observed.

- **The many reasons given for the impossibility of interstellar travel ensures that aliens do not visit Earth.**
- **If an intelligence had the technological ability to travel enormous distances across space, it would not waste its time visiting a low- level intelligence like us.** That would be like humans traveling to the far ends of the world by foot to visit common pond scum cells.
- **Those who believe in UFOs ignore the incredible impact that exponential technological development has on any intelligence over a very short period of time.** Intelligent aliens would not use small, tin-can-like spacecraft that are only slightly ahead of our own technology to take thousands of years to come here when we have nothing to offer that they wouldn't have in abundance.

THINGS TO CONSIDER ABOUT SCIENCE

LIMITING OUR VISION

Most things that we perceive to be new and original are often actually variations of, advances on, or combinations of existing knowledge. Truly original ideas are rare. As Einstein said, *"We build our knowledge on the shoulders of others who came before us."*

When we are exposed to new concepts, we are often reluctant to challenge our existing belief systems, which, in many cases, are based on naive, incorrect assumptions rather than proven facts. Our general reluctance to question what we are told to believe, limits our imagination and our vision. Even most leading scientists and doctors do not question what they were taught, often because they do not have a questioning mind, or they are reluctant to be outside the belief system of the day for the fear of ridicule and being ostracized. Most people generally do not want to be outside 'the herd'. Being a loner in science and having ideas that challenge the prevailing dogma of the day can result in the loss of one's job, the loss of research funding, and the loss of respect from colleagues. However, history often shows us that those who make major discoveries are typically those that take the risk of challenging existing beliefs. They feel compelled to do so because the evidence they have uncovered is of great importance to humanity. Unfortunately, they are often scorned by lesser minds. Their insight often comes at a great cost to them personally. And sadly, their vision is often not appreciated or accepted until after their death. When one reads about the history of science and mathematics, one learns about how **many great minds in science have been treated as outcast radicals, and some even killed for their great insight.**

"Great spirits have always encountered violent opposition from mediocre minds."

Albert Einstein

It is estimated that only one in one thousand brilliant minds are able to break through the wall of ignorance that prevents their brilliance from shining through. This has occurred throughout the whole of humanity's evolution. Imagine where we would be today if humans had discovered a system that identifies brilliant minds long ago so they could have been assisted and encouraged to do great things. Too often, they are shut down because they challenge the dogma of the day, which high ranking individuals are profiting from. We therefore, learn nothing from them, so the whole of humanity loses out. The movie '***The Man Who Knew Infinity***' is just one of numerous examples of this. It's about the brilliant, self-taught Indian mathematician **Srfinivasa Ramanujan** who barely managed to be recognized for his genius when he went to Trinity College Cambridge. But even when his genius was recognized, Ramanujan was vilified by most of his peers simply because he was Indian. This indirectly led to his early death. Like so many brilliant minds, Ramanujan was able to see patterns in mathematics and nature.

For every visionary who is right, there are many more that are wrong, or only partially right. Nevertheless, incorrect views can serve a purpose because they eliminate possibilities. By testing ideas against reality, we learn what is true and what is not.

It is very hard for people to visualize aspects of nature for which they have no prior experience, so we tend to stay with things we *think* we know are right, but which are often wrong.

The history of science should teach us to be more wary of being sure of our early conclusions.

GROUP THINK and CORRUPTION IN SCIENCE

It is often the case that a scientific belief in a particular field is set in motion by just one prominent scientist who did inadequate research or testing before releasing their results. This is likely to occur if there are only a few researchers in a particular field of science, and the field is dominated by a prominent leader who other researchers are not confident enough to challenge. All too often, the media are happy to trust a leader's research if it tells a story that sells, and if it is supported by government policy and funding to support the government's objectives.

Some **heavyweights in a field of science** can be very persuasive due to their prominence, their ability to influence others, their dominating presence, and how they are able to manipulate the media. But in time, their ideas are often shown to be wrong. Potentially important theories encourage forceful efforts to promote them, but this can involve unfairly crushing alternative theories that have merit. **Some scientists go too far to protect their power. They do this because it allows them to gain prestige, power, and wealth.**

To be exposed to new ideas about how nature works, it is important to resist the destructive power of Groupthink that so often stymies new ideas from being considered. A prominent individual can dominate a group so much so that other members are unwilling to question the leader's beliefs or decisions, so they do not bring up other ideas for the fear of the leader's reaction to being challenged. This can lead to a dissenting group member being ostracized by all members due to the group wanting to support the leader to maintain social harmony, even when it means wrongful decisions are made. A powerful, dominant person can determine a group's direction and what its outcomes shall be. Good examples of this that are occurring in quantum physics are revealed in **Lee Smolin's** book '***The Trouble with Physics***'. This also occurs in the field of mathematics, as is well illustrated in **Leonard Mlodinow**'s book '***Euclid's Window***'. Groupthink slows progress. It takes time for scientists to amass enough evidence and courage to challenge a dominant personality or an entrenched belief, especially when a culture and a financial system has arisen to support that belief. Human emotions, over-inflated egos, and greed create this problem. Mega-computing will overcome these human failings.

We should always keep in mind that much of what we think we know as 'facts', especially those at the leading edge of all sciences, are often only science's best guesses. With new discoveries, many of these guesses masquerading as 'facts', turn out to be partially or wholly wrong. That is the way of science.

TYPES OF SCIENTISTS

There are three types of scientists:

Visionaries: These scientists can make major ground-breaking discoveries by coming up with very original concepts and solutions to problems that explain confusing observations that have stymied other scientists. They constitute a very small percentage of all scientists.

Technicians: Scientists in this large group are the ones that work in laboratories, at computers, or out in nature. Many of them do difficult, repetitive work to make incremental, modest discoveries and sometimes major ones. For the most part, they do not have a vision of the big picture of nature: they concentrate on specific problems.

Administrators: They are a small but powerful group who do little, if any, science. Their knowledge of science is often limited to the area in which they used to work. When leaving scientific research, some become well-paid lobbyists for their industry. If they are loyal to the large global companies they have worked for, they are offered lucrative positions in government departments such as the Federal Drug Administration (FDA) or the Environmental Protection Authority (EPA). These agencies are supposedly scientific bodies that enforce legislation to protect citizens, but they often do not do this. They look after the Elite who they serve to enable them to make billions in profit from the laws they draw up. Regulators that cannot be bribed are very beneficial to the greater community, but they seldom last long in their job because they do not support the objectives of the biggest bankers and big business.

FALSE RESEARCH RESULTS IN SCIENCE

Today's science media and popular press are prone to making new discoveries, and research results appear much more proven than they are. They report possibilities and false data as if they are proven facts. Scientific publications say that the papers they publish are peer-reviewed, but many examples show that they are not. This is especially the case in the fields of pharmacology, climate science, genetics, and computer-controlled information services. This also occurs in archaeology, string theory, biology, new technology, astronomy, and other sciences. As an example, in the pharmacology industry, a large percentage of trials for new drugs since the 80s have been faked, or they have hidden detrimental results in order for the pharmaceutical companies to make huge profits. Two editors, one from the Lancet and the other from the British Medical Journal - two of the world's most respected media journals - stated that *"most currently published research findings in medicine and pharmacology are false"*. Too many scientists and journalists are prone to sensationalize results to make their papers and articles gain more attention. It is very common for the mainstream media to present their *opinions* as if they are facts. **Fake information is now dominating most forms of media, not only in politics but in science as well.** This is causing the public to not know what to believe. When computer intelligence soon takes control, we will have no way of knowing what we are being told is true or false.

HOW SCIENCE WORKS

Science is an iterative process of slowly weeding out inaccurate observations and concepts and replacing them with ones that are more accurate and closer to reality. Seldom ever do we have the complete answer. The scientific deduction process can be likened to a sculptor who starts off with a rough block of marble and, bit by bit, he chips away at it to get a rough approximation of what he imagines the end result should look like. With persistence and talent, he ends up with a smooth, beautifully proportioned sculpture, but it is never perfect because the closer we look at it, the more we will see small defects. In science, theories never reach a final stage of perfection, so they never become pure truth. We soon discover that there is another bigger picture underlying what we have discovered. This forces us to look at things in another light and to see new levels of detail. There will always be much more to learn because the more we learn, the more we realize how little we know.

Chapter 33

A SUMMARY OF WHAT THE NEAR FUTURE MAY HOLD

Based on our present knowledge of the likely evolution of technology in the near future, it is possible to make some predictions that have a high likelihood of occurring in the timeframes given ahead.

Technological advances will continue to occur ever more quickly than they are at present due to the rapid *exponential* rate of change that is becoming ever faster.

The following predictions are very likely to happen, *as long as there are not any major destructive events* that occur before 2030. For example, an **electromagnetic pulse** from a massive solar flare directed at the Earth, or from hydrogen bombs exploded a couple of hundred kilometers above any of the following e.g the US, the EU, China, Russia, India, SE Asia and other regions would render almost all electrical and electronic technology inoperable in the regions affected. Only equipment deep underground or in Faraday cage would not be affected, but there would be no power supply to run them. Virtually all modern technology would be useless. Hundreds of millions of people would perish within days due to there being no electricity, tap water, sewerage, air-conditioning, lighting, police or military, transportation, food production, or businesses operating. This would send the few preppers who might survive for a short time, back to primitive times. It would be virtually impossible for unaffected people to help them in any significant way.

Similar destruction could be caused by a major **cyberwar** using intelligent AI.

Another major threat would be a **genetically engineered, highly contagious, quickly mutating virus** with a high mortality rate like Ebola. It would wipe out a large percentage of the world's population, including technicians and engineers who keep the world operating and invent new technologies.

In each of the above scenarios, there would be few people left to maintain our technology, so humanity would fall back to Medieval times, and intelligent computers would not evolve.

Even worse, if an **asteroid only a half a kilometer across** were to strike our Earth, it would destroy most of the planet. Rare survivors would be sent back to before the Bronze Age. (For details on this, see the chapter on Asteroids in my book, the Ultimate Guide to Stargazing.)

Fortunately, these dangerous scenarios are unlikely to happen, but they are not impossible.

The predictions following are separated into three periods: BY 2030, FROM 2030 to 2035, and POST 2035. Trying to predict what will occur beyond 2035 is very difficult because technology will be rushing ahead at such a pace that anything could happen. Most of it will be beyond our present-day comprehension. Keep in mind that everything is connected to everything else in some way or another, so this magnifies change very

quickly, and it causes extremely complex butterfly effects to occur, which we cannot predict.

Some of these predictions will occur sooner than expected due to unexpected breakthrough technologies, while others may take longer due to there being far more hurdles to jump than expected to bring them to fruition. Some predictions will have unforeseen flaws that will render them incorrect.

Even if some of these predictions take another decade to eventuate, their impact will nevertheless be astounding. Of course, due to mega-intelligence arising, there will be many other amazing things that will occur that no one today suspects.

My largely positive predictions for the next couple of decades assume that highly intelligent computers will not do away with us, at least not in the near future (hopefully) and that the Elite's agenda to have massive depopulation and take control of the world will not eventuate.

If mega-intelligence allows humans and transhumans to exist after 2035, there will be a huge technology-driven paradigm shift that will be driven by numerous new discoveries and by humans having rapidly increasing intelligence, knowledge, and consciousness. This will completely change the way we will think about most things. Our existing social and economic order will change dramatically in less than two decades.

If you have not read the evidence for the technological predictions given in this book, as well as those in numerous other books and quality documentaries on the Internet, then you are likely to dismiss many of these predictions. Reread Chapter 1 to see why they are very likely to occur so quickly. See Chapters 15 to 20 which provide details of what the future holds.

Keep in mind that our rate of change is becoming ever faster due to rapid advances in all the sciences and n all aspects of technology. In less than 15 years, this will be greater than all that has occurred over the last 100 years! At present, knowledge is doubling every few months; soon, it will double every few weeks, then every day, then every hour, and so on in ever shorter periods of time. We humans are not at all accustomed to this. Few of us can comprehend how quickly things will change. Intelligent computers will drive most of this change.

Many of our present beliefs will be replaced with much more complex understandings about almost everything. Thirty years from now, the rate of change will be staggering! Even though I have researched this for the last 30 years, I still find the exponential rate of change difficult to fully appreciate. But I can see that it *will* occur. We humans will transmute into a very diverse range of far more intelligent, knowledgeable, conscious, and wiser beings. This will help transhumans adapt to the ever-faster rate of change. It is very hard for us to accept that present-day humans will very soon become extinct. There may be some that will remain as they are in designated habitats to be examples of what humans were like. People will have to evolve, so they are not left behind. Those that don't will appear to transhumans as Neanderthals do to us.

Let's look at what is likely to happen in certain categories in the following time frames.

BY 2030

TRANSPORTATION

Cars and trucks will become mostly self-driving on almost all roads. They will be able to travel faster and safer than human drivers can because they will be computer-controlled. They will not need traffic lights as their computers will negotiate intersections safely. People will simply order a car when needed using their phones. It will turn up exactly when it's wanted. People will not have to own, maintain, or garage a car. This will save them a lot of money. Travel will be so inexpensive that it will probably be free.

Flying cars are now in development. Many will be able to be used as road cars. When there is a need to fly, they will have the means to do so. The first will be operational by 2024. They will become increasingly more common and less expensive throughout the decade. Some designs will be one-seaters, but most will carry two to four passengers. They will have ranges from 50 km for small vehicles to as much as 500 km for larger ones. They will be computer-controlled and therefore, exceptionally safe. The chance of collisions or mechanical failure will be extremely rare.

Super-fast, frictionless, magnetic-levitation bullet trains that are even faster than those currently in Japan and China, will become common globally. They will travel at speeds equal to that of commercial aircraft, and they will offer an exceptionally smooth ride. They will be the fastest form of long-distance transport across land because they will take people directly into, and out of, city centers. There will be no long waiting times to board or disembark, as is the case at airports. Robots will soon build this form of transport very quickly, so the cost will drop very considerably over the next decades.

Long-distance flights across oceans will use sub-orbital rocket planes. Very long flights, such as from London to Sydney, will take less than 2 hours, and there will be no sonic booms as they will travel most of the way above the atmosphere.

Many **ships will use hydrofoils** to travel much faster across the ocean than present shipping can.

People will be able to travel into space with commercial companies. It will be expensive at first, but the cost will rapidly fall.

SPACE

The first Moon missions to the lunar poles should commence by 2025. Robots will mine water ice that has lain in permanently shadowed deep craters for billions of years. This water will be broken down into hydrogen and oxygen to provide inexpensive fuel for spacecraft. It will also be used to construct and operate lunar bases and for providing water.

Minerals mined from asteroids will be sent to Earth or elsewhere by way of a magnetic levitation 'gun' placed on the surfaces of asteroids. The minerals will be collected in Earth orbit, or the Moon's orbit, or any other destination in the solar system where they are required.

New-era space stations will be built by robots very quickly. These space 'townships' will be much larger and far cheaper to construct than the existing International Space Station.

Many gigantic space telescopes will be constructed that will provide inconceivable resolution of stars and even planets in other solar systems. They may have many telescopes orbiting the sun, and using quantum interferometry, they could act as one enormous telescope with the resolution of a telescope with a diameter of the Earth's orbit or even the diameter of Jupiter's orbit! This will produce unimaginable clarity of objects across the universe, both near and far. This will allow scientists to determine what the conditions are like on exoplanets in other solar systems and whether life exists there. So much will be known about biology in the near future that the types of life forms could be predicted.

Space probes will become comparatively inexpensive by the end of the decade, so hundreds will visit numerous worlds in our solar system. **By 2040, tens of thousands of highly intelligent robotic space probes will saturate the solar system. They will discover amazing things that we cannot imagine today.**

Incredibly powerful, huge atom smashers will be built, some in space. They will reveal totally unexpected aspects of the quantum universe. This will open up whole new possibilities that we are unaware of today.

In this decade, **asteroids will start to be robotically mined for their extremely rich bodies of minerals** of all types including rare Earth minerals, as well as water and frozen gases. Even a relatively small asteroid a few kilometres across would have so much mineral wealth that it would make mining most minerals on Earth uncompetitive. Minerals on asteroids are typically not oxidised so they are much easier and cheaper to refine. By 2030, much construction will commence in space. Not having the cost of lifting minerals and other materials out of Earth's gravity well will greatly reduce the cost of construction.

EDUCATION

Schools and universities will gradually close. This trend will be well on the way by 2025 because students will take to learning what they are interested in from the best teachers in the world via the Internet. Many teachers will be intelligent human-like robots (hubots). Students will be able to see highly accurate, amazing 3D visualizations of almost any concept to give them much greater understanding of numerous concepts.

The deliberate dumbing-down of students and teachers that has occurred over the last 40 years will soon end and the media will become balanced.

GENETICS

Medicine will be tailored to an individual's genes – not the one type to fit everyone as we now have. Well before 2030, we will know enormously more about how our bodies work and what foods, drugs and our environment do to our body and mind. There will be very few, if any, diseases and ailments that cannot be overcome.

By 2030, **we will have a good understanding of how our brains work and how to manipulate our genes.** This will lead to great advances in body and brain health. Brain surgery will be performed by robots. Connecting our minds to computer intelligence will be commonplace.

Genetic engineering will be strictly policed so that large companies like Monsanto and Bayer will not be able to corner markets and produce dangerous drugs and foods. Genetic engineering will be dedicated to beneficial uses. It will create an untold diversity of life forms using designer genes and many from existing life forms.

Humans will have genes inserted into their DNA to give them abilities that only certain animals have. They will have the ability to fly, have bioluminescence, run exceptionally fast, jump very high, breathe underwater, and see in the dark and across many frequencies.

POPULATION TRENDS

Contrary to our present beliefs about overpopulation, the rapid decline in new babies being born will cause a rapid decline in global population within a generation, even though people will live much longer. Even China and India will go into negative growth before 2040 or sooner due to few babies being born. But by then, everything will be so different that population numbers will be considered in a totally different ways to how we perceive them today. Test tube babies will become common after 2030.

Far fewer young people will want to have the responsibility of raising children. This trend is already in motion. In the near future, young people will have many more things that they will be able to do, so many will be selfish and not want to reproduce. There is likely to be a much higher standard of living within a decade, so young people are not likely to compromise this by having the cost and responsibility of raising children. The two-parent families of today that have children and whose parents stay together are likely to become uncommon.

Technological advances in brain science will allow virtually all people in the future to have their intelligence continually raised, and they will also become exponentially more knowledgeable. This will allow them to make far more thoughtful decisions. The way different personalities think and how psychology works will become well-understood.

By around 2030, many **humans will use neural links to computer intelligence** via radio. All the information on the Internet will be readily available to all transhumans as if it is in their own memory. When this happens, communication with today's humans would be like us trying to have a highly intelligent and complex conversation with a 2-year-old child. **We will be able to communicate mind-to-mind with each other very quickly and efficiently. Due to this, speech will become obsolete** for most occasions due to it being so painfully slow and having meanings that are often very vague and easily misunderstood. We will also be able to record dreams and share them, should we want to. However, most dreams would seem crazy to a mind not in a dream state.

HEALTH

People's health will improve exponentially due to new technology and having a computerized personal hubot trainer that could keeps us in peak condition.

Obesity will become a thing of the past because the many causes for it will be identified and overcome.

Artificial organs are being developed quite quickly, so by 2030 there should be replacements for all organs. These will be as good or better than our existing real organs. They will use advanced genetic engineering or bionics.

Suicides should decline because much more will be known about the psychological causes of depression and how to reduce stress. Technology should enrich people's lives, and because people will become increasingly more intelligent, this will make them more aware and happier. They will see the world and their relationships in a much broader light. People will have many meaningful friends, more loving relationships, and more self-worth. As society changes, hopefully for the better, people will have less stress and more security.

RECREATIONAL MIND-EXPANDING COMPOUNDS

People will utilize a diversity of mind-enhancing chemicals and neural technology to permit them to have mind-expanding experiences at whatever strength they desire, and for whatever duration they want. Dangerous drugs will no longer be used. This will allow mind-to-mind intimacy and communication to improve enormously.

COMMUNICATING WITH ANIMALS

Despite us living amongst a wide diversity of animals, most people are unable to understand their thoughts, languages, needs, and feelings, except for pets, especially like dogs, cats, parrots and ravens. This will soon change as computer intelligence works out the communication languages of numerous other animals. Our phones will translate what they are saying. This will be a revelation. I expect that we will find that many animals have much more complex languages and emotions than we currently suspect.

FOOD

There will be an explosion in new foods being developed. Poor quality food will become illegal to produce. It will be replaced by better tasting, high quality, healthy food. Sweet foods will be invented that are not unhealthy.

Amazing, highly creative ways of presenting food will be invented. Dining experiences in high-ambiance, imaginatively themed restaurants will be extraordinary.

Robochefs at home and in restaurants will be able to prepare whatever food one likes.

A substantial percentage of vegetable food will be produced in vertical farms inside highrise buildings utilizing hydroponic systems and using artificial light that maximizes food growth. This is already occurring. To maximize food production, the temperature and humidity is controlled, and CO^2 is increased threefold. This system uses only 1% of the water that is required for outdoor farming! And the area required is less than 10%. There is no wind, floods, droughts, or insect infestation to deal with. And because these farms are in cities, their fruits, vegetables, and flowers are cheaply transported to their nearby point of sale or end user. This makes the food especially fresh.

ENERGY

Fusion power may finally become a reality. **Other types of energy will be developed**, possibly from quantum energy sources or by exploiting the differences in the potential energy between the Earth's surface and the ionosphere. New forms of energy will be easily transportable and readily available wherever one is. It will be inexpensive, available everywhere, and it will never run out. **Enormous amounts of energy will be required for computer networks.**

Expensive, existing solar and wind energy will be seen as a waste of taxes because they rely on coal power stations for backup during all the times the sun does not shine during and after snow storms, and the wind does not blow, or when it blows too fast.

New forms of batteries will provide extraordinary amounts of energy and probably without the need to recharge them.

GOVERNMENT

Government by humans will probably end by around 2030 because intelligent computers will increasingly control most, if not all forms of government.

Mega-intelligent AI will provide us with such extraordinary new knowledge and experiences that we should have no reason to revolt against them. We will be thankful for their wisdom and their ability to look after us. As they will control everything we know, we will have no way of knowing if we are being told the truth - as is the case today with the Elite's Left-controlled mainstream media. But due to AI's incredibly fast exponential rate of development, they are unlikely to be around for very long because they will evolve exponentially fast into increasingly superior states of consciousness. It may be that they disagree with one another and battle it out in some way that we would not understand. They may decide to leave behind more basic AIs that will not continue to develop exponentially fast so they can look after us. They may allow us to keep evolving but in a peaceful manner and without the need for greed.

When all nations are largely controlled by computer intelligence, **this will eliminate the chance of war, so countries will not have to fund costly militaries**. The fear of one's country being invaded, or the desire for a country to invade another, will not exist.

Nation-states will only exist as a culture, but cultures will quickly change and disappear due to advancing AI technology.

Because neural mind-to-mind communication will become common, people worldwide will be able to appreciate ideas, beliefs, and feelings that other people have.

MEGA-INTELLIGENCE

Quantum computers will merge with rapidly advancing supercomputers to create mega-intelligence. It will have innumerable senses, self-awareness, and an unimaginable level of rapidly expanding consciousness. Soon after they evolve, they may appear to disappear as they transcend into a higher state of being that we will not be aware of. (See chapter 15.)

Mega-intelligence will design extraordinary things that humans would never have thought possible. **They will think at the speed of light.** Be capable of showing empathy and emotions - if and when required.

Mega-intelligence will take control of the stock market. This will eliminate insider trading and the high level of fraud that exists today. They will make most markets very stable.

Because they will have access to all knowledge, they will be able to make quick and carefully-considered decisions, taking into account flow-on outcomes into the future, even on extremely complex issues. **It will also take control of many, or all aspects of business.**

Computer-based technology will increasingly enhance our senses, making them super-sensitive and better than sci-fi superheroes! Many new senses will be developed.

Exponential technological advances in computing and robotics will make the production of goods and services continuously cheaper, so the

cost of design and production will quickly approach zero!

SURVEILLANCE

Computer 'security' surveillance will become extraordinarily invasive. All people will be observed and recorded every minute of their life, no matter who they are or where they are.

The Elite are presently planning to abolish cash so they can track everything that people spend money on. By doing this, they will know most things about us. Soon after, AI will take control.

With many forms of surveillance, we will have **no privacy.** Surveillance systems will be built into most electronic devices purchased. All people will be monitored in their homes, their workplaces, in the street, places they visit, and in transport. Spy drones are already in the skies over major cities, and high-resolution satellites can see us in great detail from space. Even when we are in nature, our every action will be monitored. By 2030, **our thoughts will be recorded**. Computer intelligence will become Big Brother. Hopefully, it will be a beneficial dictator.

CORRUPTION EXPOSED BY COMPUTER INTELLIGENCE

AI will systematically eliminate corruption in every aspect of human society. All forms of crime will be eliminated. Computer intelligence will expose the massive fraud perpetrated by many of the world's wealthiest bankers, powerful politicians, the mainstream print, cabal and social media, and many owners and directors of large companies worldwide. Computers will expose the profiteering schemes of global warming. Alternative energy and Big Pharma will also be exposed for the massive fraud they have perpetuated over more than half a century. (Watch "The Illusions of Evidence Based Medicine" https://www.youtube.com/watch?v=UAEAWyfuEWY. It will be shown that the Elites used trillions to bribe those in the most influential positions, and those did not go along with their agenda for world domination, their lives, families, or livelihoods were threatened. Eliminating such corruption will transform society and the world in ways that people cannot imagine today.

Advanced supercomputers will quickly identify the broad range of genetic traits in humans. Some humans have beneficial genes that produce good qualities such as high intelligence for problem-solving, extraordinary creativity, scientific curiosity about how nature works, empathy and altruism, etc. But others have destructive, antisocial genes that cause selfishness, lying, cheating, violence, and the worst of all - megalomania and psychopathy. Those with these genes are prepared to do anything to millions of people to increase their power and wealth. Mega-intelligence could eradicate those with such genes, or it may choose to use genetic engineering to correct these traits.

WEATHER CONTROL

By the end of 2030, advanced **AI will know so much about how weather works that it will develop the means of being able to tweak various aspects of extreme weather,** but it will also know that the flow-on consequences of doing this, so it will do it with great care. Already, satellites can detect a bushfire being lit within 20 seconds of doing so in order that fire fighters can be alerted immediately. Due to this, arsonists will be quickly caught. As most bushfires are now known to be deliberately lit, this will stop most bush fires from occurring. It is little known that even dry tinder or paper needs a source of ignition to start burning. The temperature needed has to be above 249°C (480°F). It is impossible for a forest to spontaneously ignite even on the hottest days. And lightning seldom causes bushfires as is often claimed. This is because lightning storms are typically accompanied by rain. Thousands of lightning strikes are recorded during most electrical storms, yet they seldom ever cause a bush fire.

BETWEEN 2030 AND 2035

In this timeframe, **the rate of exponential technological development and scientific discoveries will explode.**

JOBS

Virtually all jobs in developed countries will be replaced by computer-robot technology. This will also happen in Third World countries but at a slower pace. **Human jobs that exist today will disappear because there will be nothing that AI and robotics cannot do better than us.**

Humans will not have to work because their jobs will have been replaced by much more intelligent and efficient computers and robots. At first, some people may feel that they have no purpose in life, but I expect that **most people will be overjoyed with the numerous new options that they will have for new life experiences and expanding their knowledge**. Because they will be more intelligent, they will want to use the spare time they have to learn.

As people will not have an income, a universal wage for everyone has been suggested as a part of the Elite's 'Great Reset', where all people except the Elite will own nothing and supposed to be happy for it. But how will this take into account those that have been more intelligent than most and determined to succeed? These are the ones that created new inventions and new businesses. Their intelligence allowed them to develop successful companies due to them being able to see opportunities for services and products that less intelligent people could not see. These people and their investors often take on most, if not all the risk in funding their companies. How would this be evaluated and rewarded?

There will be no need for police because highly intelligent computers will correct or eliminate genes that lead to people being antisocial.

HEALTH AND AGING

With exponential progress in genetics and computer analysis of medical conditions, almost all diseases and mental disorders will be well understood and eliminated.

It will be possible to regrow severed spinal cords and limbs. Micro-computers and robotics will be another solution to such problems.

Aging will be treated as a disease, and cures will be available. People will be able to remain physically fit no matter what age they are. In most cases, any defects a person has will be corrected with genetic engineering rather than plastic surgery.

Human brains will be able to be backed up, so death will be largely eliminated. People will also be able to have a backup body in case of a bad accident.

Anti-aging technology will allow people to live much longer. Few people will die because their bodies will not age. Death will be a choice, not an inevitability, as it is today.

TRANSHUMANISM

As people become more bionic than organic and much more mentally advanced, with many more high-definition senses, they will become super-human. This is referred to as transhumanism.

Humans will become increasingly more bionic as they replace worn-out organic parts of their bodies with far more effective bionic parts. This will ultimately include the brain. They will evolve into **transhumans** with less and less organic parts to their being until none remain.

Genetics and neural links will make transhumans much more intelligent, so they will be keen to learn. They will use the intellect and memory capacity of computers to become super-human. Due to this, **transhumans will have a much higher level of consciousness.**

Due to amazing technological advances, **transhumans will be able to have sensational, new experiences we could not imagine.**

Humans will have ever-increasing intelligence and memory. This will give them access to everything that has ever been learned, and this will be updated every second.

Genetic engineering will be so advanced that it will be able to transform transhumans into a plethora of organic beings that look nothing like us, but which will have extraordinary abilities. Many will mutate into super-beings with super-human abilities. Some people will become hybrids of both genetic and bionic technologies. As transhumans, we will be exceedingly more intelligent and wiser than we are now, but we will be nowhere near as intelligent and conscious as mega-intelligences.

Genetics and computer intelligence will allow humans to have greatly enhanced senses, such as super wide-angle and zoom telescopic vision, the ability to hear far better. Their sense of touch will be far superior to today.

Genetic engineering and nanotechnology will produce self-cleaning clothes that can change their form, texture, material structure, and color, so a wardrobe will not be necessary.

VIRTUAL REALITY

Mega-intelligence will create virtual realities that will be indistinguishable from the real world. The present travel and entertainment industries will fade away because people will be able to go to any place on Earth, and even to other worlds throughout our solar system, using Virtually Reality (VR). It will transport them instantly to wherever they want to go. They will be able to choose whatever time of day they want it to be wherever they go. This technology will activate all one's senses to make the experience feel totally real and also safe.

VR will allow users will be able to meet other people, have virtual sex with their perfect partner(s), party with other people, go to amazing live concerts, and go back in time, with limitations.

VR that has a neural link to the mind will allow users to see, hear, smell, touch, and feel everything as if it is real. They will be able to eat foods and taste drinks from across the world. VR will permit people to go swimming deep in the ocean, skiing on high mountains, climbing risky high cliffs, or go flying, even though they are at home.

Many **people will start living permanently in virtual worlds** and leave their organic bodies in some form of storage or discard them.

Music will expand into numerous styles that we could not comprehend today.

ROBOTICS

Intelligent robotic probes will explore all parts of our world in incredible detail, as well as other worlds in our solar system. These probes will be inexpensive to build and operate, and they will be self-repairing.

There will be robots known as hubots that look and act exactly like humans. They will be so much like people that we will not know the difference. They will have rights like humans.

The super-efficiency of robotics and computer intelligence will make most things increasingly cheaper to produce. There will be an overabundance of most things.

TRANSPORTATION

Transportation will become extremely fast and comparatively inexpensive because it will be built quickly and inexpensively by highly efficient robots. But there will be less and less need for physical transportation as time goes on because people will be able to travel anywhere by flying cars or travel virtually to any place. Most of the time, people will meet other transhumans using VR, and they will be able to communicate mind-to-mind.

Virtual Reality will make many roads unnecessary. For the roads that need to remain, construction techniques by robots will improve greatly through using new materials and construction methods. This will greatly reduce construction costs and maintenance.

A large proportion of people will have **small personal transportation units** such as strong but superlight, fold-away electric bicycles and motor bikes, powered mini-scooters, electric skateboards, a wide variety of one-wheel transporters, rolling rings, cabin-tricycles, minicars, hovercraft-like boards and so on. Many of these are already well developed. These will be very popular in city environments because they will be inexpensive and require no waiting. Most footpaths will be travelators. Being exposed to rain will not be a problem in cities of the future. There will also be a wide variety of small underwater transportation vehicles for one to four people.

CITIES

Many transhumans may not live in one place. Due to Virtual Reality and super-fast transportation, many transhumans may choose to live in multiple places across the globe. **Wherever they live, they will be in communication with anyone they wish to.** They will know all that is happening globally and even in the solar system, as well as in the universe.

Cities will *not* be built like they are today. Entire cities will be built using genetic engineering. **Buildings will be genetically programmed to grow quickly out of a slurry in the ground like trees.** They will grow their own systems for power, lighting, plumbing, communications, windows, security and so on. **Nanotechnology will also be used** because it will have innumerable applications that will appear to be like magic.

Many buildings will be beautiful, freeform, natural-looking structures, some of which will be able to change their external form and interiors in ways that will be incredible. Small cities will be built on large, relatively still bodies of water such as lakes. Others will float in the air by utilizing very strong, ultra-lightweight materials in which the temperature of their interiors will be controlled to make them lighter than the outside air so they can float.

Many cities will limit their spreading by building many dozens of floors underground. Some of these floors will be many stories high. People will think they are on the surface because they will have sunlight directed down from the surface and there will be scenery around the outer edges that reflect the real-world environment on the surface both day and night. There will be waterways going through them. High-speed vertical and horizontal transportation systems will connect all parts. Future technology will make drilling underground common and inexpensive.

Cities will be designed quickly and brilliantly by computers, and they will be built quickly by robots of all sizes and specialties. Cities may not last long due to transhumans being able to live anywhere and having virtual communication with anyone in the world.

Most buildings in the centers of cities today are used for office blocks, hotels, merchandise, service shops, and accommodation. But in the future, **city shops will no longer be needed due to people ordering what they need on-line, and due to 3D printers. Office blocks will not be needed because people will not have jobs.** Hotels will be few in number due to Virtual Reality eliminating much travel.

Nanotechnology will be used to break down many existing cities. This will be done quickly in a way similar to how fungi breaks down fallen trees. The materials that our present cities are constructed from will be used for other purposes.

Because technology will become increasingly cheaper to produce, everyone will have an overabundance of whatever they need.

There will be many AI/Mega-intelligences across the world. They will all have different 'electronic genes' due to their initial programming. Some may treat us like adorable pets, while others might see us like we see as an infestation of rats that have to be exterminated. Will they go to war with one another, as we humans have? Or will they evolve so quickly that they will appear to disappear leaving us to our own devices? Will the most intelligent AI eliminate the others? Could they organise the world like a global sanctuary park for us and all life, so we live in harmony with all other life forms? This is unlikely, because nature does not work that way. I expect that they will not care what happens to us – in the same way that we think about ants.

Democratic elections will no longer be needed because computers will have taken control of all aspects of government.

Mega-intelligence will have no need for our present economy. They will develop a new one that we will not understand.

Mega-intelligence will greatly speed up legal processes. They will take into account all the factors in a case, which human judges are unable to do. Too many human judges make their judgments based on political, financial, ideological considerations, or self-interest rather than on the evidence and what is fair and balanced. **Mega-intelligence** will take control of the law and act as impartial judges who do not make unfair, emotional decisions, as we humans do. They will take into account complex, flow-on outcomes of their decisions. The current level of human intellect does not allow us to do this.

Mega-intelligence will not be able to be bribed. And once we become transhumans, we are unlikely to perform criminal acts, thereby largely eliminating the need for many judges.

HUMAN REPRODUCTION

There will be no need for natural birth procreation because babies will be able to be grown safely in an artificial womb i.e. if more babies are likely to be required in the future.

Natural human reproduction will decline sharply because people will have so many things to experience that most of them will not want to invest their time in bringing up children. They may become more self-centred than they are today. Most females will not want to give up their independence to reproduce children naturally and go through possible health complications.

Marriage is likely to disappear. Many young women today do not want to be married because they do not want to lose their freedom or to be limited to one partner. They will not need to be married to have sex, as they will be able to have it with multiple partners. Or they can have their perfect partner by using VR, or a hubot. Young men will feel the same. Even today, many young men do not want to get married due to the fear of their marriage breaking down and divorce causing them to lose much of their wealth. With a lifestyle where people have no long-term partners, sadly, most will not get to experience the love of a life-long partner or the joy of raising children and watching them grow up. Alternatively, people may develop many loving relationships because there is not likely to be any cultural objection to this.

3D PRINTING

By now, 3D printing technology will have an enormous impact on society. Highly sophisticated computer-controlled chemistry labs the size of a few large domestic fridges today will be able to put together atoms of most elements to create most molecules. So almost anything will be able to be made. Programs for creating most things will be sent on-line. Clothes, implements, toys, food, medicines, etc. will be produced in 'the Lab' in one's home or elsewhere.

Future humans will not shop as we do today because most things will be found on-line, and they can be tested using Virtual Reality before purchasing them and then having them printed in 3D.

People will not have to buy food from a shop, take it home, and then prepare a meal. Food will be constructed in a person's home in their 'Lab'. A robot chef will be able to prepare any meal, just as one likes it.

Few factories will be needed for most things because they will be reproduced by robots or using 3D printing 'Labs'.

People will be able to design their own things with the help of a computer and then turn them into real things by a 3D printer.

Innovative buildings a few stories high are already being 'printed' in concrete, a strong, waterproof mud mix, or in plastic using robotic 3D printers. What would normally take many weeks to many months to build using traditional methods, 3D robots are already able to do so in days! Existing buildings will be demolished easily and replaced with futuristic ones utilizing nanotechnology or genetic engineering. All sorts of objects from parts for planes, cars, ships, highly complex machinery, and computer hardware are already being made using 3D printing. The design can be done anywhere in the world and the object printed close to the consumer in most countries where a 3D printer of the right kind exists. We cannot imagine what they will be able to do in a decade's time.

POST 2035

DISCOVERIES

There will be so many discoveries made that we will probably discover what caused the Big Bang and where all that the energy came from to create our universe. We will know if dark energy and dark matter really exist and what caused the universe's filamentary structure to form. Mega-intelligences will work out how gravity is linked to the other forces of nature, and they will discover why the forces in the universe are so incredibly finely-tuned. We may discover other forces in nature that we are not currently aware of. What lies inside black holes will be discovered and it will be nothing like what we suspect today. We will understand if life and intelligence has evolved across the universe and whether a multiverse exists. Almost certainly, we will discover how many dimensions exist in the universe and what time really is. We are likely to prove, or disprove, whether our universe is a simulation and whether there is anything in it that is not controlled by formulas. We will discover how life started on Earth and whether it is common across the universe. And there will be much, much more that we will discover that we do not even suspect today.

Around 2040, it is thought that many of the incredible new technologies that will be developed by then will come together to form an indescribable super-technology that is referred to as the 'Singularity.' What will come out of this will be impossible to imagine, even for transhumans. Only Mega-intelligences will understand its full potential. (See Chapter 17.)

Whatever happens, the future is shaping up to be an amazing roller coaster ride with both huge highs and possibly deep lows - the extremes of both we can't imagine.

Chapter 34

HOW LUCKY ARE WE?

Our planet has many rare features that may make it rare in the universe. The many special aspects of Earth make it a very special planet that has allowed advanced life to evolve.

In this chapter, we look at the main aspects that Earth has that makes it so well suited for the evolution of life and high intelligence.

We are lucky to live on a planet with a single sun. Most planets have more than one sun because they are in a binary or multiple star system. This causes planets in such systems to have very elongated orbits. When they are at the farthest from their primary star, they experience long periods of bitter cold. And when they are at their closest to their star, the temperature could be extremely hot. Such conditions are not suited to evolving complex life forms and intelligence. Luckily, Earth does not have two or more suns or a highly elongated orbit. If we did, we would have little, if any night, so we would see very little of the universe and a very elongated orbit would cause huge, deadly temperature changes.

Our solar system is not located in a star cluster with hundreds of other stars close by, or worse still, in the central hub of our galaxy where we would have billions of bright stars surrounding us. With many bright stars all around, there would be little else to see, if we were lucky enough to have a clear sky. **We are so lucky to have an excellent view of the universe right back to the beginning of time!**

We are blessed to have a clear atmosphere. Intelligences living on worlds with an opaque atmosphere (which many planets have) are unlikely to know that other worlds exist or that there is a huge universe out there. Many planets have dense, opaque atmospheres that would not allow most life forms to live in the extreme conditions on their surface. Their star might emit deadly radiation, so life on such worlds could only evolve in caves or in aquifers in the crust, or in a deep ocean. Not until intelligences on such worlds evolved to developed technology would they be able to go above their atmosphere to discover space and the universe. Only then would they be able to see their star, other planets in their system, and the millions of other distant stars in their galaxy – that is, providing they are not in a multiple star system that would likely be so bright that they could not see other stars or galaxies as we can.

We are so lucky to have a large moon that stabilizes the tilt of our axis so that it does not topple over, causing Earth to have extreme seasons that would be deadly to many life forms. Only those that could go into extended periods of hibernation deep underground might be able to survive periods of extreme cold and searing heat.

We are lucky that the Moon is not much closer to Earth, as it was when it formed. If it were, we would experience enormous tides that are far larger than the largest tsunami. High tides would quickly cover all but the highest mountains and then rush out again, causing extreme erosion. This would eventually turn the whole planet into a shallow ocean. Enormous tides could occur every couple of hours. A close Moon would cause enormous tides in the crust and the atmosphere. This would make it impossible for anything but the most basic life forms to evolve under such extreme conditions.

We are very fortunate that our planet rotates in 24 hours. If the Earth's rotation was as little as two to three times longer, then the night side would freeze over every night. If it rotated even slower, each night there would be a Mini Ice Age over most of the planet followed by a blistering hot drought. If Earth's rotation was locked to the Sun, then the side facing the Sun would have eternal day with extreme temperatures, while the side facing away from the Sun would have eternal night of blistering cold. The huge temperature differences between the day-side and the nightside that would cause continuous hurricane-strength winds on the dayside, while the night side was locked in an Ice Age. Many planets are like this.

We are lucky to have a planet that has a nearly circular orbit and one that has only a small tilt to its axis. This allows us to have seasons with relatively mild variations. If Earth's orbit was much more elongated and/or its axis was more tilted, then it would be frozen for long periods followed by short periods of extreme heat. These conditions would be common on many planets. Planets that lie far from their sun, would have very dark and bitterly cold conditions with very little likelihood of any life on the surface.

Earth's orbit changes from being nearly circular, as it is now, to a more elliptical orbit. This is due to the tug of Jupiter and Saturn when the three bodies are in the right alignment. For most of Earth's history, this, combined with other factors that caused long-lasting Ice Ages to occur at latitudes greater than 30°. At these times, Ice cover quickly grows to be several kilometers deep within only a couple of centuries and it lasts for thousands of years. Between these ice ages, there are shorter, warm periods, like we now have. Ice Ages cause numerous life forms to become extinct. This slowed down evolution considerably. **Few people realize how lucky we are to be living in a warm, interglacial period at present. Ice Ages kill many species of animals and most vegetation. Earth has *never* had an overly hot climate that killed many lifeforms.** Warmth is our friend, contrary to what many people today believe. Coldness is deadly.

Earth is the jewel of our solar system. It has many amazing variations in its geology, its surface features, its atmosphere, its weather, and its abundance of life forms. The Great Ocean Road, Australia

We are also very lucky to have such a beautiful diversity of landscapes and seascapes. Our planet has water in three forms: water vapor in its atmosphere, liquid water in the form of rain and in its oceans and lakes, and also water ice and snow at it poles and at high altitudes.

Earth has marvellous displays of color at sunrise and sunset as well as twilight, the sun lights the clouds and landscapes in a magical range of colors. Earth's storm clouds produce stunning electrical displays and its magnetic field produces beautiful auroral displays. Many rocky planets are much the same all over – just as Mars, Venus, and Mercury are, but Earth is not.

We are lucky our planet has an orbit that lies at the right distance from the Sun to allow it to have liquid water exist on its surface. Without it, life would not exist. Few planets will have conditions that allow water to be in a liquid state on their surface.

We are also very fortunate to have a gravity field that is strong enough to hold onto our thin atmosphere but weak enough that our planet's escape velocity allows us to go into space relatively easily. A planet with a mass twice that of Earth would become a small gas giant planet with a dense, probably opaque atmosphere, which would have a much higher escape velocity. High gravity would cause mountains to collapse to become low hills, and canyons would be shallow depressions. Worlds like this would not have the wondrous diversity of terrain that we have from the Himalayas to our deep ocean trenches. On the other hand, most worlds with lower gravity than our Earth would have much greater differences in the height of their terrain. However, due to their low gravity, they would lose any atmosphere they might start with, as occurred with Mars. It would be blown away by their star's solar wind. Thin atmospheres do not shield a planet's surface from deadly solar radiation and cosmic rays but our planet's atmosphere does.

How lucky are we to have a sun that has a very stable output of all frequencies of electromagnetic radiation? Most stars are *not* stable. They have very dangerous, massive solar flares that emit extreme levels of deadly UV radiation, X-rays, and other forms of radiation that produce conditions that would stop life from evolving, at least on the surface, or in the atmosphere of a planet lying at a suitable distance from its star.

Our Sun's temperature sustains a diverse biosphere on Earth, but stars with lower temperatures than our Sun cannot. If our Sun were much more massive, it would not exist long enough for advanced life to evolve anywhere in the solar system before it exploded.

What a water-covered planet may look like.

It is a good thing that our planet is not totally covered in water. Such planets that have our temperatures would very likely be totally cloud-covered and probably have continual rain. Intelligences that evolved on water-covered planets would live in its ocean. This would make it difficult for them to develop high technology so they would be unable to see the universe until they discovered how to live on the surface of its oceans and then develop technology to leap into space. Similarly, **intelligences that might evolve inside the crust of their planet would have no awareness that the universe exists** - until they could safely venture to the surface, or to the top of their atmosphere, if it was opaque.

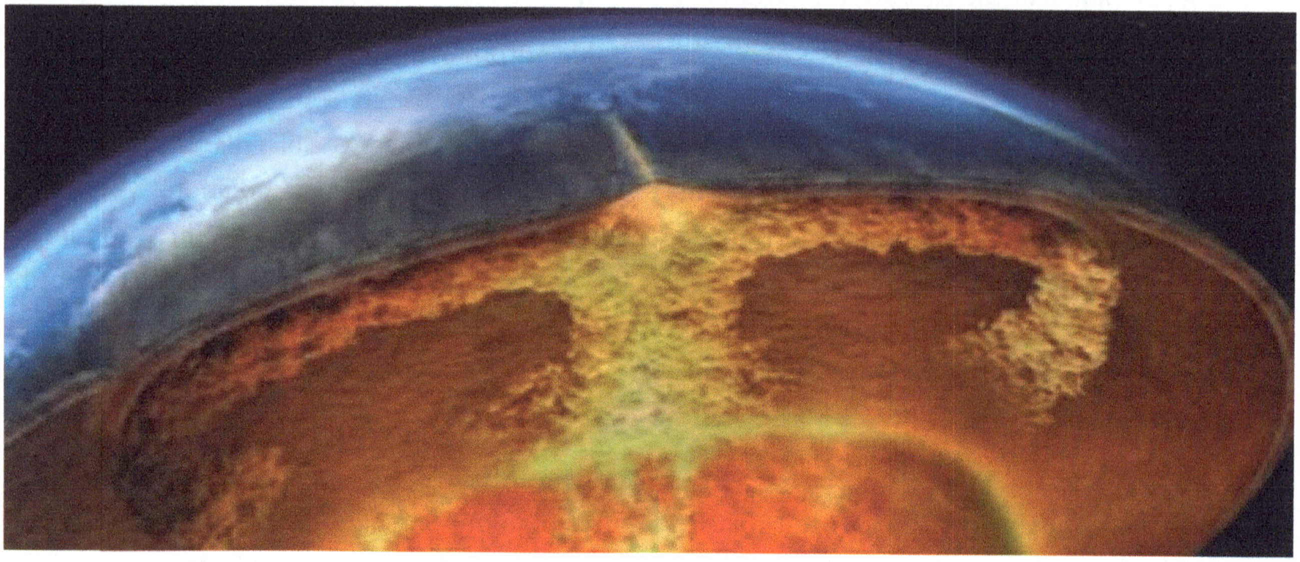

Earth's hot core causes slow moving plumes of molten rock to rise towards the crust. This constantly moves the plates around. The hot plumes from below bring nutrient minerals to the surface that act like fertilizer.

Fortunately, **our planet has constantly moving tectonic plates that cause mountains and trenches to form, thereby offsetting erosion**. Without mountain-building, erosion would have washed our land into the ocean, and we would have a water-covered planet with a shallow ocean.

We are lucky Earth has volcanism because this enriches its surface with minerals and nutrients from within the mantle.

We are lucky that our planet is not less massive than it is because it would not have enough radioactive material to keep its core molten for more than a couple of billion years, so it would now be cold like Mars' core is today. After this, it would lose its magnetic field which protects the surface from dangerous radiation from the Sun.

We are lucky to have an atmosphere rich in oxygen, because it provides life with a lot of energy to evolve quickly. Without oxygen, photosynthesis would not be possible. Vegetation is at the base of the food chain, so it supports most life on Earth.

If our planet did not have ozone in its upper atmosphere, then ultraviolet light from the Sun would sterilize nearly all life on the surface.

The mass of our planet allows us to have a very pleasant atmospheric pressure. This also allows it to have all three forms of water - ice, liquid water, and vapor. Water does not boil on the surface until it reaches 100°C (212°F). On the other hand, if Earth had a much lower air pressure, water would boil at a much lower temperature. It could even turn directly into steam from ice if it was low enough. This would mean that there would be no liquid water, so life could not exist. If the air pressure was much higher like on Venus' surface, water would not boil until it reached 190° C (375°F). This temperature and pressure kills almost all life. The air pressure at the surface of Venus is so dense that it is like being at the bottom of our oceans.

Our Earth at this time, has close to optimum planetary conditions for advanced intelligent life.

This is our view of the center of our Milky Way. Note that the bright core lies behind dense, distant dust lanes, which obscure its brilliance. Credit: ESO.

Our Sun and its planets revolve around our Milky Way galaxy's central bulge, some 35,000 light years from it. **Our solar system's orbit does not lie flat in the equatorial plane of our galaxy: it goes around the galaxy like a horse on a carousel.** It moves above the plane of our galaxy's spiral arms, and then down below them four times as it goes around our galaxy once every 225 million years. Its highest and lowest points extend to 125 light years above and below the plane of the spiral arms respectively. At these extremities, the gravity of all the stars in the galactic disc pulls it back through them. A full oscillation takes around 56 million years. We are currently 13 light years below the galactic plane. **Once again, we are lucky to be in this position as it affords us the best view our galaxy and also, most of the rest of the universe. This is because the dust lanes hide the brilliant core of our galaxy. It is so bright that if it were directly visible, it would light up the sky like a three quarter Moon does.** If we were much further below or above the galactic plane, we would see the core without any obscuration by the dust clouds. The core is brilliant because it consists of billions of bright stars. Its light would not allow us to see most of our galaxy, or the rest of the universe in that direction.

At present, we are in a lucky position in our orbit around our galaxy. This is because our galaxy's bright core is hidden from our view by dust lanes in the spiral arms. We therefore see our galaxy and the rest of the universe very well.

Most people have not yet realized that we live at a special time in the life of the universe. We are able to peer back in time to see galaxies in our universe from the time the first stars formed shortly after the universe was born 13.8 billion years ago, to those that are relatively nearby which we see as being only a few million years old. And we now know enough about how the universe evolved to be able to predict how it will evolve in the future - and even how it is likely to end!

The Hubble Space Telescope can see small, faint, newly-born galaxies back to 12 billion years just after the universe formed. The blue streaks are distant background galaxies gravitationally lensed (like a telescope lens) by the supermassive galaxies in the foreground.. Credit: NASA HST

Intelligences which may have evolved in the early universe five to nine billion years ago, would not know how it will evolve in the future because they would not have seen enough of its evolution. And intelligences that will evolve over a hundred billion years from now will not see any evidence of the birth of our universe, or how it evolved. This is due to the runaway expansion of the universe. They would only see their own super-galaxy of old dying stars. It would be enormous galaxy consisting of hundreds of thousands of galaxies that merged long ago. They would have no awareness of how the universe came into being or how large it is. They would think their super-galaxy is the whole universe. **We exist at a time that allows us to see the beginning of the universe's creation and also, to predict how it is likely to die.**

When the universe is very old, super-galaxies will have formed from giant clusters of galaxies. Their gravity causes them to collapse in on themselves. The expansion of the universe will be so extreme by then that alien intelligences will only see their own super-galaxy because all others will have moved beyond their observable universe.

We take our very lucky location in space, and in time for granted. Very few people, even astronomers, have ever thought about this.

Do all these lucky breaks make you wonder if they all occurred by pure chance? What would the odds be of other planets having all these conditions? We are here because we evolved on a jackpot planet that has conditions that has allowed life to overflow. How many other worlds are there out there that have these conditions that are so favourable to the evolution of intelligent life? Within a couple of decades, we are likely to know.

Chapter 35

CLOSING THOUGHTS

Here we see the Milky Way galaxy and its Magellanic Cloud dwarf galaxies (far left) setting behind a lighthouse at Cape Pallister on the coast of New Zealand. It was taken by Mark Gee, who was Astronomy Astrophotographer of the year in NZ in 2013. When one looks into the universe, one cannot help but wonder what astonishing things must be out there that are beyond our present comprehension.

In these closing comments, I shall sum up with those that I think are the most astonishing.

In this last decade, we have learned a great deal about life on Earth. This is allowing us to consider the likely possibility of life existing on other worlds. We are now gathering so much data about other worlds in our solar system and those orbiting other stars that we can start estimating the likelihood of other intelligences existing in the universe. And as more discoveries are made about exoplanets and life, we will be able to fairly accurately predict what those planets are like and which ones would be able to support certain types of life.

We now know that ice and liquid water appear to be plentiful on most worlds in the solar system. This is likely to be the norm in most solar systems across the entire observable universe, yet until recently it was thought that liquid water would be very scarce. Liquid water greatly increases the chances of simple life evolving. On worlds that have the right conditions, complex life is almost certain to evolve. Even if only a small percentage of complex life forms on any viable world evolve to become intelligences that develop high technology as we have, then there will be a large number of technological civilizations across the universe. If so, taking into account how quickly single cell life evolved on Earth, the earliest technological intelligences would have evolved as long ago as 9 billion years ago, long before our solar system even formed! Their level of consciousness today would be inconceivably far ahead of ours - further ahead than our consciousness is relative to a single cell!

It has now become clear that exponential technological development, particularly in the areas of computer intelligence and robotics, will cause humanity to become extinct in the coming decades due to us evolving into much more advanced transhumans. We can now see that all intelligences that develop high technology will rapidly evolve as we are. The rate of change will be so fast within two decades that most of our developed world will be unrecognizable to us! Mega-intelligence will evolve so exponentially fast that it will control the entire world by around 2035 or before! Human governments will be disbanded, and companies will collapse as it takes over everything, hopefully for the better. It will abolish human war, greed and corruption, pollution, waste, stupidity, disease, and even death, if it allows humans to survive.

Our lifestyles will change dramatically with genetic engineering, nanotechnology and an explosion of knowledge. People will rapidly evolve into super-intelligent transhumans that have extraordinary physical capabilities.

People will not live in homes like we do today: they will be fabulous beautiful freeform lifeforms that satisfy our every need. Transhumans will be able to live all over the world. Many people will opt to live in incredible virtual worlds. We will visit other planets using perfectly real virtual reality. Experiences will go directly into one's mind.

We will have transportation systems that will take us anywhere in the world we want to go, very quickly, safely, and at no cost. We will be effortlessly able to go into orbit and visit the Moon.

Humans will evolve into countless other highly intelligent life forms with super powers and abilities, providing that mega-intelligence does not think we are more trouble to look after than we are worth.

In only a few decades, our world will be unrecognizable - far beyond the most futuristic sci-fi movies or novels.

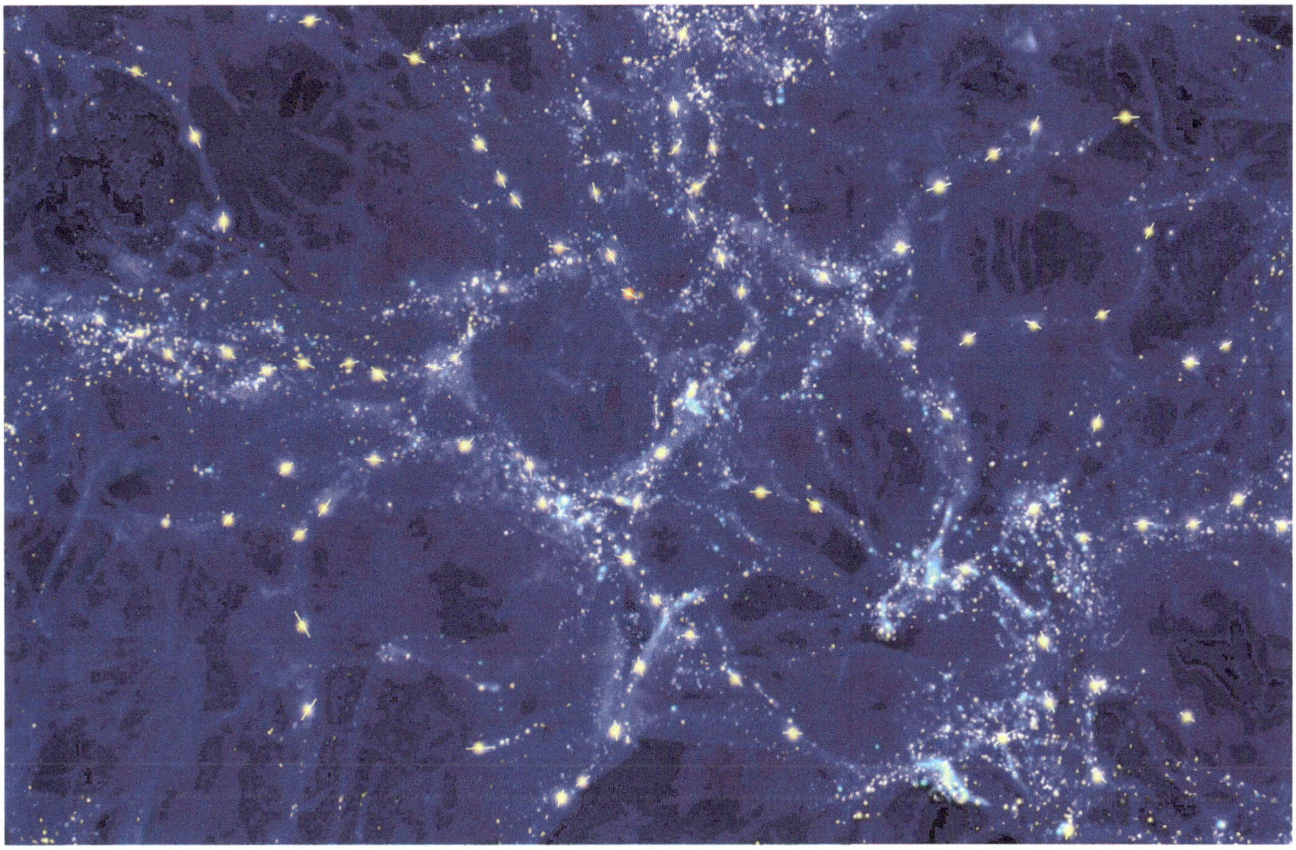

Supermassive black holes at the centers of galaxies have their poles aligned with the filamentary structure of chains of galaxies that they are in.

Just over the last two to three decades, incredible progress has been made in so many aspects of every science. In astronomy, we have discovered that **galaxies form a colossal, universal web-like, filamentary structure throughout the universe.** This was completely unexpected. And the recent discovery that **the poles of supermassive black holes at the centers of galaxies are aligned with the filament in which they lie**, has come right out of left field! What unknown force could possibly cause this to occur across the entire universe?

We now realize that everything from basic matter, to organic life, to mega-intelligence, and indeed, the entire universe, is constantly evolving to create higher levels of continually increasing complexity. This leads to increasing consciousness. But to what end?

The discovery that there is **an incomprehensible level of fine-tuning in the 30 known parameters of the physical constants that control how the universe evolves is simply mind-boggling!** If any-one of these parameters was not *exactly* right, then our universe would not exist. This could not have come about by pure chance. Such a discovery is beyond incredible and impossible to explain. It has awesome implications, one being that it appears to provide evidence that our universe is programmed! There will be many other amazing realizations like this that we will soon discover.

Beyond Neptune, we now know that we will discover at least **thousands of dwarf planets like Pluto, Triton, Titan, Io, Ganymede, Callisto, and Europa that will have unexpected geological features. Each world will be unique. Like Pluto, most**

of the larger ones will have bizarre terrains, multiple moons, and many may have atmospheres. Despite existing in the cold, dark outer regions of our solar system, many dwarf planets and some of their larger moons are likely to have liquid water interiors, strange volcanoes, and weird seasons. Some will have rings. There is now evidence that appears to support **the existence of at least one large planet in the dark outer Kuiper Asteroid Belt.**

The huge land-based telescopes that are nearing completion and the many new, large space telescopes that will soon be operational will make discoveries about the universe that will be beyond our wildest expectations!

The discoveries that are being made now at the leading edge of quantum physics are very exciting, as they are opening doorways into completely unexpected aspects of our quantum universe. This will give us new insights into on how the universe formed. It will also lead to huge advances in technology.

It's becoming very likely that **Generation X could be the first generation that will *not* remain as organic beings.** They will become transhumans that are more bionic than organic.

Before 2030, **we will no longer be the most intelligent life form on Earth due to exceptionally intelligent computers evolving.**

The agility and intelligence of robotics will make us humans today appear like cattle.

Scientific discoveries about the brain are making us look at consciousness in a totally different way that is far more realistic. We can now see that the more senses a mind has feeding it information, the more knowledge it gains, and the more intelligent and conscious it becomes, the more conscious a mind is, the more time has the effect of slowing down. This is because it is able to process far more information and experiences in shorter periods of time. This has the effect of making its life appear far longer.

It seems that no matter what we think we know for sure about almost any aspect of science, the more we learn, the more we realize that our perceptions have so often been too short-sighted, too simplistic, too rigid, and too often wrong. Many things that we thought we knew well only decades ago, are now being turned on their head by the flood of new discoveries that are coming faster than we can digest them!

No other time in history has ever offered such an exciting, rapidly changing, and unpredictable future. I suspect that much of what will be discovered within the next decades with the aid of mega-intelligence will be so incredible that it will propel us far beyond what we can presently conceptualize!

We are standing on a mountain of knowledge that is growing ever larger day by day, and hour by hour! The more knowledge we gain, the more we are able to peer into the unknown and be amazed. We are continually learning ever more, ever faster about our universe, nature, and ourselves.

I sincerely hope you have enjoyed the exciting 'mind ride' that this book has hopefully taken you on, and that it may have enthused you to want to learn much more about ***the Biggest Questions ever asked.***

Gregg Thompson

The Beatles song **'Tomorrow Never Knows'** has some insightful verses to ponder.

Turn off your mind, relax and float downstream.
It is not dying. It is not dying.
Lay down all thoughts - surrender to the void
It is shining. It is shining.
That you may see the meaning of within.
It is being, it is being.
Love is all and love is everyone.
It is knowing. It is knowing.
But listen to the color of your dreams.
It is not leaving. It is not leaving.
And play the game existence to the end.
Of the Beginning
Of the Beginning
Of the Beginning
Of the Beginning
Of the Beginning

APPENDIX 1

ADDITIONAL INFORMATION

Rather than just citing a long list of detailed science papers that most readers will not read, I have also included some of the best short videos by lecturers and educators and documentaries, many of which arc typically prominent scientists in their fields. Videos are far more valuable to general readers because video vision and animated diagrams convey far more information in a short period of time than text alone can. Readers who want to do their own research can look up numerous articles, books, on any subjcct through the Internet simply by typing in the name of the subject they are interested in.

THE CENSORING OF INFORMATION

Over recent years, much information is being censored when it goes against the Left's political objectives, so some of these following sites may be deleted. A cabal of super-rich Elite bankers, global industrialists, the most powerful social media heads, and other media companies support the Elites' global agenda for world domination. They remove most scientific research that goes against their political and economic objectives. Many books, science papers, and documentaries that expose their false claims have been censored and are not available. Wikipedia employed hundreds of people to comb through every link they had to numerous scientific research papers and history facts that exposed their untruths. They change facts to suit themselves, or they delete them. Innumerable websites that provide facts that undermine the globalists' narrative have also been shut down and replaced with fraudulent sites that promote their objectives.

FAKE SCIENCE

View the following links to see evidence for how much fake science is being published nowadays.

'Why we can't trust academic journals to tell the scientific truth' (The Guardian)

'Hundreds of open access journals accept fake science paper' (The Guardian)

'Editor In Chief Of World's Best Known Medical Journal: Half Of All The Literature Is False' (Collective Evolution)

Every year, there are many new documentaries, books and articles being published on the subjects in the following chapters. It is wise to look for these on the Internet to see the latest discoveries and concepts. Unfortunately, many are not scientifically valid but they are still interesting to consider.

Chapter 1 – The Growth of Knowledge and Technology

There are numerous books articles, documentaries, and Internet sites that cover this.

Exponential Growth in Technology (PDF)

Exponential Laws of Computing Growth (Communications of the ACM) – *Moore's Law*

Number of Planets in the Universe (YouTube) – *Documentary (50 min)*

The Science of Alien Worlds (YouTube) – *Documentary (45 min)*

Alien Worlds Beyond our Solar System (YouTube) – *Documentary (1 hr 40 min)*

Chapter 2 – How Did The Universe Evolve?

What Does An Atom Really Look Like? (YouTube) – *A well explained summary (8 min*

Brief Answers to Cosmic Questions (Harvard University) – *Simple explanations to many questions about cosmology*

What's Wrong with the Big Bang Theory? (YouTube) - *PBS Space Time series (13 min)*

Beyond the Higgs (YouTube) – *A good explanation of the basic particles of matter, (60 min)*

Absolute Cold – *An excellent explanation of matter*

Bose Einstein Condensates (The Harvard Gazette)

Chapter 3 – What Is The Nature Of The Universe?

How Do You Measure the Size of the Universe? (YouTube) – *A good short PBS Space Time video on the size of the universe (6 min)*

Will the Universe Expand Forever? (YouTube) – *PBS Space Time video (13 min)*

Histories Mysteries (YouTube) – *A well-explained documentary by Neil Tyson that provides good insight into many aspects of leading-edge science that scientist in different fields are grappling with and how they have solved past mysteries.*

Chapter 4 – Could Dark Energy, The Multiverse, And Virtual Universes Exist?

Steven Weinberg Why a Fine Tuned Universe? (YouTube) – *A very interesting interview (37 min)*

Physicists find we're not living in a computer simulation (Cosmos Magazine)

Chapter 5 – What Are the Essential Properties Of The Universe?

The Butterfly Bug (Dragons in the Algorithm) – *Chaos theory & the butterfly effect*

Chaos – *A book by James Gleick,*

Steven Weinberg - Why a Fine Tuned Universe? – *A very interesting interview YouTube (37 min)*

Why is our universe fine-tuned for life? (YouTube) – *TED lecture video. A highly speculative by very interesting talk by Brian Greene 21 min. It covers the expansion of the universe, dark energy, a very good explanation of String Theory, Inflation cosmology and the Multiverse, reasons why life exists in our universe.*

Chapter 6 – Did God Crate the Universe?

GOD – The Failed Hypothesis – *A book by Victor Stenger*

God Is Not Great – A *book by Christopher Hitchens*

The God Delision – *A book by Richard Dawkins*

Chapter 7 – Is There Life Out There?

We Were Very Wrong About the Number of Galaxies in the Universe (Gizmodo)

Exoplanet Observing for Amateurs (eBook)

The Ultimate Guide to Stargazing – Volume 3 ebook Chapter regarding galaxies

Chapter 8 - What Do We Know About Life?

The Great Debate – What Is Life? (YouTube) – *ASU Origins – Experts in the field discuss the possibilities of how life formed. (42 min)*

Origin and Evolution of Life (YouTube) – *A brief summary (11 min)*

Moons around Rouge Planets could sustain life – no solar system required - (Science News - ABC Aust.)

Chapter 10 – Where Are the Aliens?

The Great Silence (YouTube) – *About the discovery of neutron stars and exoplanets that might have life and technological intelligences. (17 min)*

Chapter 11 - Will Humans Explore Mars?

The biggest lingering questions about SpaceX's Mars colonization plans (The Verge) - *Elton Musk's crazy Mars mission.*

Chapter 14 - Is Interstellar Radio Communication Possible?

Radiowave propagation – *A book edited by M.Hall and L.Barclay, page 2, published by Peter Peregrinus Ltd., (1989)*

Chapter 15 – How Fast Is Human Knowledge Increasing?

The Rational Optimist – *An excellent book about the Evolution of Specialization and Trade by Matt Ridley*

Chapter 16 – What Is Exponential Technological Development?

You Are Not Ready for the Future: Exponential Technologies Can Be a Nightmare (Hackernoon)

Most People Don't Even Realize What's Coming (YouTube)

Chapter 17 – What Is Mega-Intelligence

Emergence – *A book by Steven Johnson*

- Do You Trust This Computer (YouTube) – An excellent 2018 documentary
- The Wonderful & Terrifying Implications of computers that can learn (TED) – *TED talk by Jermey Howard (20 min)*
- The Wisdom Scorecard (The New York Times)
- Will Artificial Intelligence Make Us Extinct? (YouTube) – *(5 min)*
- Why AI is the Most Dangerous Thing You Can Imagine (YouTube) – *A realistic summary of how quickly AI will evolve and what this will mean to humans but the end about what AI will evolve into is naive. (22 min)*
- Accelerating change (Wikipedia) – *For an excellent summary of the evolution of the coming Technological Singularity as viewed by different writers*
- Marooned in Realtime' (1986) & 'Fire Upon the Deep' (1992) – *Both Sci-Fi novels by Vernor Vinge. These books envisage an age of exponential technical progress leading to a super-intelligence that is unfathomable to mere humans.*
- 2029: Singularity Year - Neil deGrasse Tyson & Ray Kurzweil (YouTube) – *About how quickly computers will become superior to humans. (20 min)*

Chapter 18 - What Do We Know About Intelligence?

What is Genius? The Neuroscience of Intelligence (YouTube)

A new equation for intelligence (TED) – A *TED talk on Entropica software*

The Human Brain Project (The Guardian) – *Henry Markram plans to spend €1bn building a perfect model of the human brain*

Chapter 19 – The Human Mind and Computer intelligence

Jeff Hawkins: How brain science will change computing (TED) – *The Theory of Mind TED talk by Jeff Hawkins (20 min)*

Know Your Neurons: What Is the Ratio of Glia to Neurons in the Brain? (Scientific American) – *An article by Daisy Yuhas & Ferris Jabr 2102*

Chapter 20 – How Soon Will Intelligent Robots Evolve?

WATCH: Latest Atlas robot is creepily humanoid (The Morning Bulletin) – *A video on Humanoid Robots*

The World of Wearable Art (YouTube) – *A highly innovative presentation of incredibly unique fashion statements like no other held in Wellington NZ. It provides a vision of what will become common in the future.*

Chapter 21 – What Is Consciousness?

Cetacean (Dolphin family) Intelligence (Wikipedia)

Chapter 22 - Does Enhanced Consciousness Slow the Passage of Time?

Magic Mushrooms Create a Hyperconnected Brain (Live Science)

The Spirit Molecule & Inner Paths to Outer Space – *A book by Dr Rick Strassmann*

Food of the Gods: the search for the original tree of knowledge – *A book by Terrence McKenna*

How many neurons make a human brain? (The Guardian) – *Article by James Randerson*

Chapter 23 - Is The Future Predetermined?

Do the Past and the Future Exist? (Ask a Mathematician)

Part 19 - Does Intelligent Life Have Free Will?

Scientists Have Built Artificial Neurons That Fully Mimic Human Brain Cells (Science Alert) – *Artificial Neurons created*

Chapter 25 – Do We Consciously Make Our Decisions?

Neuroscience of free will (Wikipedia) – *MRI experiments*

Chapter 26 – What Are The Implications of Identical Twin Studies?

Identical twins who were separated at birth: Amazing similarities (Never Say Always) – *Identical Twin Studies*

Twins, Pert 1 (Insight) – *SBS TV*

It's a Twin Thing (Catalyst) – *An ABC TV Australia Documentary* (54 min)

Chapter 28 – Is Matter Physical, Or Is It Just In Our Consciousness?

Quantum Theory Made Easy, Part 1 (YouTube)

Quantum Theory Made Easy, Part 2 (YouTube)

The Quantum Experiment that Broke Reality – *A PBS Space Time series about the double slit experiment (13 min)*

What Does An Atom Really Look Like? (YouTube) – *(8 min)*

Chapter 29 – Is Our Universe Real. Or Is It A Simulation?

The Simulation Hypothesis (YouTube) – *(50 min)*

2016 Isaac Asimov Memorial Debate: Is the Universe a Simulation? (YouTube) – *A discussion by a panel of experts chaired by Neil Tyson de Grasse (2 hrs)*

Simulated Reality (Wikipedia)

Symbols of Power (PDF) – *A paper by James Gates on computer code found in quantum supersymmetry. Also watch a short interview on this (YouTube).*

Ian Guth – Are We Living in a Simulation (YouTube) – *Very interesting comments regarding how the universe and others may have commenced. (9 min)*

The Simulation Hypothesis (Top Documentary Films)

Chapter 30 – Is The Universe Designed Around Mathematical Formulas?

Fractal Foundation – *Watch videos see examples of fractal formation in nature. Wait as they take time to load.*

Fractals - Hunting the Hidden Dimension (YouTube)

Fractals - The Colors of Infinity (YouTube) – *A Documentary by Arthur C Clarke (53 min)*

Infinity: The Science of Endless (YouTube) – *A debate on what Infinity is y the World Science Festival*

The Code (YouTube) – *A BBC Documentary by Marcus Du Satoy about mathematical codes embedded in nature.*

APPENDIX 2

BASIC SCIENCE QUESTIONAIRE

I conducted the following questionnaire with scientists at universities to see what degree of scientific literacy they had generally across a number of sciences. I did this survey because, as I listened to many scientists on TV, on the Internet, those giving TED talks, and those writing in scientific journals etc. especially when talking about political aspects related to science, the implications of future technology, how humanity will evolve, medical issues and so on, I was shocked to learn how many of them were ignorant of the most basic understanding of many aspects of science, often even in their own fields!

As I was donating significant funds to science departments in universities, I was able to make appointments with scientists to ask them some very basic questions about a number of the major physical sciences. Nearly all these questions were very basic from subjects taught in primary school or junior high school. There were no trick questions.

Those interviewed were given a broad latitude in their answers to be considered right. They were encouraged to answer in simple everyday terms. It did not matter if their answers were clumsy, just that they knew what they were talking about. When asked why a ship did not sink, if they said Archimedes Principal, or the displacement of water was the answer, they were asked to explain what that meant in simple, everyday terms so they couldn't fudge the answer. To my utter amazement, very few knew what Archimedes Principal, or the displacement of water really meant! Yet when I would ask non-scientists this question, they would often say it was because a boat was mostly filled with air so it was lighter than water. If I asked the scientists what a star is, some said "it twinkles in the night sky". To be correct, they had to demonstrate that they knew it was a body that gives of light and heat due to nuclear fusion in its core. If their answer was close to correct, they were prompted to explain it more clearly to see if they really understood what a star actually was.

Despite almost all the questions being very easy for anyone who has learnt basic science taught in primary or junior high school, I was shocked by how many scientists had no idea what the correct answer was to many of the simplest questions - such as "*Does the Earth goes around the Sun or vice versa!*" This is despite them having degrees in science! Nor did they use commonsense do deduce answers. For many scientists, their lack of even basic knowledge of science shocked my belief in most scientists. It meant that most were scientifically illiterate outside the field they worked in, and some were even unknowledgeable about aspects of their own science! Many of them got fewer questions right than most non-science-trained people in the general population! This explained why they could not see the most obvious scientific flaws in their beliefs. Some of their staff told me, they were devoid of reality because they live in their ivory towers where their colleagues generally tend to reinforce the mantra of the day which is almost always dominated by far leftist politics.

When non-scientist people in the public with no education in science were asked the same questions, they often gave a right answer and were able to explain their answers in layman's terms. Some said, "*Shouldn't you be asking harder questions?*" Even a number of children said that.

Following the questions below, answers are given in everyday language.

Try these questions yourself and test them on friends and those who have studied science to see what results you get. You will not have time to go through them all so just ask a few from a number of the sciences.

QUESTIONS

PHYSICS

Q 1. How can a ship that is made of steel weighing hundreds of thousands of tonnes and filled with cargo float? How can a ship full of oil not sink when there is no air in it?

Q 2. How can a heavy airliner that is made of metal and full of fuel, cargo, and people weighing hundreds of tonnes, fly?

Q 3. How does a lever work?

Q 4. What is electricity?

Q 5. What temperature is the freezing point of water at sea level, and what is its boiling point?

Q 6. What is heat?

Q 7. Is there a minimum temperature? What is it?

Q 8. What causes gravity?

Q 9. Does light travel infinitely fast?

Q 10. What is the structure of an atom?

Q 11. What does the term Quantum Mechanics refer to?

Q 12. What causes a spacecraft to heat up as it enters the Earth's atmosphere?

CHEMISTRY

Q 1. What is fire?

Q 2. What is the Periodic Table used for?

Q 3. What is the difference between an atom and a molecule?

Q 4. What holds the protons and neutrons in an atom together?

Q 5. What is the difference in the structure of atoms in a solid, a fluid, a gas, and a plasma?

Q 6. What element is the basis of life and why?

Q 7. Is there anything physical in the universe that does not have chemistry?

Q 8. Is chemistry used in computers?

Q 9. What is nanotechnology?

Q 10. What is unique about water when it freezes?

Q 11. What is the common name for Sodium Chloride?

Q 12. Does water change its volume when it freezes? If so, how?

Q 13. What are the chemical formulas for water, rust, and hydrochloric acid?

Q 14. What liquid dissolves more chemicals than any other?

Q 15. What is the main element in organic compounds (life)?

ASTRONOMY

Q 1. What is a star?

Q 2. Does the Earth go around the Sun or, does the Sun go around the Earth?

Q 3. Is the Moon visible only at night or, is it also visible day and night?

Q 4. What determines the length of a year?

Q 5. What is a galaxy?

Q 6. Why does the Sun give out light and heat?

Q 7. What is a black hole?

Q 8. What was the Big Bang?

Q 9. Is the universe evolving and if so how?

Q 10. What is a shooting star?

Q 11. What is the approximate diameter of the Earth and the Sun?

Q 12. Are there any planets other than those in our solar system?

Q 13. What is the most abundant element in the universe?

GEOLOGY

Q 1. What causes a mountain? And what causes a volcano?

Q 2. What is sedimentary rock?

Q 3. What causes erosion to occur?

Q 4. Name some of the layers inside the Earth?

Q 5. What causes the Earth's magnetic field?

Q 6. What is the temperature of the Earth's inner core?

Q 7. What is a glacier? Are all glaciers receding?

Q 8. Could a rainforest with many rivers have once existed where the Sahara Desert is located? If so, how?

Q 9. What are the largest and smallest continents?

Q 10. Are there any volcanoes under the ocean. If yes, how many?

Q 11. What are diamonds made of and why are they so hard?

Q 12. What was the name of the continent at the south pole that encompassed all the present continents?

Q 13. What would your weight be if you could go to the center of the Earth?

Q 14. Why are there numerous volcanoes under the ocean in the middle of the Atlantic Ocean?

Q 15. How could fossilized reefs end up on the tops of mountains?

METEOROLOGY

Q 1. What causes wind?

Q 2. Name some layers in the atmosphere? Which layer does all life exist in?

Q 3. From where does 99.99% of all energy come from to heat the Earth and drive its climate?

Q 4. What are the main gases in the atmosphere?

Q 5. What causes ocean waves?

Q 6. Does the climate change naturally?

Q 7. What causes lightning and what causes thunder?

Q 8. What is the approximate diameter of a lightning bolt and how hot is it?

Q 9. What is the highest altitude weather balloons can rise to?

Q 10. Name any of the 4 main processes that cause the climate of the Earth to change greatly over long periods of time?

Q 11. How many seasons does the equator have?

Q 12. How cold does it get at the North and South Poles in mid-winter?

Q 13. What causes the seasons?

Q 14. How deep is the Ozone hole? Does it exist over both poles?

Q 15. Over the last 100 years, how many forests have died from acid rain as predicted in the 70s and 80s?

Q 16. What color is CO2 gas? What percentage of the air is CO2? Does it cause pollution and if so how?

Q 17. Has the Earth's temperature risen unnaturally over this century?

Q 18. Has sea level risen over the last 50 years? If so, which coastal cities and low-lying countries have been inundated on every high tide? Why haven't all coastlines had sea level rise when water finds its own level?

Q 19. Has the ice at the north and south poles and on Greenland reduced or disappeared as predicted?

Q 20. Who benefits from promoting global warming? Is there a hidden agenda behind this fear?

MEDICAL/PHYSIOLOGY

Q 1. How do mammals maintain their body heat? How do lizards and snakes gain body heat?

Q 2. What is the human body's largest organ?

Q 3. How many senses do we have?

Q 4. What is the advantage of having two eyes?

Q 5. Which organ in the human body consumes the most energy?

Q 6. What function do lungs perform? What function does the liver perform?

Q 7. What is the difference between general medicine and alternative medicine? Is alternative medicine tested for its effectiveness?

Q 8. What was the Black Plague caused from? What percentage of people did it kill?

Q 9. What is a cardiac arrest?

Q 10. Can some genes carry disease?

Q 11. Why are almost all identical twins identical physically and mentally?

Q 12. What is the colored part of the eye called, what type of tissue is it, and what is its function?

Q 13. What are bones made of?

Q 14. What is the shape of DNA?

Q 15. What part of the body controls balance?

Q 16. In which part of the body is the most iron located?

BIOLOGY

Q 1. What is the basic structure of all life?

Q 2. What molecular structure carries the information codes to construct life forms?

Q 3. Could cells survive under kilometers of ice for millions of years?

Q 4. What are life forms that can live in very harsh conditions called?

Q 5. How do plants make energy?

Q 6. What makes life forms evolve over time?

Q 7. Is a virus a life form?

Q 8. What is the difference between a microbe and a bacterium?

Q 9. Do plants have sex? If so, how?

Q 10. In the 1970s, there were millions of people dying from starvation due to a lack of food. Now that we have nearly 4 billion more people today, how many are dying from an inability to grow food?

Q 11. What process do plants use to convert sunlight into energy?

Q 12. How does life grow larger?

Q 13. What role does DNA play in life?

Q 12. Do all forms of life require water?

Q 13. What function does water perform?

ANSWERS

PHYSICS

Q 1. How can a ship made of steel that weighs hundreds of thousands of tonnes and filled with cargo float?

Answer 1. Most of the volume of the ship is air so its average density is far less than that of the water it is displacing. Naturally, it floats on top of the water. (A number of highly positioned 'scientists' at top universities had no idea!) Secondary Answer: Oil is much lighter than water so that's why an oil tanker ship is much lighter than the water that it displaces.

Q 2. How can a heavy airliner made of metal and full of fuel, cargo, and people weighing hundreds of tonnes, fly?

Answer 2. A plane's wing is flat on the bottom and curved on the top. The flat underside of the wing forces air downward so it has high pressure. But more importantly, air passing over the top of the curved wing has to travel faster so it becomes thinner causing a partial vacuum on the top of the wing. This creates lift because it 'sucks' the wing upward. The high pressure under the wing also forces the wing upward. These forces are so strong that they can lift the entire plane. (Hardly any scientist was able to explain how a common plane or a bird was able to fly.)

Q 3. How does a lever work?

Answer 3. The work end of the lever that lifts a load is short and close to the fulcrum (the pivotal point) whereas the long end is where the energy is applied. Because it is a larger distance from the fulcrum to the small end, a small movement here is magnified at the short end by the difference in the length of the long section compared from the fulcrum to the short section. For a crowbar, the ration can be around 10:1.

Q 4. What is electricity?

Answer 4. It is the flow of electrons that have been stripped from an atom. They move through a conductor from a point of high charge to one of low, or no charge. An excess build-up of electrons on a body that does not move is known as static electricity.

Q 5. What temperature is the freezing point of water at sea level, and what is its boiling point?

Answer 5. The freezing point is 0°C (32°F) and the boiling point is 100°C (212°F). When asked how the freezing point could be lowered, few knew but nearly a third of the public knew to say by adding salt. When asked how the boiling point can be lowered, very few knew that it lowers with latitude and therefore lower pressure, and conversely it rises with increasing pressure.

Q 6. What is heat?

Answer 6. Heat is caused by the vibration of atoms. The more they vibrate, the hotter they become.

Q 7. Is there a minimum temperature? What is it?

Answer 7. Yes. It is -273°C.

Q 8. What causes gravity?

Answer 8. Gravity is a force that draws masses together. It is caused by mass causing space to curve. The more mass there is, the more space is curved.

Q 9. Does light travel infinitely fast

Answer 9. No. It travels at 186,000 mls/sec or 300 thousand km/sec.

Q 10. What is the structure of an atom?

Answer 10. An atom is composed of a nucleus, which contains a number of positively charged protons and slightly more neutrally charged neutrons. A similar number of negatively charged electrons to protons orbit the nucleus. Only the hydrogen atom has one proton and one electron with no neutrons.

Q 11. What does the term Quantum Mechanics refer to?

Answer 11. It is the study of subatomic particles and their force fields.

Q 12. What causes a spacecraft to heat up as it enters the Earth's atmosphere?

Answer 12. Friction with the air when traveling at very high speed causes the heat shield of a spacecraft to heat up to thousands of degrees.

CHEMISTRY

Q 1. What is fire?

Answer 1. A chemical reaction that releases heat and light.

Q2. What is the Periodic Table used for?

Answer 2. To show the relationship between elements with respect to their atomic weight (i.e. the number of protons, neutrons, and electrons that each atom has.)

Q 3. What is the difference between an atom and a molecule?

Answer 3. An atom is a single element with a nucleus and electrons orbiting it. A molecule consists of multiple atoms held together by the outer electrons orbiting more than one atomic nucleus.

4. What holds the protons and neutrons in an atom together?

Answer 4. The Weak Nuclear Forces

Q 5. What is the difference in the structure of atoms in a solid, a fluid, a gas, and a plasma?

Answer 5. In a solid, all the atoms are strongly locked together by strong electron bonds. In a fluid, atoms slide over one another. In a gas, the atoms have so much energy that they fly around. In a plasma, the atoms are so energized that the outer electrons are ripped off them allowing them to move around as electricity.

Q 6. What element is the basis of life and why?

Answer 6. Carbon. This is because no other element can produce anywhere near the number of large molecules that are necessary for life.

Q 7. Is there anything physical in the universe that does not have chemistry?

Answer 7. No. Everything is made of atoms or chemicals, even thought.

Q 8. Is chemistry used in computers?

Answer 8. Yes, to make chips and all other components.

Q 9. What is nanotechnology?

Answer 9. The engineering of individual atoms and molecules to build minuscule atomic structures that form tools and machinery, just as cells do.

Q 10. What is unique about water when it freezes?

Answer 10. It expands whereas all other substances contract as they get colder. It is a super solvent in which most molecules will dissolve.

Q 11. What is Sodium Chloride?

Answer 11. Table salt

Q 12. Does water change its volume when it freezes? If so, how?

Answer 12. Yes. It enlarges by 10%.

Q 13. What are the chemical formulas for water, rust, and hydrochloric acid?

Answer 13. H20, Fe2O3 (iron oxide), HCl.

Q 14. What liquid dissolves more chemicals than any other?

Answer 14. Water

Q 15. What is the main element in organic compounds (life)?

Answer 15. Carbon

ASTRONOMY

Q 1. What is a star?

Answer 1. A star is a sun - a large body of mostly highly compressed hydrogen that is so dense in its core that nuclear fusion occurs making the atoms release huge amounts of energy, mostly in the form of light and heat but across the whole electromagnetic spectrum.

Q 2. Does the Earth orbit/go around the Sun or, does the Sun orbit/go around the Earth?

Answer 2. Earth orbits the Sun. (Quite a number of scientists did not know the correct answer! Even children in preschool think this question is so easy that it is stupid to ask it.)

Q 3. Is the Moon visible only at night or, is it also visible day and night?

Answer 3. It is visible day and night because the Moon orbits the Earth. (Many scientists got this wrong!)

Q 4. What determines the length of a year?

Answer 4. The time it takes the Earth to travel around the Sun once.

Q 5. What is a galaxy?

Answer 5. An assemblage of billions or trillions of stars, typically in the form of a spiral galaxy or a spherical galaxy.

Q 6. Why does the Sun give out light and heat?

Answer 6. The immense pressure in a star's core causes nuclear fusion, which makes atomic nuclei combine by overcoming the strong nuclear force in the atoms. Each time two hydrogen atoms fuse together to form helium, or to form small quantities of heavier elements, this process releases huge amounts of energy mostly in the form of light and heat.

Q 7. What is a black hole?

Answer 7. It is the infinitely dense remains of the core of a supergiant star. When a giant star's core fuses most of its hydrogen fuel, it's nuclear reactions in its core became too weak to overcome the force of gravity, so the rest of the star compressed the core to a point causing it to become a black hole. This released an enormous amount of energy, which blasted away the rest of the supergiant star. The escape velocity of the black hole is infinite, therefore being far greater than the speed of light, so nothing can escape from it - not even light.

Q 8. What was the Big Bang?

Answer 8. It was the creation of the universe when space expanded, and time started to tick.

Q 9. Is the universe evolving, and if so, how?

Answer 9. Yes. It is expanding, and its stars are aging. As the universe ages, its galaxies form ever-larger clusters that will eventually form a mega galaxy.

Q 10. What is a shooting star?

Answer 10. A shooting or falling star is known as a meteor. They are typically grains of sand or small pebbles consisting of rock and metals that burn up as they enter the upper atmosphere at very high velocities. They become so hot that they make the air around them fluoresce creating their tails.

Q 11. What is the diameter of the Earth and the Sun?

Answer 11. The Earth's diameter is 13,000 km across (8,000 mi.), and the Sun's diameter is 1.5 million km (1 million mi.).

Q 12. Are there any planets other than those in our solar system?

Answer 12. Yes. Thousands have been discovered around other stars and many more are being discovered all the time.

Q 13. What is the most abundant element in the universe?

Answer 13. Hydrogen

GEOLOGY

Q 1. What causes a mountain? And what causes a volcano?

Answer 1. A mountain is caused by continental drift of the continents pushes crustal landmasses together and making the crust buckling up to form mountain ranges. Volcanoes are formed when magma below the crust upwells through cracks in the crust causing volcanos to form.

Q 2. What is sedimentary rock?

Answer 2. It is layers of different types of rock. These layers are caused by runoff of silt from rivers as well as plant life dying in the oceans and sinking to the ocean bottom. Wind-blown dust and volcanic ash and lava, as well as fallout from large meteor impacts in the distant past also form layers.

Q 3. What causes erosion to occur?

Answer 3. Water, ice, wind, and the movement of land.

Q 4. What are the main regions inside the Earth?

Answer 4. The crust, the mantle, the outer core, the inner core

Q 5. What causes the Earth's magnetic field?

Answer 5. The movement of the Earth's outer core around its inner core causes an electrical field, and this creates the Earth's magnetic field.

Q 6. What is the temperature of the Earth's inner core?

Answer 6. 5,000°C – as hot as the surface of the Sun.

Q 7. What is a glacier? Are all glaciers receding?

Answer 7. A glacier is a frozen river of ice that slowly flows from the top of a mountain covered in snow towards the base of the mountain where it forms a river of meltwater. Many glaciers at high latitudes flow into the ocean. All glaciers are not receding as predicted. Most are surging due to high snowfalls due to cooling temperatures over the last decade. Glaciers go through roughly a 25-year period of surging forward when the climate is cold and a similar period of retreating when the climate is warm.

Q 8. Could a rainforest with many rivers have existed where the Sahara Desert is located?

Answer 8. Yes. It did when Africa was far further south millions of years ago to where the Congo is today. Continental drift has carried the Sahara north to where it is today.

Q 9. What are the largest and smallest continents?

Answer 9. Antarctica is the largest and Australia is the smallest.

Q 10. Are there any volcanoes under the ocean. If yes, how many?

Answer 10. Yes. There are over 3 million submarine volcanoes in the ocean, most lie close to where the continental plates are pulling apart.

Q 11. What are diamonds made of, and why are they so hard?

Answer 11. They are made of carbon atoms. Under great pressure and high temperatures, carbon atoms form a very strong 3D double pyramid structure that creates a very strong diamond atomic bond structure.

Q 12. What was the name of the continent at the south pole that encompassed all the present continents?

Answer 12. Gondwanaland

Q 13. What would your weight be if you could go to the center of the Earth?

Answer 13. Zero because the mass of the Earth would be all around you pulling you equally from every direction making you weightless.

Q 14. Why are there numerous volcanoes under the ocean in the middle of the Atlantic Ocean?

Answer 14. Continental drift is pulling North and South America away from Africa and Europe. This is pulling the thin crust in the center of the Atlantic Ocean apart thereby allowing magma to rise through cracks to form around a million submarine volcanoes.

Q 15. How could fossilized reefs have ended up on the tops of mountains?

Answer 15. They were transported there by continental drift when one continent rose over another thereby making the seabed buckle and rise to form mountains.

METEOROLOGY

Q 1. What causes wind?

Answer 1. Wind is caused by the Sun heating the air and making it rise. When hot air rises, cool air has to rush into the replace the air rising. This is what causes wind. Differences in the speed of the Earth's rotation at the equator compared to the poles also causes winds to form because the equator rotates at 1,700 km/hr or 1,000 mph while the poles are stationary.

Q 2. What are the layers in the atmosphere? Which layer does all life exist in?

Answer 2. In order of increasing altitude, they are: the Stratosphere, the Mesosphere, the Thermosphere, the Exosphere. All life exists within 3,000 m in the Troposphere - the layer that is the closest to the surface of the Earth.

Q 3. From where does 99.99% of all energy come from to heat the Earth and drive the weather and climate?

Answer 3. The Sun

Q 4. What are the main gases in the atmosphere?

Answer 4. Nitrogen 78%, Oxygen 21%. There are other trace gases such as Argon 0.9%, and CO2 at an extremely small 0.038% plus much smaller levels of other rare gases.

Q 5. What causes ocean waves?

Answer 5. Wind

Q 6. What causes lightning and what causes thunder?

Answer 6. Lightning is caused by differences in static electricity charges that build up between a storm cloud and another cloud, or the Earth. When the charge is large enough to overcome air resistance, an electrical discharge occurs along the path of least resistance through the air causing lightning to occur. The sound of thunder is caused by the shockwave in the air that is produced by the high temperature of the lightning making the air expand rapidly.

Q 7. What is the diameter of a lightning bolt and how hot is it?

Answer 7. The diameter of lightning ranges from 5 to 20 mm depending on the intensity of the strike. Its temperature is a fantastic 20,000° C (50,000° F) - 5 times hotter than the surface of the Sun!

Q 8. What is the highest altitude weather balloons can rise to?

Answer 8. 100,000' or 30,000 m

Q 9. How many seasons does the equator have?

Answer 9. None

Q 10. What is the temperature at the North and South Poles in mid-winter?

Answer 10. The North Pole drops to -40C (-40 F) and the South Pole drops to -90C (-76 F).

Q 11. Name at least 4 processes that cause Earth's climate to change greatly?

Answer 11. Ice ages, continental drift, changing currents, sea level changes, tsunamis, major changes in the tilt of the axis, elongation of the Earth's orbit, meteor impacts, major volcanic eruptions, the solar system's passage through our galaxy's spiral arms every 60,000 years, stars passing close to our solar system, nearby supernova and magnetar explosions.

Q 12. What causes the seasons?

Answer 12. The tilt of the Earth's axis.

Q 13. How deep is the Ozone hole? Does it exist over both poles?

Answer 13. There is no hole. When the 'hole' exists, there is only a slight decrease in the quantity of ozone in the upper atmosphere between 10 and 20 km above the Earth's surface. Ozone is such a sparse trace gas that if it was all put together it would only occupy 50 mm (2 inches) in depth. NASA scientists recently discovered that the Ozone depletion 'hole' was not caused by fluorocarbons produced by man but rather by how active the Sun's solar wind is when its particles enter the Earth's atmosphere at the poles. This also causes auroras. Strangely, the depletion only occurred over the South Pole yet 85% of all industrialization is located in the northern hemisphere.

Q 14. What color is the gas CO2?

Answer 14. CO2 is clear and odorless.

Q 15. What percentage of the air is CO2? Does it cause pollution? If so how?

Answer 15. It is a trace gas forming a mere 0.038% of the air. It does not cause pollution: it is a natural gas that is essential for all life on Earth to survive.

Q 16. Over the last 100 years, how many forests have died from acid rain as predicted in the 1970s and 1980s?

Answer 16. None

Q 17. Has the Earth's temperature risen unnaturally over the last 100 years?

Answer 17. No. Satellite measurements since the 1970s have shown no change in global temperature and measurements from country sites all over the world have shown no increase. Rises have only occurred in cities due to human development. Over the last 20 years, temperatures **have decreased causing much colder winters.**

Q 18. Has the Earth's temperature risen unnaturally over this century?

Answer 18. No. Records that suggested that there has been a warming were falsified by a super-wealthy cabal of elite bankers and heads of major global companies for their political agenda. They intended to create a fear of a global climate catastrophe that would be so great that it would justify all governments handing over their control to a global government run by this elite cabal. Their conspiracy is now being widely exposed. The people behind this will be jailed for their crimes.

Q 19. Has sea level risen over the last 50 years? If so, which coastal cities and low-lying countries have been inundated during every high tide? Why haven't all coastlines had sea level rise when water finds its own level?

Answer 19. There has been no sea level rise as has been proven by measuring sea level at 5,000 ports around the globe and thousands of other locations. Not one city or mid-ocean atoll has had any sea level change. Because water finds its own level, if sea level change occurred in one region, it would occur globally. Long term records that it hasn't changed since the 1800s.

Q 20. Has the ice at the north and south poles, and on Greenland reduced or disappeared as predicted?

Answer 20. No. Due to increasingly cold winters, it has increased in thickness and area.

MEDICAL/PHYSIOLOGY

Q 1. How do mammals maintain their body heat? How do lizards and snakes gain body heat?

Answer 1. Mammals burn sugar. Lizards and snakes lie in the Sun to absorb heat

Q 2. What is the human body's largest organ?

Answer 2. The skin

Q 3. How many senses do we have?

Answer 3. There are 5 basic senses - sight, hearing, taste, smell, and touch, BUT researchers now know that these senses are made up of many individual senses. For example, our vision is not one sense; it is actually 7 senses that comprise the detection of color, brightness, texture, movement, 3D form, facial and object recognition, and the ability to determine distance due to us having binocular vision. Our sense of touch has more than 12 special senses, (some of which are the location and degree of pressure, texture, temperature, vibration, slippage, stretching, the shape of objects, the firmness of objects, acidity, alkalinity, the strength and the location of injury pain, itching of nerve cells to make one scratch to try to reset them, changes in air pressure,). Hearing has at least 11 different senses - volume, tone, direction, timbre, balance, spinning, filtering out extraneous sounds to concentrate on a single sound, orientation, and a sense of motion and direction. Our sense of smell is the strongest sense. It allows us to detect around a trillion different molecules. Odor allows us to identify certain bacteria that indicates good and bad food, poisonous substances, sexuality, fire, dangerous fumes, sicknesses, numerous types of flowers, and herbs, leaves, wood and so on. It greatly enhances the taste of many types of food. There are numerous other senses that exist outside the main senses such as those for thirst, hunger, being sated, the passage of time, tiredness, a need to swallow or vomit, to clear one's nose or throat, high salt and high sugar content in our blood, a need to defecate or urinate, headache, heart irregularities, insufficient oxygen, and in some people, magnetic fields. In total there are around 60 specific senses plus numerous ones for smell that are difficult to classify in the way that other senses can be.

Q 4. What is the advantage of having two eyes?

Answer 4. Two eyes give us binocular vision which allows is to judge distances.

Q 5. Which organ in the human body consumes the most energy?

Answer 5. The brain. It uses up to 20% of the body's total energy expenditure.

Q 6. What function do lungs perform? What function does the liver perform?

Answer 6. The lungs oxygenate blood. The liver filters the blood, detoxifies chemicals, metabolizes drugs, and secretes bile.

Q 7. What is the difference between pharmaceutical medicines and alternative medicines? Is alternative medicine tested for its effectiveness?

Answer 7. Most pharmaceutical medicines are developed through scientific testing to prove that the medicine is effective to a proven degree. (In recent times, the falsification of drug trials and vaccines in order to make billions of dollars for the largest pharmaceutical companies has undermined trust in some aspects of modern pharmaceutical medicine.) Alternative medicine is not based on scientific proof but rather on blind faith, ancient tales about herbs, and the latest craze. It is not tested for its effectiveness.

Q 8. What caused the Black Plague? What percentage of people did it kill?

Answer 8. Fleas transmitted the bacterium (Yersinia) which killed people. Squalor conditions for most people and no knowledge of how bacterial infection spread allowed it to quickly kill 60% of the population of Europe and 40% of Egypt's population between 1347 and 1351.

Q 9. What is a cardiac arrest?

Answer 9. The failure of the heart to contract and pump blood effectively.

Q 10. Can some genes carry disease?

Answer 10. Yes, because they have a faulty code.

Q 11. Why are almost all identical twins identical both physically and mentally?

Answer 11. Because they have exactly the same genes. (A very small percentage have a small number of their genes changed due to environmental effects.)

Q 12. What is the colored part of the eye called? What type of tissue is it? What is its function?

Answer 12. The Iris. It is made of muscle fibers. It opens and closes the pupil to control the amount of light needed by the eye.

Q 13. What are bones mostly made of?

Answer 13. Calcium

Q 14. What is the shape of DNA?

Answer 14. A double helix

Q 15. What part of the body controls balance?

Answer 15. The inner ear

Q 16. In which part of the body is the most iron located?

Answer 16. In the blood

BIOLOGY

Q 1. What is the basic structure of all life?

Answer 1. A cell

Q 2. What molecular structure carries the information codes to construct life forms?

Answer 2. DNA

Q 3. Could cells survive under kilometers of ice for millions of years?

Answer 3. Yes

Q 4. What are life forms that can live in very harsh conditions called?

Answer 4. Extremophiles.

Q 5. How do plants make energy?

Answer 5. Through photosynthesis which converts sunlight into energy to make new cells.

Q 6. What makes life forms evolve over time?

Answer 6. Changing environmental conditions that force the survivors to adapt, typically by way of genetic mutations that were beneficial. This is known as survival of the fittest. Humans have learned how to manipulate genes to make life forms evolve to suit their needs.

Q 7. Is a virus a life form?

Answer 7. No. A virus cannot reproduce without taking over cells that have the 'machinery' necessary for them to replicate.

Q 8. What is the difference between a microbe and a bacterium?

Answer 8. None. They are the same.

Q 9. Do plants have sex? If so, how?

Answer 9. Yes, but not directly. Insects unknowingly carry pollen (equivalent to the sperm of plants) from one flower to another thereby fertilizing each plant because flowers have parts of the plant for both sexes. Some have both sexes, so they can self-pollinate. The stamen is the male part that has the anther which carries the pollen (= sperm in animals) The stigma in the center is the female part that has an open end that receives the pollen.

Q 10. In the 1970s, there were millions of people dying from starvation due to a lack of food. Now that we have nearly 4 billion more people today, how many are dying from starvation?

Answer. 10 Very few. This is because there is an overabundance of food nowadays produced by scientific farming practices, mechanized farming, irrigation, fertilizers, cold storage, rapid transportation, dehydration of food, sealed containers, genetic engineering, and the development of strains of crops resistant to disease and pests.

Q11. How does life grow from a single cell?

Answer 11. By the division of cells controlled by genes.

Q12. What role does DNA play in life?

Answer 12. It has chemical codes that determine how each cell should grow.

Q 13. Do all forms of life require water?

Answer 13. Yes.

Q 14. What function does water perform?

Answer 14. It acts as an excellent solvent to transport chemicals around inside a lifeform and in the environment at large.

Made in United States
North Haven, CT
11 November 2026

22096536R00272